Building Systems for Interior Designers

BUILDING SYSTEMS
FOR INTERIOR
DESIGNERS

THIRD EDITION

Corky Binggeli, ASID

WILEY

Published by John Wiley & Sons, Inc., Hoboken, New Jersey.

Published simultaneously in Canada.

For general information about our other products and services, please contact our Customer Care Department within the United States at (800) 762–2974, outside the United States at (317) 572–3993 or fax (317) 572–4002.

Wiley publishes in a variety of print and electronic formats and by print-on-demand. Some material included with standard print versions of this book may not be included in e-books or in print-on-demand. If this book refers to media such as a CD or DVD that is not included in the version you purchased, you may download this material at http://booksupport.wiley.com. For more information about Wiley products, visit www.wiley.com.

Cover image: © Matejay/iStockphoto
Cover design: Wiley

Library of Congress Cataloging-in-Publication Data:

Binggeli, Corky, author.
  Building systems for interior designers / Corky Binggeli.—Third edition.
     pages cm
  Includes bibliographical references and index.
  ISBN 978-1-118-92554-6 (hardback)
   ISBN 978-1-118-92555-3 (ePub)—ISBN 978-1-118-92556-0 (Adobe PDF)—ISBN 978-1-118-92554-6 (hardback) 1. Buildings—Environmental engineering. 2. Buildings—Mechanical equipment—Design and construction. 3. Buildings–Electric equipment–Design and construction. I. Title.
  TH6014.B56 2016
  696—dc23
                          2015031808

Printed in the United States of America

SKY10030901_102521

Contents

PART IV
WATER AND WASTE SYSTEMS

CHAPTER 9
Water Supply Systems 145

CHAPTER 10
Waste and Reuse Systems 164

CHAPTER 11
Fixtures and Appliances 176

Preface

The first edition of *Building Systems for Interior Environments* arose from my need for a textbook to use in teaching interior design students that gave them the understanding and skills to work with architects and engineers. This third edition of *Building Systems for Interior Environments* updates the role of interior designers as part of the building design team, and addresses their special needs for information on today's building system design and equipment.

Interior designers today work closely with other design and construction professions to provide functional, sustainable, and healthy buildings. Sophisticated digital control systems permit design that supports varied occupancy and work styles and schedules. This results in more individualized control of the interior environment, which is conducive to worker satisfaction and productivity.

Sustainable design supports a holistic approach to building systems, where the older territorial distinctions between the various building system and architectural/engineering disciplines are opening up. Existing buildings are being adapted and reused for new purposes. The impact of energy efficiency and water conservation in buildings is widely recognized. This third edition of *Building Systems for Interior Environments* reflects these changes.

In addition, the third edition benefits from the comments of experienced educators as to the best ways to organize and focus the information. The updated contents are arranged to flow smoothly in an order that facilitates the teaching process.

Part I looks at environmental conditions and the site, the building envelope and the building design process, sustainable design, the interaction of the human body with the built environment, and how building codes protect us. Part II addresses building forms, structures, and elements including floor/ceiling assemblies, walls, stairs, windows, and doors. Part III introduces acoustic design principles and architectural acoustics. Part IV addresses water supply, waste, and reuse systems, as well as plumbing fixtures and appliances. Part V covers principles of thermal control, indoor air quality, ventilation, and moisture control, and heating and cooling. Part VI explains electrical system basics and electrical distribution, as well as lighting systems. Part VII concludes the book with coverage of fire safety design, conveyance systems, and communications, security, and control equipment.

More than 40 percent of the third edition's text is new. This edition contains over 485 illustrations, approximately 260 of which are new and 160 redrawn or revised. There are also over 175 tables, 125 of which are new or significantly revised. References to related materials in other chapters are included. Tips indicating material of assistance and interest to interior designers are also included. Key terms are defined in the text and indicated in **bold**. Quotations from architectural and engineering sources aid in understanding the perspective of other design professionals.

Interior designers need to understand the viewpoints and respect the expertise of other design professionals. *Building Systems for Interior Designers* provides the information they need to do this, without delving into engineering calculations. It focuses on the parts of the building design process that most affect the occupants' functional needs, and provides a technical but readily understandable foundation for the design of interior spaces. Residential as well as commercial and institutional spaces are included.

Building Systems for Interior Designers is listed by the National Council for Interior Design Qualification (NCIDQ) as a reference for preparation for the NCIDQ exam. The third edition is supplemented by online materials including an Instructors Manual with knowledge areas, topics for discussion, and definitions of key terms, PowerPoint presentations for each chapter, and a Test Bank with sample questions. Supplemental material is available at www.wiley.com/go/bsid3e.

Corky Binggeli, ASID
Arlington, MA

Acknowledgements

The authorship of a book involves a great amount of solitary work and persistence. Turning the manuscript and illustrations into a published book involves the efforts of many others. I have now published seven books—plus second and third editions of some of them—with the team at John Wiley & Sons, and I want to thank them once again for their professionalism, support, and good advice.

I especially want to thank my editors, Paul Drougas, Lauren Poplowski, and Seth Schwartz, their able assistants Michael New and Melinda Noack, and production editor Amy Odum.

The feedback of my colleagues and students is invaluable in assuring that this edition meets their needs. I especially benefitted from the review team of Dr. Jane L. Nichols, IDEC, NCIDQ, High Point University; Ji Young Cho, LEED AP, Kent State University; and Brian Sweny, AIA, IIDA, LEED AP, IDEC, SEED, Savannah College of Art and Design.

As is the case with all my books, I am indebted to my husband, Keith Kirkpatrick, for his support and assistance. He puts up with my obsessive focus, makes sure I am properly fed, and gives solid advice when I'm struggling with a decision. Keith also reviewed and commented on all of the illustrations. Thank you, Keith!

PART

I

THE BUILDING, THE ENVIRONMENT, AND HEALTH AND SAFETY

Interior designers today work closely with other design and construction professionals to provide functional, sustainable, and healthy buildings. Sustainable design is supporting a holistic approach to building systems, and older territorial distinctions between various architectural and engineering disciplines are opening up. Existing buildings are valued for the materials and energy they embody, and many projects involve the renovation of building interiors.

Interior designers are increasingly working as part of environmentally aware design teams. Sustainable design involves interior designers observing the impact of a building's site, climate, and geography on its interior spaces. Building interiors are increasingly open to natural settings and views, and the interior designer's work may bridge interior and exterior spaces. Wise energy use dictates awareness of how sun, wind, and heat or cold affect the building's interior.

While focusing on building elements that affect interior designers, *Building Systems for Interior Designers, Third Edition* addresses this multidisciplinary approach to building design. We begin our study of building systems in Part I by looking at the relationships among the environment, the building, and human health and safety.

Chapter 1, "Environmental Conditions and the Site," looks at climate change, energy sources and consumption, and how site conditions affect building design.

Chapter 2, "Designing for the Environment," investigates the building envelope and the role of insulation in heat flow. Energy efficient design, the building design process, and sustainable design are introduced.

Chapter 3, "Designing for Human Health and Safety," addresses the interaction of the human body with the built environment and how building codes protect us.

1

A common thread ... is the attitude that buildings and sites should be planned and developed in an environmentally sensitive manner, responding to context and climate to reduce their reliance on active environmental control systems and the energy they consume. (Francis D.K. Ching, *Building Construction Illustrated* (5th ed.), Wiley, 2014, Preface)

Environmental Conditions and the Site

Buildings evolved from our need for shelter. In addition to shelter, we depend on buildings for sanitation, visual and acoustic environments, space and means to move, and protection from injury.

> A building's form, scale, and spatial organization are the designer's response to a number of conditions—functional planning requirements, technical aspects of structure and construction, economic realities, and expressive qualities of image and style. In addition, the architecture of a building should address the physical context of its site and the exterior space. (Francis D. K. Ching and Corky Binggeli, *Interior Design Illustrated* [3rd ed.], Wiley 2012, page 4)

We depend on the building's site to provide clean air and to help control thermal radiation, air temperature, humidity, and airflow. Building structures rely on site conditions for support and to help keep out water and control fire. The site can also play a role in providing clean water, removing and recycling wastes, and providing concentrated energy.

Once these basic physical needs are met, we turn to creating conditions for sensory comfort, efficiency, and privacy. We need illumination to see, and barriers that create visual privacy. We seek spaces where we can hear others speak clearly, but which offer acoustic privacy. The building's structure gives stable support for all the people, objects, and architectural features of the building.

The next group of functions supports social needs. We try to control the entry or exit of other people and of animals. Buildings facilitate communication and connection with the world outside through windows, telephones, mailboxes, and computer and video networks. Our buildings support our activities by distributing concentrated energy to convenient locations, primarily through electrical systems.

Finally, a building capable of accomplishing all of these complex functions must be built without excessive expense or difficulty. Once built, it must be able to be operated, maintained and changed in a useful and economical manner. The building should be flexible enough to adapt to changing uses and priorities. Eventually, the building's components may be disassembled and returned to use in other construction.

The design of a building that incorporates all these functions requires coordination between building systems' designers, builders, and users. The building's environmental conditions and its site generate complex factors for architects, engineers, and other design professionals. They, along with landscape architects, examine the site's subsoil, surface water levels, topsoil, and rocks with regard to excavations, foundations, and landscaping. Hills, valleys, and slopes affect stormwater drainage and soil erosion, and the location of roads and paths. Shelter from the wind or exposure to sunlight help determine where the building can be built and the type of landscaping. Nearby buildings create shade, divert wind, and change the natural drainage patterns; they can also result in a lack of acoustic and/or visual privacy.

INTRODUCTION

In Chapter 1, we begin to examine the design of the building and its site. Interior designers benefit from a general understanding of both passive systems and mechanical systems that meet the environmental requirements of buildings. Awareness of building systems provides interior designers with the terminology and basic requirements to ask intelligent questions of architects, engineers, and contractors.

This awareness starts with a basic understanding of environmental and site conditions. Climate affects how buildings are designed in different places and how they relate to their sites. An understanding of energy sources and their history helps put their use in buildings in perspective. As interior designers seek to open building interiors to their surroundings, it is important that they understand what opportunities and challenges are involved.

Throughout history, buildings have looked both out towards the surrounding site and environment, and in towards the people, activities, and objects they contain. (See Figure 1.1) Although interior designers are primarily concerned with the building's interior, their work is often influenced by the building's exterior construction and site.

The building's form and orientation on the site are major concerns of the building's architect. A building's climate and surrounding natural and built features are priorities for the architect, landscape architect, and engineers. During the last decades of the twentieth century, architects began to expand their view of architecture to include areas of social concern, including accessibility and sustainable design, both of which are important to interior designers.

The design of the building, including its massing, configuration, and orientation, generate the relationship between the interior space and the exterior environment. In order to be active and responsible members of the building design team, interior designers must understand the roles and concerns of the architects, engineers, and other consultants who make up the design team. In turn, the rest of the design team will benefit from an awareness of the concerns they share with interior designers.

The interior designer's concern about climate change and renewable energy sources leads to caring about how a building responds to its site and climate and how it fuels its operation. Although they are not directly responsible for deciding on the site and building energy sources, interior designers can play a major role in selecting interior materials that support the conservation of energy and the use of energy sources available on-site.

The interior layout can support or block solar radiation to help keep the interior warm or cool. Selecting thermally massive interior materials can aid passive solar design. In many instances, the project is in an existing building, and most of the work is interior design. In addition, interior designers may be involved in the design of outdoor spaces such as patios adjacent to the building.

CLIMATE CHANGE

Interior designers' concern for how a building responds to its site and climate, and how it fuels its operation are involved in their selection of interior materials that do not contribute to greenhouse gases and that support the conservation of energy and the use of sustainable energy sources.

According to the 2014 report of the Intergovernmental Panel on Climate Change (IPCC), the warming of the climate is unequivocal, and mostly caused by human-created greenhouse gases. (See Figure 1.2) The atmosphere and ocean have warmed, the amounts of snow and ice have diminished, and sea level has risen.

> Continued emission of greenhouse gases will cause further warming and long-lasting changes in all components of the climate system, increasing the likelihood of severe, pervasive and irreversible impacts for people and ecosystems. Limiting climate change would require substantial and sustained reductions in greenhouse gas emissions which, together with adaptation, can limit climate change risks. (IPCC Fifth Assessment Synthesis Report, Climate Change 2014 Synthesis Report Summary for Policymakers, www.ipcc.ch/)

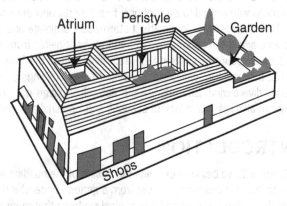

Figure 1.1 Roman residence

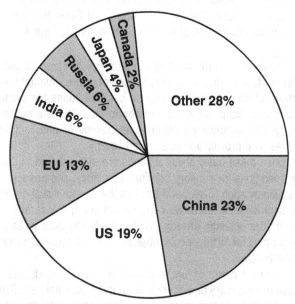

Figure 1.2 Global greenhouse gas emissions, 2008

Small increases in global temperatures are already resulting in hotter summers, changes in precipitation patterns, and rising sea levels. More droughts are occurring in some areas, with floods in others. Warm climate diseases such as malaria are likely to spread. Species extinction is anticipated.

Damaging results include the melting of permafrost in northern Canada, Alaska, and Russia, which can cause huge amounts of organic material to decompose, giving off **carbon dioxide (CO_2)** and methane. According to the US **Environmental Protection Agency (EPA)**, permafrost melting is already causing sinking land that can damage buildings and infrastructure.

ENERGY SOURCES

According to the EPA, US buildings account for 41 percent of total energy use and 65 percent of all electricity consumption. The sources of the energy that is used to construct and operate buildings are important to interior designers concerned about global climate change and energy conservation. As part of the building design team, interior designers can support the use of sustainable energy sources as well as efforts toward passive design. Conserving energy and using clean and renewable energy sources in buildings reduces the amount of air pollution produced by electric power plants and by burning fuels in buildings.

Energy sources are often categorized as renewable or non-renewable. **Renewable energy** is energy that comes from naturally occurring resources such as sunlight, wind, or geothermal heat that are naturally replenished on a human timescale.

All of our energy sources are derived from the sun, with the exception of geothermal, nuclear, and tidal power. Before 1800, solar energy was the dominant source for heat and light, with wood used for fuel. Wind was used for transportation and processing of grain. Early industries located along rivers and streams utilized waterpower. In the 1830s, the earth's population of about one billion people depended on wood for heat and animals for transportation and work. Oil or gas were burned to light interiors. By the 1900s, coal was the dominant fuel, along with hydropower and natural gas. Around the beginning of the nineteenth century, mineral discoveries led to the introduction of portable, convenient, and reliable coal, petroleum, and natural gas fuels to power the industrial revolution.

Fossil fuels such as coal, petroleum, and natural gas were formed from decaying plant and animal matter over vast periods of time beneath the earth's surface. Although fossil fuels continue to be formed, their timeframes are such that they are not replaceable at anything like the rates that they are being used, and they are not considered to be sustainable materials. Remaining world fossil fuel reserves are limited, with much of it expensive and environmentally objectionable to remove. Buildings being built today could outlive fossil fuel supplies used at current rates.

As the world's supply of fossil fuels diminishes, buildings must use non-renewable fuels conservatively if at all, and look to on-site resources such as daylighting, passive solar heating, passive cooling, solar water heating, and **photovoltaic (PV)** electricity.

Some types of energy, such as solar energy, can be used directly by a building for heating and cooling. Others, such as electricity, are produced from another fuel source.

Electricity

Today's buildings are heavily reliant on electricity because of its convenience of use and versatility. Electricity is considered a high-quality energy source; however, only one third of the source energy (often coal) used to produce electricity actually reaches its end use, with most of the rest wasted during production and transmission. As of 2009, consumption of electricity has begun to decline for the first time since World War II, with reductions in all world regions except Asia and the Middle East.

Electric lighting produces heat, which in turn increases air conditioning electrical energy use in warm weather. The use of daylighting is an important sustainable design technique. However, daylight is dependent on weather and time of day, so electric lighting continues to have an important role.

For more information on daylighting and electric lighting, see Chapter 17, "Lighting Systems."

The use of electricity for space heating employs a high-quality source for a low-quality task. Passive or active solar heating design uses this unlimited, free source of energy to heat building interiors.

See Chapter 14, "Heating and Cooling," for more information on solar heating design.

Renewable Energy Sources

Renewable sources include solar (heat, light, and electricity), wind, hydroelectric, geothermal, and biomass. Electricity produced by solar or wind energy can in turn be used to generate hydrogen, a high-grade fuel, from water. The above are all considered to be renewable resources because they can be constantly replenished, but our demand for energy may exceed the rate of replenishment. Some, such as hydroelectric power, can have negative impacts on the environment.

SOLAR ENERGY

The sun acts on the earth's atmosphere to create climate and weather conditions. The earth's rotation determines which part of the earth faces the sun, controlling day and night. Plant life depends on the sun's energy for growth, and humans and other animals depend on plants for food and shelter. Solar energy is the source of almost all of our energy resources. It does not produce air, water, land, or thermal pollution, and is decentralized and very safe to use. It is used for space heating, hot water heating, and photovoltaic electrical energy.

During the day, the sun's energy heats the atmosphere, the land, and the sea. At night, much of this heat is released back into space. The warmth of the sun moves air and moisture across the earth's surface and generates seasonal and daily weather patterns.

SOLAR ENERGY HISTORY

The sun has long been used to heat and illuminate buildings. The Romans incorporated glass windows into their buildings around 50 BCE to bring in daylight and solar heat, and wealthier Romans often added sunrooms to their villas.

In *The Ten Books on Architecture*, Roman architect and engineer Marcus Vitruvius Pollio wrote:

> If our designs for private houses are to be correct, we must at the outset take note of the countries and climates in which they are built. . . . This is because one part of the earth is directly under the sun's course, another is far away from it, while another lies midway between these two. (Translated by Morris Hickey Morgan, Harvard University Press 1914, republished by Dover Publications, Inc., 1960, page 170)

Italian Renaissance architect Andrea Palladio (1508–1580), author of *The Four Books of Architecture*, was influenced by Vitruvius. Palladio placed summer rooms on the north side and winter rooms on the south side of his buildings to take advantage of the sun.

In the seventeenth century, solar heating was revived in Northern Europe for growing exotic plants in greenhouses. Improved glassmaking techniques led to the popularity of greenhouses (conservatories) attached to upper-class residences.

American modernist architect George Frederick Keck (1895–1980) designed the "House of Tomorrow" for the Century of Progress exhibition in Chicago in 1933. His realization that the all-glass house was warm on sunny winter days, even without a furnace led to his designing solar houses in the 1930s and 1940s. (See Figure 1.3)

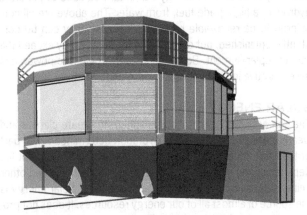

Figure 1.3 George Frederick Keck "House of Tomorrow," 1933

SOLAR RADIATION

Solar radiation drops with the distance from the sun, as solar rays spread out. The path of solar rays through the atmosphere is longer in the morning and evening than at noon, and longer at noon at the poles than at noon at the equator. (See Figure 1.4)

The **electromagnetic spectrum** of radiation emitted by the sun includes wavelengths ranging from extremely short x-rays to very long radio waves. (See Figure 1.5) Radiation is reflected, scattered, and absorbed by dust, smoke, gas molecules, ozone, carbon dioxide, and water vapor in the earth's atmosphere. Radiation that has been scattered or re-emitted is called diffuse radiation. The portion of the radiation that reaches the earth's surface without being scattered or absorbed is referred to as direct radiation.

Ultraviolet (UV) wavelengths make up only a small percentage of the sun's rays that reach sea level, and are too short to be visible by the human eye. UV radiation triggers **photosynthesis** in green plants, producing the oxygen we breathe, the plants we eat, and the fuels we use for heat and power. During photosynthesis, plants take carbon dioxide from the air and give back oxygen. Humans and other animals breathe in oxygen and exhale carbon dioxide. Plants transfer the sun's energy to us when we eat them, or when we eat plant-eating animals. That energy goes back into the environment when animal waste decomposes and releases nitrogen, phosphorus, potassium, carbon and other elements into the air, soil, and water. Animals or microorganisms break down dead animals and plants into basic chemical compounds, which then reenter the cycle to nourish plant life.

UV radiation kills many harmful microorganisms, purifying the atmosphere and eliminating disease-causing bacteria from

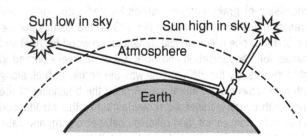

Figure 1.4 Sun's path through the atmosphere

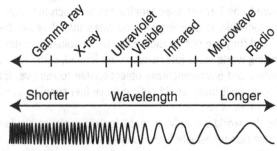

Figure 1.5 Electromagnetic spectrum

sunlit surfaces. It also creates vitamin D in our skin, which we need to utilize calcium. Photosynthesis also produces wood for construction, fibers for fabrics and paper, and landscape plantings for shade and beauty.

Infrared (IR) radiation, with wavelengths longer than visible light, carries the sun's heat. The sun warms our bodies and our buildings both directly and by warming the air around us.

The distance that radiation must travel through the earth's atmosphere, as well as atmospheric conditions, largely determine the amount of solar radiation that reaches the earth's surface. The distance varies with the angle of the earth's tilt toward or away from the sun. The angle is highest in the summer, when direct solar radiation strikes perpendicular to the earth's surface. The angle is lowest in the winter, when solar radiation travels a longer path through the atmosphere. Consequently, the greatest potential solar gain (in the Northern Hemisphere) for a south-facing interior space occurs during the winter. (See Figure 1.6) Nearer to the equator, the sun remains more directly overhead throughout the year. (See Figure 1.7)

The sun illuminates the indoors through windows and skylights during the day. Direct sunlight is often too bright for comfortable vision. When daylight is scattered by the atmosphere or blocked by trees or buildings, it offers an even, restful illumination. Under heavy clouds and at night, artificial light provides adequate illumination.

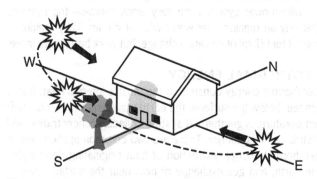

Figure 1.6 Sun angles in northern latitudes

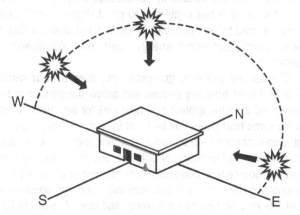

Figure 1.7 Sun angles in tropical latitudes

Sunlight can also be destructive. Most UV radiation is intercepted by the high-altitude ozone layer, but enough gets through to burn our skin painfully. Over the long term, exposure to UV radiation may result in skin cancer. Sunlight contributes to the deterioration of paints, roofing, wood, and other building materials.

Fabric dyes may fade, and many plastics decompose when exposed to direct sun. This is an issue for interior designers when specifying materials.

Photovoltaic (PV) technology converts solar energy directly into electricity at a building's site. PV collectors provide energy for heating water or for electrical power. PV cells are often made of silicon, the most common material in the earth's crust after oxygen. PV cells are very reliable and have no moving parts. They produce no noise, smoke, or radiation.

See Chapter 15 Electrical System Basics for more information on photovoltaic (PV) technology.

The use of solar energy for heating requires consideration of shading to avoid overheating. Greek and Roman buildings used porticoes and colonnades for both shade and protection from rain. Following their example, Greek Revival architecture in the southern United States adopted large overhangs supported by columns, as well as large windows for increased ventilation and white exterior colors for maximum solar reflectance.

WIND POWER

Wind power derives from currents created when the sun heats the air and the ground. Wind power uses a turbine to convert the energy of wind flow into mechanical power that a generator can turn into electrical energy.

By 200 BCE, windmills were used in China to pump water. By the eleventh century CE, windmills were used in the Middle East to grind grain. Their use declined as steam engines dominated during the Industrial Revolution. Larger **wind turbines** were developed in Denmark by 1890 to generate electricity. In the 1930s, wind turbines brought low-cost electrical power to rural areas of the United States. (See Figure 1.8)

Wind energy is plentiful in most of the United States. Wind turbines require a windy site, and raising the turbine as high as possible accesses higher wind speeds. (See Figure 1.9) Although wind is an intermittent source, turbines can be connected to the electrical grid for steady power. Stand-alone systems require battery storage. Hybrid systems combine wind with photovoltaics, with wind power dominating in less sunny, windier winters and solar power providing electricity in summers. The noise produced by small wind turbines is generally not objectionable to most people, and larger turbines are being engineered to reduce noise levels.

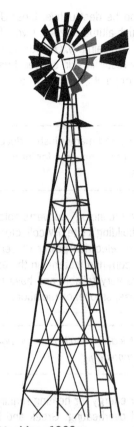

Figure 1.8 Wind turbine, 1930s

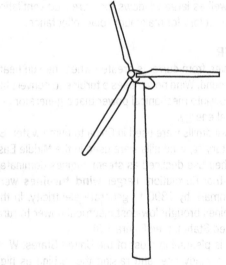

Figure 1.9 Wind turbine today

HYDROELECTRIC POWER

Hydroelectric power (hydropower) is energy that is produced when water stored behind a dam is released at high pressure. This energy is transformed into mechanical energy, which is used by a turbine to generate electricity. In the United States, about 5 percent of energy is produced by falling water.

Figure 1.10 Water-powered machinery for boring gun barrels

Source: Reproduced from Diderot's Encyclopedia (1777), in T.K. Derry and Trevor I. Williams, *A Short History of Technology*, Oxford University Press 1960, republished by Dover 1993, page 149

Hydroelectric power has a long history. (See Figure 1.10) The world's first hydroelectric power plant began operation on the Fox River in Appleton, Wisconsin, in 1882. By 1907, hydropower was providing 15 percent of US electrical generation.

Today hydroelectric power is almost exclusively used to generate electricity. The dams needed typically require flooding large areas of land to produce storage lakes. This disturbs the local ecology and can prevent fish from reaching their spawning grounds. Some outdated, dangerous, or ecologically damaging dams in the United States are being demolished.

Micropower systems are very small hydroelectric systems that rely on running river water without a dam. They require at least 3 feet (1 m) of elevation change, but work better with more.

GEOTHERMAL ENERGY

Geothermal energy consists of the earth's internal heat. About ten feet below the surface, the earth maintains a fairly constant temperature. A geothermal system collects, concentrates, and distributes this energy. There are two common applications of geothermal energy: extraction of heat originating deep within the earth, and geo-exchange of heat near the surface using a heat pump.

Geothermal energy can be extracted where sufficient heat is brought near the surface by conduction, bulging magma, or ground water that has circulated to great depths. Geothermal energy is used to heat buildings in Iceland and Japan. In Boise, Idaho, direct geothermal energy heats over 65 downtown businesses.

The second process, geo-exchange, uses a **heat pump** to extract heat from the ground just below the surface in the winter, and uses the ground as a heat sink for summer cooling, so the same heat pump can be used for both heating and cooling. Geo-exchange heat pumps can significantly reduce energy consumption and emission of pollution and greenhouse gases. A ground-source heat pump offers much greater efficiency than an air-source heat pump, is between three and four times as efficient as electric resistance heating, and uses 70 percent less energy than standard air conditioning equipment.

For more information on heat pumps, see Chapter 14, "Heating and Cooling."

BIOMASS ENERGY

Biomass is the organic matter of plants. Photosynthesis provides the materials for biomass conversion, which includes the combustion of firewood, crop waste, and animal wastes. Biomass can replace chemicals made from fossil fuels to generate electricity and as fuel for transportation vehicles.

Biomass makes use of energy from two types of sources: plants grown for their energy content, and organic waste from agriculture, industry, or garbage. When biomass decomposes, it creates food for new plants; converting it into energy diverts it from this use. Biomass can be considered to be carbon neutral, removing the same amount of carbon dioxide as it grows that is returned to the atmosphere when it is burned.

Biofuels derived from biomass include ethanol alcohol, biodiesel, and methane. Biomass conversion into fuel may require more energy than is obtainable from the product itself, in which case it is not a sustainable process.

HYDROGEN

The most abundant element on earth, hydrogen is found in many organic compounds as well as water. Although not occurring naturally as a gas, it can be separated from other elements and burned as a fuel. Used in fuel cells, hydrogen combines electrochemically with oxygen to produce electricity and heat, with only water vapor emitted in the process. When used as a fuel, hydrogen is nonpolluting, producing only water when it is burned, and does not contribute to global warming.

Chemical bonds must be broken to free the hydrogen locked within compounds such as water. The most practical method involves production from water by electrolysis, which breaks water into hydrogen and oxygen by passing electrical current through water, using wind or PV-generated electricity.

Hydrogen can be used to generate pollution–free electricity in fuel cells, or to power automobile engines. Hydrogen must be stored in heavy and expensive high-pressure tanks. If stored as a liquid, it must be cooled to −423°F (−253°C).

STORING RENEWABLE ENERGY

Both wind and solar energy are not easily stored. With any energy storage option, some energy is lost. Batteries lose some of the electrical energy they store as heat, and it takes a large volume of batteries to store a lot of energy.

Wind and hydropower generate electricity by mechanical means, and their energy can be stored before it is converted into electricity. Hydrogen produced by electrolysis can be stored and later recombined with oxygen to recover energy. Hydrogen fuel cells can produce a controlled release of stored energy.

Connecting a photovoltaic system to the existing electrical grid allows the grid to supply electricity when the PV system is inactive at night. Extra PV energy is sent onto the grid. Using a special PV electrical meter, **net metering** only charges the user for electricity used in excess to what they produce.

Non-Sustainable Energy Sources

Petroleum and natural gas split the energy market about evenly by 1950. The United States was completely energy self-sufficient, thanks to relatively cheap and abundant domestic coal, oil and natural gas.

Beginning in the 1950s, the United States experienced steadily rising imports of crude oil and petroleum products. In 1973, political conditions in oil-producing countries led to wildly fluctuating oil prices, and high prices encouraged conservation and the development of alternative energy resources. The 1973 oil crisis had a major impact on building construction and operation. Unstable political conditions led to an emphasis on reducing imported oil. Between 2005 and 2011, the amount of oil imported by the United States dropped by 33 percent.

Our most commonly used fuels—oil, gas, and coal—are fossil fuels. We started using fossil fuels around 1850. Although limited supplies still remain, it is becoming continually more difficult to access them without causing environmental damage. Burning fossil fuels produces most of the air pollution and smog we experience, plus acid rain and global climate change. These resources are clearly not renewable in the short term, and are not sustainable resources.

OIL

Petroleum is a liquid mixture of hydrocarbons that is present in certain rock strata and can be extracted and refined to produce fuels including gasoline, kerosene, and diesel oil. It is often called "oil," especially when used as a fuel or lubricant. Oil is used to heat buildings and to make lubricants, plastics, and other chemicals, as well as to power vehicles.

The first oil wells were dug in China starting around the fourth century CE. Oil was first distilled into kerosene for lighting in the mid-nineteenth century. In 1859, the first oil well was drilled in Titusville, Pennsylvania. Fuel oil began to replace coal for building heating in the 1920s.

New wells are deeper, underwater, or in almost inaccessible locations. Oil shale is becoming a more common source of oil. It requires huge amounts of energy to extract oil from tar sands, resulting in a high cost to mine and a very high cost to the environment.

GAS

The first well intentionally drilled to obtain natural gas was drilled in 1821 in Fredonia, New York. It was originally used as a fuel for streetlights. In 1855, Robert Bunsen invented the Bunsen burner so that gas could be used to provide heat for cooking and warming buildings. However, there were very few pipelines for natural gas until the 1940s.

Natural gas is used to generate electricity, and for industrial, residential, and commercial uses. More than half of US commercial establishments and residences are heated using gas.

Natural gas is primarily composed of methane, and produces carbon dioxide when burned. It is delivered to most parts of the United States and Europe by an extensive network of pipelines. Most of the easily obtained natural gas in North America has already been taken out of the ground, with limited supplies available from deep wells.

Natural gas recovered from shale now supplies 30 percent of US natural gas. Supplies of natural gas from shale are relatively clean, compared to the coal used to generate electricity. However, the process of **hydraulic fracturing (fracking)** poses significant environmental risks, including the contamination of well water and the increased possibility of earthquakes. Extraction and transportation can emit methane, a greenhouse gas.

Liquid natural gas (LNG) is odorless, colorless, non-toxic, and non-corrosive. LNG is condensed and shipped in tankers at −260°F (−162°C) over long distances. High costs of production and expensive cryogenic tank storage have limited its widespread commercial use. Concerns about cost and safety have limited development of US terminals.

COAL

Surface mining of coal and its household use is documented by archeological evidence in China from about 3490 BCE. The process of mining coal developed in sixteenth-century Scotland, with advances in the seventeenth century. The Industrial Revolution in eighteenth-century Britain led to extensive use of coal in Britain to drive steam engines. (See Figure 1.11)

Since the 1990s, coal use in buildings has declined, with many large cities limiting its application. Coal is the largest source of energy for the generation of electricity worldwide. It is also used for industrial processes such as refining metals. Currently in the United States, most coal is used for electric generation and heavy industry.

Coal is inconvenient to transport, handle, or use. Because it is dirty to burn and may cause acid rain, its use is restricted to large burners, where expensive equipment is installed to reduce air pollution. Modern techniques scrub and filter out sulfur ash

Figure 1.11 Early nineteenth-century Bradley coal mine, Staffordshire, England

Source: Reproduced from T.K. Derry and Trevor I. Williams, *A Short History of Technology*, Oxford University Press 1960, republished by Dover 1993, page 470

from coal combustion emissions, and older coal-burning plants that still contribute significant amounts of airborne pollution are under governmental pressure to improve. Even with this equipment, burning coal still produces CO_2 and contributes to global warming.

The United States has enough coal to last well over a century. However, deep mining exposes miners to the risk of explosions and cave-ins, as well as severe respiratory ailments by exposure to coal dust. Strip mining is damaging to the surface of the land, and reclamation, although possible, is expensive. The western United States, a location where water for reclamation is a scarce resource, is the site of much of the current strip mining.

NUCLEAR ENERGY

Nuclear energy is produced by fission. The introduction of nuclear power promised an energy source that used resources very slowly.

The first nuclear power plant started operation in 1957. In 2014, 62 operating nuclear plants with 100 reactors, located in 31 US states. Nuclear power produces around 20 percent of electricity consumed in the United States, according to the EPA.

Despite originally being hailed as the answer to our energy problems, nuclear fission has become one of the most expensive and least desirable ways to produce electricity. The narrowly averted disaster in 1979 at Three Mile Island, Pennsylvania; the 1986 explosion at Chernobyl, Ukraine; and the Fukushima Daiichi nuclear disaster in 2011 in Japan have realized safety fears.

Nuclear plants contain high pressures, temperatures, and radioactivity levels during operation, and have long and expensive construction periods. The public has serious concerns over the release of low-level radiation over long periods of time, the risks of high-level releases, and problems with disposal of radioactive fuel. Nuclear reactors consume huge amounts of cooling water, and heat up rivers. Added to this is the threat of nuclear materials falling into the hands of terrorists or unreliable governments. Civilian use has been limited to research and generation of electricity.

Global Climate Change

Global warming and cooling are part of the earth's natural cycle. However, the global warming that is currently occurring is due in large part to the actions of human beings. The presence of certain gases in the atmosphere allows the sun's ultraviolet (UV) radiation to pass through, but blocks the infrared (IR) radiation emitted by sources on earth. The increasing production of some gases, most commonly carbon dioxide, is resulting in a warming trend called the greenhouse effect.

GREENHOUSE EFFECT

The greenhouse effect is a natural phenomenon that helps regulate the temperature of our planet, protecting the earth's surface from extreme differences in day and night temperatures. **Greenhouse gases** are pollutants that trap the earth's heat, especially emissions of carbon dioxide from burning fossil fuels. As greenhouse gases accumulate in the atmosphere, they absorb sunlight and IR radiation and prevent some of the heat from radiating back out into space, trapping the sun's heat closer to the earth. If all of these greenhouse gases were to suddenly disappear, our planet would be 60°F (15.5°C) colder than it is, and uninhabitable. Significant increases in the amount of these gases in the atmosphere cause global temperatures to rise.

Human activities contribute substantially to the production of greenhouse gases. Activities including building construction and operation are adding greenhouse gases to the atmosphere at a faster rate than at any time over the past several thousand years, accelerating global climate change. According to the US Department of Commerce National Oceanic and Atmospheric Administration (NOAA) Annual Greenhouse Gas Index for 2013, global warming has increased 34 percent since 1990.

Greenhouse gases include carbon dioxide, methane (which is pound for pound over 20 times greater than carbon dioxide over a 100-year period), nitrous oxide, ozone, **chlorofluorocarbons (CFCs)** and other gases. (See Table 1.1) Most are produced by burning fossil fuels, which include coal, oil, and natural gas. Water vapor can also be considered a greenhouse gas, as it absorbs IR radiation reradiated from the earth.

The earth's current average temperature is around 61°F (16°C). The polar ice caps are already melting, and as they do, sea levels will rise, leading to coastal flooding. Related changes in climate can affect agricultural production and the habitability of some regions.

We can help control global warming by reducing the use of fossil fuels through energy conservation and alternative fuel sources. Designers can design for energy conservation and the use of clean and renewable energy sources.

Interior designers can specify materials and equipment that avoid fuel combustion and environmentally damaging refrigerants, and select insulation, upholstery, and other products made with environmentally benign materials.

OZONE

A blanket of ozone gas screens the earth from harmful UV radiation from the sun. Depletion of ozone in the stratosphere results in more UV radiation reaching the earth's surface. We need to preserve the high-altitude ozone layer that intercepts most UV radiation before it can reach earth.

The thinning of the ozone layer has been in large measure due to the use of chlorofluorocarbon (CFC) refrigerants in building air conditioning systems. CFCs escaped from air conditioners, released as spray can propellants, or released from other refrigerant sources slowly migrated to the upper atmosphere, where they can continue to deplete the protective ozone layer for an estimated 50 years.

Production of CFC refrigerants has been banned, but previously released chemicals continue to thin the ozone layer. The US Clean Air Act of 1990 allows the sale of new refrigeration equipment that uses **hydrofluorochlorocarbons (HCFCs)** until 2020; after that, only service of existing systems will be permitted. In the United States, production levels of HCFCs are scheduled to be frozen in 2015 and banned entirely in 2030. As of 2014, the stratospheric ozone layer is recovering, but complete recovery is not expected until 2050 or later. Refrigerants including CFCs, HCFCs, and **hydrofluorocarbons (HFCs)** are also greenhouse gases that contribute directly to global warming.

See Chapter 12, "Heating and Cooling Systems," for more information on refrigerants.

Energy Consumption by Buildings

According to the National Academies, 84 percent of the energy used in the United States comes from fossil fuels. With less than five percent of the world's population, the United States consumes nearly 25 percent of primary energy consumption. In 2010, buildings accounted for 41 percent of this use.

TABLE 1.1 GREENHOUSE GASES

Greenhouse Gas	Man-made Sources	Comments
Carbon dioxide (CO₂)	Burning fossil fuels, cement production, deforestation	After water vapor, the most common greenhouse gas
Methane (CH₄)	Landfills, rice farming, cattle raising, burning fossil fuels, landfills	Next most common greenhouse gas after CO₂
Nitrous oxide (N₂O)	Nylon production, nitric acid production, agriculture, automobile engines, biomass burning	Third most common greenhouse gas after N₂O
Hydrofluorocarbons (HFCs)	Fire suppressants, refrigerants	Replacement for CFCs, do not affect ozone layer but add carbon dioxide
Perfluorocarbons (PFCs)	Production of aluminum from alumina	Very potent, very long-lived
Sulfur hexafluoride (SF₆)	Electrical equipment manufacturing, window filling inert gas, magnesium casting	Extremely potent but low amount in atmosphere

The sun's energy arrives at the earth at a fixed rate, and the supply of solar energy stored over millions of years in fossil fuels is limited. As the population keeps growing, people continue using more energy. We do not know exactly when we will run out, but we do know that wasting the limited resources we have is a dangerous way to go. Through careful design architects, interior designers, and building engineers can help make these finite resources last longer.

The design of the building's interior can help keep the interior warm or cool. Thermally massive interior materials support **passive solar design**, which relies on the design of the building itself, rather then on fuel-consuming mechanical equipment.

Building designers and owners now strive for energy efficiency to minimize costs and conserve resources. United States building codes include energy conservation standards. The United States is increasingly taking energy use and global climate change seriously. The conservation of resources and use of environmentally friendly energy sources have become standard practice for building designers.

BUILDING SITE CONDITIONS

Early in the design process, site-planning decisions affect the options considered for building lighting and control systems, as well as the amount and type of energy used. The use of on-site resources such as sun, water, wind, and plant life can replace or supplement the building's dependence on non-renewable fuels. Site planning considers existing site conditions, climatic conditions, and intended building use. It is important for interior designers to be informed and involved in the building design process from the beginning.

Building Placement

Where and how a building is positioned on the earth affects its structure, supply and retention of water, collection and retention of heat from the sun and the earth, cooling and ventilation by winds, exposure to fire, and level of acoustic quiet or noise. Each of these conditions shapes the building's design; the result can reflect and communicate a sense of place.

CONNECTIONS TO SURROUNDINGS

Buildings bring people, vehicles, materials, and sounds of activity to a site. The site is connected to utilities including electricity, water, and natural gas. The building changes how the sight is illuminated with electric lighting at night. Heat flows both to and from the building through openings and the building envelope. Landscaping changes the flow of water, and liquid and solid wastes are often moved offsite for treatment or disposal.

When a building fills an entire site, on-site resources are limited, and wind and sunlight may be blocked. Less heat and water are able to be absorbed, and the roof may be the only area capable of nurturing plants.

Climates

Climate views the weather statistically over long periods of time, including such criteria as temperature, **relative humidity**, solar radiation, and wind speed. Climates vary with the earth's position in relation to the sun, and with latitude and longitude. The characteristics of a climate include the amount of sun, humidity and precipitation, and air temperature, motion, and quality.

Relative humidity (RH) is the amount of water vapor present in air expressed as a percentage of the amount needed for saturation at the same temperature.

Designing for local climate conditions reduces the fuel needed to operate mechanical heating and cooling equipment. Interior shading reduces the need for mechanical cooling, while operable window treatments allow in the sun's warmth when desired.

See Chapter 6, "Windows and Doors," for more information on windows and window treatments.

LOCAL CLIMATES

Local temperatures vary with the time of day and the season of the year. Because the earth stores heat and releases it at a later time, a phenomenon known as **thermal lag**, afternoon temperatures are generally warmer than mornings. The lowest daily temperature is usually just before sunrise, when most of the previous day's heat has dissipated. Although June experiences the most solar radiation in the Northern Hemisphere, summer temperatures peak in July or August due to the long-term effects of thermal storage. Because of this residual stored heat, January and February—about one month past the winter solstice—are the coldest months. It is usually colder at higher latitudes, both north and south, as a result of shorter days and less solar radiation.

MICROCLIMATES

Sites may have **microclimates**, different from surrounding areas, which result from the interaction of the larger climate with site-specific characteristics such as topography, vegetation, elevation, proximity to large bodies of water, views, and wind patterns. Buildings influence microclimates by redirecting rainwater to nourish plants, blocking or channeling wind, and moderating or maintaining hotter temperatures by storing heat in massive materials. A site should be selected with its microclimate advantages in mind.

HEAT ISLANDS

Cities sometimes create their own microclimate **heat islands** with relatively warm year-round temperatures produced by heat sources such as air conditioners, furnaces, electric lights, car engines, and building machinery. The amount of added heat in a heat island varies widely depending on its location, the season and time of day, and the buildings it contains.

Cities are often more cloudy than surrounding areas, and because of their accumulated heat, tend to have more rain instead of snow. Wind is channeled between closely set buildings. High vertical walls and narrow streets diminish solar radiation. Sun is absorbed and reradiated off massive surfaces, and less is given back to the obscured night sky. Highly reflective glass surfaces intensify glare and add to summer cooling loads of adjacent buildings. The convective updrafts created by the buildings in large cities can affect the regional climate.

CLIMATE TYPES

Environmentally sensitive buildings are designed in response to the climate type of the site. Indigenous architecture, which has evolved over centuries of trial and error, provides models for building in the four basic climate types: cold, temperate, hot arid, and hot humid.

COLD CLIMATE DESIGN

Cold climates feature long cold winters with short, very hot periods occurring occasionally during the summer. Cold climates are generally found around 45 degrees latitude north or south; in the United States, North Dakota is an example.

Buildings designed for cold climates emphasize heat retention. Minimizing the surface area of the building reduces exposure to low temperatures. The building is oriented to absorb heat from the winter sun. Passive solar heating is often used to encourage heat retention without mechanical assistance. Cold climate buildings may have fewer windows to limit heat loss. Wind protection may be necessary.

Setting a building into a protective south-facing hillside reduces the amount of heat loss and provides wind protection, as does burying a building in earth. (See Figure 1.12)

TEMPERATE CLIMATE DESIGN

Temperate climates have cold winters and hot summers. Temperate climates are found between 35 degrees and 45 degrees latitude, in Washington, DC, for example.

A temperate climate favors a design that encourages air movement in hot weather while protecting against cold winter winds. Buildings designed for temperate climates employ winter heating and summer cooling, especially where it is humid. In the Northern Hemisphere, south-facing walls are exposed to winter sun, with summer shade for exposures on the east and west sides and over the roof. Deciduous shade trees that lose their leaves in the winter help to protect the building from sun in hot weather and allow the winter sun through. (See Figure 1.13)

Adding materials that retain heat inside a well-insulated building helps even out solar heat gains and losses through windows and internal heat gains from activities; it also moderates high summertime daytime temperatures.

HOT ARID CLIMATE DESIGN

Hot arid climates have long hot summers and short sunny winters, and the daily temperatures that range widely between dawn and the warmest part of the afternoon. They may have very cold winters. Arizona is an example of a hot arid climate in the United States.

Buildings in hot arid climates feature heat and sun control. They often try to increase cooling and humidity by taking advantage of any wind and rain, and make the most of the cooler winter sun.

Designs for hot arid climates provide summer shade to the east and west and over the roof. They use massive walls to create a time lag as heat moves slowly from outside to inside. (See Figure 1.14) Strategies include shading small windows and outdoor spaces from the sun, with light interior colors to diffuse the limited **daylight.**

Enclosed courtyards offer shade and encourage air movement, and the presence of a fountain or pool and plants increases

Figure 1.12 Pioneer dugout home near McCook, Nebraska, 1890s

Figure 1.13 House for temperate climate, 1930s

Figure 1.14 Taos Pueblo, New Mexico, 1880

humidity. In warm climates, sunlit surfaces should be a light color, to reflect as much sun as possible.

HOT, HUMID CLIMATE DESIGN

Hot, humid climates have very long summers with slight seasonal variations and relatively constant temperatures. The weather is consistently hot and humid, as in New Orleans.

Buildings designed for hot, humid climates take advantage of shading from the sun to reduce heat gain and cooling breezes with many large windows, overhangs, and shutters. They minimize east and west exposures to reduce solar heat gain, although some sun in winter may be desirable. High ceilings can accommodate large windows and to help stratify air, with cooler air at the level of occupants and warmer air exiting the space above. Floors may be raised above ground (See Figure 1.15) with crawl spaces for air circulation.

Additional climate design examples are found in Chapter 2, "Designing for the Environment," and Chapter 12, "Principles of Thermal Comfort."

Figure 1.15 Treehouses in Buyay, Mount Clarence, New Guinea

Site Conditions

Architects analyze the local conditions of the building site, including sun and wind patterns in the summer and winter, water runoff patterns, and microclimate conditions. They look at how privacy and site access, views, sound, heat, light, air motion, and water change with the vertical distance from the surface, and apply this information to design the functions of building spaces to match the horizontal and vertical layers of the building. (See Figure 1.16)

The climate of a particular building site is determined by the sun's angle and path, the air temperature, humidity, precipitation, air motion, and air quality. Building designers describe sites by the type of soil, the characteristics of the ground surface, and the topography of the site. The presence of water on the site affects the plants and animals found there. People living on the site are exposed to and alter its views, heat levels, noise, and other characteristics.

Building structures depend on the condition of the soil and rocks on the site. The construction of the building may remove or use earth and stone or other local materials. Alterations destroy, alter, or establish habitats for native plants and animals. Elevating a structure on poles or piers minimizes disturbance of the natural terrain and existing vegetation. Setbacks provide access to daylight and fresh air, and may be required by code to meet height restrictions. (See Figure 1.17)

Typically, public utilities connect distribution systems at the building boundary. These include piping for water and gas, and electric service wiring. Buildings contribute to air pollution directly through fuel combustion, and indirectly through the electric power plants that supply energy and the incinerators and landfills that receive waste. The presence of people has a major environmental impact.

The interior of the building responds to these surrounding conditions by opening up to or turning away from views, noises, smells, and other disturbances. Interior spaces connect to existing on-site

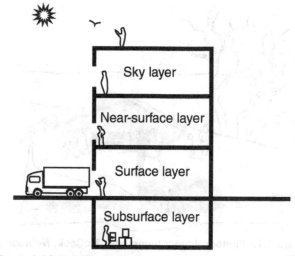

Figure 1.16 Building use layers

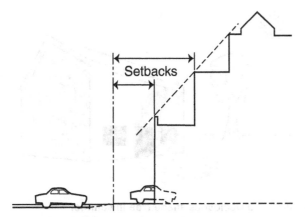

Figure 1.17 Building setbacks

Source: Redrawn from Francis D.K. Ching, *Building Construction Illustrated* (5th ed.), Wiley 2014, page 1.25

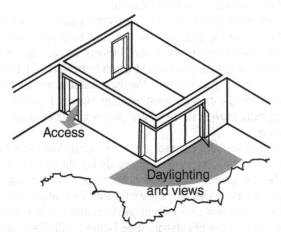

Figure 1.18 Connecting interior to outdoors

Source: Redrawn from Francis D.K. Ching and Corky Binggeli, *Interior Design Illustrated* (3rd ed.), Wiley 2012, page 28

walks, driveways, parking areas, and gardens. (See Figure 1.18) The presence of wells, septic systems, and underground utilities influence the design of residential bathrooms, kitchens, and laundries as well as facilities in commercial buildings.

The hard surfaces and parallel walls in cities intensify noise. Mechanical systems of neighboring buildings may be very noisy, and are hard to mask without reducing air intake, although newer equipment is usually quieter. Plants only slightly reduce the sound level, but the visually softer appearance gives a perception of acoustic softness, and the sound of wind through the leaves may help to mask noise. Fountains also provide helpful masking sounds.

PREVIOUS USES OF SITES

Effective planning can create efficient, environmentally sustainable urban forms while minimizing urban sprawl. Siting buildings near mass transit avoids consumption of fossil fuels and pollution from automobiles.

Reusing land that has already been impacted by human activities rather than undeveloped properties supports land conservation by bringing land back to productive use. (See Table 1.2) Land reuse can also promote economic and social revitalization in distressed areas.

WIND AND THE SITE

Winds are usually weakest in the early morning and strongest in the afternoon, and can change their effects and sometimes their directions with the seasons. Evergreen shrubs, trees, and fences slow and diffuse winds near low-rise buildings.

The wind patterns around buildings are complex, and localized wind conditions between buildings often increase wind speed and turbulence just outside building entryways. The flow of wind typically returns to its original flow pattern after encountering an obstacle. (See Figure 1.19) Less-dense windbreaks such as fences and plants tend to reduce the returning velocity more than thicker building materials.

TABLE 1.2 LAND REUSE

Type	Description	Comments
Greenfields	Undeveloped natural properties that have experienced little or no impact from human activities. Can include agricultural land without activities other than farming.	Loss of prime farmland should be avoided. Support biodiversity by siting the building to avoid encroachment on animal habitats.
Brownfields	Abandoned, idled, or underused industrial and commercial facilities for which expansion or redevelopment is complicated by real or perceived environmental contamination.	Hazardous substances, pollution, or contamination may complicate reuse. Brownfields near preexisting infrastructure and a potential workforce can be valuable.
Grayfields	Blighted or obsolete building sites on land that is not necessarily contaminated.	May be valuable due to scarcity of available urban land, preexisting infrastructure, government incentives.
Blackfields	Include properties such as abandoned coal strip mines and subsurface mines.	Surface water may have very low pH levels, may be contaminated with iron, aluminum, manganese, sulfates.

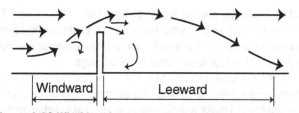

Figure 1.19 Wind barrier
Source: Redrawn from Francis D.K. Ching, *Building Construction Illustrated* (5th ed.), Wiley 2014, p. 1.22

WATER AND THE SITE

Large bodies of water moderate air temperature both between day and night and throughout the year. The evaporation of smaller waterbodies helps cool summer air temperatures.

Water appears on the site due to precipitation of rainwater or snow, or as groundwater and soil moisture. Some sites offer **potable** (drinkable) water. Treatment or removal of wastewater occurs on some sites.

Fountains, waterfalls, and trees tend to raise the humidity of the site and lower the temperature. Large bodies of water, which are generally cooler than the land during the day and warmer at night, act as heat reservoirs that moderate variations in local temperatures and generate offshore breezes. They are usually warmer than the land in the winter and cooler in the summer.

Rainwater falling on steeply pitched roofs with overhangs is usually collected by gutters and downspouts to be carried away as surface runoff, or underground through a storm sewer. Drain leaders are pipes that run vertically within partitions to carry the water down through the structure to the storm drains. Even flat roofs have a slight pitch, with the water collecting into roof drains that pass through the interior of the building.

Sites and buildings should be designed for maximum on-site rainfall retention. Roof ponds and cisterns hold water that falls on the roof, giving the ground below more time to absorb runoff.

See Chapter 9, "Water Supply Systems," for more information on using rainwater.

PLANT AND ANIMAL LIFE

Building sites provide environments for a variety of plant and animal life. Grasses, weeds, flowers, shrubs, and trees trap precipitation, prevent soil erosion, provide shade, and deflect wind. Grassy areas are cooler than paved areas, both day and night. Plants play a major role in food and water cycles, and their growth and change through the seasons help us mark time. They can help keep unwanted solar heat and light out of buildings during the warmest part of the year. Vegetation absorbs moisture during the day for release at night. Plants improve air quality by trapping particles on their leaves, to be washed to

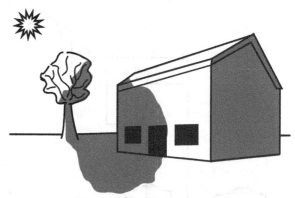

Figure 1.20 Deciduous shade tree in summer

the ground by rain; photosynthesis assimilates gases, fumes and other pollutants.

Plants increase our sense of enjoyment and enhance privacy. Plants frame or screen views, moderate noise, and visually connect the building to the site.

Deciduous plants grow and drop their leaves on a schedule that responds more to the cycles of outdoor temperature than to the position of the sun. In the Northern Hemisphere, where the sun reaches its maximum strength from March 21 through September 21, plants provide the most shade from June to October, when the days are warmest. (See Figures 1.20 and 1.21) Evergreens provide shade all year and help reduce snow glare in winter.

In the Northern Hemisphere, a deciduous vine on a trellis over a south-facing window grows during the cooler spring, shades the interior during the hottest weather, and loses its leaves in time to welcome the winter sun. Vertical vine-covered trellises work well on east and west facades, while horizontal ones work in any orientation. (See Figure 1.22) The vine also cools its immediate area by evaporation.

Bacteria, mold, and fungi break down dead animal and vegetable matter into soil nutrients. Bees, wasps, butterflies, and birds pollinate plants, but are kept out of the building. Termites may attack the building's structure. Building occupants may welcome cats, dogs, and other pets into a building, but want to exclude nuisance animals such as mice, raccoons, squirrels, lizards, and stray dogs.

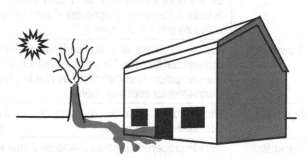

Figure 1.21 Deciduous shade tree in winter

Figure 1.22 Trellis with vine

Termite damage is a problem in warmer climates. It is advisable to recommend an inspection before beginning a remodeling project if there are termites in the area.

SHADE

The ability of trees to provide shade depends on their orientation to the sun, their proximity to the building or outdoor space, their shape, height and spread, and the density of their foliage and branch structure. In the Northern Hemisphere, the most effective shade is on the southeast in the morning and the southwest during late afternoon, when the sun has a low angle and casts long shadows.

Air temperatures in the shade of a tree are about 5° to 11°F (3° to 6°C) cooler than in the sun. A wall shaded by a large tree in direct sun may be 20° to 25°F (11° to 14°C) cooler. This temperature drop is due to the shade plus cooling evaporation from the surface area of the leaves. Shrubs right next to a wall produce similar results, trapping cooled air and preventing drafts from infiltrating the building. Neighborhoods with large trees have maximum air temperatures up to 10°F (6° C) lower than those without.

BUILDING SITING AND ORIENTATION

The site location, building orientation and geometry, and local climate conditions affect the design of the building and its systems. Site selection and building orientation benefit from the involvement of architects and their consultants as early as possible. By close coordination of an interdisciplinary design team, the architect can optimize the use of the site and its integration with its local environment. Each member of the design team—including the interior designer—can determine how the site affects his/her discipline and how to best further sustainable design goals.

The interaction of building orientation and interior layout has a long history. The Roman architect Vitruvius wrote about the design of public baths that "the warmest possible situation must be selected; that is, one which faces away from the north and northeast. The rooms for the hot and tepid baths should be lighted from the southwest, or, if the nature of the situation prevents this, at all events from the south, because the set time for bathing is principally from midday to evening." (*The Ten Books on Architecture*, Dover Publications, Inc., 1960, page 157)

Building orientation considerations include solar orientation, topography or adjacent structures, prevailing winds, available daylight and shading, views, and landscaping and irrigation needs. The building's orientation, form, and compactness have significant impacts on heating, cooling, and lighting systems, and energy conservation. The orientation of the building and its width and height determine how the building will be shielded from excess heat or cold or open to ventilation or light. For example, the desire to provide daylight and natural ventilation to each room limits the width of multi-story hotels.

Buildings that minimize east and west exposures are generally more energy efficient, especially where extensive glazing absorbs heat during summer months. Orienting the main elevations with operable windows perpendicular to prevailing breezes aids natural ventilation. A rectilinear building with its length oriented in an east/west direction will present its longer south façade to the sun for maximum winter solar gain.

When designing a building to take advantage of solar heating, provisions must be made to prevent overheating in warm weather. Roofs provide a barrier to excess summer solar radiation, especially in low latitudes where the sun is directly overhead. The transmission of solar heat from the roof to the interior of the building can result in high ceiling temperatures. High ceiling temperatures can be reduced with thermally resistant materials, materials with high thermal capacity, or ventilated spaces in the roof structure.

Orienting building entrances away from or protected from prevailing cold winter winds, and buffering entries with airlocks, vestibules, or double entry doors dramatically reduces the amount of interior and exterior air change when people enter.

Locating an unheated garage, mudroom, or sunspace between the doors to a conditioned interior space is a very effective way to control air loss in a building.

Interior Layout

To ensure overall compatibility, the layout of a building's interior should be considered while the building is being located on its site, and while its rough shape, shading, and orientation are being established.

In the Northern Hemisphere, spaces with maximum heating and lighting needs should be located on the building's south face. Buffer areas, such as toilet rooms, kitchens, corridors, stairwells, storage, garage, and mechanical and utility spaces need less light and air conditioning, and can be located on a north or west wall. The areas with the greatest illumination level needs should have access to natural lighting. Conference rooms, which need few or no windows for light and views, can be located farther away from windows. Spaces that need a lot of cooling due to high internal heat gains from activities or equipment should be located on the north or east sides of the building.

Energy for mechanical heating and cooling can be conserved by locating spaces that accommodate cooler temperatures on the north side of the building. Buffer spaces such as garages on the north or west protect the building interior from winter cold or, for the western exposure, summer sun.

For more information on daylighting and layout, see Chapter 17, "Lighting Systems."

Openings in the building are the source of light, sun, and fresh air. Building openings provide opportunities for wider personal choices of temperature and access to outdoor air. On the other hand, they limit control of humidity, and permit the entry of dust and pollen.

Planning the layout of interior spaces to maximize the use of standard sizes of materials and products minimizes construction waste.

Existing Buildings

Existing buildings are often demolished and replaced with new products in new buildings. Building reuse is a more sustainable practice. **Demolition by hand salvage**, in which the building is taken apart and its constituent pieces are reused, is an alternative that is labor-intensive (which may provide job training experience) but energy-wise.

Historic preservation and adaptive reuse represent the highest form of recycling. Building reuse reduces the demand for new land, recycles existing buildings, uses fewer materials, and reduces the amount of construction demolition and waste going to landfills. By keeping older buildings in usable condition and protecting their original use or finding a new one, communities create a sense of continuity and cultural richness.

Construction work in occupied existing buildings requires separation and protection of occupied areas from construction areas. Noise control measures must be taken. Ducts and airways must be protected from dust, moisture, particulates, chemical pollutants, and microbes during demolition and construction. There should be increased ventilation and exhaust air at the construction site.

This first chapter has introduced climate change, energy sources, and site conditions. In Chapter 2, we explore how the building envelope—the interface between the outside and the inside—controls building heat flow. The basics of the building design process and how sustainable design promotes energy efficiency are also introduced.

2

Designing for the Environment

Both the architecture of the building and its environmental control systems are used to control the interior thermal environment. The building itself and its environmental systems should work together to promote the thermal comfort of its occupants in an energy-efficient manner. Architects and their consultants seek to find the balance between the passive environmental controls of the building's enclosure and the active mechanical equipment.

> It is best to restrict the role of active devices insofar as possible to that of fine tuning the internal environment to optimum conditions, once the heavy work of environmental modification has been performed by the passive devices of the site and the building enclosure. . . . Machines, no matter how powerful, can never substitute fully for the good design judgment of the architect who sites and configures the building, for it is the enclosure of the building, not the mechanical equipment, that must create the basic conditions for human satisfaction, both spiritual and physical. (Edward Allen, *How Buildings Work* (3rd ed.), Oxford University Press 2005, page 253)

INTRODUCTION

In Chapter 2, we investigate the building envelope and the role of insulation in heat flow. Energy efficient design, the building design process, and sustainable design also are introduced.

The experience of the people who inhabit and use the buildings designed by architects, landscape architects, engineers, and interior designers can be best enhanced by designs that are both aesthetically pleasing and socially meaningful. This requires both sensitivity to composition and form, plus an understanding of science and technology.

Architects seek to interpret ideas through physical form for human habitation. They aim to design an environment that nurtures human endeavor without imposing excessive external stress.

As part of the design team, interior designers are involved in supporting, rather than obstructing, the flow of energy through the building envelope. They have a role in the design of passive systems and should care about energy use. Sustainable design and the use of environmentally responsible materials benefit from the skills of interior designers, and they play an important role in the **LEED** building certification process for both residential and commercial spaces.

Leadership in Energy & Environmental Design (LEED) is a green building certification program that certifies building projects that satisfy prerequisites and earn points. It was developed by the US Green Building Council (USGBC).

BUILDING ENVELOPE

Like our skins, a building is the interface between our bodies and our environment. The building envelope is the point at which the inside comes into contact with the outside, the place where energy, materials, and living things pass in and out. The building's interwoven structural, mechanical, electrical, plumbing, and other systems create an interior environment that supports our needs and activities, and responds to exterior weather and site conditions. In turn, the building itself and its site affect and are affected by the earth's larger natural patterns.

The building envelope encloses and shelters space. It furnishes a barrier to rain and protects from sun, wind, and harsh temperatures. Entries are the transition zone between the building's interior and the outside world.

The building envelope admits or excludes heat gain, contains internal heat, and dissipates excess internal heat. The building's surface affects the comfort of the occupant primarily through surface temperatures. This in turn can modify air temperature as air moves across warm surfaces. Air motion and relative humidity are important to cooling. In addition, air quality is important for both heating and cooling for most building occupancies.

History

The design of buildings throughout history has shown many ways to design shelters that respond to local environments without the use of mechanical equipment.

Sod houses were built primarily by German and Scandinavian immigrants during the settlement of the US and Canadian prairies, which have cold winters and hot summers. (See Figure 2.1) The thick, tough roots of prairie grass allowed it to be cut into rectangles that were then laid crosswise into walls two feet (0.6 meter) thick. The south side had a door and two or more windows, often with a window at each end. A variety of roofing materials were used. Sod houses were well insulated but damp and vulnerable to rain damage, and required frequent maintenance. Canvas or plaster lined interior walls. Sod houses were warm in winter and very cool in summer, and were not liable to be burned by fire.

In the hot arid climates of Middle Eastern deserts, the Arab tent historically served as a nomadic house. (See Figure 2.2) The tent was made of a light wood pole frame with a handmade textile covering. It was a lightweight, modular, and mobile shelter that provided shade and encouraged air movement.

In traditional buildings, shelter from the weather was provided through passive systems for heating, cooling, and daylighting.

Figure 2.2 Traditional Arab tent

Careful design of roofs, walls, and windows, along with interior surfaces, can maintain comfortable interior temperatures for most of the year in most North American climates. Scheduling can allow us to avoid the most uncomfortable hours, as the tradition of a midday siesta demonstrates.

With the advent of mechanical building systems, the building envelope came to be regarded as a barrier separating the interior from the outdoor environment. Architects created an isolated environment, and engineers equipped it with energy-using devices to control conditions. The desire to control air moisture, motion, and pollution content resulted in a tendency to seal the building to exclude outdoor air except through controlled mechanical equipment intakes. This led to exclusion of daylight, view, and useful solar heat.

Dynamic Building Envelope

A building envelope does not have to be a barrier, but can dynamically enclose the interior space. A traditional Native American tipi exemplifies a traditional dynamic building envelope. (See Figure 2.3)

Today, the need to conserve energy encourages us to see the building envelope as a dynamic boundary interacting with the

Figure 2.1 Sod house, Anselmo, Nebraska
Source: Wikimedia Commons, Anselmo, Nebraska sod house.JPG by Ammodramas, released to public domain

Figure 2.3 Sioux tipi

Figure 2.4 Building envelope
Source: Redrawn from Francis D.K. Ching and Corky Binggeli, *Interior Design Illustrated* (3rd ed.), Wiley 2012, page 9

external natural energy forces and the internal building environment. (See Figure 2.4) The envelope is sensitively attuned to the resources of the site: sun, wind, and water. The boundary is manipulated to balance the energy flows between inside and outside.

Dynamic elements of the building envelope include operable windows, window shading devices, and insulating shutters. A dynamic building envelope can be sensitive to changing conditions and needs, letting in or closing out the sun's warmth and light, breezes, and sounds. Openings and barriers may be static, like a wall; allow on/off operation, like a door; or offer adjustable control, like venetian blinds. The appropriate architectural solution depends on the range of options desired, the local materials available, and local style preferences.

For more on how the building affects thermal comfort, see Chapter 12, "Principles of Thermal Comfort."

The more flexibility is built into the building envelope, the more important it is to provide adequate controls. Modifications are often made for aesthetic as well as practical reasons. Occupants of residential buildings routinely modify the building envelope to accommodate changes in sun angle and outside temperature as well as to invite in breezes while excluding insects. Thermal shades and other window treatments can block the passive solar collection of warming infrared (IR) radiation on a cold, sunny day. Other user controls for daylighting and solar control devices include awnings, opaque draperies, and translucent curtains.

For more information on window treatments, see Chapter 6, "Windows and Doors."

A dynamic envelope demands that the user understand how, why, and when to make adjustments. This requires communication from the designer to the appropriate building personnel, or automated controls.

Opaque components of the building envelope, such as solid walls, floors, and roofs, are typically intended as fixed barriers to heat, light, air, and noise. Their ability to block heat transmission varies, depending on their construction, orientation, and material selection.

Building Envelope and Codes

Today, building designs are expected to conform to energy codes and standards of client-established energy criteria. As a result, building designers must be able to evaluate the performance of the building envelope. As new concepts and new energy-saving products are introduced to the market, the designer needs to analyze their claims.

Building code requirements for thermal envelope performance of non-residential buildings in North America are usually those set forth in *ANSI/ASHRAE/IES Standard 90.1-2013—Energy Standard for Buildings Except Low-Rise Residential Buildings.* Residential buildings are usually subject to the requirements of the *International Energy Conservation Code* or *ASHRAE Residential Energy Standard 90.2.*

ASHRAE® is a global society advancing human well-being through sustainable technology for the built environment. It focuses on building systems, energy efficiency, indoor air quality, refrigeration, and sustainability through research, standards writing, publishing, and continuing education.

The standards in these codes are intended as minimum requirements for building envelopes. The **US Green Building Council's (USGBC) Leadership in Energy & Environmental Design (LEED)** rating system requires that envelopes be designed to exceed these requirements by between 15 and 60 percent. Buildings that meet LEED standards usually employ elements that control daylighting and solar heat.

Exterior Walls

Exterior walls are part of the building envelope. They support vertical loads as well as horizontal wind loads. Exterior walls control the passage of heat, air, sound, moisture, and water vapor in and out of the building. They are constructed for durability to the weathering of sun, wind, and rain, and for fire resistance. Rigid exterior walls serve as **shear walls** to transfer lateral wind and earthquake loads to the building's foundation. A great variety of construction details are used for exterior walls. (See Figure 2.5)

Vertical structural support for a building is usually provided either by load-bearing walls or by a framework of columns and

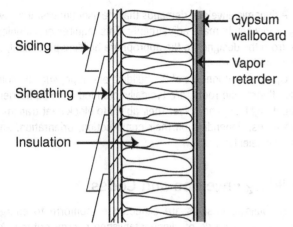

Figure 2.5 Sample exterior wall section

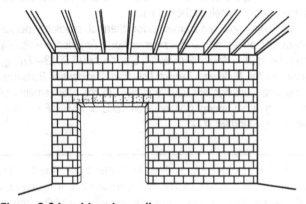

Figure 2.6 Load-bearing wall

Source: Redrawn from Francis D.K. Ching and Corky Binggeli, *Interior Design Illustrated* (3rd ed.), Wiley 2012, page 153

beams. Load-bearing walls are commonly built parallel to each other. (See Figure 2.6)

For more information on loadbearing walls, see Chapter 5, "Floor/Ceiling Assemblies, Walls, and Stairs."

Curtain walls are nonload-bearing exterior surface treatments. (See Figure 2.7) The framing or panels of the curtain wall is supported by either the columns alone or by columns and spandrel beams or by the edges of floor slab. The structural support for a curtain wall may be located in front of, within, or behind its surface plane.

SMART FAÇADES

Smart façades are also known as double-skin façades or climate walls. They have an additional glass skin that allows solar control and **natural ventilation**. A smart façade can integrate passive solar collection, shading, daylighting, increased thermal resistance, and mechanical systems.

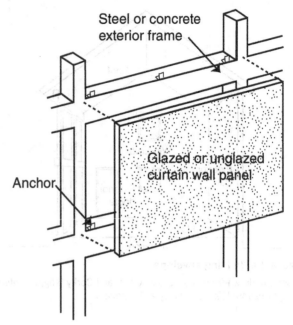

Figure 2.7 Curtain wall

Source: Redrawn from Francis D.K. Ching, *Building Construction Illustrated* (5th ed.), Wiley 2014, page 7.24

A smart façade usually comprises a double-glazed unit adjacent to a 6" to 30" (152 to 762 mm) deep air space, with a layer of **safety glazing** (glazing that meets Consumer Product Safety Commission test requirements) on the outside. Venetian blinds and operable windows make the façade dynamic. The design of a smart façade supports comfortable ventilation and night-flush cooling while preventing the entry of rain, noise, and excessively high air speeds.

Roofs

Roofs control thermal radiation, temperature, humidity, and airflow. They can provide access to views and create visual privacy. Roofs are part of the building's structural support, block the entry of living creatures, and serve to keep out water and control fire. They can also be a source of clean water and aid acoustic privacy.

Roofs are exposed to extreme weather conditions. Hot roofs are a summer heating problem. Radiant losses on a clear night can lower the roof temperature below that of the outdoor air, which may be good in summer, but is bad in winter.

Architects may view roofs as a primary sheltering element for the interior spaces of a building. The roof's form and slope must be compatible with the type of roofing. Roof construction also controls the passage of moisture vapor, **infiltration** (leaking) of air, and the flow of heat and solar radiation. The roof may be required to be fire-resistant.

The roof system of a building shelters the interior space. (See Figure 2.8) The roof structure must be strong enough to

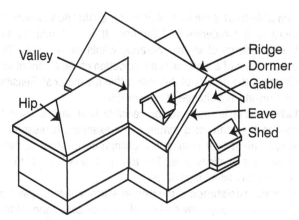

Valley
Hip

Ridge
Dormer
Gable
Eave
Shed

Figure 2.8 Roof terminology

Source: Redrawn from Francis D.K. Ching, *Building Construction Illustrated* (5th ed.), Wiley 2014, page 6.16

span the space, carry its own weight, plus any attached equipment and the weight of snow or rain. In addition, the roof must resist lateral and **seismic** (earthquake) forces and the uplifting of wind forces. The roof's structural system must correspond to that of the column and bearing wall systems. This influences the layout of interior spaces and the type of ceiling that the roof structure may support. Long roof spans open up a more flexible interior space while shorter roof spans lend themselves to more finely defined spaces.

ROOF FORMS

The form of a roof structure has major impact on a building's appearance. (See Table 2.1) They are either flat or made up of one or more slopes. Sloping roofs shed water to eave gutters. The interior space under a sloping roof may be usable.

TABLE 2.1 ROOF FORMS

Form	Description
Flat roof	Slope usually ¼" per foot (1:50 mm), often leading to interior drains
Pitched roof	One or more slopes
Shed roof	A single pitched slope
Gable roof	Slopes down in two directions from a central ridge, forming a triangular gable at each end
Gambrel roof	A pitched roof with a shallower upper slope and steeper lower slope on each side
Hip or hipped roof	Sloping sides like a gable roof, but ends also slope in to meet central ridge
Mansard roof	Resembles a shallow hipped roof set on top of a steeper lower part

ROOF INSULATION AND ATTICS

Buildings can gain a great deal of heat through the roof. This gain can be reduced by using additional insulation and by reflecting the sun's radiation. A white roof can reflect half of the heat that would have been absorbed by a black roof.

The insulation for a flat roof is placed on top of the roof deck to avoid the penetration of the thermal envelope by structural members, lighting fixtures, or air ducts. The insulation should follow the rafters or top cord of trusses on sloped roofs. It is important for the ductwork to be on the indoor side of the thermal envelope.

In cold climates, an uninsulated attic beneath a roof is sometimes separated from the living space of a house by an insulated ceiling. Although the attic is not insulated, it may still be warmer than the outdoor temperature. Venting the attic to the outdoors will allow it to disperse moisture that manages to migrate through the insulated ceiling.

GREEN ROOFS AND ROOF GARDENS

A **green roof**, also called vegetated roofing, is a natural roof covering typically consisting of vegetation planted in lightweight soil or growing medium over a waterproof membrane. (See Figure 2.9) The natural covering protects the roof membrane from daily temperature fluctuations and ultraviolet (UV) radiation. Green roofs control the volume of stormwater runoff and improve air and water quality. They help reduce the heat island effect in urban areas. A green roof can help stabilize indoor air temperatures and humidity, and reduce heating and cooling costs. (See Table 2.2)

Green roofs are valued for creating wildlife habitats, and making buildings and cities more humane. A small green roof can be designed as a roof garden available for use by building occupants.

A green roof must be built on a frame strong enough to support it, with carefully applied waterproofing, as it is difficult to locate leaks once the growing medium is in place. The plantings require ongoing care.

TABLE 2.2 GREEN ROOF TYPES

Type	Description
Intensive	Require minimum 12" (305 mm) soil depth for roof garden with larger trees, shrubs, grasses, plus irrigation and drainage systems. Roof deck usually concrete.
Extensive	Low maintenance. Typically use 4" to 6" (102 to 152 mm) deep lightweight growing medium to grow small, hardy plants and thick grasses. Concrete, steel, or wood roof deck.
Modular block	Anodized aluminum containers or recycled polystyrene trays with 3" to 4" (76 to 1–2 mm) of engineered soil supporting low-growing plants. Pad fastened to bottom of each block protects roof surface and controls drainage.

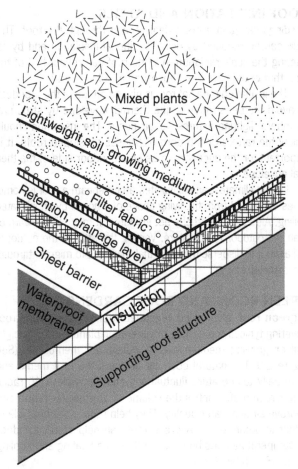

Figure 2.9 Green roof

Source: Redrawn from Francis D.K. Ching, *Building Construction Illustrated* (5th ed.), Wiley 2014, page 7.09

See Chapter 4, "Building Forms, Structures, and Elements," for information on building foundations.

HEAT FLOW AND THE BUILDING ENVELOPE

Now that we have looked at the physical parts of the building envelope, it is time to investigate how heat flows through the envelope. This involves taking a look at some terminology and the basics of thermodynamics.

Terminology

There are several terms that are useful when discussing energy flow. These will be revisited when we consider the basics of electrical and thermal design, but an introduction now will help us understand the building envelope and its relationship to sustainable design.

- **Power** is the instantaneous flow of energy at a given time.
- **Energy**, in the context of buildings, is power usage over time.

- **Sensible heat** is the form of heat energy that flows whenever there is a temperature difference. It is apparent as the internal energy of atomic vibration within all materials. The temperature of a material is an indication of the extent of this vibration, the density of the heat within the material. Sensible heat results in a change of temperature.
- **Latent heat** is the term for sensible heat that is used to change the state of a material, for example, to evaporate water. Latent heat results in a change in moisture content, often in the humidity of air. The total heat flow is equal to the sensible and latent heat flows.
- **Thermal resistance** is the rate at which a building gains or loses heat through any portion of its enclosure under stable indoor and outdoor temperatures. Roughly, the thermal resistance of a material is inversely proportional to its density.

Thermodynamics

Thermodynamics is the branch of science that deals with the relationship between heat and other forms of energy. A basic grasp of thermodynamic principles is important in understanding the energy efficiency of building envelope. We will keep this simple.

LAWS OF THERMODYNAMICS

The laws of thermodynamics are fundamental laws of physics. They define the physical qualities of temperature, energy, and **entropy** (the tendency toward running down or disorder). The first and second laws of thermodynamics are important to understanding how thermal energy (heat) flows from one object to another.

The **first law of thermodynamics** states that energy can be neither created nor destroyed. The total amount of energy in an environment remains constant, and can be accounted for as it transfers from one place to another and from one form to another. It is possible to use the same energy over and over again for different purposes in buildings; this leads to energy efficiency.

The **second law of thermodynamics** expresses the tendency toward disorder as part of the normal nature of things. It states that, as time goes by, energy will tend toward lower and lower quality. Energy is lost every time it is converted from one form to another. Ideally, high-quality energy would be reserved for high-quality uses, such as lighting and running motors. Once degraded to lower quality ones, it could be used for space heating or water heating.

These physical laws relate to the maintenance of building materials and systems, which requires the regular addition of energy and materials in order to resist the forces of nature. The laws of thermodynamics have specific applications to the design of heating, ventilating, and air conditioning (HVAC) systems.

Another law also applies; the **Zeroth law of thermodynamics** helps define the concept of temperature. It states that if two systems are in thermal equilibrium with a third system, they must be in thermal equilibrium with each other. Heat will only flow from a higher to a lower temperature, so heat flow requires a difference in temperature.

Heat Flow and Building Envelope

The flow of heat through a building depends on the season (to the outside in the winter, and to the inside in the summer), and its path either through building envelope materials or via air exchange with the outdoors. The amount of heat that flows either into a building from outside, or out of a building from inside, depends on the difference in temperature between the inside and outside, the thermal resistance of the building envelope materials, and the ability of the building envelope to store heat.

How much heat the building envelope gains or loses is influenced by the construction of the outside of the building envelope, along with the wind velocity outside the building. Each layer of material making up the building's exterior shell helps resist the flow of heat into or out of the building. The amount of resistance depends on the properties and thickness of the materials making up the envelope. Heavy, compact materials usually have less resistance to heat flow than light ones.

Some parts of walls and roofs—such as metal framing studs—transmit heat more rapidly than others. (See Figure 2.10) These pieces of the construction are called **thermal bridges**, and they can increase heat loss significantly in an otherwise well-insulated assembly. When a thermal bridge exists in a ceiling or wall, the cooler area can attract condensation, and the water can stain the interior finish. Wood studs conduct less heat through the wall than metal studs. (See Figure 2.11)

Architects increase thermal resistance by adding insulation or reflective sheets, or by creating more air spaces. Air enclosed between two surfaces in the building envelope will only effectively resist airflow if it lacks substantial air circulation. The thickness of the air space is not usually critical, but the number

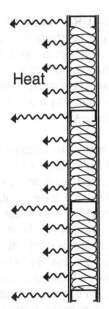

Figure 2.11 Thermal bridge through metal stud wall

of air spaces makes a difference. Highly efficient insulation materials like fiberglass batt insulation, which hold multiple air spaces within their structure, are better than empty air spaces alone. High levels of insulation maintain comfortable interior temperatures, control condensation and moisture problems, and reduce heat transmission through the envelope.

The use of tighter construction and house wraps have reduced infiltration significantly. Insulating sheathing that creates a continuous unbroken thermal envelope over the entire building greatly improves performance.

THERMAL CAPACITY

The ability of a material to store heat is called its **thermal capacity**. A building envelope constructed with materials with high thermal capacity may be able to reduce the heat flow by storing heat during the daytime, and emitting it later at night. This slows down and may diminish excess heat that makes it into the interior of a building. Thermal capacity is roughly proportional to mass.

INTERNAL AND SKIN LOAD DOMINATED BUILDINGS

Building form also affects how heat moves into and out of a building. Tall thick buildings shelter larger amounts of floor space from outside climate. The heat generated by their electrical lighting can be more than enough to keep the building warm in winter. These buildings are called **internal load dominated**, and require air conditioning throughout the year.

The interior spaces in thinner buildings almost all have an exterior wall that needs to be heated in cold weather and cooled in hot weather. Electric lighting does not provide as much heat, as daylight suffices for most daytime needs. These buildings are called **skin load dominated**.

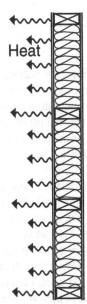

Figure 2.10 Thermal bridge through wood stud wall

HEAT AND MOISTURE FLOW PROCESSES

When an object is at a temperature different from its surroundings, heat flows from hotter to cooler. Moisture flows from areas of greater concentration to those of lesser concentration. Both of these occur within a building envelope.

Sensible heat, which results in a change of temperature, is gained or lost in three ways:

- **Convection** is the exchange of heat between a fluid (typically air) and a solid. The motion of the fluid, which is due to heating or cooling, plays a critical role in the extent of the heat transfer.
- **Conduction** involves the transfer of heat directly from molecule to molecule. It can occur within or between materials, with the proximity of the molecules (the material density) playing a critical role in the extent of the heat transfer.
- **Radiation** is the process by which heat flows via electromagnetic waves from hotter surfaces to detached, colder ones. Radiation can occur across empty space and great distances, such as between the sun and the earth.

Another process that can take place in the building envelope is **evaporation**, which carries heat away from wet surfaces. When water evaporates, it changes from a liquid state to a vapor, and loses latent heat. Evaporation is much less important for buildings than for the human body; moisture flow through envelope assemblies and by way of air leakage are the principal means of latent heat gain and loss in a building.

Heat transmission through the building envelope is affected by its surface area, construction materials, thickness, orientation to the sun, shading, exterior color, temperature of its surroundings, and temperature of the interior space.

Heat is conducted through solid envelope layers, and is transmitted by radiation and convection through the envelope's air pockets. Interior air currents bring air molecules into contact with room surfaces, thus transferring heat between the air and room surfaces.

Once it moves inside the solid material of a wall, heat moves by conduction. In the open spaces between studs framing a wall, heat flows by convection or by radiation.

As heat is transferred to the air, a film of air next to the wall rises in temperature; it then continues to rise due to the buoyancy of warm air. Heat transferred from the air to an interior surface cools, becomes heavier and denser, and falls, reversing the circulation pattern. Heat radiates between the inside surfaces and room contents, and between outside surfaces and building surroundings.

Heat transfer through the building envelope results in sensible heat losses and heat gains within the building. Latent heat gains result from the presence of people, which add humidity to the air, and ventilation control equipment, which can add to the latent cooling load if it is humid outdoors.

THERMAL MASS

Massive construction results in **thermal lag**, which tends to produce more stable conditions. Adobe structures in the US

Figure 2.12 Interior of adobe house, Pie Town, New Mexico, 1940
Source: Library of Congress Prints and Photographs Online Catalog

Southwest are a good example. (See Figure 2.12) Massive construction on a building's west and east sides and on its roof minimizes solar heat gain in summer.

Thermal lag time is useful when the time of maximum solar heat differs from the time of maximum internal heat need. As the temperature swing for passive solar heating is controlled by the sun and building occupants rather than by a thermostat, there are typically daily variations in indoor temperatures.

Thermal mass also passively keeps room temperatures from rising too high during hot summers. In the winter, sun-warmed surfaces can help create a space that is comfortable at 5° to 10°F (3° to 6°C) lower than normal room temperatures.

See Chapter 12, "Principles of Thermal Comfort," for more information on thermal mass.

HEAT LOSS

Buildings lose heat through the building envelope by transmission, infiltration, and ventilation. The amount of loss depends on the exposed area, the temperature difference between indoors and outdoors, and the thermal resistance of the building skin. Building heat loss can be minimized by compact design, the use of common walls between buildings, and the application of insulation.

Heat loss through infiltration depends on how much cold outside air is able to enter the building, and the temperature difference between the indoor and outdoor air. It is calculated by either the air change method or crack method, neither of which are very precise, or more accurately by a blower-door test. (See Figure 2.13) Blower-door tests are frequently used to assess the need for insulating existing residential buildings. They are relatively simple to perform, and useful when considering renovation work.

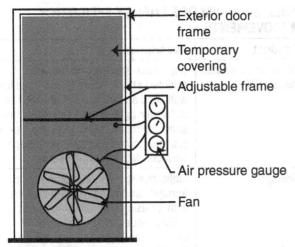

Exterior door frame
Temporary covering
Adjustable frame
Air pressure gauge
Fan

Figure 2.13 Blower-door test

HEAT GAIN

Buildings gain heat by transmission, infiltration, ventilation, and the effect of the sun on thermal mass. They also gain heat from internal heat sources, including people, lighting, and mechanical equipment.

By specifying light-colored finishes, an interior designer can reduce heat gain generated by solar radiation passing through glazing and heating opaque surfaces.

See Chapter 13, "Indoor Air Quality, Ventilation, and Moisture Control," for more information on infiltration and ventilation.

U-FACTORS AND R-VALUES

U-factors (sometimes called U-values) are expressions of the steady-state rate at which heat flows through architectural envelope assemblies. They are used in codes and standards, and by engineers to specify envelope thermal design criteria.

The U-factor includes all elements in a building envelope assembly and all sensible modes of heat transfer (convection, conduction, and radiation). This includes heat flow through windows and skylights, where the situation becomes more complicated due to differences in heat flow rates between various parts of the window or skylight unit. The National Fenestration Rating Council (NFRC) U-factor combines these variations into a single value for an entire unit. The lower the U-factor (it is usually less than 1), the lower the heat flow for a given temperature difference.

Fenestration is defined as the arrangement of windows and doors on the elevations of a building.

TABLE 2.3   R-VALUES FOR INTERIOR MATERIALS

Category	Material	R-Value
Interior Finishes	Drywall 1/2"	0.45
	Drywall 5/8"	0.56
	Paneling 3/8"	0.47
Flooring	Plywood 3/4"	0.93
	Particle board underlayment 5/8"	0.82
	Hardwood flooring 3/4"	0.68
	Tile, linoleum	0.05
	Carpet, fibrous pad	2.08
	Carpet, rubber pad	1.23
Air films	Interior ceiling	0.61
	Interior wall	0.68
Air spaces	1/2"–4" approximate	1.00

R-values measure the thermal resistance of a given material. The R-value is the reciprocal of the U-factor. It represents the degree of resistance to heat flow, and thereby, the insulating ability of a building envelope element. The lower the U-value, the higher the R-value and the insulating value, so the higher the R-value, the more time is required for heat transfer through the material. It is determined experimentally.

The materials and construction assemblies used in a building's envelope affect its R-value. (See Table 2.3) The structure's orientation to the sun and exposure to strong winds also influence the amount of heat that will pass through the barrier. By knowing the R-value along with the desired indoor temperature and outdoor climate conditions, an engineer can estimate the building envelope's ability to resist thermal transfer and to regulate indoor conditions for thermal comfort.

See Chapter 6, "Windows and Doors," for U-factors and R-values for doors and windows.

Moisture Flow through Building Envelope

Water moves through building envelope assemblies in both liquid and vapor states. Air contains water vapor, and water vapor will move from an area of greater concentration to one of lower concentration. In the summer, moisture will typically flow into an air-conditioned building, increasing humidity. This usually requires dehumidification, often through removal of the latent heat of condensation of the added moisture. In the winter, it may be necessary to add water vapor to the air to achieve the desired relative humidity. This is often achieved by evaporating water by adding the latent heat of vaporization.

VAPOR PRESSURE

Vapor pressure is the pressure of a vapor in contact with its liquid or solid form. Differences in vapor pressure drive the flow of moisture through the components of a building envelope assembly. Where there are gaps in the assembly, water vapor is carried by airflow. Water vapor condensing within the building envelope construction can make insulating materials ineffective, damage wood structural elements, and harbor mold and mildew. Water will readily permeate many building materials, including gypsum board, concrete, brick, wood, and some types of insulation, along with most interior finish materials.

VAPOR RETARDERS

Vapor retarders are materials that resist the flow of water vapor through the envelope assembly. They are very thin membranes that take up virtually no space inside the building envelope. Vapor retarders are sometimes referred to as vapor barriers, but this is a less accurate term.

The placement of vapor retarders within a wall, roof, or floor is critical, and varies with the type of construction and the local climate. The vapor retarder should stop the flow of water vapor before the vapor can come in contact with its **dew point** (the temperature below which droplets begin to condense and dew can form) within the envelope assembly.

The vapor retarder should be installed below the finish material on the warm side, but the warm side can change. The warm side in a cold climate is inside. In an air-conditioned building in a hot climate, the warm side is outside. In climates where never very hot or very cold, vapor retarders not needed. When adding insulation to an existing building, designers should consider the effect on any vapor retarder type and location.

Plastic films in the proper location within the envelope assembly provide better protection from migrating water vapor than vinyl wallpaper or vapor retarder paints on interior surfaces. They also help to block airflow through the envelope. In existing older buildings with vapor problems, it is often not feasible to install a proper vapor retarder. An alternative may be to plug air leaks in walls and apply good coat of paint to warm-side surface; special vapor–retarding paints are available.

Envelope Thermal Performance

Twentieth-century architecture often ignored the energy efficiency of the building envelope and relied on mechanical equipment fueled by cheap oil to create the desired interior environment. The 1973 oil crisis changed that, and energy efficient design became a priority.

There have been major increases in envelope component energy efficiency since the 1970s. (See Table 2.4) Thicker and better wall and roof insulation have improved R-values and U-factors. Better window materials and construction have improved performance.

TABLE 2.4 ENVELOPE ENERGY EFFICIENCY IMPROVEMENTS

Product	Description
Structural insulated panels (SIPs)	Rigid foam sandwiched between oriented strand board (OSB). Used for walls and as structural roof sections. Better air tightness and insulation than site-assembled framing systems, using insulated cores and thermal storage surfaces with less wall thickness.
Aerogel	Light, transparent, and porous silica aerogel can be foamed into cavities without using chlorofluorocarbons (CFCs). Adding carbon absorbs IR radiation and increases its R-value.
Gas-filled panels	Sealed plastic bags for insulation, enclosing honeycomb baffles of thin polymer films and low-conducting argon, krypton, or xenon gas.

INSULATION MATERIALS

Insulation reduces heat gain and loss and reduces drafts. Indoor comfort is dramatically increased when insulation is added to exterior walls, ceilings, and especially windows. Insulation also helps control interior mean radiant temperature.

Mean radiant temperature (MRT) is used by engineers to give a value to the way an interior space and its furnishings radiate and emit heat to a human body in a given location. MRT measures the temperature of each surface in a space and defines the specific spot in the space where the MRT is to be measured. It takes into account how much heat each surface emits, and how the surface's location relates to the point where the MRT is being measured. The MRT is derived by a detailed analysis and complex calculations. An MRT is quite abstract and cannot be directly measured.

Insulation reduces energy consumption otherwise used to heat interior spaces by increasing the MRT. Insulation is the primary defense against heat loss transfer through the building envelope. It is important to insulate a building from top to bottom. In a cold climate, a house should be insulated wherever a heated area comes in contact with an unheated area. (See Figure 2.14)

You can check whether an existing building's walls are insulated by removing an electrical outlet cover and looking inside, or by drilling two 1/4" (6mm) holes above one another about 4" (102 mm) apart in a closet or cabinet along an exterior wall, and shining a flashlight in one while looking in the other. An insulation contractor can blow cellulose or fiberglass insulation into an existing wall.

To add insulation to an unheated attic without flooring, add a layer of unfaced batts about 12" (305 mm) deep across the joists. To insulate a space with a finished cathedral ceiling,

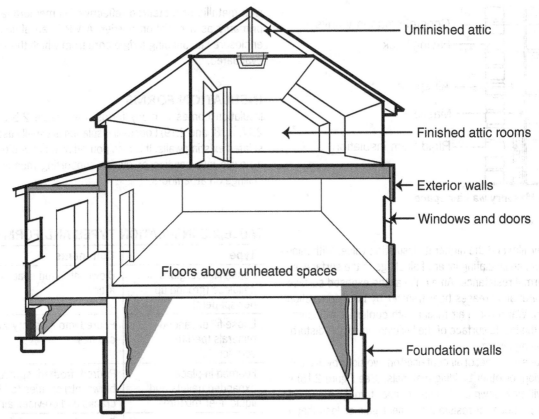

Figure 2.14 Where to insulate a house

Source: Redrawn from "Where to Insulate in a Home", Energy.gov

Labels on figure:
- Unfinished attic
- Finished attic rooms
- Exterior walls
- Windows and doors
- Floors above unheated spaces
- Foundation walls

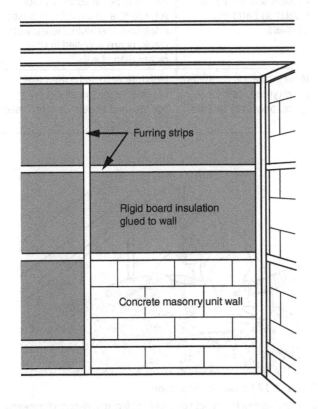

Figure 2.15 Insulating basement walls

Labels on figure:
- Furring strips
- Rigid board insulation glued to wall
- Concrete masonry unit wall

either the interior drywall is removed to install insulation, or a new insulated exterior roof is built over the existing roof.

Uninsulated foundation walls can be responsible for as much as 20 percent of a building's heat loss. A concrete or masonry basement wall can best be insulated with panels of closed-cell polystyrene foam on the exterior wall. (See Figure 2.15) To insulate the inside of the basement, the walls must first be kept dry from the outside. Polystyrene foam insulating panels can then be directly adhered to the inside wall surfaces, with wood strips laid on them to create a ventilated airspace. The finish is nailed or screwed to wood strips with ventilation gaps at the floor and ceiling. It is important to correct any drainage problems before insulating the basement.

The surfaces of most insulating materials should not be exposed either indoors or outdoors.

Air Films and Air Spaces

An air film is a thin layer of stagnant air attached on the exposed surface of a construction assembly. Air films on interior or exterior surfaces contribute substantially to the insulating capability of some construction assemblies. The rougher the surface, the

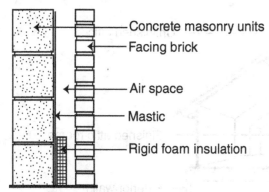

Figure 2.16 Masonry wall air space

- Concrete masonry units
- Facing brick
- Air space
- Mastic
- Rigid foam insulation

thicker the air film and the higher its insulation value. With minimal air motion, an insulating layer of air clings to the surface, increasing thermal resistance. An air film at the exposed surface of a solid material increases heat transfer by both convection and radiation. Warm room air mixing with cooler air will gently flow against the inside surface of the building envelope, disturbing the insulating air film.

An air space is a sheet of air contained on two sides by drywall, brick, insulation, or other building materials. (See Figure 2.16) A layer of air will slow down the transfer of heat through a building envelope. The amount of resistance to heat gain or loss that it provides is determined by its width, position (horizontal, vertical, or tilted), and surrounding materials.

AIR BARRIERS

An air barrier may be required by building codes in walls and roofs to avoid air leaks. The type of air barrier depends on the type of construction. Many types of framed walls use continuous sheet material wrapped around building exterior just before exterior finish material is applied. The air barrier must be airtight, but allow water vapor to pass. In some situations, where installed on the warm side of a wall, the air barrier also acts as a vapor barrier.

Insulation Types and Forms

When deciding on a type of insulation, its performance specifications and any complications from material thicknesses should be considered. Characteristics of insulation include R-value, moisture resistance, fire resistance, potential for toxic smoke, physical strength, and stability over time.

Inorganic fibrous or cellular materials include fiberglass, rock wool, slag wool, perlite, and vermiculite. Organic fibrous or cellular materials include cotton, synthetic fibers, cork, foamed rubber, and polystyrene.

Most insulating materials owe their effectiveness to very small air spaces that slow down heat transfer. Two exceptions are reflective insulation and vacuum insulation panels (VIPs). Reflective insulation comprises larger air spaces faced

with metallic or metalized reflective foil membranes and acts primarily as a radiation barrier. A VIP is an almost gas-tight enclosure surrounding a rigid core from which the air has been evacuated.

INSULATION FORMS

Insulation comes in many forms. (See Table 2.5 and Figures 2.17, 2.18, and 2.19) Loose-fill insulation is usually used for uninsulated existing walls. It is also poured by hand or blown through a nozzle into cavities or over a supporting membrane above ceilings on attic floors.

TABLE 2.5 INSULATION TYPES AND FORMS

Type	Comments
Loose-fill fiberglass or cellulose (ground up newspapers)	Blown into stud spaces and attics
Loose-fill expanded minerals (perlite, vermiculite)	Poured into masonry wall cavities
Foamed-in-place expanded pellets and liquid-fiber mixtures	Poured, frothed, sprayed, or blown into cavities to fill corners, cracks and crevices airtight
Foams	Sprayed into cavities or on surfaces such as basement walls
Flexible and semi-rigid insulation batts and blankets	Batt may be faced with vapor retardant; also used for acoustic insulation, ineffective when wet; blankets are supplied in rolls rather than sheets
Rigid insulation (extruded polystyrene, cellular glass, polyisocyanurate)	Moisture resistant, often used outside; comes in blocks, boards, and sheets or preformed for use on pipes

Figure 2.17 Loose-fill insulation

Source: Redrawn from photo by Dennis Schroeder, National Renewable Energy Lab, Energy.gov

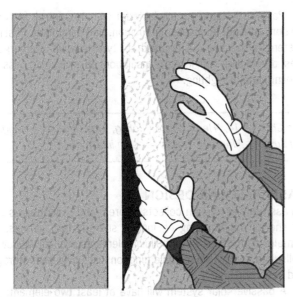

Figure 2.18 Batt insulation

MOVABLE INSULATION

Movable insulation provides extra insulation on winter nights. It also eliminates the black hole effect at skylights and windows and can provide extra insulation and shading in the summer. In addition, some types of movable insulation have aesthetic value. Rigid panels have high R-values, but are complex to install and use.

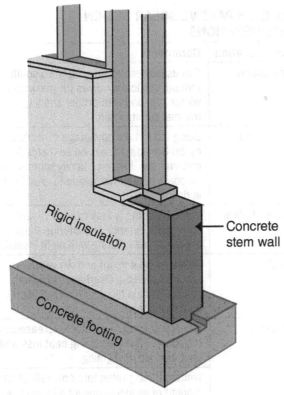

Rigid insulation

Concrete stem wall

Concrete footing

Figure 2.19 Rigid insulation

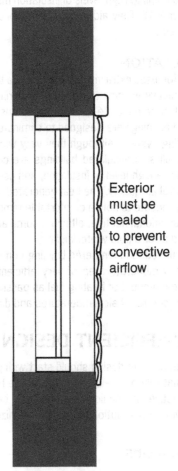

Exterior must be sealed to prevent convective airflow

Figure 2.20 Exterior insulating shutter

Wind tends to limit thermal performance of exterior shutters, which need to be sealed prevent short-circuiting convection of conditioned air near the window. (See Figure 2.20)

Drapery with thermal liners of insulating foam or reflective films increase R-value up to three R-units. Edges must be sealed to prevent convection. Thermal-lined drapery should extend from ceiling to floor and have magnetic strips or Velcro™ for good edge seals. It is advisable to also use a vapor barrier to reduce condensation on windows.

Venetian blinds with a reflective coating or insulating louvers (slats) control daylight as well as heat gain and loss. The louvers adjust to admit or limit passage of light and heat.

Night insulation over direct-gain solar glazing conserves indoor heat, and can also reject sun on summer days. It can also provide privacy control and eliminate the black hole appearance of bare glazing.

ADDING INSULATION

The thermal resistance of the building envelope is increased by adding more insulation. Insulating materials with slow conduction rates provide high thermal resistance. By reducing heat transmission through the envelope, they reduce energy requirements for heating and cooling. High levels of insulation maintain a comfortable interior MRT. They aid in controlling condensation and moisture problems.

SUPERINSULATION

Superinsulation uses extra insulation plus extra thermal mass, along with insulative window treatments. Superinsulated buildings use about twice the code-required minimums of insulation. Superinsulated buildings are designed to eliminate the need for a central heating system. Although they vary with climate and building style, all superinsulated buildings are constructed to be airtight, have a high level of insulation, and use a ventilation system to control air quality. The heat generated by people and equipment is adequate to hold a comfortable temperature overnight without additional heating, although some additional heat may be needed after a cold winter night.

A superinsulated, sun-tempered building combines thermal insulation with a moderate area of very efficient windows. It uses about same amount of heating fuel as passive solar building, with better control of air temperatures and direct sunlight.

ENERGY-EFFICIENT DESIGN

Energy efficient building design should start with passive building systems that rely on the architecture of the building itself. Fuel-efficient active mechanical systems can be considered once the building's contribution has been maximized.

Passive Systems

The design of passive systems requires a change in the perspective of the design team and the way that energy codes and standards are evaluated. The majority of energy efficiency standards have dealt only with on-site energy usage, neglecting off-site energy such as that used to transport fuel and losses during electrical generation.

Passive design begins early in the design process, and requires early and continuous architectural attention. It also requires the sincere involvement and cooperation of the building's occupants. Passive solar systems tend to create a more pleasant indoor environment than active solar systems. The building itself can be used to teach its users and visitors about how it works.

Passive design solutions let nature do the work, and usually employ renewable energy resources. The approach is to think small, simple, and local.

Passive systems work with the rules of physics, and often look to nature for biological design models. Passive systems shape building form to support solar heat, daylighting, and natural ventilation. Passive design avoids the use of high-grade resources for low-grade tasks.

Highly integrated passive system components relate to the whole rather than to isolated parts. An increased awareness of multidisciplinary, multifunctional system capabilities allows first costs to be spread over multiple features.

Passive systems are available for climate control, fire protection, lighting, acoustics, circulation, and sanitation. Passive systems also have an increased ability to survive in emergencies, due to their high levels of insulation, passive solar, passive cooling, and daylighting.

See Chapter 14, "Heating and Cooling," and Chapter 17, "Lighting Systems," for more passive design information.

PASSIVE SOLAR DESIGN

A passive solar system collects, stores, and redistributes solar energy without using equipment such as fans, pumps, or complex controllers. Basic building elements such as windows, walls, and floors have multiple functions including heat storage and heat radiation.

A passive solar system will have at least two elements: a collector with south-facing glazing (Northern Hemisphere), and energy storage, which is usually a thermally massive material such as rock or water. Passive solar design is usually the second step of a three-tier approach. First, heat retention is provided, then passive solar design, and finally, mechanical heating (if needed).

General considerations for passive solar systems include orientation, interior plan, slope of glazing, shading, and reflectors. (See Table 2.6)

TABLE 2.6 PASSIVE SOLAR DESIGN CONSIDERATIONS

Consideration	Comments
Orientation	Sun usually enters building from south through vertical windows for maximum winter sun, and with proper shading, minimal summer sun.
Interior plan	Designed to take advantage of sun's daily cycle. Breakfast areas on east side for morning sun. Living or family room spaces on south or southwest side for sun later in day.
Slope of glazing	Vertical glazing is less expensive and safer, both exterior and interior shading are easier, as is fitting with night insulation.
Shading	Reflected heat in hot and dry areas requires shading. Overhangs (awnings, balconies) help block diffuse radiation in humid regions.
Reflectors	Exterior specular reflectors increase solar collection while minimizing heat loss and gain, and aid daylighting.
	White diffusing reflectors only reflect small amount of incident sunlight into window.

DIRECT GAIN SYSTEMS

Direct-gain passive solar systems introduce solar energy directly into the interior space through ordinary fenestration. The space itself is used as a solar collector and for storage of excess daytime heat for later release at night and on overcast days. Solar collection depends on south-facing glazing (Northern Hemisphere) located strategically to strike thermal mass within the interior space. Depending on the specifics, a properly designed direct-gain system can be 30 to 75 percent efficient for the capture and use of incident solar energy.

Direct gain systems utilize thermal mass materials such as masonry or water to store heat. Floors, walls, and/or ceiling are typically made of thermally massive materials that store heat well. Massive furnishings or other objects within the space can also provide thermal storage.

Many interior contents, such as drywall, furniture, and books, can act as thermal mass. Medium to dark colors aid in heat absorption. Thermal storage floors must not be covered by carpets.

Direct gain systems require careful design of site conditions and window treatments to prevent glare indoors. Transparent or light-diffusing glazing may be used. Adjustable window insulation inhibits the loss of solar heat to the outside through glazing at night. Smaller or fewer windows reduce summer heat gain problems. South-facing windows (Northern Hemisphere), clerestories, and skylights facilitate the greenhouse effect.

Glass helps to filter UV radiation but with direct gain systems, enough is transmitted to bleach paints, interior furnishings, and other building materials. It is advisable to select colors and materials that resist fading.

Direct-gain systems tend to produce relatively large daily space temperature fluctuations, with 10° to 30°F (6° to 17°C) swings typical. Even with the addition of a conventional heating system, some temperature fluctuation can be expected.

INDIRECT GAIN SYSTEMS

Indirect gain systems place thermal storage mass between the sun and the occupied space. Sunlight strikes the thermal mass, where it is absorbed and stored, then slowly transferred to the occupied space. The thermal storage material may be masonry or water. Three basic types of indirect gain systems include thermal storage walls, roof ponds, and greenhouses and sunspaces.

THERMAL STORAGE WALLS

Trombe wall systems and water thermal storage walls are both indirect gain thermal storage walls. Trombe walls were developed by engineer Felix Trombe and architect Jacques Michel

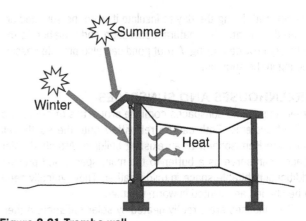

Figure 2.21 Trombe wall
Source: Redrawn from Energy.gov

in the 1960s. A Trombe wall consists of thermal mass, often around 12" (305 mm) thick, just inside a sheet of south-facing glazing. (See Figure 2.21) It requires heavy masonry construction and is usually made of solid materials such as concrete, brick, stone, or adobe, or consist of steel containers of water. Tubes of translucent or transparent plastic that allows some light to pass are also used.

Sunlight is absorbed and conducted through the wall to the occupied space. Heat produced and longwave radiation emitted by the wall is trapped between the glazing and the wall. The heat gradually migrates through the wall, producing low-grade but effective radiant heat inside at night.

Trombe walls usually have a direct gain opening within wall for daylight and view. They are thermally stable, with a large portion of their heat delivered by radiation to space. Keeping the air space between the Trombe wall and the glass covering it clean can be a problem. Objects on the wall's interior surface that interfere with radiant heat must be minimized. Trombe walls can make a living space feel very enclosed, but may be appropriate for bedrooms, where enclosure and darkness are more welcome.

Water thermal storage walls operate similarly to masonry ones, except that, since the heat transfers through the wall by convection rather than conduction, it moves to the interior much more rapidly. They are made with corrugated galvanized steel culverts, steel drums, or opaque fiberglass-reinforced plastic tubes.

ROOF PONDS

Roof ponds are indirect gain solar systems that consist of water contained in large plastic bags located on the roof of a single-story building. They are usually supported by a metal deck roof structure, which is used as the finished ceiling of the room below, with heat conducted from the storage radiated into the interior space. Insulated panels over the bags of water are manipulated with electric motors.

In the winter, the roof pond is exposed to sunlight in the daytime, and covered with insulation at night. In the summer, it is covered during the day to insulate it from the sun, and uncovered at night, when natural convection and radiation to the night sky provide cooling. A roof pond can also provide passive cooling in the summer.

GREENHOUSES AND SUNSPACES

Greenhouses and sunspaces combine properties of direct and indirect systems. Both can be retrofitted onto the south wall (Northern Hemisphere) of an existing building. Attached solar spaces can serve as a buffer to the main space, and provide additional habitable space in mild weather. They typically need to be shaded and vented in warm weather.

Greenhouses are directly heated by solar radiation. A thermal wall (masonry or water) between the greenhouse and the occupied space receives direct sunlight and transmits heat to the adjacent space. Heated greenhouse air may also be vented into the occupied space. Fans are sometimes used to extract additional heated air from a greenhouse to heat adjoining spaces. Heat from the thermal wall is released at night, which can aid plant life. Thermal mass moderates temperatures during the day. The efficiency of a solar greenhouse can be as high as 60 to 75 percent. When the greenhouse is used to grow plants, the temperature range and fluctuations must be considered as they relate to plant needs.

A **sunspace** is used as an extension of the occupied space only when the weather permits. Heat built up in the sunspace transfers to the occupied space to meet heating needs, and the sunspace is not heated at night. They can be pleasant living spaces, but keeping comfortable temperatures in the main space is achieved with very wide temperature swings in the sunspace. A row of water containers to store heat takes substantial floor area.

Sunspaces require venting to the outside to prevent overheating; openings can be smaller if a fan is used. They also require openings such as doors, windows, or vents in the common wall between the main building and the sunspace to heat the interior in the winter. To be effective, all these openings must add up to at least 16 percent of the glazing area.

Active Solar and Hybrid Systems

Typically, solar heated buildings use hybrid systems that combine active and passive strategies. Active solar space-heating systems use mechanical equipment to collect and store solar energy. The design criteria for collection and storage are somewhat complex. The large arrays of solar collectors require pipes or ducts to distribute heat for space heating.

THE DESIGN PROCESS

Historically, architects addressed the environmental needs of a building themselves. However, in the twentieth century, this role was delegated to engineers, who responded with mechanical and electrical equipment. Today, the design process is more collaborative and holistic.

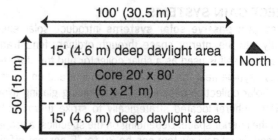

Figure 2.22 Locating functions for daylighting

At the initial conceptual design stage, the architect and interior designer group similar functions and spaces with similar needs close to the resources they require, consolidating and minimizing distribution networks. (See Figure 2.22) The activities that attract the most frequent public participation belong at or near ground level. Closed offices and industrial activities with infrequent public contact can be located at higher levels and in remote locations. Spaces with isolated and closely controlled environments, like lecture halls, auditoriums, and operating rooms, are often placed at interior or underground locations. Mechanical spaces that need acoustic isolation and restricted public access, or that require access to outside air, should be close to related outdoor equipment like condensers and cooling towers, and must be accessible for repair and replacement.

Large buildings are broken into zones. Perimeter zones are immediately adjacent to the building envelope, usually extending 15' to 20' (4.6 to 6 m) inside. Perimeter zones are affected by changes in outside weather and sun. In small buildings, the perimeter zone conditions continue throughout the building. Interior zones are protected from the extremes of weather, and generally, require less heating, as they retain a stable temperature. Generally, interior zones require cooling and ventilation.

The spaces in a mixed-use building are divided between perimeter zones that are most affected by exterior conditions and interior zones that remain relatively stable. (See Figure 2.23) Varying uses as well as scheduling and internal heating and cooling requirements within these zones could require subdivision into additional zones.

The Design Team

In the past, architects were directly responsible for the design of the entire building. Heating and ventilating consisted primarily of steam radiators and operable windows. Lighting and power systems were also relatively uncomplicated.

Today, the architect typically serves as the leader and coordinator of a team of specialist consultants, including structural, mechanical, and electrical engineers, along with fire protection, acoustic, lighting, and elevator specialists, and interior designers. (See Figure 2.24) Energy conscious design requires close coordination of the entire design team from the earliest design stages.

On larger projects, the design team can include a wide range of professionals. Interior designers work directly for the

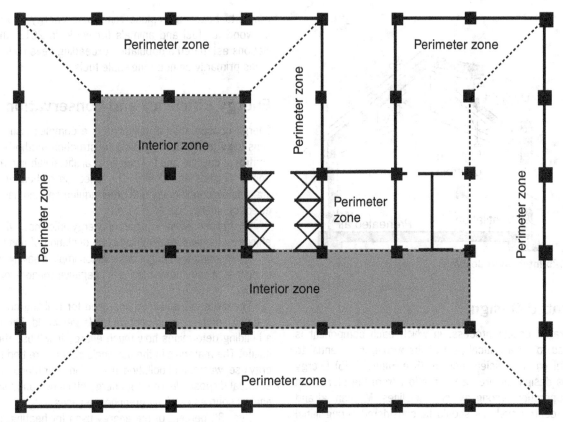

Figure 2.23 Perimeter and interior building zones
Source: Redrawn from Walter T. Grondzik and Alison G. Kwok, *Mechanical and Electrical Equipment for Buildings* (12th ed.), Wiley 2015, page 245

architect as part of the architectural team, serve as consultants to the architect, or in some cases, work independently on building interiors.

The interior designer often meets with the architect and engineers in the preliminary stages of the design process to coordinate the interior design with new and existing plumbing,

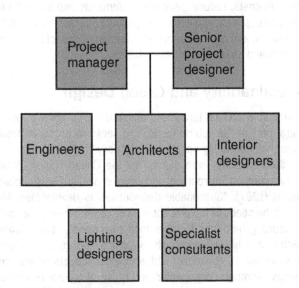

Figure 2.24 Project team

mechanical, and electrical system components. The location of plumbing fixtures, sprinklers, fire extinguishers, air diffusers and returns, and other items covered by plumbing and mechanical codes must be coordinated with interior elements. The plumbing, mechanical, and electrical systems are often planned simultaneously, especially in large buildings. Vertical and horizontal **chases** for distribution are integrated into building cores and stairwells. Suspended ceiling and floor systems house mechanical, electrical and plumbing components. Their locations affect the selection and placement of finished ceiling, wall, and floor systems.

The design process establishes the design intent of the project by defining characteristics of a proposed building solution. These become focal points for the efforts of the design team. Example design intents include:

• Providing outstanding comfort for its occupants
• Accommodating the latest in information technology
• Focusing on indoor environmental quality and sustainable design
• Using primarily passive systems to accomplish design goals through the building itself rather than through adding equipment
• Providing a high degree of flexibility for its occupants, so that changes in use or other conditions are readily accommodated

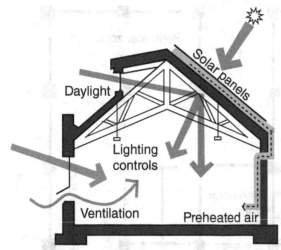

Figure 2.25 Integrated design

Integrated Design

An integrated design process, in which each component is considered to be a critical part of the whole, is essential to successful energy-efficient design. (See Figure 2.25) Energy conscious design requires a team effort from the start, with close cooperation among all the disciplines. Mechanical and electrical control systems should be considered during initial planning, as they strongly affect building form, site location, and orientation.

SUSTAINABLE DESIGN

Today, architects have accepted the goals of sustainable design and the need for collaboration with other design professionals. They support organizations working to educate architects, including the US Green Building Council (USGBC) and its Leadership in Energy and Environmental Design (LEED) programs.

> Sustainability…calls for a holistic approach that considers the social, economic, and environmental impacts of development and requires the full participation of planners, architects, developers, building owners, contractors, manufacturers, as well as governmental and non-governmental agencies. (Francis D. K. Ching, *Building Construction Illustrated* (5th ed.), Wiley 2014, page 1.03)

Energy efficiency and conservation and the use of sustainable materials are important parts of sustainable design. Interior designers are also supporters of these goals and contributors to this process.

The growth of human population generates much of the work for building design professionals; it is also the source of great problems. According to the United Nations (UN) Department of Economic and Social Affairs (2013), around 1930, the world's population reached approximately two billion. Around that time in the United States, kerosene and other petroleum products began to replace renewable energy sources such as wood for fuel and animals for work. In 2013, the United Nations estimated a population exceeding seven billion, which relies primarily on non-renewable fuels.

Energy Efficiency and Conservation

Energy conservation in buildings is a complex issue involving sensitivity to the building site, construction methods, use and control of daylight and the design of artificial lighting, and selection of finishes and colors. The selection of heating, ventilating, and air conditioning and other equipment can have a major effect on energy use.

Energy conservation is part of energy efficiency. While energy efficiency focuses on minimizing the depletion of non-renewable energy resources, energy conservation implies saving energy by using less. It sometimes carries the negative connotation of doing without.

The materials and methods used for building construction and finishing have an impact on the larger world. The design of a building determines how much energy it will use throughout its life. The materials in the building's interior are tied to the energy use, waste, and pollution involved in their manufacture and eventual disposal. Increasing energy efficiency and using clean energy sources can limit greenhouse gases.

Over 80 percent of the energy used for heating, refrigeration, and air conditioning buildings comes from non-renewable sources. In the United States, about 67 percent of electricity is produced by the consumption of fossil fuels; this also contributes carbon dioxide to the atmosphere. Reducing the amount of energy used requires energy conservation and reduces both the cost and environmental impact.

Global warming problems can be minimized by designing for energy conservation. Passive heating, ventilating, and air conditioning (HVAC) strategies and the use of renewable energy technologies help. Energy-efficient buildings lower electrical and fuel costs, reduce peak power demands and the need for new power plants, and reduce air pollution, carbon dioxide (CO_2) emissions, and other negative environmental impacts from production and distribution of fossil fuels.

Sustainability and Green Design

Sustainable architecture looks at human civilization as an integral part of the natural world, and seeks to preserve nature through encouraging conservation in daily life.

Sustainability was defined by the United Nations World Commission on Environment and Development in *The Brundtland Report* (1987): "Sustainable development is development that meets the needs of the present without compromising the ability of future generations to meet their own needs." **Sustainable design** is a holistic approach to building design that reduces negative social, economic, and ecological impacts on the environment through conservation and reuse of natural resources, energy, water, and materials.

The goal of sustainability is no net negative environmental impacts. This involves avoiding non-renewable resource consumption, keeping the rate of consumption of renewable resources below that of regeneration, and limiting any pollution produced.

Sustainable design is also called **green design**. The US Environmental Protection Agency (EPA) has defined green building as:

> [T]he practice of creating structures and using processes that are environmentally responsible and resource-efficient throughout a building's life-cycle from siting to design, construction, operation, maintenance, renovation and deconstruction. This practice expands and complements the classical building design concerns of economy, utility, durability, and comfort.

The potential effects of green design performance can be measured in terms of depletion of fossil fuels and other non-renewable resources, water use and protection of water bodies, global warming, stratospheric ozone depletion and smog creation, accumulation of acids in the environment, and release of toxic materials.

ASHRAE 189.1—*Standard for the Design of High-Performance Green Buildings Except Low-Rise Residential Buildings* provides a total building sustainability package for those who strive to design, build, and operate green buildings. The 2012 ICC-700 National Green Building Standard® (NGBS) for residential construction is used to certify single- and multi-family homes and residential developments.

Architect William McDonough and chemist Michael Braungart have proposed three principles to guide sustainable design:

1. Waste equals food: Produce everything so that, when its useful life is over, it becomes a healthy source of raw materials to produce new things.
2. Respect diversity: Design everything to respect the region, culture, and materials of a place.
3. Use solar energy: Buildings must be designed to be responsive to this non-polluting and renewable energy source.

REGENERATIVE DESIGN

Supporters of sustainable design are familiar with the saying, "reduce, reuse, recycle." Today, many designers are adding a fourth work, "regenerate."

The goal of energy efficiency is to reduce net negative energy impacts. The goal of green design is to reduce net negative environmental impacts. The goal of **regenerative design** is to produce positive environmental impact, leaving the world better off in terms of energy, water, and materials. (See Figure 2.26)

The John T. Lyle Center for Regenerative Studies at Cal Poly Pomona views the development of regenerative systems as the most promising method for ensuring a sustainable future, going beyond conservation of critical natural resources to enhancing them over time. They teach that regenerative design

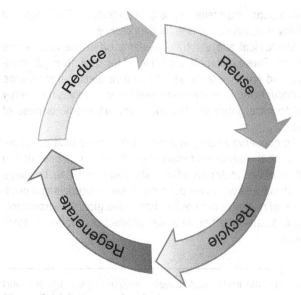

Figure 2.26 Regenerative design

emphasizes the development of community support systems that can be restored, renewed, revitalized, or regenerated through the integration of natural processes, community action, and human behavior.

Energy and Materials

Most buildings have a functional lifetime of 50 years, and up to 100 years of structural life. Designing for the future is difficult, especially when considering how a building system may relate to a changing global environment. Our dependence on non-renewable energy sources, often from distant locations, means that consumers typically have no direct contact with energy sources and systems that are typically designed to remain unseen. Most mechanical and electrical systems use non-renewable materials (primarily metals and plastics).

Once non-renewable resources are exhausted, they cannot be replaced in a timeframe that relates to the human species. Renewable resources such as solar energy or sustainably managed forests, on the other hand, arrive at rates controlled by nature; they are available indefinitely.

Off-site sources such as natural gas or oil, the electrical grid, and water and sewer lines are usually subsidized by society at large, and often entail significant environmental costs. Using sustainable on-site resources such as daylighting, solar heating, cooling, water heating, and photovoltaic electrical production as well as rainwater retention and on-site wastewater treatment, can supplement or replace off-site sources.

EMBODIED ENERGY AND CONSTRUCTION MATERIALS

Embodied energy is the energy that is used to obtain, process, fabricate, transport, and dispose of a unit of building material. Embodied energy considers the impact of building energy consumption, maintenance, and replacement. Each type of building material contains embodied energy.

Mechanical and electrical systems often use metals and plastics. These materials are selected for their strength, durability, and fire resistance as well as their electrical resistance or conductivity. Their environmental impact involves the energy cost to mine, fabricate, transport, and ultimately, dispose of them.

To conserve energy and materials, and to limit pollution, it is wise to reduce materials used for building construction and to utilize materials efficiently. Reuse existing buildings and structures wherever possible. Select materials and products that minimize the destruction of the global environment, and evaluate whether material production produces toxic wastes.

Reducing materials use lowers environmental impact and minimizes waste from material handling and construction. Use durable materials made from renewable resources, and avoid non-renewable sources or materials. Use materials with recycled and recyclable content. Local building materials reduce the embodied energy due to transportation.

Setting Sustainability Goals

Environmentally conscious interior design is a practice that attempts to create indoor spaces that are environmentally sustainable and healthy for their occupants. Sustainable interiors address their impact on the global environment. To achieve sustainable design, interior designers must collaborate with architects, developers, engineers, environmental consultants, facilities and building managers, and contractors.

The professional ethics and responsibilities of the interior designer include the creation of healthy and safe indoor environments. The interior designer's choices can provide comfort for the building's occupants while benefiting the environment, an effort that often requires initial conceptual creativity rather than additional expense.

It is often possible to use techniques that have multiple benefits, spreading the cost over several applications to achieve a better balance between initial costs and benefits. For example, a building designed for daylighting and natural ventilation also offers benefits for solar heating, indoor air quality, and electric lighting. This approach cuts across the usual building system categories and ties the building closely to its site. We will be discussing many of these techniques in this book, crossing conventional barriers between building systems in the process.

Sustainable Design Strategies

Sustainable design strategies are available for just about any building system. (See Table 2.7) Building designers should look

at the building envelope, HVAC system, lighting, equipment, and appliances as well as renewable energy systems as a whole. **Energy loads**—the amount of energy the building uses to operate—are reduced by integration with the building site, use of renewable resources, the design of the building envelope, and the selection of efficient lighting and appliances. Energy load reductions lead to smaller, less expensive, and more efficient HVAC systems that in turn use less energy.

Buildings, as well as products, can be designed for reuse and recycling. A building designed to easily adapt to changed

TABLE 2.7 SUSTAINABLE BUILDING SYSTEM STRATEGIES

Building Systems	Strategies
Site	Use on-site energy sources. Site buildings to enhance daylighting and reduce use of electric lighting.
Building Envelope	Use insulation to reduce heat transfer through the envelope.
Building Construction	Use materials with less embodied energy. Use products made with recycled materials and ones that can be recycled.
Fenestration	Employ user-operated controls such as window shades or operable windows.
Acoustics	Reduce noise pollution through building siting, acoustic absorption, space planning, materials selections, and efficient equipment selection.
Water	Use rainwater retention for irrigation and flushing toilets. Specify water-conserving and waterless plumbing fixtures.
Waste	Control construction and demolition waste. Recycle and reuse materials and water.
Fixtures, appliances	Specify energy efficient fixtures and appliances.
IAQ, ventilation, moisture control	Use natural ventilation. Improve indoor air quality with proper ventilation and by avoiding pollutants.
Heating, cooling	Use passive solar heating and cooling as well as shading.
Electrical Power	Use photovoltaic energy or fuel cells.
Lighting	Design for daylighting with supplemental energy-efficient electric lighting.
Communication and Controls	Use integrated and intelligent building controls for fresh air ventilation, sunlight and shading, and electrical lighting.

uses reduces the amount of demolition and new construction and prolongs the building's life. Products that do not combine different materials allow easier separation and reuse or recycling of metals, plastics, and other constituents than products where diverse materials are bonded together. The use of removable and reusable demountable building parts leaves the building structure intact, but discourages integration of mechanical and structural systems, and is more susceptible to energy leaks.

It takes more energy to produce electric heating (3 units heat are equal to one unit of electricity) than combustion heating (up to 90 percent efficient, so 1.1 to 1.7 units of fuel produce 1 unit of heat).

SUSTAINABLE STRATEGIES FOR RESIDENCES

Interior designers are often in a primary position to support sustainable design strategies for residences. Daylighting and energy-efficient light sources should be used throughout the home. Locate energy-efficient windows for passive solar heating and minimal winter heat loss as well as to minimize cooling-season heat gain.

Using standard-sized products and materials helps to minimize construction waste. Recycle construction and demolition waste, and donate serviceable cabinetry, appliances, and fixtures for reuse.

Residential kitchen appliances are responsible for about 29 percent of the energy use in a home. When working with clients to design a kitchen, rather than developing an ever-expanding mandate, encourage them to think small, and encourage minimalist design. Specify environmentally healthy building and interior finish materials from local sources, including salvaged and repurposed products. Support the use of energy efficient products and appliances and water-conserving fixtures.

LEED System

As introduced above, Leadership in Energy & Environmental Design (LEED) is a green building certification program. To receive LEED certification, building projects satisfy prerequisites and earn points to achieve different levels of certification. Prerequisites and credits differ for each rating system, and teams choose the best fit for their project.

LEED v4 was introduced in December 2013. This version emphasizes material transparency, requiring a better understanding of the products being used in a building and their sources. Prerequisites such as metering and recording a building's energy and water use were also introduced. LEED v4 supports integrated building systems.

USGBC continues to revise LEED to meet current needs. Check for updates at http://www.usgbc.org/.

LEED CRITERIA AND RATING SYSTEMS

Each LEED rating system groups requirements that address the unique needs of building an project types on their path toward LEED certification. (See Table 2.8) Project teams use the credits appropriate to their chosen rating system to guide design and operational decisions.

LEED rating systems include Integrative Process requirements that promote reaching across disciplines to incorporate diverse team members during the pre-design period.

There are four levels of LEED certification. The lowest is Certified, then Silver, Gold, and Platinum. The number of points a project earns determines the level of LEED certification.

LEED Professional Accreditation recognizes an individual's qualifications in sustainable building. Interior designers are among those becoming LEED-accredited professionals by passing the appropriate LEED professional accreditation exam. The LEED Green Associate Exam is the first level. Experience working on a LEED-registered project is strongly recommended before taking the LEED AP exam for one of the five project categories.

In addition to LEED, there are other sustainable design programs throughout the world. (See Table 2.9)

High Performance Buildings

High performance buildings integrate and optimize all major high-performance building attributes, including energy efficiency, durability, life-cycle performance, and occupant productivity. The design of the role of mechanical systems in high-performance buildings should clarify the role of the system in reaching net-zero energy or carbon-neutral outcomes. Historically, passive systems have been used, but these have some limitations. More recently, active systems with fewer limitations have dominated, but these may have serious environmental impacts.

Net zero buildings are buildings that become self-sufficient and operate entirely on renewable energy.

TABLE 2.8 LEED V4 RATING SYSTEMS

Rating System	Building Types Covered
Building Design and Construction (BD+C)	New construction, core and shell, schools, retail, hospitality, data centers, warehouses and distribution centers, healthcare, homes
Interior Design and Construction (ID+C)	Commercial interiors, retail, hospitality
Building Operations and Maintenance (O+M)	Existing buildings, schools, retail, hospitality, data centers, warehouses and distribution centers
Neighborhood Development (ND)	Neighborhood development plan

TABLE 2.9 OTHER SUSTAINABLE DESIGN PROGRAMS

Program	Description
Canada Green Building Council	The Canada Green Building Council has a version of LEED adapted for the Canadian climate, and construction practices and regulations.
2012 National Green Building Standard®	Residential green building rating system approved by American National Standards Institute (ANSI). Developed by International Code Council (ICC) and National Association of Home Builders (NAHB).
Architecture 2030 Challenge	Goal is to reduce greenhouse gas emissions of buildings to 0 by the year 2030.
Living Building Challenge	Rigorous international sustainable building certification program created by International Living Future Institute to promote advanced measurement of sustainability in the built environment.
Green Globes™	The Green Building Initiative's (GBI) online assessment tool for new or existing buildings, sustainable interiors, or building intelligence.
R-2000	Canadian voluntary technical performance standard for energy efficiency, indoor air tightness quality, and environmentally responsible home construction administered by Natural Resources Canada (NRCan).
BREEAM	The Building Research Establishment's Environmental Assessment Method (BREEAM) in the United Kingdom, Hong Kong, and Canada. Environmental assessment and rating system for sustainable design, construction, operation.

The term **passive house** (the German term *Passivhaus* is often used) refers to a voluntary standard for energy efficiency in a building that reduces its ecological footprint so that it requires little energy for space heating or cooling. The standard has been applied to office buildings, schools, and other building types in addition to residential buildings.

The characteristics of a house that uses no more energy than it produces include superinsulated walls, roof, and floor, and airtight construction with a heat recovery unit for ventilation. High-performance windows are oriented properly for solar and climate requirements, and fully shaded in summer. Passive solar is used for space heating; and active solar, for domestic hot water. Appliances, electric lighting, and heating and cooling systems are all designed for high-efficiency. Photovoltaics provide any additional electrical needs.

This chapter has explored the building envelope and introduced energy efficient design, the building design process, and sustainable design. In Chapter 3, we look at how the human body interacts with the built environment as well as how building codes protect us.

Designing for Human Health and Safety

3

Architects, engineers, and interior designers are all concerned about how the human body interacts with the built environment. Engineers design systems that maintain thermal equilibrium and other environmental requirements. All three professions use building codes to satisfy minimum health and safety requirements.

Members of the American Institute of Architects (AIA) and other licensed architects with mandatory continuing education requirements are required to complete a minimum number of hours of Health, Safety, and Welfare (HSW)-related training. AIA defines HSW in architecture as "anything that relates to the structural integrity or soundness and health impacts of a building or building site. Courses must intend to protect the general public." (Continuing Education System, The American Institute of Architects, http://www.aia.org/education/ces/AIAB089080, accessed July 4, 2014)

According to the American Society of Interior Designers (ASID):

> Protecting health, safety and welfare is the professional responsibility of every interior designer. Every decision an interior designer makes in one way or another affects the health, safety and welfare of the public. Those decisions include specifying furniture, fabrics and carpeting that meet or exceed fire codes and space planning that provides proper means of egress. Additionally, designers deal with accessibility issues, ergonomics, lighting, acoustics and design solutions for those with special needs. (Health and Safety, www.asid.org/content/health-and-safety, accessed July 4, 2014)

INTRODUCTION

The work of interior designers is directly concerned with the health, safety, and welfare of building occupants. Aspects of sustainable design such as indoor air quality and sustainable materials have direct impacts on human health. Requirements for health and safety differ between residential and commercial spaces.

Building safety has been a code issue since around 1772 BCE, when the Babylonian Code of Hammurabi dictated consequences for improper building construction. Concern about building-related illnesses arose in the late 1970s, due to indoor air problems in newly constructed office buildings; by the 1990s, the focus became the chemical contents of building materials.

HUMAN BODY AND THE BUILT ENVIRONMENT

Buildings provide environments where people can feel comfortable and safe. To understand the ways building systems are designed to meet these needs, we must first look at how the human body perceives and reacts to interior environments.

Maintaining Thermal Equilibrium

Our perception that our surroundings are too cold or too hot is based on many factors beyond the temperature of the air. The season, the clothes we are wearing, the amount of humidity and air movement, and the presence of heat given off by objects in the space all influence our comfort. Contact with surfaces or moving air, or with heat radiating from an object, produces the sensation of heat or cold. There is a wide range of temperatures that will be perceived as comfortable for one individual over time and in varying situations. We can regulate the body's heat loss with three layers of protection: the skin, clothing, and buildings.

HUMAN BODY HEAT PRODUCTION

The human body operates as an engine that produces heat. (See Figure 3.1) The fuel is the food we eat, in the form of proteins, carbohydrates, and fats. The digestive process uses chemicals, bacteria, and enzymes to break down food. Useful substances are pumped into the bloodstream and carried throughout the body. Waste products are filtered out during digestion and stored for elimination.

The normal human internal body temperature is around 98.6°F (37° C). This temperature cannot vary by more than a few degrees without causing physical distress. Our body turns only about one-fifth of the food energy we consume into mechanical work, with the remaining energy given off as heat or stored as fat. The amount of heat our body produces depends on what we are doing. This is why a room full of people doing aerobic exercise quickly heats up. The body requires continuous cooling to give off all of this excess heat.

An individual's metabolism sets the rate at which energy is used; it is based primarily on our level of muscular activity. (See Figure 3.2) Our **metablolic rate** follows a normal daily cycle, and is influenced by what, when, and how much we eat. The metabolic rate varies with body surface area and weight, health, sex, and age. Our metabolic rates are highest at around age 10, and lowest in old age. Pregnancy and lactation increases the rate by about 10 percent.

The amount of clothing a person is wearing and the surrounding thermal and atmospheric conditions also influence the metabolic rate. It increases when we have a fever, during continuous activity, and in cold conditions if we are not wearing warm clothes. The weight of heavy winter clothing may add 10 to 15 percent to the metabolic rate.

The metabolic rates of specific groups of people (children, older adults) may influence the thermal qualities of the spaces we design for them.

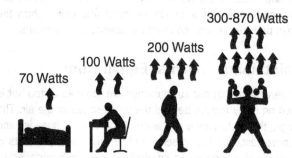

Figure 3.1 Activity and body heat

70 Watts
100 Watts
200 Watts
300-870 Watts

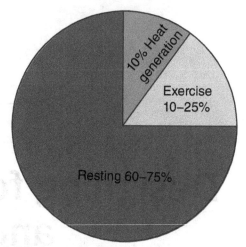

Figure 3.2 Sedentary metabolic rates

10% Heat generation

Exercise 10–25%

Resting 60–75%

The set of conditions that allows your body to stay at the normal body temperature with the minimal amount of bodily regulation is called **thermal equilibrium**. We feel uncomfortable when the body works too hard to maintain its thermal equilibrium.

As designers of interior spaces, our goal is to create environments where people are neither too hot nor too cold to function comfortably and efficiently. We experience **thermal comfort** when heat production equals heat loss. Our mind feels alert, our body operates at maximum efficiency, and we are at our most productive.

SKIN AND INTERNAL BODY INTERACTIONS

Our skin is our primary interface with our environment, and the most important regulator of heat flow. We sense pressure and pain, heat and cold when our skin touches surfaces or moving air or senses heat by radiation. The brain's hypothalmus receives these signals from the skin, as well as core body temperatures, and responds with changes in blood distribution.

If the brain senses that we are too cold, the rate of body heat loss in decreased by reducing blood flow from the body core to the skin surface. In addition, sweat glands force less water to the skin surface, resulting in less evaporation and less heat loss. Colder temperatures result in goose bumps, our body's attempt to create insulation by fluffing up our scant body hair. Shivering tries to increase the body's metabolic rate to burn more fuel and produce more heat.

When we are too hot, blood flow toward the surface is increased. Sweat glands secrete water and salt (which lowers the water vapor pressure) to the skin surface, increasing heat loss by evaporation. (See Figure 3.3)

EFFECTS OF HEAT AND COLD

Our skin surface provides a layer of insulation between the body's interior and the environment that is about equal in effect to putting on a light sweater. When we are cold, we lose too much heat too quickly, especially from the back of the neck, the head, the back, and the arms and legs. (See Figure 3.4)

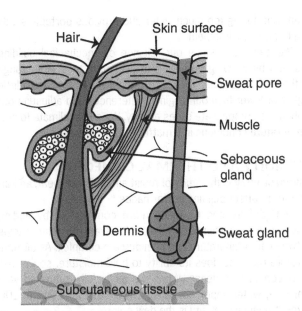

Figure 3.3 Skin and sweat glands

Source: Redrawn from www.naturallyhealthyskin.org

Hair
Skin surface
Sweat pore
Muscle
Sebaceous gland
Dermis
Sweat gland
Subcutaneous tissue

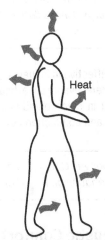

Heat

Figure 3.4 Body heat loss areas

This puts an increased strain on the heart, which pumps an increased amount of the blood directly to the skin and back to the heart, bypassing the brain and other organs, and we can become lethargic and mentally dull. Ultimately, when deep body temperatures fall, we experience hypothermia, which can result in a coma or death. The slide toward hypothermia can be reversed by exercise to raise heat production, or by hot food and drink, and a hot bath or sauna.

When we get too hot, the blood flow to the skin's surface increases, sweat glands secrete salt and water, and we lose body heat through evaporation of water from our skin. Water constantly evaporates from our respiratory passages and lungs; the air we exhale is saturated with water. In high humidity, evaporation is slow and the rate of perspiration increases as the body tries to compensate. When the surrounding air approaches body temperature, only evaporation by dry, moving air will lower our body temperature.

Overheating increases fatigue and decreases our resistance to disease. If the body is not cooled, deep-body temperature rises and impairs metabolic functions, which can result in heat stroke and death. (See Table 3.1)

Conduction, convection, radiation, and evaporation were introduced in Chapter 2, "Designing for the Environment." We will look at these again in Chapter 12, "Principles of Thermal Comfort."

At a normally comfortable temperature around 70°F (18°C), we lose approximately 72 percent of our body heat per hour by radiation, convection, and conduction. (See Figure 3.5) Evaporation from our skin surface results in 15 percent lost, and evaporation from air exhaled from our lungs adds another 7 percent. The warming of air inhaled into our lungs accounts for 3 percent, and the remaining 3 percent is heat expelled in feces and urine.

When air and surface temperatures approach body temperature, radiation, convection, and conduction are not effective, and evaporation is the only process that effectively cools us. Access to dry moving air helps evaporation do its job.

TABLE 3.1 SURPLUS BODY HEAT TRANSFER MECHANISMS

Mechanism	Description	Primary Variables
Conduction	Heat is transferred through direct contact with cooler surfaces	Surface temperatures
Convection	Heat from the body is absorbed by air molecules	Air temperature, air motion, humidity
Radiation	Heat is transferred to cooler surfaces without physical contact	Surface temperature, orientation to the body
Evaporation	Heat is drawn from the body's surface to provide energy to turn liquid water into water vapor	Humidity, air motion, air temperature

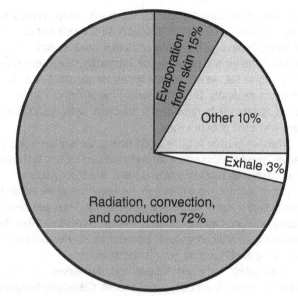

Figure 3.5 Body heat loss by type

CLOTHING AND ACCLIMATION

Over time, we have increasingly asked our buildings to do more to maintain thermal equilibrium and our bodies to do less. This leads to greater energy use. However, our personal thermal equilibrium is something we can mostly control ourselves with clothing. (See Figure 3.6)

Clothing is an additional layer that protects our bodies and helps maintain thermal equilibrium. In cold environments, clothing acts as an insulating layer. In a hot humid environment, we expose our skin to moving air to increase heat loss, while seeking shade to protect our skin from burning. Clothing helps us

keep from losing too much water and provides portable shade in a hot arid environment.

We can also reduce our reliance on mechanical building systems by allowing ourselves to acclimate to a wider range of thermal conditions. Tolerating temperatures that are a little higher or lower than our original preferences and adjusting our clothing to compensate saves energy. We can acclimate to new temperature conditions in a matter of days or weeks.

HUMIDITY AND THERMAL COMFORT

Psychrometry is the study of moist air. Moisture, heat, and air interact to affect building performance.

Hot, dry air can be made more comfortable by adding moisture to raise humidity levels. However, the relationships between air temperature and humidity are complex. As air temperature rises, so does its ability to hold moisture, so warmer air becomes less dense. When the air is fully saturated with moisture, water vapor condenses; this saturation line of 100 percent relative humidity is the dew point.

In Chapter 2, "Designing for the Environment," we looked at how condensation can occur within insulation due to falling air temperatures and the dew point.

Humidity is most effective in influencing heat transfer in hot dry environments, where evaporative heat loss dominates. It has less of an impact in cold conditions, where heat loss by convection, radiation, and conduction dominate.

See Chapter 13, "Indoor Air Quality, Ventilation, and Moisture Control," for more information on humidity.

Visual and Acoustic Comfort

Visual comfort covers a range of situations, including provision of adequate illumination for the task at hand, controlling glare, and providing views and connections to the outdoors.

Our eyes can be damaged if we look even quickly at the sun, or for too long at a bright snow landscape or light-colored sand. Direct glare from lighting fixtures can blind us momentarily. Low illumination levels reduce our ability to see well. The adjustment to moderately low light levels can take several minutes, an important consideration when designing entryways between the outdoors (which may be very bright or very dark) and the building's interior.

Interior designers should avoid creating strong contrasts that can make vision difficult or painful, for example a very bright object against a very dark background or a dark object against light. Lighting levels and daylighting are important parts of interior design.

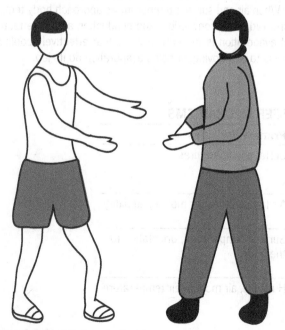

Figure 3.6 Clothing and climate

The buildings we design should help us use our senses comfortably and efficiently. Loud sounds can damage our hearing, especially over time. We have trouble hearing sounds that are much less intense than the background noise. The art and science of acoustics addresses how these issues affect the built environment.

See Chapter 7, "Acoustic Design Principles"; Chapter 8, "Architectural Acoustics"; and Chapter 17, "Lighting Systems" for more information on acoustics and visual comfort.

Other Human Environmental Requirements

Building systems meet our needs for water and waste removal, and assure a supply of fresh air. They also help protect us from bodily harm.

WATER AND WASTE REMOVAL
We need a regular supply of water to move the products of food processing around the body. Water also helps cool the body. We need food and drinking water that is free from harmful microorganisms. Contaminated food and water spread hepatitis and typhoid. Building systems are designed to remove body and food wastes promptly for safe processing.

We will look at these issues in Chapter 9, "Water Supply Systems," and Chapter 10, "Waste and Reuse Systems."

FRESH AIR
We must have air to breathe for the oxygen it contains, which is the key to the chemical reactions that combust (burn) the food-derived fuels that keep our body operating. (See Figure 3.7)

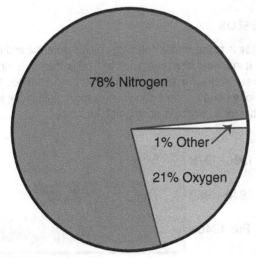

Figure 3.7 Oxygen in air

When we breathe air into our lungs, some oxygen dissolves into the bloodstream. We exhale air mixed with carbon dioxide and water, which are produced as wastes of combustion. Less than one-fifth of the air's oxygen is replaced by carbon dioxide with each lungful, but a constant supply of fresh air is required to avoid unconsciousness from oxygen depletion and carbon dioxide accumulation. Building ventilation systems assure that the air we breathe indoors is fresh and clean.

PROTECTION
The human body is attacked by a very large assortment of bacteria, viruses, and fungi. Our skin, respiratory system, and digestive tract offer a supportive environment for microorganisms. Some of these are helpful or at least benign, but some cause disease and discomfort. Our buildings provide facilities for washing food, dishes, skin, hair, and clothes to keep these other life forms under control. Poorly designed or maintained buildings can be breeding grounds for microorganisms. These are issues for both the design of building sanitary waste systems and indoor air quality.

See Chapter 13, "Interior Air Quality, Ventilation, and Moisture," for more information on indoor air quality and ventilation systems.

Our buildings exclude disease-carrying rodents and insects. Pests can spread typhus, yellow fever, malaria, sleeping sickness, encephalitis, plague, and various parasites. Inadequate ventilation encourages tuberculosis and other respiratory diseases. Adequate ventilation carries away airborne bacteria and excess moisture. Sunlight entering the building dries and sterilizes our environment.

Our soft tissues, organs, and bones need protection from hard and sharp objects. Smooth floor surfaces prevent trips and ankle damage. Our buildings help us move up and down from different levels without danger of falling, and keep fire and hot objects away from our skin.

Interior designers must always be on the alert for aspects of a design that could cause harm from falling objects, explosions, poisons, corrosive chemicals, harmful radiation, or electric shocks. By designing spaces with safe surfaces, even and obvious level changes, and appropriately specified materials, we protect the people who use our buildings. Our designs help prevent and suppress fires, and facilitate escape from burning buildings.

Buildings give us space to move, to work, and to play. Our residential designs support family life with places for the reproduction and rearing of children, preparing and sharing food with family and friends, studying, and communicating verbally, manually, and digitally. We provide spaces and facilities to pursue hobbies and to clean and repair the home. Our designs create opportunities to display and store belongings, and to work at home. The spaces we design may be closed and private at times, and open to the rest of the world at others. We design buildings that are secure from intrusion, and provide ways to communicate

Figure 3.8 Movement
Source: Redrawn from Francis D.K. Ching and Corky Binggeli, *Interior Design Illustrated* (3rd ed.), Wiley 2012, page 198

both within and beyond the building's interior. We provide stairways and mechanical means of conveyance from one level to another for people with varied levels of mobility. (See Figure 3.8)

For more information on stairs, see Chapter 5, "Floor/Ceiling Assemblies, Stairs, and Ramps." See Chapter 19, "Conveyance Systems," for information on elevators and wheelchair lifts.

Our designs also support all the social activities that occur outside of the home. We provide power to buildings so that workshops, warehouses, markets, offices, studios, barns, and laboratories can design, produce, and distribute goods. These workplaces require the same basic supports for life activities as our homes, plus accommodations for the tasks they house. We gather in groups to worship, exercise, play, entertain, govern, educate, and to study or observe objects of interest. These communal spaces are even more complex as they must satisfy the needs of many people at once.

HAZARDOUS MATERIALS

Much of the work of interior designers takes place in existing buildings. Repurposing or recycling demolition waste is a more sustainable alternative to sending it to a landfill. However, some materials require special handling.

Renovation Considerations

Existing buildings may contain materials that are hazardous to human health. Some of these can be exposed or disturbed during demolition. Demolition work can reveal the remains of dead animals and insects within walls, attics, and other spaces. Mold is a common problem that can affect health in buildings where moisture accumulates.

Local regulations should be reviewed for proper handling of suspect demolition waste. The demolition area should be isolated from occupied areas of the building. This should

include the heating, cooling, and ventilating ductwork and equipment. In addition, any areas containing dead animals should be isolated.

Lead

Lead is a neurotoxin that accumulates in the body and is especially damaging to fetuses, infants, and young children, causing learning disabilities, nausea, neurological damage, and death. Particles are suspended in the air or settle on surfaces. Children ingest and inhale lead from paint chips through playing on floors and other dusty surfaces and then putting their hands in their mouths.

Lead-based paint is found in three quarters of pre-1975 US homes. (See Figure 3.9) Until 1985, pipes and solder contained lead; old pipes and solder should be replaced.

The 2010 Lead Renovation, Repair, and Painting Rule in the United States applies to all housing built in 1978 or earlier. All contractors performing renovations, repairs, or painting must be trained and certified to follow lead-safe work practices, including containment of the work area, minimizing dust, and cleaning up thoroughly. The contractor must provide information on lead-safe practices to the property owner.

Interior designers should verify whether lead paint is present before beginning renovation work.

A professional licensed contractor should perform lead-paint abatement. Occupants must be out of the building during the process, and workers must be properly protected. Lead paint should not be sanded or burned off. Moldings and other woodwork should be replaced or chemically treated. Wood floors must be sealed or covered, belongings should be removed or covered, and dust should be contained during the process. The final cleanup should be done using a **high-efficiency particulate air (HEPA) filter** vacuum.

Asbestos

Asbestos is found in many buildings built before the end of the 1970s. It may have been combined with other materials, or used as a preservative for wood. The inhalation of asbestos fibers over a long period of time can cause cancer, fluid in the lungs, and asbestosis, a fibrous scarring of the lungs.

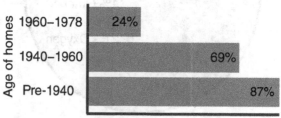

Figure 3.9 Lead in homes (EPA data)

Figure 3.10 Asbestos

Source: "Anthophyllite asbestos SEM" courtesy of US Geological Survey (www.usgs.gov)

Asbestos is white, light gray, or light brown, and looks like coarse fabric or paper; it may appear as a dense, pulpy mass of light gray, stucco-like material applied to ceilings, beams, and columns. (See Figure 3.10) Up until 1975, asbestos was widely used for steam pipe and duct insulation and in furnaces and furnace parts. Before 1980, acoustic tiles, and fiber-cement shingles and siding contained asbestos. Vinyl floor tiles made from the 1940s to the 1980s may contain asbestos as may their adhesive. Asbestos fibers may still be found in existing construction, especially in the insulation on heating system components and other equipment, acoustic ceiling and vinyl floor tiles, and drywall joint-finishing material and textured paint purchased before 1977.

Most asbestos can be left undisturbed as long as it does not emit fibers into the air. If it is not crumbling, it can be sealed with a special sealant and covered with sheet metal. If it remains in place, it must be dealt with later during renovation or demolition.

Disturbing asbestos material by drilling holes, hanging materials on walls or ceilings, causing abrasion, or removing ceiling tiles below asbestos-containing materials should be avoided.

Wrapping can repair asbestos-covered steam lines and boiler surfaces, but asbestos in walls and ceilings usually cannot be repaired as it is difficult to keep airtight. It is possible to encapsulate asbestos by enclosing it in areas with low ceilings or small areas that are unlikely to be disturbed or damaged by water, or where the asbestos is unlikely to deteriorate. However, encapsulation may cost more than removal.

Removal is the only permanent solution, but if done improperly, can be more dangerous than leaving the asbestos in place. Removal must be done by a properly certified and licensed expert.

Areas from which asbestos is being removed must be isolated using airtight plastic containment barriers, and kept under negative pressure with special HEPA filtration. The work site should be inspected and its air quality tested after the work is done.

Mold

Dampness in buildings supports the growth of bacteria, fungi including **mold**, and insects. Dampness in buildings results from internal sources including leaking pipes or external sources like rainwater. It becomes a problem when materials such as rugs, walls, and ceiling tiles become wet for extended periods of time. Excessively high relative humidity and flooding can also produce dampness.

> Research studies have shown that exposures to building dampness and mold have been associated with respiratory symptoms, asthma, hypersensitivity pneumonitis, rhinosinusitis, bronchitis, and respiratory infections. Individuals with asthma or hypersensitivity pneumonitis may be at risk for progression to more severe disease if the relationship between illness and exposure to the damp building is not recognized and exposures continue. (National Institute for Occupational Safety and Health [NIOSH] Alert: *Preventing Occupational Respiratory Disease from Exposures Caused by Dampness in Office Buildings, Schools, and Other Nonindustrial Buildings*, DHHS [NIOSH] Publication No. 2013-102, November 2012)

See Chapter 13, "Indoor Air Quality, Ventilation, and Moisture Control," for more information on mold.

BUILDING CODES AND STANDARDS

Governments respond to concerns for safety by developing building codes, government mandated documents establishing minimum acceptable building practices. These codes dictate both the work of the interior designer and architect, and the manner in which the building's mechanical, electrical, plumbing and other systems are designed and installed.

Standards set minimum requirements for an aspect of building design. Standards are developed by a recognized authority, usually by a consensus process with substantial external review and input. Codes often refer to standards.

Building Codes

Codes define the minimum that society deems acceptable. Most of the codes in the United States are **prescriptive codes**, which mandate that something be done in a certain way. Prescriptive codes define the means and methods by which the code is to be carried out. **Performance codes** are codes that state the objective that must be met, and may offer options for compliance.

The jurisdiction of a project is determined by the location of the building. A **jurisdiction** is a geographical area that uses the same codes, standards, and regulations. A jurisdiction may be as small as a township or as large as an entire state. The authority having jurisdiction for a particular project location—for example, the building department or health department—enforces the code requirements; there may be several authorities for a single project.

Most jurisdictions have strict requirements as to who can design a project and what types of drawings are required. Often, drawings must be stamped by a licensed architect or licensed engineer registered within the state. In some cases, interior designers are not permitted to be in charge of a project, and may have to work as part of an architect's team. Some states may allow registered interior designers to stamp drawings for projects in buildings with limited numbers of stories and square feet. Working out the proper relationships with the architects and engineers on your team is critical to meeting the code requirements.

The *ICC International Performance Code for Buildings and Facilities (IPC)* is a model building code that attempts to unify code requirements across geographic barriers. Introduced by the International Codes Council (ICC) in 2002, the *IPC* has been adopted statewide or in some localities in all US states.

Some states have statewide codes based on a model code, while others have local codes, and sometimes both state and local codes cover an area. Not every jurisdiction updates its codes on a regular basis, which means that in a particular jurisdiction, the code cited may not be the most current edition of that code. When codes are changed, one or more yearly addenda are published with the changes, and incorporated in the body of the code when the next full edition of the code is published.

The interior designer must check with the local jurisdiction for which codes to follow. Professional organizations and government agencies offer continuing education programs when major code changes are introduced.

In addition to the basic building code, jurisdictions issue plumbing, mechanical, and electrical codes. On projects with a major amount of plumbing or mechanical work, registered engineers will take responsibility for design and code issues. On smaller projects, a licensed plumber or mechanical contractor will know the codes.

Interior designers are not generally required to know or to research most plumbing or mechanical code issues. However, the interior designer needs to be aware of some plumbing and mechanical requirements, such as how to determine the number of required plumbing fixtures.

CODE OFFICIALS

The codes department is the local government agency that enforces the codes within a jurisdiction. A code official is an employee of the codes department with authority to interpret and enforce codes, standards, and regulations within that jurisdiction.

The plans examiner is a code official who checks plans and construction drawings at both the preliminary and final permit review stages of the project. The plans examiner checks for code and standards compliance, and works most closely with the designer.

The fire marshal usually represents the local fire department. The fire marshal checks drawings with the plans examiner during preliminary and final reviews, looking for fire code compliance.

The building inspector visits the project job site after the building permit is issued, and makes sure all construction complies with the codes as specified in the construction drawings and in code publications.

Standards and Organizations

Codes cite standards developed by government agencies, trade associations and standard-writing organizations as references. A standard may consist of a definition, recommended practice, test method, classification, or required specification.

The nonprofit **National Fire Protection Association (NFPA)** was formed in 1896 to reduce the burden of fire and other hazards on the quality of life by providing and advocating codes and standards, research, training, and education. The NFPA develops and publishes more than 300 standards intended to minimize the possibility and effects of fire and other risks. *NFPA 101® Life Safety Code®* and *NFPA 70: National Electric Code®* are both NFPA publications that provide guidelines for fire safety. The NFPA establishes testing requirements covering everything from textiles to firefighting equipment to the design of means of egress.

The American **National Standards Institute (ANSI)** originated in 1918, and strives to assure the safety and health of consumers and the protection of the environment. The Institute oversees the creation, promulgation, and use of thousands of norms and guidelines that directly impact businesses in nearly every sector, including acoustical devices, construction equipment, energy distribution, and many more.

ASTM International, formerly known as the American Society for Testing and Materials (ASTM), dates to 1898. ASTM International develops over 12,000 ASTM voluntary consensus standards used around the world to improve product quality, enhance safety, facilitate market access and trade, and build consumer confidence.

ASHRAE®, formerly known as the American Society of Heating, Refrigeration, and Air-Conditioning Engineers (ASHRAE), was formed in 1959 to sponsor research projects and to develop performance level standards for HVAC and refrigeration systems. Mechanical engineers and refrigeration specialists and installers use ASHRAE standards. *ANSI/ASHRAE/IES Standard 90.1—Energy Standard for Buildings Except Low-Rise Residential Buildings* provides minimum requirements for energy efficient designs for commercial and multi-story residential buildings. It is frequently updated to respond to new technologies.

Figure 3.11 UL label

UL, formerly Underwriters Laboratories (UL), is a global independent safety science company working since 1894 to innovate safety solutions from the public adoption of electricity to new breakthroughs in sustainability, renewable energy, and nanotechnology. Dedicated to promoting safe living and working environments, UL helps safeguard people, products, and places. The organization certifies, validates, tests, inspects, audits, and advises and trains users of products, systems, and materials. UL lists all the products it tests and approves in product directories.

Interior designers may find UL's *Building Materials, Fire Protection Equipment,* and *Fire Resistance* directories useful. Codes require UL testing and approval for certain products. UL tags appear on many household appliances as well as lighting and other electrical fixtures. (See Figure 3.11)

Federal Codes and Regulations

The federal government regulates the building of federal facilities, including federal buildings, Veterans Administration hospitals, and military establishments. The construction of federal buildings is typically not subject to state or local building codes and regulations. The federal government issues regulations for government built and owned buildings, similar to the model codes. On a particular project, the authorities involved may opt to comply with stricter local requirements, so the designer must verify what codes apply.

There are over one thousand separate codes and a wide variety of federal regulations. In an effort to limit federal regulation, the Consumer Product Safety Commission encourages industry self-regulation and standardization, and industry groups have formed hundreds of standards writing organizations and trade associations representing almost every industry.

Congress can pass laws that supersede all other state and local codes and standards. They are collected in the *Code of Federal Regulations,* which is revised annually. The *Occupational Safety and Health Act (OSHA), Fair Housing Act (FHA),* and *Americans with Disabilities Act (ADA)* are examples of congressionally passed laws with wide implications for interior designers and architects.

OSHA

The **Occupational Safety and Health Administration (OSHA)** was established in 1970 to assure safe and healthful working conditions for working men and women by setting and enforcing standards and by providing training, outreach, education, and assistance. OSHA adds to code requirements by regulating the design of buildings and interior projects where people are employed. Contractors and subcontractors on construction projects must strictly adhere to OSHA requirements. Interior designers should be aware that these regulations exist and affect the process of building construction and installation of equipment and furnishings.

AMERICANS WITH DISABILITIES ACT

The term **accessible** in building codes refers to handicapped accessibility as required by codes, the *Americans with Disabilities Act (ADA)*, and other accessibility standards. The Departments of Justice and Transportation developed the provisions of the ADA, which was passed by Congress in 1990. In addition, some states also have their own accessibility standards.

The ADA is a comprehensive civil rights law with four sections. Title I protects individuals with disabilities in employment. Title II covers state and local government services and public transportation. Title III covers all public accommodations, defined as any facility that offers food or services to the public. It also applies to commercial facilities, which are non-residential buildings that do business but are not open to the general public. Title IV deals with telecommunications services, and requires telephone companies to provide telecommunications relay services for individuals with hearing and speech impairments.

We will be referring to ADA provisions throughout this book. ADA Titles III and IV affect the work of interior designers most directly.

The regulations included in Title III have been incorporated into the **2010 ADA Standards for Accessible Design**, which gives helpful information on interpretation and compliance. (See Figure 3.12) The ADA deals with architectural concerns such as accessible routes and the design of restrooms for wheelchair access. Communication issues covered include alarms systems and signage for people with vision and hearing impairments.

All new buildings with public accommodations and/or commercial facilities must conform to specific ADA requirements. This includes a wide range of project types, including lodging, restaurants, hotels, and theaters. Shopping centers and malls, retail stores, banks, places of public assembly, museums, and galleries are also covered. Libraries, private schools, day-care centers, and professional offices are all included. State and local government buildings and one- and two-family dwellings are not required to conform.

ADA requirements are most stringent for new buildings or additions to existing buildings. The requirements are not as clear concerning renovation of existing buildings and interiors. When an existing building is renovated, specific areas of the building must be altered to conform to ADA requirements. These alterations are limited to those deemed readily achievable in terms of

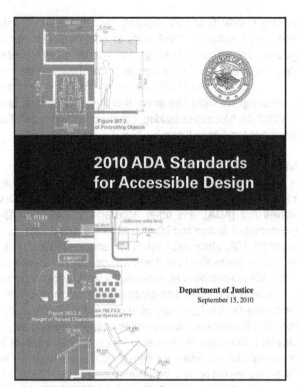

Figure 3.12 2010 ADA Standards for Accessible Design
Source: Reproduction of this document is encouraged by ADA

structure and cost. Exemptions may be made for undue burden as a result of the difficulty or expense of an alteration. These situations are determined on a case-by-case basis by regulatory authorities or the courts, and often involve difficult judgment calls.

Energy Efficiency Requirements

Energy efficiency requirements mandate the minimum performance that is considered acceptable, rather than an optimal performance level.

Introduced in 1998 by the international Code Council, the **International Energy Conservation Code (IECC)** addresses energy efficiency, including cost savings, reduced energy usage, conservation of natural resources, and the impact of energy usage on the environment. In addition, some states have their own energy codes. These codes cover virtually every building system, including lighting and electrical distribution.

The most generally used standard for commercial and institutional buildings in the United States is *ANSI/ASHRAE/IES Standard 90.1—Energy Standard for Buildings Except Low-Rise Residential Buildings* This standard provides the minimum requirements for energy-efficient design of most non-residential buildings. It offers the minimum energy-efficient requirements for design and construction of all or portions of new buildings and their systems, and new systems and equipment in existing buildings as well as compliance criteria. It is an indispensable reference for engineers and other professionals involved in the design of buildings and building systems.

Codes and standards for residential energy efficiency include the *International Energy Code* and *ANSI/ASHRAE Standard 90.2—Energy Efficient Design of Low Rise Buildings*. These residential energy requirements focus on minimum requirements for the building envelope (walls, floors, roofs, doors, windows) and mechanical equipment performance for heating, cooling, and domestic hot water.

The 2012 **International Residential Code (IRC)** incorporated the residential provisions of the International Energy Conservation Code into its sections on energy efficiency. This divides the United States and Canada into eight climatic zones designated by state, province, county, and territory, with many code requirements specific to these zones. These include requirements for windows, exterior doors, and insulation.

Chapters 1 through 3 have introduced the environmental and governmental criteria that affect the design of buildings. In Part II, "Building Components," we see how the building's form and structure, interior architectural elements, and windows and doors support, protect, and shape its interior spaces.

PART

II

BUILDING COMPONENTS

Although the work of interior designers is concerned with interior spaces, interior designers benefit from an understanding of the way buildings are built, why they stand up or fall down, and how different building techniques affect the shaping and utilization of interior space.

Part II, Building Components, comprises three chapters:

Chapter 4, "Building Forms, Structural Principles, and Elements," looks at how the building's form interacts with site conditions, affects building energy use, and meets the needs of its occupants.

Chapter 5, "Floor/Ceiling Assemblies, Walls, and Stairs," introduces the floor systems that provide horizontal support, and the wall systems, stairs, and ramps that give vertical support and facilitate movement.

Chapter 6, "Windows and Doors," addresses how windows and skylights control thermal radiation, offer daylight and views, and sometimes provide ventilation, and how doors admit entry and help control fire.

As in Part I, there are multiple interconnections between the topics presented.

Almost every component of a building serves more than one function, with some components commonly serving ten or more simultaneously, and these functions are heavily interdependent. For example, if we decide to build the partitions in a school building of thin sheets of gypsum wallboard over a framework of steel studs, instead of bricks and mortar, we will affect the thermal properties of the building, its acoustical qualities, the quality and quantity of light in the classrooms, how the piping and wiring are installed, the usefulness of the wall surfaces, the deadweight that the structure of the building must support, the fire resistance of the building, which trades will construct it, and how it will be maintained. (Edward Allen, *How Buildings Work* (3rd ed.), Oxford University Press 2005, page 31)

4

Building Forms, Structures, and Elements

A building gives expression to its architectural form through its structural elements. The form, in turn, affects how a building reacts to the conditions of its site. These reactions shape the way that building systems use energy to meet the needs of occupants. This is especially true of buildings that use passive systems.

Chapter 4 looks at building forms and structural systems, introduces the basics of building structural loads and the elements that carry them, and surveys the basic types of building structures.

INTRODUCTION

The building's structure provides support to the building and makes adjustments to movement. It is designed to withstand fire and aid in its containment. Building structures can be designed to control thermal radiation, air temperature, the thermal qualities of surfaces, humidity, and airflow. How the structure is designed affects visual and acoustic privacy, and the entry of living creatures and materials.

> We must configure every building in such a way that it will support its own dead load plus a live load equal to the worst combined total of people, furnishings, snow, wind, and earthquake that may reasonably be expected. (Edward Allen, *How Buildings Work* (3rd ed.), Oxford University Press 2005, page 172)

The building's architects and engineers must estimate the magnitude of these loads. They must select a structural system appropriate to the site, the uses of the building, and the expected loads. This process involves determining the exact configurations and necessary strengths and sizes of components of the structural system, including all fastening devices used to hold larger components together.

History

Early buildings were built of perishable materials such as branches and animal hides. These were succeeded by more durable stone, clay, wood, and eventually manmade materials.

Buildings became higher and capable of spanning greater distances over time. Stronger materials and greater knowledge of building materials aided this process.

Most of the less durable early buildings did not make it into the archeological record. Some stone buildings such as those at Skara Brae in Scotland's Orkney Islands (occupied around 3180 to 2500 BCE) survived to provide evidence of their construction today.

HISTORIC PRESERVATION

Many interior design projects are in existing buildings. Some projects involve minimal changes in use and layout, while others require extensive demolition and rebuilding. Buildings with historic value deserve special attention and care.

The federal government sets standards to assure that federally funded work does not adversely affect buildings that are either on or eligible for the National Register of Historic Places (NRHP). State and local governments and many private organizations also use these standards. These standards are mandatory only when federal funding or other preservation related funding incentives are used; otherwise, the property owner is only required to meet applicable codes and zoning ordinances.

BUILDING FORM

Building form affects the design of building systems and the amount of energy they use. As Chapter 2 indicates, engineers designate some buildings as internal load dominated and others as skin load dominated.

The enclosures of thick, tall buildings keep most of their interior space away from climatic influences. Heat-producing electric lighting added to the heat generated by the occupants and their equipment create internal load dominated buildings that require mechanical cooling through all seasons.

Thinner buildings need heating in cold weather and cooling in hot weather to counter the effects of the weather on the building envelope. Daylight is able to reach a skin load dominated building's center, making electric lighting mostly unnecessary by day.

In the Northern Hemisphere, the sun is welcomed into the building for heat through glazed openings on the south wall of the building. The amount of large areas of cold glass may need to be limited to retain as much heat and daylight as possible. The building's designers may find themselves looking for surplus heat elsewhere in the building to warm colder perimeter spaces.

STRUCTURAL SYSTEM

The **structural system** of a building is designed and constructed to support the loads applied to the building, and to transmit them safely to the ground without damage to the building. The components of the building structure protect the building's occupants and contents. Some structural systems, such as heavy timber structures, are based on a single material. Others combine more than one material; structural steel framed buildings often have horizontal steel beams and vertical steel columns, along with horizontal floor planes of steel and concrete. A building may have more than one structural system, as when a concrete **foundation** supports a light wood framed **superstructure**. (See Figure 4.1) In other cases, what appears to be the structural material, such as brick, is actually only an exterior facing material, with the real **structural load** carried by a concrete structural system.

The underground part of the building is in direct contact with soil, rock, and groundwater. The above-ground superstructure is affected by wind, rain or snow, and sun. The design of buildings can blend into or shut out environmental conditions, and these design choices have a direct impact on our world.

The building's vertical extension above ground is called its superstructure, and includes the columns, beams, and load-bearing walls that support its floors and roof structures. The building's superstructure rests on its foundation.

The foundation, in turn, is supported by the earth below and surrounding it. Soil conditions, the presence of water below the surface, and the characteristics of bedrock below are all taken into account in the design of the foundation.

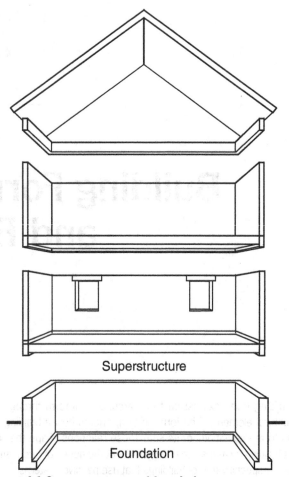

Superstructure

Foundation

Figure 4.1 Superstructure and foundation
Source: Redrawn from Francis D.K. Ching and Corky Binggeli, *Interior Design Illustrated* (3rd ed.), Francis D.K. Wiley 2012, page 8

Foundations

The building rests on a foundation made of concrete, concrete block, or stone that supports the floor structure and anchors the rest of the building. A foundation may be deep enough for a basement, only leave room for a crawl space, or consist of a slab-on-grade set directly on the ground.

A foundation consisting of a grid of **piers** or poles can be used to elevate the building's superstructure well above ground level to prevent damage from flooding, to accommodate a steeply sloping site, or to allow cooling air to circulate below the building.

Foundations also serve to keep the rest of the building above wet earth and to keep water out of the building. Although the soil surrounding the foundation will help to hold building heat in, foundations are often given an exterior layer of insulation as well. Although architects and engineers take great pains to provide a stable base for the building's foundation, it is considered normal for the structure of a building to settle slightly, gradually subsiding as the soil below compacts under the building's loads.

Building load

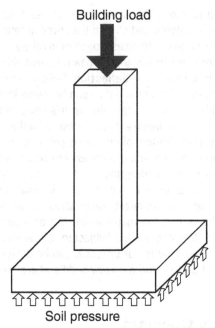

Soil pressure

Figure 4.2 Footings spread load

Figure 4.3 Cast-in-place concrete foundation

At the base of the foundation, footings spread the load over a wider area. (See Figure 4.2) Some buildings on unsuitable or unstable soil require especially deep foundations. Pile and caisson foundations rest on layers of rock or on dense sands and gravels farther below the surface.

Foundation walls are usually made of concrete or of concrete masonry units, although older buildings may have stone foundations. Concrete foundation walls are **cast-in-place** within forms. (See Figure 4.3) They have fewer joints that could admit groundwater than CMU walls. **Concrete masonry units (CMUs)** are small units 8" (203 mm) high and wide, and 16" (457 mm) in length. They can be handled easily and do not require formwork.

SLABS-ON-GRADE

A **concrete slab-on-grade** is supported directly on the earth, and made thick enough to carry the wall and column loads for a one or two-story building. A concrete slab can conduct heat well, so the temperature at the outside edge of the slab may be significantly lower than the interior ambient air temperature when the slab is covered with carpet on a foam pad.

The lower perimeter temperature of a concrete slab can result in interior condensation and moisture problems as a result of thermal transfer, rather than foundation leaks.

Unless the building is closed and air conditioned or dehumidified, moisture is likely to condense on top of a cool slab in humid weather. A thick drainage layer of crushed rock under the slab keeps water from building up. A continuous membrane between the crushed rock and the slab will block water vapor migration through the slab.

It is best not to insulate the inside of a foundation wall, as this defeats the benefit of its thermal mass. Instead, the earth side of the foundation wall should be insulated all the way down to the footing.

FOUNDATION WALL SYSTEMS

Insulating concrete form (ICF) systems comprise preformed blocks or panels with plastic ties that are designed primarily for use below grade. They are cast as formwork for concrete and steel reinforcing rods, and remain in place as insulation.

Structural insulated panels (SIPs) are sandwiched prefabricated panels that are quickly connected with splice plates (and without thermal bridges). They allow much less infiltration than standard construction. The facing boards of their stressed skin construction carry most of the structural load.

BASEMENTS

Basement walls are typically foundations walls, and may be partly above and partly below grade. Basement floors are typically concrete slabs.

Basements remain within a narrow temperature range all year. The greatest problem for basements and other earth-sheltered building components is keeping water out.

The depth of the basement affects how much heat will flow through basement walls and floors. With insulation, heat loss can be reduced significantly.

CRAWL SPACES

A **crawl space** is created under the superstructure by a continuous foundation wall or piers. A crawl space may be literally that, and be too low to stand or even sit. Crawl spaces give access to electrical, plumbing, and mechanical equipment.

Crawl spaces should be ventilated to deal with moisture that may migrate from the interior and from the ground. When moisture disperses into a crawl space, it can produce mold and mildew.

BUILDING LOADS

The term **building load** refers to any of the forces to which a structure is subjected. The weight of the building materials is part of the building load, as are any furnishings or equipment inside the building. Building loads also take into account people and things that move into and out of the building. In addition, building structures are designed to accommodate winds (including storms), the weight of accumulated snow, and seismic forces (earthquakes).

Types of Building Loads

Dead loads are static loads acting vertically downward on a structure, including the weight of the structure itself and the weight of building elements, fixtures, and equipment permanently attached to it.

Static loads are applied slowly and steadily to a structure until reaching their maximum. A structure responds slowly to a static load and is affected the most when the static force is at its greatest. Static loads include the weight of the structure itself, the weight of building elements, fixtures, and equipment permanently attached to it, and movable or moving live loads.

Live loads change over time, but generally do so gradually. They include the weight of the building's occupants, any mobile equipment and furnishings, and any collected snow and water.

Dynamic loads are applied suddenly to a structure, often with rapid changes in the size of the force and the point to which it is applied. Earthquakes and the loads caused by winds are examples of dynamic loads.

Compression, Deflection, and Tension

Building structures are subject to forces that produce compression, deflection, and tension. (See Figure 4.4)

Compression is the shortening or pushing together of a material resulting in a reduction in its size or volume. For example, if you press down on a firm cushion, the material inside

is squeezed together. This squeezing may cause the shape of the cushion to deform, and stretch the covering fabric in some places while causing it to squish together in others.

The perpendicular distance a beam is bent down when a load is placed on it is called **deflection**. Deflection increases as the load becomes heavier and the span becomes longer. When a beam deflects under a load, the beam's material under the load at the top of the beam gets compressed. At the same time, the material at the bottom of the load is pulled apart.

Tension is the reaction of a material to stresses stretching or pulling on a material along the direction of its length. A beam that bends and deflects is subject to an internal combination of compressive and **tensile** (pulling) stresses. An interesting effect of this is that the material that lies between the compressed and stretched areas is subject to relatively little stress. This is why steel beams, for instance, can be designed with a wide top and bottom and a narrow middle where there is little work to do.

Spanning Openings

Structural loads are transferred through the building's structural system to the ground. Loads are transferred across openings by systems of beams and columns.

There are quite a few ways to carry a building load across an opening, including beams, trusses, lintels, arches, and corbels. Vaults and domes cover an area with a three-dimensional structure. Cantilevers reach out over a space with support at only one end.

BEAMS

In a **beam**—a horizontal structural member that is longer than it is wide or deep—compression may cause the beam to bend downward. The distance between the beam's supports is called its **span**. If the beam's load is applied at the center of a beam that is supported at both ends, the center will bend down from the horizontal.

Wood beams have a long history, although most have not lasted a long time. The pagoda of the Horyu-ji in Ikaruga, Nara Prefecture, Japan is an exception. Constructed in 607 CE, it is widely considered to be one of the oldest existing wooden buildings in the world. The original building was probably destroyed by lightning and burned to the ground. It was reconstructed around 711, and repaired and reassembled in the twelfth century, in 1374, and in 1603, and restored in 1954, preserving 15 to 20 percent of the original seventh century materials.

There is a variety of types of wood beams. (See Figure 4.5) Beams cut from large trees are used in heavy timber construction. Light-wood frame construction uses **dimensional lumber** beams that are uniformly cut and designed to carry loads across specific spans. Spaced beams are made with blocking and open spaces to keep weight light. Laminated beams are made from pieces of wood glued together; they can be very strong. There are also a number of types of built-up beams.

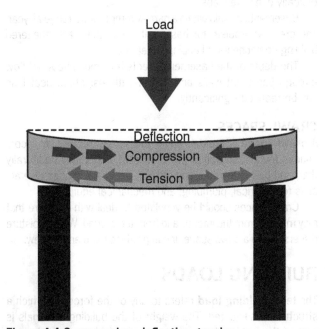

Figure 4.4 Compression, deflection, tension

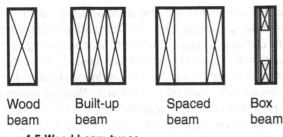

Wood beam | Built-up beam | Spaced beam | Box beam

Figure 4.5 Wood beam types

Source: Redrawn from Francis D.K. Ching, *Building Construction Illustrated* (5th ed.), Wiley 2014, page 4.35

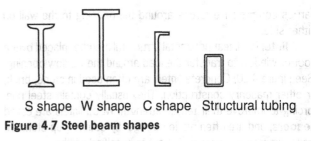

S shape W shape C shape Structural tubing

Figure 4.7 Steel beam shapes

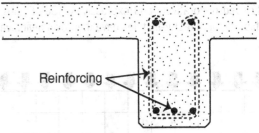

Reinforcing

Figure 4.8 Reinforced concrete beam

Source: Redrawn from Francis D.K. Ching, *Building Construction Illustrated* (5th ed.), Wiley 2014, page 4.04

TRUSSES

Another way to span an opening is a **truss.** (See Figure 4.9) A truss could be thought of as a type of built-up beam that utilizes the inherent stability of a triangle to spread and support a load. Unlike a beam with its combination of tension and compression, in a truss all the structural members are in tension.

Some trusses look like big triangles with smaller triangles inside. Others are rectilinear—or even curved on top—but still made up of smaller triangles.

Steel trusses are composed of structural steel angles and tees bolted or welded together into a triangle-based framework. Steel trusses can take on many angled, curved, flat, and triangular forms.

Laminated wood trusses are made up of smaller pieces of wood glued together into large trusses. Laminated wood trusses can be designed to be stronger than solid wood trusses, and are easier to shape into curves.

LINTELS, ARCHES, AND CORBELS

Openings weaken the structure of a rigid wall. Building a lintel or arch above a door or window opening supports the load and

Figure 4.6 Stone beams and column, Pemberton Mill, Lawrence, Massachusetts, 1967

Source: Library of Congress Prints and Photographs, HABS, Robert M. Vogel, Photographer

Stone beams are good in compression but relatively poor in tension. (See Figure 4.6) They are also very heavy, and limited to shorter spans than steel beams.

Steel beams are strong and good in both tension and compression. (See Figure 4.7) They are also relatively easy to assemble into grid patterns with steel columns, making it possible to create interiors with large open spaces.

Reinforced concrete beams incorporate lengthwise and web steel reinforcement. (See Figure 4.8) They are almost always formed as part of the slab they support.

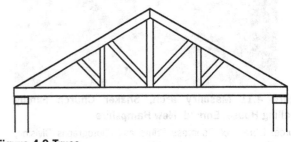

Figure 4.9 Truss

carries compressive forces around the opening to the wall on either side.

A **lintel** is a linear horizontal structural member placed over a door or window to transfer the load around the window opening. (See Figure 4.10) Concrete lintels are often used in stone, brick, or other masonry construction. They usually contain steel reinforcing to improve their tensile strength. Wood lintels are called **headers**, and are often made stronger by doubling up standard wood dimension lumber that is cut to specified widths.

Arches, like columns, are structural members that are frequently used decoratively and expressively. (See Figure 4.11) Many historic architectural styles have their own definitive type of arch. Some arches are able to stand alone because they transfer the load in both directions from each part of the arch.

An arch uses a curved structure to span an opening. Arches are designed to support a vertical load mostly by compression along its axis and to transfer the load to load-bearing surfaces adjacent to the two sides of the arch. Masonry arches are constructed of individual wedge-shaped stones or bricks. Rigid arches of curved, rigid timber, steel, or reinforced concrete have additional ability to carry some bending stress.

A Roman arch is semicircular. (See Figure 4.12) Each of the stones in the arch (voissoirs) presses against its neighbor, and the keystone at the top holds them all in place.

Corbels consist of masonry units stacked with each row extending past the row below. (See Figure 4.13) They are held in place by the weight of the construction on top of them. A corbelled arch is made from masonry units such as bricks or stone blocks, with each row extending from the side of the opening farther into the opening; the opening gets smaller as it becomes higher. Corbels are not able to stand alone; they depend on the weight of masonry above the extending units to keep them from falling into the arch. Corbels are generally inferior to arches for spanning openings.

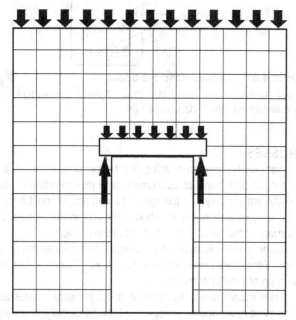

Figure 4.10 Lintel transferring load around opening

Source: Redrawn from Francis D.K. Ching, *Building Construction Illustrated* (5th ed.), Wiley 2014, page 2.17

Figure 4.12 Roman arch

Figure 4.11 Masonry arch, Shaker Church Family Dwelling House, Enfield, New Hampshire

Source: Library of Congress Prints and Photographs Division, Aubrey P. Janion, photographer, 1959

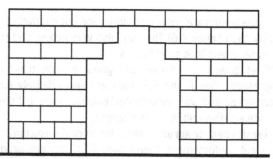

Figure 4.13 Corbelled arch

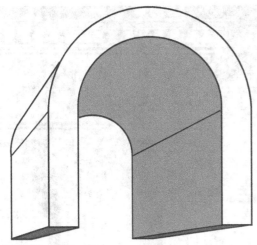

Figure 4.14 Barrel vault

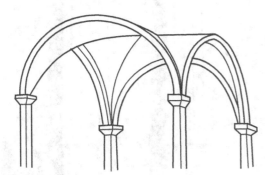

Figure 4.15   Groin vault

VAULTS

The ceiling over a hall or room can be spanned by a **vault**, which is an arched structure of stone, brick, or reinforced concrete. A vault functions like a three-dimensional arch. The supporting walls along the length of a vault are buttressed against outward stresses, like the sides of an arch.

A **barrel vault** is essentially a series of Roman arches lined up along an axis. (See Figure 4.14) It is semicircular in cross section.

Groin (cross) vaults are formed by the perpendicular intersection of two vaults, creating the appearance of two arches intersecting diagonally. (See Figure 4.15) When the arches are elaborately curved, the vault takes on an ornate three-dimensional form.

DOMES

Domes span circular openings with a spherical surface structure consisting of stacked blocks, reinforced concrete, or short linear elements as in a geodesic dome. (See Figure 4.16) One way to visualize a dome is as an arch spun on its vertical axis. Domes can cover large open interior spaces. They are supported by bearing walls, columns and arches, or piers.

Figure 4.16 Interior of dome, Hudson County Courthouse, Jersey City, New Jersey

Source: Library of Congress Prints and Photographs Division

CANTILEVERS

It is possible to support a horizontal structure from one end only, with the structure extending out past the support; this is called a **cantilever**. (See Figure 4.17) The weight of the cantilevered structure and any load on it is transferred back to the supporting wall. A simple example of a cantilever is a diving board. Balconies and overhangs are often cantilevers.

Vertical Supports

Some of the vertical structural members available for use in buildings include columns, pilasters, piers, posts, and piles.

COLUMNS

Compression and tension both occur in a **column**—a vertical structural member that is longer than it is wide or deep. When a load is placed on top of a column, the material in the column will be compressed. Sometimes the load is great enough to compress the column until its material crumbles; this can happen to stone columns. (See Figure 4.18) With other materials, the column will bend (deflect) along its length, much like a beam would. One side of the deflected area will be in tension and the other in compression. Thin wood columns tend to fail in this way.

The goal in designing with columns and beams is to carry the load safely to the ground. A load on a beam will be transferred to the support (column) at each end, so that each column carries only half the load. The columns then carry the load down to the ground. Well-designed and properly loaded columns and beams will deflect only within very well-defined parameters; they are designed to withstand loads greater than those anticipated in normal use.

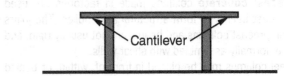

Figure 4.17 Cantilevers

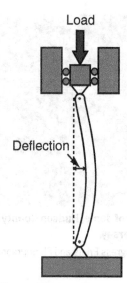

Figure 4.18 Buckling column
Source: Redrawn from Buckled column.png by Peter Schwartz on Wikimedia Commons, released to public domain

Wood columns may be either solid wood or built out of wood pieces solidly glue-laminated or mechanically fastened together. Spaced wood columns are made of multiple structural members with blocking and spaces inside. When a wood column fails under a compression load, it is because the wood fibers are crushed. Slender wood columns also fail by buckling.

Masonry columns and pilasters are made up of small, usually uniform units that are strong in compression. The bricks or blocks are stacked up nominally a minimum of 12" (305 mm) in both width and length, and a maximum of thirty times as long as they are wide. A row of columns connected by arches is called a **colonnade**.

A **pilaster** looks like a column protruding from a wall; it may stick out on one or both sides. (See Figure 4.19) The pilaster buttresses the wall, making it stronger and less likely to fall over when lateral forces are applied.

Stone columns can be carved from a single giant block of stone or built of large blocks stacked one upon another. They rely almost entirely on compression to carry a load to the ground. Stone columns with stone beams are laid out close together. For example, the classic Egyptian column was often assembled in a densely packed hypostyle hall.

The reinforcing in concrete columns helps resist applied forces. Concrete columns are laid out on a regular grid in order to facilitate the economical forming of beams and slabs. The columns are designed as continuous units from the foundation to their tops. When used with a grid of steel or timber beams, concrete columns are connected to the other components with steel connectors.

Precast concrete columns made in factories are used with precast beams to form structural assemblies. The joints between precast columns and beams are not usually rigid, and they are normally assembled with shear walls.

Steel columns may be placed in front of, within, or behind the exterior wall plane of the building. Steel columns are most

Figure 4.19 Pilasters, county courthouse, Springfield, Ohio
Source: Wikimedia Commons, Architecture_pilasters.jpg, photographer Derek Jensen, 2005, released to public domain

commonly made in the wide-flange or W shape, but are also available as round pipes and square or rectangular tubes. Some concrete columns have structural steel columns inside.

PILLARS AND POSTS

Pillars and posts are linear vertical structural members much like columns. They may function like columns to carry a building load to the ground. Pillars can also be free-standing as monuments. The term **post** is often used for wood columns made from a single tree trunk. Shorter wood or metal vertical pieces used in fences and other construction are also called posts.

LOAD-BEARING WALLS

A **load-bearing** wall is one designed to spread a load over its entire surface. In the case of a structural building frame, filling the opening with a load-bearing wall makes the area that carries the compressive forces to the ground much greater.

More information on load-bearing walls is found in Chapter 5, "Floor/Ceiling Assemblies, Walls, and Stairs."

Lateral Forces

If you support the ends of a simple beam on two columns, and then push one of the columns from the side, it will all tend to topple over. This structural arrangement does not resist **lateral** (perpendicular to the direction of the support) forces very well. The situation can be improved by bracing the corners where the columns meet the beam.

Lateral stability is the ability of a structure to resist lateral forces without sliding, overturning, buckling, or collapsing. (See Figure 4.20) Lateral support can be added to a frame to support building surfaces, collect surface loads, and conduct them to columns or bearing walls. The frame then carries the vertical loads and transfers them to the earth via the building's foundations. This can be accomplished by using rigid joints, diagonal bracing, or shear panels. In addition, floor and roof planes add lateral stability, acting like very deep horizontal beams.

In our beam and columns example, corner bracing will make the columns and beam into a rigid frame. Imagine a picture frame without a picture in it; the frame can easily shift out of alignment. When you put a stiff rectangular piece of cardboard in the frame, it will regain its right-angle corners. The flat surface of the cardboard will also help to spread the stresses evenly and help to keep the frame straight.

BUTTRESSES

A wall will be strongest along the direction of its surface. If the load is perpendicular to the plane of the wall, the wall will need extra support for lateral stability. This support can be provided by buttressing the wall with pilasters, adding cross walls or transverse rigid frames, or inserting horizontal slabs between pairs of walls. A **buttress** is a thickening in a wall that provides additional structural support. It may be a relatively simple pilaster, a heavy exterior buttress, or a more complex flying buttress. All of these serve to increase the area that helps to carry the load, and to resist lateral forces.

CROSS WALLS

Another way to accommodate lateral loads is by locating other walls—**cross walls**—perpendicular to the bearing wall. These walls become part of the building's structural system and are necessary to the building's stability. Even if they are not carrying loads from above, they help the load-bearing walls do their work.

Lateral stability can also be increased by building pairs of walls connected by horizontal slabs. The slabs help to even out the load between the walls and resist sideways forces.

Shearing Forces

A **shearing force** describes what happens when two parallel surfaces move in opposite directions. (See Figure 4.21) This is why scissors are called shears: the parallel blades move past each other.

A **shear wall** is a wood, concrete, or masonry wall that transfers a load to the foundation and resists changes in shape. Shear walls are designed to deal with shearing forces. In our earlier example, the rigid cardboard in the picture frame acted as a shear wall, to keep the frame's corners at right angles.

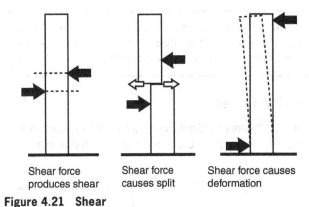

Shear force produces shear

Shear force causes split

Shear force causes deformation

Figure 4.21   Shear

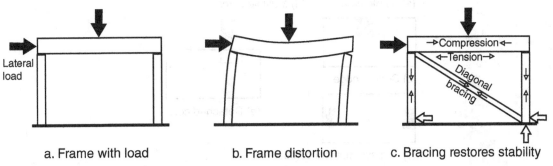

a. Frame with load   b. Frame distortion   c. Bracing restores stability

Figure 4.20 Lateral stability

TWISTING FORCES

A **braced frame** is composed of timber or steel with diagonal bracing members. The diagonal members pick up the loads that would otherwise twist the frame, and prevent the corners from moving closer together or farther apart. Diagonal bracing members can save both materials and weight as compared to sheets that fill the entire frame.

As a building becomes taller and thinner, it is more likely to twist or sway. High-rise buildings require added bracing with diagonal bracing and/or a rigid core. Many high-rise buildings employ tube structures, consisting of bracing systems at the perimeter of the building connected by rigid floor planes.

Grid Frameworks

Large and complex structural forms can be made of smaller structural units. Linear column and beam frameworks create a grid that can be added to both horizontally and vertically with ease. Within that framework, designers are free to add non-bearing walls that do not follow the grid.

A **structural grid** creates regular open spaces between structural elements. Interior designers are then free to design non-bearing partitions of varying heights and configurations to shape the interior space.

Structural grids can be modified by adding additional columns or by eliminating supporting elements. In the latter case, the support for the load must be transferred away from the missing column.

Removing a column may be a complex process, and an interior designer should consult an architect or structural engineer before removing existing structural elements.

Service Cores

In most multi-story buildings, the stairs, elevators, toilet rooms, and supply closets are grouped together in a **service core**. The mechanical, plumbing, and electrical chases, which carry wires and pipes vertically from one floor to the next, are also often located in service cores, as are the electrical and telephone closets, service closets, and fire protection equipment. Often, the plan of these areas varies little if at all from one floor to the next.

Service cores may have different ceiling heights and layouts than the rest of the floor. Mechanical equipment rooms may need higher ceilings for big pipes and ducts. Some functions such as toilets, stairs, and elevator waiting areas benefit from daylight, fresh air, and views, so access to the building perimeter can be a priority.

There are several common service core layouts. (See Figure 4.22) Central cores are most frequently used. In high-rise office buildings, a single service core provides the maximum amount of unobstructed rentable area and creates efficient distribution patterns. Locating cores at the building's edge provides access to perimeter daylight. Two symmetrical cores can provide lateral bracing and shorter distribution runs, but reduces layout flexibility. Multiple cores are used in apartment buildings and structures made of repetitive units, with the cores located between units along interior corridors. Detached cores are located outside the body of the building to save usable floor space, but require long service runs.

Service cores can take up a considerable amount of space. Along with the entry lobby and loading docks, service areas may nearly fill the ground floor as well as the roof and basement. Their locations must be coordinated with the structural layout of the building. In addition, they must coordinate with patterns of space use and activity. The clarity and distance of the circulation path from the farthest rentable area to stairs in the service core have a direct impact on the building's safety in a fire. (See Figure 4.23)

For more information on service cores, see Chapter 10, "Waste and Reuse Systems."

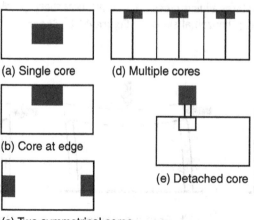

(a) Single core

(b) Core at edge

(c) Two symmetrical cores

(d) Multiple cores

(e) Detached core

Figure 4.22   Service core layouts

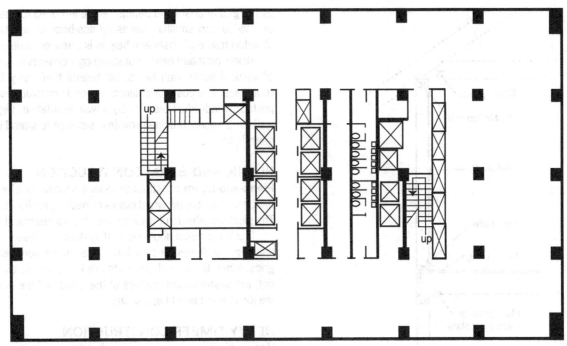

Figure 4.23 Plan with service core

STRUCTURAL TYPES

The remainder of this chapter will introduce basic structural construction types, and look at how they affect the interiors of buildings. These include light frame, post-and-beam and heavy timber, masonry, concrete, steel structures, and other structural types.

Light Frame Structures

Framing is the process of fitting and connecting relatively slender members to shape and support a structure. **Light frame construction** forms the structural elements from closely spaced members made of dimension lumber or light-gauge steel. Light frame structures are often built on-site from local materials.

LIGHT WOOD FRAMING
Due to the availability of lumber from extensive forests, wood-frame construction traditionally has been used to build private residences and small public buildings in North America. The building's frame, a skeleton of relatively slender members, gives support and shape to the superstructure.

A frame house is constructed with a wood skeleton that is usually sheathed with siding. **Sheathing** is a protective covering of boards, plywood, or other panels applied to the frame as a basis for siding, flooring, or roofing. **Siding** such as shingles, boards, or sheets of metal surfaces the exterior walls of frame buildings.

Strength is given to corners by assembling two or three studs at the intersection of two framed walls. Corners are further reinforced by diagonal corner braces set into the studding.

In **platform framing**, a wooden building is built of stacked floors made with studs only a single story high. (See Figure 4.24) Each story rests of the top plates of the story below, or on the sill plates of the foundation wall.

LIGHT STEEL FRAMING
Light-gauge steel studs are manufactured as channels or C-shapes, and are usually pre-punched to allow piping, wiring, and bracing to be run through them. Light-gauge steel joists are lighter, more dimensionally stable, and can span longer distances than wood joists. Light-gauge steel studs are cold-formed from sheets or strips of steel. Light-gauge studs are easy to cut and assemble in the field, and are frequently used for lightweight, non-combustible, and damp-proof wall structures. (See Figure 4.25) They are used for load-bearing walls that support light-gauge steel joists, and for non-bearing partitions. Like wood studs, light-gauge steel stud walls have cavities for insulation and utilities, and can be finished with a wide array of materials.

Post-and-Beam and Heavy Timber

Post-and-beam and plank-and-beam construction both use vertical posts and horizontal beams to carry structural loads. Heavy timber construction uses large wood structural members, which may be combined with brick walls. Pole construction uses wood poles to raise parts of the building above the ground.

POST-AND-BEAM CONSTRUCTION
In **post-and-beam** construction, vertical posts and horizontal beams carry floor and roof loads. The interior spaces formed

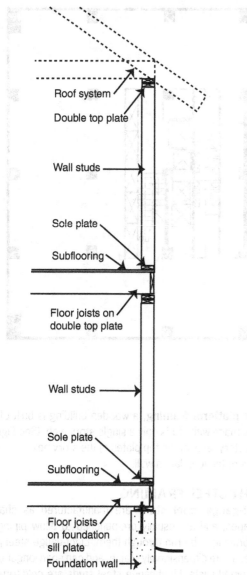

Figure 4.24 Platform framing
Source: Redrawn from Francis D.K. Ching, *Building Construction Illustrated* (5th ed.), Wiley 2014, page 5.42

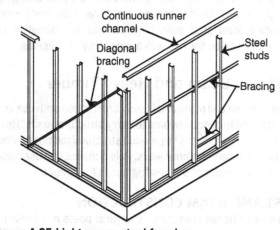

Figure 4.25 Light-gauge steel framing

by the grid of posts and beams can be left open to one another or divided into smaller spaces by non-bearing partitions. The skeleton frame of posts and beams is often left visible.

Wood post-and-beam construction connects a framework of vertical posts with horizontal beams that carry floor and roof loads. The posts or columns in wood construction are supported by individual piers or by a wall foundation. Rigid shear walls or diagonal bracing provide resistance to lateral wind and seismic forces.

PLANK-AND-BEAM CONSTRUCTION

Plank-and-beam construction uses a framework of wood timber beams to support wood planks or decking for floors or roofs. The planks or sheets of decking span the framework of beams.

Plank-and-beam floor and roof systems are used with post-and-beam wall systems to form three-dimensional structural grids. When these grid elements are left exposed on the interior, the aesthetic appearance of the wood and the detailing of the joints must be of high quality.

HEAVY TIMBER CONSTRUCTION

Heavy timber construction with large wooden posts and beams is more resistant to fire than light wood-frame construction. Non-combustible, fire-resistive exterior walls combine with wood members and decking that meet minimum size requirements specified in the building code.

Heavy timber construction combined with brick is called **mill construction**, due to its use in early mill buildings in North America. The combination of fire-resistant brick walls, large open floor spaces, and daylight streaming through large windows provided an ideal setting for New England's textile industry. Many of these buildings still stand, and have been converted into office spaces, housing and studios for artists, and museums.

POLE CONSTRUCTION

Pole construction uses a vertical structure of pressure-treated wood poles firmly embedded in the ground. This pier foundation supports the building above the surface of land or water. Pole construction allows buildings to be built on steeply sloping land without removing all the trees and grading the area to a flat surface.

Masonry Structures

Masonry construction dates back to the earliest civilizations. Building units of natural or manufactured products such as stone, brick, or concrete block are typically held together with mortar. Masonry walls are durable, fire-resistant, and structurally efficient for compression loads.

Masonry construction has developed its own specific vocabulary over the centuries. (See Figure 4.26) A single horizontal row of masonry units is called a **course**. Multiple courses stacked together are referred to as a **field**. The continuous vertical section of a masonry wall one unit in thickness is called a **wythe** (or withe).

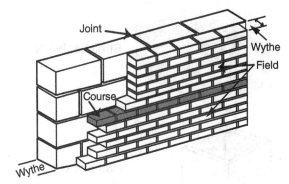

Figure 4.26 Masonry terms

Masonry walls are typically constructed in parallel sets to support steel, wood, or concrete spanning systems. They are often spanned by open-web steel joists, timber or steel beams, or site-cast or concrete slabs.

MASONRY WALL CONSTRUCTION

Masonry units can be assembled into solid walls, cavity walls, or veneered walls with joints filled with **mortar**, which is usually composed of **portland cement**, sand, and water. Solid masonry walls are constructed of either solid or hollow masonry units. **Composite walls** are made with more than one masonry type.

In grouted masonry walls, two wythes each one unit thick are bonded into a single mass with **grout** (thin mortar). Where the facing and backing units of a masonry wall are constructed separately, the result is called a **cavity wall**. The space between the two parts of the cavity wall keeps water from penetrating through the entire wall into the interior. The air-filled cavity serves to insulate the wall.

A wall that has a non-structural facing of stone, brick, concrete, or tile bonded to a supporting structure is known as a **veneered wall**. The surface veneer obscures the actual structural material, which may be concrete or masonry.

Masonry walls can be unreinforced or reinforced. Reinforced masonry walls have steel reinforcing bars inside grout-filled cavities and joints.

Masonry bearing or shear walls are required to have a minimum thickness of 8" (203 mm). With reinforcing, this can be reduced to 6" (152 mm). Solid six-inch masonry walls in single-story buildings are limited to nine feet in height.

Because masonry is composed of small units, it can be assembled into curving or irregular forms. Masonry walls provide texture and color to interior spaces.

MASONRY OPENINGS

Openings in masonry walls are spanned by arches or with stone or concrete lintels. Precast concrete lintels are available for brick and concrete masonry walls. Concrete masonry lintels rest on masonry on each side of the opening. Paired steel angles can support the facing and back up wythes across an opening. Reinforced brick lintels are built four to seven courses

high, with steel reinforcing bars embedded in grout in the center of the brick construction.

MOVEMENT JOINTS

Masonry materials expand and contract with changes in temperature and moisture content. Clay masonry units absorb water and expand, and concrete masonry units shrink as they dry after manufacture.

Movement joints are incorporated into masonry walls to control these changes, with expansion joints designed to close slightly when masonry materials expand, and control joints constructed to open slightly as concrete masonry shrinks. Movement joints are located each 100 to 125 feet (30 to 38 m) along unbroken lengths of masonry walls, at changes in wall height or thickness, at columns, pilasters, and wall intersections, and near corners. They are also installed on both sides of openings greater than six feet (1.8 m) wide, and on one side of openings less than six feet.

STONE MASONRY

Masonry of natural stone is durable and weather resistant. (See Table 4.1) Stones may be simply laid in mortar as double-faced walls, or stone used as a facing veneer tied into a concrete or masonry backup wall.

Structural stone masonry patterns are sometimes visible on interior walls. Interior walls built with stone add scale, color and texture. In other instances, stone is used to face walls constructed of other materials.

BRICK MASONRY

Brick is a rectangular masonry unit made of clay and hardened in the sun or by firing in a **kiln** (a furnace or oven). **Common brick** is used for general building purposes and is given no special color or texture. **Facing brick** (also called face brick) is used on visible surfaces. Facing brick is made of special clays or treated to create desired colors and textures.

Bricks will absorb water and are graded for durability when exposed to weather. They are made by molding clay to produce

TABLE 4.1 COMMON BUILDING STONE TYPES

Stone Type	Description
Granite	Very hard, strong and durable, good weathering and abrasion resistance
Marble	High compressive strength, most durable in dry climates or where protected from precipitation
Limestone	Softer and more porous than granite, becomes harder when exposed to weather, durability greatest in dry climates
Sandstone	Highly porous, easy to work but relatively low durability
Slate	Splits into slabs easily, extremely durable

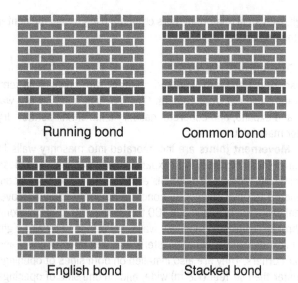

Running bond   Common bond

English bond   Stacked bond

Figure 4.27 Sample masonry bond patterns

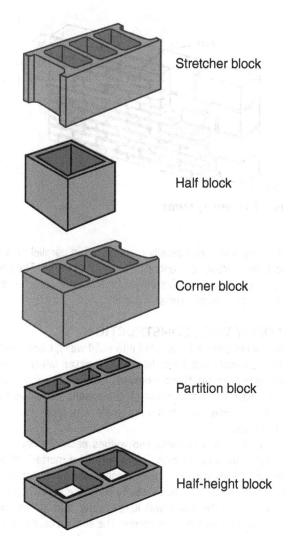

Stretcher block

Half block

Corner block

Partition block

Half-height block

Figure 4.28   Sample concrete masonry unit shapes

a variety of surface textures and densities. Flashed bricks are fired alternately with too much and then too little air to produce a varied face color. Firebrick is made of high-temperature resistant refractory clay; it is used for lining furnaces and fireplaces.

Bricks are made in a variety of standard sizes, which are given in **nominal dimensions**. These are larger than the actual brick dimension; the added size accounts for the thickness of a mortar joint. The dimensions and proportions of bricks affect the scale and appearance of brick walls. The types and thicknesses of mortar joints are chosen for appearance and for their ability to shed water.

The orientation of each brick in a wall affects the wall's solidity and appearance. The arrangement of masonry units into regular bond patterns takes many forms. (See Figure 4.27)

CONCRETE MASONRY UNITS

A concrete masonry unit (CMU) is a precast masonry unit made of portland cement, fine aggregate and water. CMUs are available in many shapes and styles. (See Figure 4.28)

Concrete blocks are solid or hollow concrete masonry units; they are often wrongly referred to as cement blocks.

Concrete blocks are assembled into walls in a single wythe, with reinforcing embedded in the mortar of the horizontal joints. The vertical cells of CMUs may be filled with steel reinforcing embedded in grout. Steel reinforcing increases the wall's ability to carry vertical loads and its resistance to buckling and lateral forces.

GLASS BLOCK WALLS

Translucent, hollow glass blocks have clear, textured, or patterned faces. Inserted into openings in exterior walls, they let in light while limiting views. As interior partitions, they contribute sparkle, translucency, and texture.

UNFIRED EARTH CONSTRUCTIONS

Sun-dried bricks made of mud and water have been used for thousands of years. Although unfired stabilized earth has low tensile strength and is unsuited for tall structures, its high compressive strength, ability to store heat, low cost, and ability to be produced locally on a small scale have made it widely used in areas of little rainfall throughout the world.

Adobe construction is traditionally used without insulation in hot dry climates without cold winters. Adobe is made of sun-dried clay masonry mixed by hand or machine and cast in wood or metal forms. Straw is sometimes added to the mix. Adobe bricks vary in size, although they are commonly 10" by 14" (254 by 305 mm) and 2" to 4" (51 to 102 mm) thick. Adobe bricks are very heavy. They are usually plastered on both the inside and outside. Openings are traditionally spanned by rough-hewn wood beams called vigas. Both adobe and rammed earth constructions have little resistance to earthquakes, and may require a separate structural frame.

Rammed-earth construction is an unfired stabilized earth technology consisting of a stiff mixture of clay, silt, sand, and water compressed and dried within wall forms. The damp soil is compacted by hand or machine into lifts or layers 6" (152 mm) high before being set onto the previously placed layers.

Both adobe and rammed earth walls vary from 8" (203 mm) for interior non-bearing walls to 18" (457 mm) thick exterior walls capable of supporting two stories up to 22 feet (6.7 m) high.

Concrete Structures

Concrete has been used as a structural material since the Roman Empire. It is made by mixing cement that has been **calcined** (heated to a high temperature without melting or fusing) and mineral aggregates, plus water.

Cement is made from finely pulverized clay and limestone, and is used as an ingredient in concrete and mortar; the term is often used incorrectly for the term "concrete."

Aggregates added to concrete are hard, inert mineral materials such as sand and gravel. (See Table 4.2) Aggregates make up 60 to 80 percent of the volume of concrete, and greatly affect its strength, weight, and fire-resistance. Other additives (also called admixtures) are added to concrete to provide specific desired properties.

Concrete is either cast-in-place where it will be used by a concrete mixer or agitator truck, or precast off-site in factory-controlled conditions. Concrete used for construction is usually reinforced with steel reinforcing bars or with welded-wire fabric.

TABLE 4.2 CONCRETE ADDITIVES

Additive	Description
Air-entraining agent	Disperses tiny air bubbles in concrete or mortar mix; improves workability, produces lightweight insulating concrete
Accelerator	Increases rate of setting and strength development
Retarder	Slows setting, allowing more time working time
Surface-active agent (surfactant)	Helps water wet and penetrate the mix, or aids in distributing other additives
Water-reducing agent (superplasticizer)	Reduces amount of water required to keep concrete workable, generally increasing concrete strength
Coloring agent	Pigment or dye to alter concrete color

CAST-IN-PLACE CONCRETE

Cast-in-place concrete is cast into formwork that is usually made of prefabricated reusable panels. Additional forms can be added to create linear recesses and other patterns on the concrete's surface.

Cast-in-place concrete beams are usually formed and cast with the slabs they support. This structural integration adds to the strength of the building.

Cast-in-place reinforced concrete floor slabs are integrated into their supporting concrete beams. Precast concrete slabs or planks may be supported by beams or by load-bearing walls.

TILT-UP CONSTRUCTION

Tilt-up construction consists of reinforced concrete wall panels cast directly on the floor slab, without formwork, then tilted up with a crane into their final vertical position. Tilt-up panels are designed to withstand the stresses of being lifted and moved, which may be greater than in-place loads. Tilt-up construction panels have good structural strength, insulating value, and thermal mass, and are weather and fire resistant.

PRECAST CONCRETE

Precast concrete structural elements are cast and steam-cured off-site, transported to the site, and set in place with cranes. Precast building components offer consistent strength, durability, and finish quality, and eliminate on-site formwork. They may be reinforced or prestressed for extra strength or reduced thickness.

Prestressed concrete is reinforced by putting tension on high-strength steel tendons inside concrete. (See Table 4.3) The tensile stresses in the tendons are transferred to the concrete, which deflects in a direction opposite to the deflection that will take place when the concrete is subjected to a load. The two forces, from the prestressed concrete and cables and from the load, cancel each other out, allowing the prestressed concrete to span greater distances with heavier loads.

Precast concrete wall panels are used for bearing walls and columns that support site-cast concrete floors or steel floor and roof systems. (See Table 4.4)

TABLE 4.3 PRECAST PRESTRESSED CONCRETE STRUCTURAL ELEMENTS

Type	Description
Solid flat slab	Plank used for sort spans, uniformly distributed loads
Hollow-core slab	Plank with hollow cores to reduce weight; uniformly distributed loads over medium to long spans
Single and double tees	T-shaped planks with single or double stem and broad, flat slab
Ledger beam	L or inverted T shaped beam with projecting ledges to support the ends of joists or slabs

TABLE 4.4 PRECAST CONCRETE COLUMNS

Column Size	Approximate Area Supported
10" by 10" (255 by 255 mm)	2000 square feet (185 square meters)
12" by 12" (305 by 305 mm)	2750 square feet (255 square meters)
16" by 16" (405 by 405 mm)	4500 square feet (418 square meters)

Metal Structures

Archeological evidence dates the beginnings of iron production to Anatolia around 1200 BCE. By around 650 BCE, steel was being produced in large quantities in Sparta, Greece. The introduction of the Bessemer process in the nineteenth century industrialized the process of removing impurities from molten iron, and cast iron, wrought iron, steel and stainless steel became available for building construction.

CAST IRON AND WROUGHT IRON

Cast iron is a brittle iron-based alloy that contains carbon and small amounts of silicon. It is cast in a mold and machined to make building and ornamental products. Cast iron is strong in compression but weak in tension and bending; it becomes structurally unreliable in fires.

Cast iron columns have been used since the nineteenth century to support glass-enclosed buildings such as London's Crystal Palace. Their lightweight and elegant appearance lives on in historic examples, although today most architectural cast iron is purely decorative.

Wrought iron is tough, malleable and relatively soft. It is easy to work with, and is often forged into beautiful linear designs. Wrought iron is used for interior products, gates and fences, and for other architectural components with relatively light loads.

STEEL FRAMING

Steel is an alloy of iron with carbon; sometimes other materials are added for specific properties. The strength, hardness, and elasticity of steel depend on its carbon content and heat treatment. Adding carbon to iron increases its strength and harness, but decreases it ductility and ability to be welded.

Structural steel framing is made of hot-rolled steel beams and columns spanned by **open-web joists** and metal decking. (See Table 4.5) Steel framing is most efficiently used in a regular grid pattern of girders, beams, and joists. Columns are spaced according to the spans of girders or beams. Lateral wind or seismic loads are resisted by shear planes, diagonal bracing, or rigid framing that uses special connectors.

Rolled structural steel is generally as strong in tension as in compression, making it an excellent structural building material. However, steel rusts when exposed to water and air. Because steel can deform at high temperatures, steel structures must be

TABLE 4.5 SAMPLE STRUCTURAL STEEL FORMS

Form	Description
W-shape (wide flange)	H-shaped form with wide parallel flanges; often used for beams and columns
S-shape (American standard beam)	I-shaped section with sloped inner flange surfaces
American standard channel	Rectangular C-shaped section with sloped inner flange surfaces
Angle (angle iron)	L-shaped section with equal or unequal leg lengths; a pair of angles joined back to back form a double angle
Bar	A long, solid piece with a square, rectangular or other simple cross-sectional shape
Structural tubing	Hollow structural shape with a square, rectangular or circular cross section; pipe is round structural tubing.
Plate	A thin, flat sheet of uniform thickness; may have a distinctive waffle surface pattern
Sheet metal	A thin sheet or plate of metal; corrugated for decking use.
Open-web steel joist (bar joist)	Lightweight steel joist supported by upper chord; web is often a zigzag steel bar.

coated with fire-resistant materials or combined into assemblies that are rated for fire resistance.

The outer walls of a steel structure may consist of exterior panels rigidly connected to the frame, which stiffen the structure and resist wind loads. Curtain walls are exterior walls supported by the structural frame and carrying no load other than their own weight and the wind load.

Curtain walls were introduced in Chapter 2, "Designing for the Environment."

A rigid steel frame is frequently used for single-story light industrial buildings, warehouses, and recreational facilities. The frame consists of two columns and a beam or girder fastened together with rigid connections. A row of long-span beams or girders supported by pairs of exterior columns, called a one-way beam system, generates long, narrow, column-free spaces.

Structural steel frames are well adapted to buildings with open interior spaces.

Metal decking is corrugated to increase its stiffness, which in turn allows for increased spans. Horizontal metal decks are welded to steel joists or beams that they rest on, and screwed, welded, or seamed to each other along their lengths.

Other Structural Types

Shell, membrane, and cable structures are other ways to cover an area with three-dimensional forms.

SHELL STRUCTURES

A **shell structure** is a thin, curved plate structure, designed to transmit applied forces by compression, tension, and shear. Shells can withstand relatively large uniformly applied forces, but lack the bending resistance necessary to accommodate concentrated loads. Thin shells structures are constructed of reinforced concrete. A shell surface may be circular, elliptical, or parabolic.

CABLE STRUCTURES

Cable structures use flexible wire rope or metal chains with high tensile strength as a means of support. Suspension structures support applied loads with cables suspended and prestressed between compression members. Cable-stayed structures employ cables extending from vertical or inclined masts to support parallel or radially arranged horizontal spanning members.

MEMBRANE STRUCTURES

A **membrane structure** is a thin, flexible surface that transmits loads primarily along lines of tensile stress. There are several types of membrane structures. (See Table 4.6)

TABLE 4.6 MEMBRANE STRUCTURE TYPES

Type	Description
Tent structure	Prestressed by external forces; completely taut under all anticipated load conditions; usually sharply curved in opposite directions
Net structure	Surface of closely spaced cables rather than fabric material
Air-supported structure	A single membrane supported by internal air pressure; anchored and sealed along perimeter
Air-inflated structure	A double-membrane structure supported by pressurized air within building elements

MIXED STRUCTURAL TYPES

Many buildings are made of a mix of materials and some combine different structural types. By understanding the nature of structural materials, it is possible to determine that a brick-covered building is actually a concrete structure. It is not uncommon for a building to have a structural frame of one material, with infill panels of another material forming the exterior walls.

There is a great deal more that could be said about building structural types; perhaps this short survey has wet your curiosity to learn more about historic and innovative buildings. In Chapter 5, we focus on the horizontal elements of building interiors, and at the building systems that allow movement from one layer of the building to another.

5

Floor/Ceiling Assemblies, Walls, and Stairs

INTRODUCTION

In addition to their structural roles, a building's floors, walls, and stairs define interior spaces and facilitate movement through the building's interior.

Floors, walls, and ceilings do more than mark off a simple quantity of space. Their form, configuration, and pattern of window and door openings also imbue the defined space with certain spatial or architectural qualities. (Francis D.K. Ching and Corky Binggeli, *Interior Design Illustrated* [3rd ed.], Wiley 2012, page 6)

Stairs provide means for moving from one level to another and are therefore important links in the overall circulation scheme of a building. Whether punctuating a two-story volume or rising through a narrow shaft, a stairway takes up a significant amount of space. Safety and ease of travel are, in the end, the most important considerations in the design and placement of stairs. (Francis D.K. Ching, *Building Construction Illustrated* (5th ed.), Wiley 2014, page 9.02)

The construction of floors, walls, stairs, and ramps affect interior design and life safety. Their configurations and finishes are very important issues in the event of fire.

HORIZONTAL STRUCTURAL UNITS

The basic horizontal structural units used in buildings are either reinforced concrete slabs or grids of **girders** (large primary beams), beams, and **joists** (smaller parallel beams) that support planks or decking. Rigid planar structures spread loads in many directions along the plane. The stresses from the loads generally take the shortest and stiffest routes to the vertical supports.

Rigid floor structures act as flat, deep beams, transferring lateral loads to rigid frames, shear walls, or braced frames. Rigid frames are usually constructed of steel or reinforced concrete with rigid joints. Rigid frames are used in low and medium-rise buildings.

Floor/Ceiling Assemblies

Floors are the flat, level base planes of interior space. As the platforms that support our interior activities and furnishings, they must be structured to carry the resulting loads safely. (Francis D.K. Ching and Corky Binggeli, *Interior Design Illustrated* (3rd ed.), Wiley 2012, page 150)

FLOOR FUNCTIONS

A floor must be designed to withstand the use intended for that specific space. Floors interact with numerous other building systems. They provide structural support, control fire and the thermal qualities of surfaces, and provide useful surfaces. Floors can help control humidity, airflow, acoustic privacy, and the entry of living creatures. They sometimes control thermal radiation and air temperature, and help keep out water.

The ceiling portion of the assembly plays a role in controlling thermal radiation, air temperature, humidity, fire, and visual and acoustic privacy. It also helps with flow of air and adjusts to movement, and sometimes provides structural support.

FLOOR/CEILING ASSEMBLY COMPOSITION

Floor systems fall into two general categories: a series of linear beams and joists topped with a plane of sheathing or decking, or a slab of reinforced concrete.

The term **floor/ceiling assembly** describes how the floor and ceiling are built. Each floor/ceiling assembly differs from others in thickness, components, and fire and acoustic ratings.

The depth of the floor/ceiling assembly affects the height from interior ceiling to floor as well as total building height. The depth of the floor system is related to the size of the structural bays that it spans and the strength of the materials used. The design of the assembly often has to accommodate mechanical, plumbing, and electrical equipment.

The way that the floor structure's edges connect to the supporting foundation and structural wall systems affects the structural integrity of the building and its physical appearance. The details of these connections determine the building's ability to control airborne and structure-borne sound, and affect the fire-resistance rating of the assembly. A relatively stiff assembly with some elasticity helps support moving loads. Too much deflection and vibration in the floor/ceiling assembly can be a problem for finish materials as well as for human comfort.

For information on how floor systems are involved in blocking noise and fire, see Chapter 8, "Architectural Acoustics," and Chapter 18, "Fire Safety Design."

EXPOSED CEILINGS

The maximum possible floor to ceiling height can be achieved by exposing the architectural structure as the finished ceiling. Exposed ceilings reveal mechanical equipment, plumbing, electrical conduits, sprinkler systems, lighting fixtures, and other items that are ordinarily hidden above a ceiling, in the process allowing access to these components.

Exposing the structure eliminates the cost of installing a ceiling. The cost of designing mechanical, electrical, and plumbing systems to present a code-compliant and finished appearance may offset or even exceed the savings. In addition, exposed ceilings and their equipment may need to be painted.

BETWEEN FLOORS AND CEILINGS

An enclosed portion of the building structure that is designed to allow the movement of air, forming part of an air distribution system, is commonly called a **plenum**. Although the term plenum is specifically used for the chamber at the top of a furnace (also called a bonnet) from which ducts emerge to conduct heated or conditioned air to the inhabited spaces of the building, it is also commonly used to refer to the open area between the bottom of a floor structure and the top of the ceiling assembly below. In some cases, air is carried through this space without ducting, and then it is called an open plenum.

Building codes limit where open plenum systems can run in a building, prohibit combustible materials in plenum spaces, and limit the types of wiring permitted. The area between the floor above and ceiling below is usually full of electrical, plumbing, heating and cooling, lighting, fire suppression, and other equipment. (See Figure 5.1) Equipment in the plenum sometimes continues vertically down a structurally created shaft (chase). The open plenum must be isolated from other spaces so that debris in the plenum and vertical shaft is not drawn into a return air intake.

Interior designers are often concerned with locating lighting or other design elements in relation to all the other equipment in the plenum.

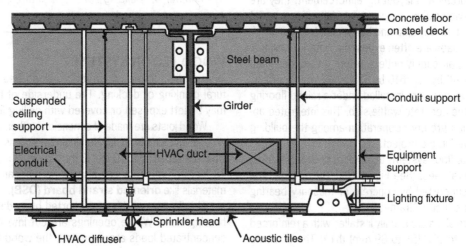

Figure 5.1 Space between floor and ceiling

Interior Design Concerns

During remodeling projects, it is important for the interior designer to know what components are in the existing floor and to avoid damaging them. Floors may be weak or uneven due to poor construction, age, or sagging from a settling foundation. Uneven floors make it difficult to install ceramic tile without breaking tiles and cracking grout. In addition, uneven floors often squeak.

Floor joists should be examined to determine how a floor is constructed, whether damage exists, and the size of the framing members. Concealed air vents and plumbing running through the floor/ceiling assembly may be difficult and expensive to relocate.

Water damage from leaking toilets or pipes may weaken the floor structure. Prior to renovating a bathroom, the floor covering should be lifted or removed, or the ceiling below checked for leaks. Severe damage may require replacement of the **subflooring** (the structural material used to span floor joists) or reinforcement of floor joists. Floors over a concrete slab should be checked for moisture and cracks requiring sealing.

Appliances require stable, even floors in order to fit with each other or with cabinetry. The built-in leveling legs of residential kitchen appliances are difficult to adjust properly, so level floors should do the job.

Some equipment is very heavy, and may require additional structural support. Exercise equipment, for example, requires a floor with enough support for the equipment and enough stiffness for the jumping and pounding that occurs with its use.

Floor Systems

Floor systems are constructed from concrete slabs, wood, or steel. Floating floors and access floors require special construction methods.

CONCRETE SLABS

Concrete slab floors are reinforced with steel and either poured in place or delivered to the site as precast planks. Depending on their thickened areas and layout of reinforcement, they are categorized as one-way slabs, one-way joist slabs, two-way flat plates or slabs, and two-way slab and beam constructions.

Concrete waffle slabs are often exposed as interior ceilings, with lighting fixtures, air supply outlets, return grilles, designed to fit into the 24" by 24" by 18" (610 by 610 by 457 mm) coffers. Ductwork, pipes, and wires are installed under a raised flooring system resting on the concrete waffle slab. This integrated approach requires much greater cooperation among the building professionals involved in the project.

Precast concrete floor systems use precast concrete slabs, beams, and structural tees. (See Figure 5.2) These structural members may be supported by concrete or masonry bearing walls, or by steel or concrete frames.

Precast concrete slabs are usually installed with a reinforced concrete topping 2" to 3½" (51 to 89 mm) thick. The concrete topping may be eliminated with smooth-surfaced concrete slabs

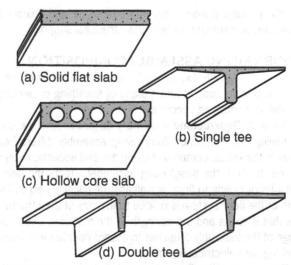

(a) Solid flat slab

(b) Single tee

(c) Hollow core slab

(d) Double tee

Figure 5.2 Precast concrete floor elements

when they are finished with carpet and pad. The undersides of precast slabs can be caulked and painted and left exposed, or a ceiling finish may be applied or suspended below the slab.

Subflooring material must be smooth, level, and suitable for installation of the floor's finish material. Sometimes an existing concrete floor receives a new subfloor or the application of a leveling material to smooth out irregularities.

Concrete slab-on-grade floors are built in direct contact with the ground. The temperature of the ground is often different from that of outdoor air, and earth conducts heat more readily than air. A few inches of insulation can make a significant difference in the heat flow from the slab. Concrete slab joints allow for contraction and expansion. (See Figure 5.3)

See Chapter 4, "Building Forms, Structures, and Elements," for more on slab-on-grade floors.

WOOD FLOOR SYSTEMS

In wood light-frame construction, wood beams support structural planking or decking. The underside of the floor structure may be left exposed or covered with ceiling finish material.

Wood joists are made of dimension lumber, which is easy to cut on site with simple tools. Nominal dimensions range from 2 × 6, which can span up to 10 feet (3 m), through 2 × 12, which can span 18 feet (5.5 m). They can also be prefabricated from composite materials like **oriented strand board (OSB)**.

The floor assembly is supported by girders, posts, or load-bearing walls. Where openings are cut into the floor or where concentrated loads are expected, the wood framing members are doubled up. Such is the case where non-bearing partitions

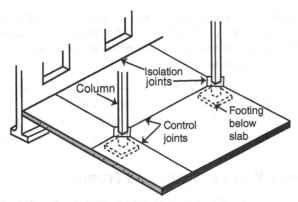

Figure 5.3 Concrete slab joints

Source: Redrawn from *Construction and Home Repair Techniques Simply Explained*, Naval Education and Training Command Management Support Activity, republished by Dover Publications, Inc., 1999, page 180

run parallel to the length of the joists. The load from non-bearing partitions can be distributed across many joists when the partitions run perpendicular to them. Load-bearing walls must be supported from below.

Wood plank and beam floor systems transmit sounds of objects striking them, such as footfalls. The optional installation of a ceiling below provides a concealed space for thermal and acoustic insulation, piping, wiring, and ductwork. Wood frame construction is combustible, and requires fire-resistant ceiling and floor finishes for fire resistance.

Subflooring serves as a working platform during construction and provides a smooth base for the floor finish. Combined with joists, the subfloor can form a structural diaphragm that transfers lateral forces to shear walls. (See Figure 5.4) Plywood is usually used for subflooring, but OSB, waferboard, or particleboard is also sometimes used.

Underlayment is an additional layer, often plywood or hardboard, added on top of the subflooring. Underlayment helps spread out impact loads on the floor, and prepares the surface for direct application of finish materials that require a smooth surface. The underlayment may be a separate layer over the boards or panels of the subfloor, or may be combined with the subfloor itself as a compound material.

Subflooring and underlayment panels are nailed into place over at least two open spans with their length perpendicular to the joists and their ends staggered. When glued and nailed to the floor joists, combined subfloor/underlayment panels act together with the joists as large beams, increasing the stiffness of the floor assembly and reducing floor creep and squeaking.

Prefabricated joists and trusses are best used for long spans across simple floor plans. They can be factory assembled to the engineer's design and shipped to the building site. Prefabricated, pre-engineered wood is lighter and more **dimensionally stable** (able to maintain its original dimensions during use) than sawn lumber, and can be manufactured in greater depths and lengths to span longer distances.

Wood beams are made of solid sawn lumber or glued-together smaller pieces of wood. They are supported and

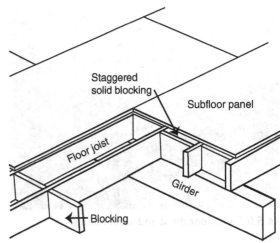

Figure 5.4 Wood floor framing

Source: Redrawn from *Construction and Home Repair Techniques Simply Explained*, Naval Education and Training Command Management Support Activity, republished by Dover Publications, Inc., 1999, page 264

joined with metal connections that are manufactured in many designs. When left exposed, as is often the case, the quality of the wood and its construction is important to the appearance of the interior.

Wood plank and beam framing is usually used to support moderate, evenly distributed loads. The beams support a structural floor plane of wood decking, plywood, or prefabricated stressed-skin panels. Most partitions in this type of framing are nonbearing, and can be placed on a soleplate perpendicular to the planks. When aligned with the direction of the floor decking, nonbearing partitions may require extra support.

STEEL FLOOR SYSTEMS

Steel-framed buildings usually employ steel decking, precast planks, or concrete slab floors. The steel beams that support the decking or precast planks are in turn supported by girders, columns, or load-bearing walls, which are usually parts of a steel skeleton frame system. (See Figure 5.5)

Light gauge or open-web joists are also used to support floors. They may be supported by masonry or reinforced concrete bearing walls, or by steel beams or joist girders (heavier versions of open-web joists).

FLOATING FLOORS

Floating floors are floors that are not nailed or glued to the subfloor. Relatively simple floating floors are used with laminate flooring and other finishes.

Suspended ceilings or raised floors help designers deal with ceiling noise and equipment clutter, but decrease the floor-to-ceiling height of the finished space.

Figure 5.5 Steel floor deck and steel beams

Acoustic floating floors, used where the transmission of sound must be strictly controlled, are more complex. The floor must spread loads in a way that does not crush the acoustical padding that prevents the noise produced by an impact from being transmitted. The total construction must be airtight so no sound is transmitted. The baseboards at the walls should not bridge the gap between the floor slab and the finish. Special care must be taken at walls and penetrations, and consistent construction should be used throughout.

ACCESS FLOORING SYSTEMS

Access flooring systems are typically used in office and institutional spaces to provide flexible placement of furnishings and equipment as well as access to mechanical and electrical equipment below the floor. The system consists of removable floor panels supported on adjustable pedestals. The pedestals can raise the finished floor height up to 30" (455 mm); minimum heights vary.

The steel, aluminum, or lightweight reinforced concrete floor panels are 24" (610 mm) square. They may be finished with carpet tile, resilient tile, hardwood, terrazzo, or porcelain tile, or other finishes. Finishes with fire ratings and designed with electrostatic discharge control are also available.

For more information about access flooring systems, see Chapter 16, "Electrical Distribution," and Chapter 14, "Heating and Cooling."

WALL SYSTEMS

Walls are vertical building elements that enclose, separate, and protect interior spaces. Walls are often designed to support loads transferred from floors and roofs, but are sometimes a framework of columns and beams with nonstructural panels attached to or filling in between them.

Walls, like floors, serve many building functions. They control thermal radiation, air temperature, humidity, airflow, acoustic and visual privacy, and the entry of people and animals. Walls

provide structural support, keep out water, and control fire. They can provide useful surfaces, control their thermal qualities, and adjust to movement. They also house the wires and pipes of other building systems.

The pattern of load-bearing walls and columns should be coordinated with the layout of the interior spaces of a building.

Load-Bearing Walls and Frames

Load-bearing walls are often constructed of concrete and masonry, and are classified by building codes as noncombustible construction. They are heavy and solid, and support loads in compression by the mass of their material. They are strong in compression but require reinforcing to accommodate tensile forces. Their lateral stability is affected by their height-to-width ratios. Their design and construction requires proper placement of expansion joints.

In order to accommodate tensile stresses, concrete and masonry walls can be reinforced with steel. They are designed to carry relatively heavy loads across 20 to 40 foot (6 to 12 m) spans.

Interior surfaces of concrete and masonry walls are often left exposed, creating texture and pattern.

Load-bearing walls can also be built with metal or wood **studs.** Stud wall frames can be assembled on-site or panelized off-site. The relatively small, lightweight components are easily worked into a variety of forms.

The cavities between the studs are used for thermal insulation, vapor retarders, mechanical distribution, and mechanical and electrical service outlets.

It is important to ascertain whether a wall is load-bearing before contemplating its demolition. The direction of ceiling joists can be checked above the space, in the attic, or using an access panel. Load-bearing walls tend to run perpendicular to the joists. Joists running in the same direction as the wall probably indicate that the wall is not load-bearing. However, the advice of a qualified professional is strongly recommended.

STRUCTURAL FRAMES

Structural frames support and accept a variety of nonbearing or curtain wall systems. Concrete frames are typically rigid, qualifying as noncombustible, fire-resistive construction. Timber frames span shorter distances and carry lighter maximum loads than concrete or steel. They require diagonal bracing or shear planes for lateral stability. Timber frames may qualify as heavy timber construction if they meet certain requirements.

Interior Walls and Partitions

Interior walls and partitions subdivide the space within a building. They may or may not be load bearing. Their construction should be able to support finish materials, provide acoustical separation, and accommodate connections to mechanical and electrical distribution services.

Interior walls are commonly made with metal or wood studs spaced 16" or 24" (406 or 610 mm) on center that carry vertical loads. Due to their relatively small pieces and the variety of fastening techniques available, they are quite flexible in form. Cavities in the wall frame accommodate insulation and mechanical and electrical distribution.

The finish materials determine the fire-resistance rating of the wall assembly.

WALL OPENINGS

Openings such as doors and windows in walls are constructed so that any vertical loads from above are distributed around the openings. The location and size of door and window openings are influenced by the structural system and by modular dimensions of wall materials.

SOUND CONTROL

To prevent sound transmission between interior spaces, walls can be insulated with fiberglass or other acoustic insulating material. Building a double-studded wall, especially with staggered studs, on two separate plates is effective, but costs more for increased labor and materials. Double-studded and staggered stud walls also may negatively affect the running of plumbing, wiring, and ductwork.

For more information on walls and sound control, see Chapter 8, "Architectural Acoustics."

EXISTING WALLS

Existing walls may hide wiring, plumbing, drain and vent pipes, and heating, air conditioning, and air return ducts. It is important to check for these before demolition work begins on a renovation.

Uneven walls and the absence of square corners in both new and older construction make it difficult to install finishes, appliances, and cabinets. Some walls may not have been initially installed properly. Old plaster walls often have cracks, and may be distorted when undergoing repairs. Drywall that has been moisture damaged or not properly repaired can soften and sometimes bow. As a result, tile and stone finishes may not lie flat on misshapen walls, and tile lines may be uneven. Mirrors may not contact the wall evenly, and the edges of a glass door may not fit flush and flat.

When an uneven wall is to remain, using surface treatments like liners or wall panels can create a surface smooth enough to work with.

Where walls or corners are off square, preformed showers and other fixtures may not fit corners well, and gaps will have to be filled. It is especially difficult to install cabinets or appliances in uneven corners without gaps. Cabinet fillers can often be used to fill gaps at walls produced by uneven construction, but the problem may require a more complex solution.

STAIRS AND RAMPS

In buildings with multiple floors, it is necessary to provide ways for people and materials to move vertically inside the building. Stairs and ramps are important links in the design of a building's circulation. Stairs and their landings take up a significant amount of space. Ramps take up much more floor space than stairs to rise to a given height.

Stairs

Stairs not only connect the levels of a building, but they are also important forms of spatial transition between rooms. (See Figure 5.6) The design of the stair shapes how we approach a

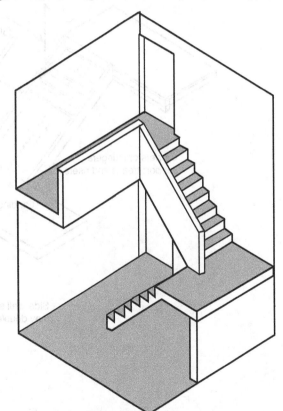

Figure 5.6   Stairway form
Source: Redrawn from Francis D.K. Ching and Corky Binggeli, *Interior Design Illustrated* (3rd ed.), Wiley 2012, page 31

stairway, the pace and style of our ascent and descent, and what we are able to do along the way. The stairway's form can fill and provide focus for a space and offer engaging views. It may run along a room's edge or wrap around it. Stairs can help people escape from a burning building, but can also invite danger.

STAIRS AND CODES

The design of stairs is strictly regulated by building codes and by the Americans with Disabilities Act (ADA). Stairs that are part of building exits have additional fire-safety requirements. Codes also regulate the elements of stairs that make them safe to support weight and to move up and down.

Building codes may allow limited use of curved, winder, spiral, switchback and alternating tread stairs. This depends upon the occupancy classification, number of occupants, use of the stairs, and the dimension of the treads.

Stairs require support from stringers, beams, or side walls. (See Figure 5.7) Handrails help to steady users and prevent falls. Landings break up long runs of stairs and provide resting places.

Stringers and trim may project a maximum of 1½" (38 mm). They must be continuous without interruptions by a **newel post** supporting the handrail or another obstruction.

Doors must swing in the direction of egress, and cannot reduce the landing to less than one-half of its required width. When fully open, doors must not intrude into the required stairway width by more than 7" (178 mm).

The 2015 International Building Code (IBC), 2015 International Residential Code (IRC), and 2010 ADA Standards for Accessible Design requirements for stairs vary and have exemptions that affect their application. Check current applicable codes.

RISERS, TREADS, AND NOSINGS

A stair designed with proper proportions should be comfortable and absolutely uniform. **Treads** are the horizontal part of the stair that we step on, and span the distance between the two sides of the stair. Treads need to be level and secure. **Risers** are the vertical boards that close off the space between treads and add rigidity to the assembly. The riser heights for

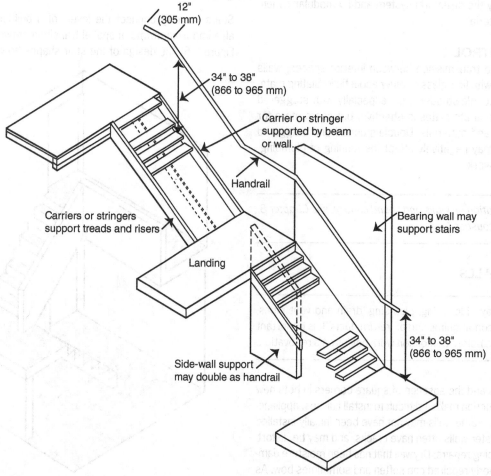

Figure 5.7 Stair parts
Source: Redrawn from Francis D.K. Ching and Corky Binggeli, *Interior Design Illustrated* (3rd ed.), Wiley 2012, page 204

the stair are established by dividing the overall height into small increments that assure a comfortable and safe ascent and descent. Treads and risers in a flight must be uniform within a small tolerance. Nosing dimensions, the slope of risers, and the projection of treads past risers are all covered by code.

Code requirements for treads, risers, and **nosings** where the tread and riser meet include: (See Figure 5.8)

- A minimum of three risers per flight is recommended to prevent tripping.
- The ADA sets a minimum tread depth at 11" (279 mm), and a riser height between 4" and 7" (102 and 178 mm). In public occupancies, codes tend to require that riser heights not exceed 7" (178 mm).
- The 2015 International Residential Code (IRC) limits residential risers to 7¼" (196 mm) with a minimum 10" (254 mm) tread depth. The 2015 International Building code IBC sets riser heights at 7" (178 mm) maximum and 4" (102 mm) minimum.
- Risers and treads must maintain uniform dimensions.
- Open risers are not permitted, with some exceptions. Where permitted, open risers must not allow a 4" (102 mm) diameter sphere to pass through.

Codes usually mandate that a stair have a minimum of three risers, although some jurisdictions require four risers. The concern is that, with fewer than three or four risers, a user may not recognize the level change, especially when viewed from above. Single risers are especially difficult to see, and may cause trips and falls.

To determine the actual riser and tread dimensions for a set of stairs, divide the total rise or floor-to-floor height by the desired riser height. Round off the result to arrive at a whole number of risers. The total rise is then re-divided by this whole number to arrive at the actual riser height.

The calculated riser height must be checked against the maximum riser height allowed by the building code. If necessary, the number of risers can be increased by one and the actual riser height recalculated.

In any flight of stairs, there is always one less tread than the number of risers. Once the actual riser height is set, the tread run can be determined using one of the following comfortable proportioning formulas:

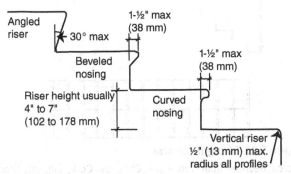

Figure 5.8 Risers, treads, and nosings

- Tread in inches + 2 times riser in inches = 24 to 25
- Riser in inches times tread in inches = 72 to 75

STAIRWAY WIDTHS

Most stairs must have a clear minimum width of 44" (1118 mm). With an occupant load under 50, a 36" (914 mm) minimum width may be permitted.

The IRC requires a residential stair to have a minimum width of 36" (914 mm). The IBC requires stairs as part of a means of egress to have a minimum width of 44" (1118 mm). Stairs 48" (1219 mm) wide allow two people to pass in opposite directions, although a width 60" (1624 mm) is recommended where the stair is intended for people to use in both directions at the same time.

HEADROOM AND LANDINGS

People—especially those with visual impairments—must be protected from low headroom in open spaces below stairs. The 2015 IBC requires a minimum headroom clearance above a stair and its landing to be 80" (2032 mm), measured from the nosing. This may present a design challenge involving changes in texture and material as well as **guards** located at or near the open sides of elevated walking surfaces to prevent falls. When the usable space under stairs is enclosed, it is required by code to be 1-hour fire-rated construction, except in certain residential occupancies.

A flight of stairs may not rise more than 12' (3.7 meters) between floors or landings without intermediate landings. Each run of stairs requires a landing at the top and bottom, and a flat floor or landing on each side of a door. The landings of a stairway should be designed to integrate with the structural system to avoid framing complications.

Landings must be at least as wide as the stairway width, and a minimum of 44" (1118 mm) in length measured in the direction of travel. In dwelling units, a minimum of 36" (914 mm) is allowed. According to the 2015 IBC, landings serving straight-run stairs need not be longer than 48" (1219 mm).

HANDRAILS

A **handrail** is a barrier made of horizontal rails supported by uprights called **balusters** (or banisters). A railing with supporting balusters is called a **balustrade**. A newel (or newel post) is a post that supports one end of a handrail at the top or bottom of a flight of stairs. Handrails should be continuous without interruption by a newel post or other obstruction.

The 2015 IBC requires handrails for stairs in all covered buildings, with some exceptions. The maximum height is required to be no less than 34" (864 mm) and no greater than 38" (965 mm) measured directly above the front nose of the treads. Two handrails are required for most stairs. A single handrail is allowed within dwelling units and for spiral stairways. Intermediate handrails are required for each portion of a stair 88" (2235 mm) or more wide. Check specific uses with applicable codes.

Most codes do not consider the required handrail projection into a stair as a reduction in the stair's width. According to the

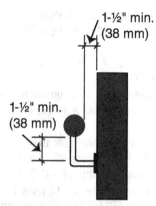

Figure 5.9 ADA handrail projection and clearance

Source: Redrawn from 2010 ADA Standards for Accessible Design, Figure 505.6

2015 IBC, handrails may project a maximum of 4½" (115 mm) into the required width; stringers and trim may project a maximum of 1½" (38 mm). The handrail should extend a minimum of 12" (305 mm) horizontally beyond the top riser of a stair flight, and extend at the slope of the stair run for a horizontal distance of at least one tread depth beyond the last riser nosing of the flight. (See Figure 5.9)

Handrails should be free of sharp or abrasive elements. Handrail ends should return smoothly to the wall or the walking surface, or continue to the handrail of an adjacent stair flight.

The circular cross section of a handrail should be between 1¼" (32 mm) and 2" (51 mm) in diameter; other shapes with similar properties may be permitted if their maximum cross section does not exceed 2¼" (57 mm). Circular rails minimum 1 ¼" diameter, maximum 2". Rectangular handrail cross-sections are not permitted, as they are difficult to grasp.

Code requirements for handrails are complex. Check current codes for the specific use and jurisdiction.

GUARDS

A **guard** (guardrail) is required to protect the open or glazed sides of stairways, ramps, porches, mezzanines, and unenclosed floor and roof openings. Stairs open on one or more sides must have guards on the open sides. A stair landing or any platform that is open and 30" (762 mm) or more above the adjacent level must have guards. The 2015 IBC requires guards to be 42" (1067 mm) high. The IRC requires guards in dwelling units to be a minimum of 36" (915 mm) high.

Some jurisdictions allow reduced height along the side of a stair, so the guard can be used as a handrail. Glazed guards may have special requirements.

A guard must be configured so that a sphere with a 4" (102 mm) diameter cannot pass through any opening in the railing from the floor up to 34" (865 mm). From 34" to 42" (865 to 1070 mm), the pattern may allow a sphere up to 8" (205 mm) in diameter to pass. This is intended to prevent the head of an infant or

small child from passing through. Spindles or balusters forming a stair rail must be set three per tread to meet this requirement. To avoid a 4" (102 mm) sphere from passing below the stair rail, a 10" (254 mm) tread should have a riser no greater than 6-½" (165 mm) high, and an 11" (179 mm) tread's riser should not exceed 6-¼" (159 mm).

STAIR TYPES

A stairway may be approached or departed either axially or perpendicularly to the stair run. Stairs create opportunities for dramatic three-dimensional openings and forms, inspiring vistas, and intriguing details. All stairs require careful planning for safety and support. (See Figure 5.10) Openings must be protected to prevent falls, unauthorized climbing of railings, and entanglements (especially by children). Openings for stairways in wood-framed structures must be double-framed to transfer loads around the open stairwell.

Stairs are made in many configurations. A flight of straight-run stairs has no turns (See Figure 5.11). Quarter-turn and half-turn stairs change direction in the course of their run. (See Figures 5.12 and 5.13)

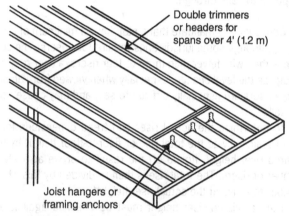

Figure 5.10 Floor framing for stair

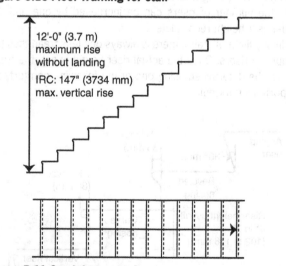

Figure 5.11 Straight-run stair

Source: Redrawn from Francis D.K. Ching and Corky Binggeli, *Interior Design Illustrated* (3rd ed.), Wiley 2012, page 202

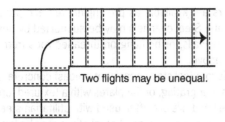

Figure 5.12 Quarter-turn stair
Source: Redrawn from Francis D.K. Ching and Corky Binggeli, *Interior Design Illustrated* (3rd ed.), Wiley 2012, page 202

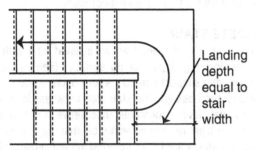

Figure 5.13 Half-turn stair
Source: Redrawn from Francis D.K. Ching and Corky Binggeli, *Interior Design Illustrated* (3rd ed.), Wiley 2012, page 202

WINDERS, CURVED, AND SPIRAL STAIRS

Quarter-turn and half-turn stairs sometimes may use **winders** rather than a landing to conserve space when changing directions. (See Figure 5.14) The 2015 IBC does not permit winder treads in means of egress stairways except within dwelling units. Where allowed, winder treads typically require a minimum 11" (279 mm) tread depth at a walk line 12" (9.5 mm) from the interior curve, with at least 10" (254 mm) within the clear stair width.

The 2015 IBC requires curved stairs with winder treads to follow requirements for winders. (See Figure 5.15) The smallest radius is required to be at least twice the minimum width or required capacity of the stairway.

The 2015 IBC permits the use of spiral stairs as a component in the means of egress only within dwelling units or from a space not more than 250 square feet (23 m²) in area and serving not more than five occupants, with some exceptions. (See Figures 5.16 and 5.17) Headroom clearance is required to be 78" (1981 mm) minimum, with riser height no more than 9½" (241 mm). The minimum stairway clear width at and below the handrail is required to be 26" (660 mm).

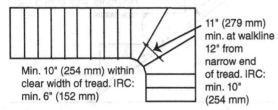

Figure 5.14 Winding stair
Source: Redrawn from Francis D.K. Ching and Corky Binggeli, *Interior Design Illustrated* (3rd ed.), Wiley 2012, page 203

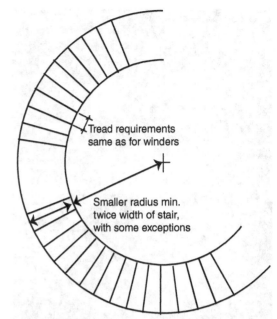

Figure 5.15 Curved stair
Source: Redrawn from Francis D.K. Ching and Corky Binggeli, *Interior Design Illustrated* (3rd ed.), Wiley 2012, page 203

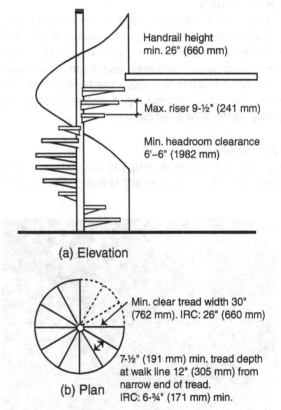

Figure 5.16 Spiral stair
Source: Redrawn from Francis D.K. Ching and Corky Binggeli, *Interior Design Illustrated* (3rd ed.), Wiley 2012, page 203

Figure 5.17 Spiral stair designed by Frank Gehry, Stata Center, Massachusetts Institute of Technology

WOOD STAIRS

Wood stairs are supported by stringers or carriages that serve as inclined beams. **Stringers** are the sloping finish members that run along the sides of the staircase, providing endpoints for treads and risers. **Carriages** running under the treads and risers may be attached to a supporting beam, header, or wall framing. The number and spacing of carriages depends on the spanning ability of the tread material.

STEEL STAIRS

Steel stairs are similar in design to wood stairs, with steel channels serving as carriages and stringers. (See Figure 5.18)

Figure 5.18 Steel stair

Steel stairs are available as pre-engineered and prefabricated components. Steel channels may be supported by beams, rest on masonry walls, or be hung on threaded rods from the floor structure above.

Treads for steel stairs consist of precast concrete, concrete-filled pans, bar grating, or flat plates with a textured top surface. Metal pipe handrails are often used with utilitarian steel stairs.

The ends and underside of steel stairs may be left exposed to view, or finished with gypsum board or metal lath and plaster. A stair can function like a sloped plane connecting one floor with another. Metal stairs may be noisy in use, and can be finished with resilient treads to dampen impact noise.

CONCRETE STAIRS

A concrete stair is designed as an inclined slab with steps formed as an integral part of its surface. (See Figure 5.19) Load, span, and support conditions for concrete stairs require careful structural design. The stair slab thickness is related to the stair's span, its horizontal distance between slab supports. Concrete stairs are reinforced with steel that may extend into a side wall. Handrails are either inserted in cast-in-place sleeves or attached with brackets to the stairs or a low wall.

Concrete stairs are finished with slip-resistant nosings and treads. These may include cast metal nosing with an abrasive finish; metal, rubber, or vinyl treads with a grooved surface, or stone treads with an abrasive strip.

EXIT STAIRS

For more information on fire stairs, see Chapter 18, "Fire Safety Design." Refer to the applicable building code to verify exit stair requirements.

Stairs are an important part of the fire-safety provisions for a building. In a fire, elevator shafts can become conduits for fire and smoke. Firefighting ladders reach only to about the seventh floor of a building, and occupants of tall buildings must usually rely on stairs to get down from upper floors. This process can take hours in large and heavily occupied buildings.

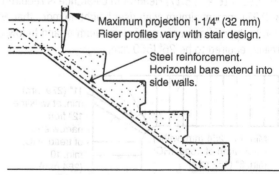

Maximum projection 1-1/4" (32 mm) Riser profiles vary with stair design.

Steel reinforcement. Horizontal bars extend into side walls.

Figure 5.19 Concrete stair

Source: Redrawn from Francis D.K. Ching and Corky Binggeli, *Interior Design Illustrated* (3rd ed.), Wiley 2012, page 205

Some stairs do not meet the strict requirements to be considered part of an exit, and are not designated as exit stairs, even though they may be very prominently located. In an emergency such as a fire, properly designed exit stairs provide safe ways out of the building and allow firefighters access.

Exit stairs, as defined by code, are the most common type of exit. An exit stair includes the stair enclosure, any doors opening into or exiting out of it, and the stairs and landings within the enclosure. The stair enclosure must meet fire rating requirements. All doors in an exit stair must swing in the direction of discharge.

The occupant load determines the required width of an exit stairway. It is based on the use group and the floor area served.

Exit stair requirements are similar to those for other stairs, but stricter. The 2015 IBC requires that exit stairways be enclosed in fire-rated construction, and lead directly to the building's exterior or via an exit passageway, with some exceptions. The IBC requires that exit stairways shall not decrease in width along direction of exit travel.

Code requirements for exit stairs are complex. Verify requirements for specific projects and jurisdictions with current codes.

Codes require that when fully open, the door must not intrude into the required width by more than 7" (180 mm). Riser heights are required to be a minimum of 4" (102 mm) and a maximum of 7" (178 mm) for new stairs. A minimum tread depth for new stairs is 11" (279 mm). The minimum headroom is 6'-8". The minimum required headroom clearance is 80" (2.3 meters), with 12' (3.7 meters) maximum vertical distance between landings.

The minimum dimension of the landing in direction of travel is required to be equal to the width of stair, but need not exceed 4 feet (1.22 meters) when the stair has a straight run. Intermediate landings shall not decrease in width along the direction of exit travel. Doors may not open onto stairs without a landing at least the width of the door.

Ramps

Within a building, a **ramp** is a sloping floor or walk connecting two levels. (See Figure 5.20) A stepped ramp is a series of ramps connected by steps. A curved ramp is technically known as a **helicline.**

Ramps are designed to make travel from one level to another smooth and relatively unobstructed. Ramps require relatively long runs. If they are too steep or continue for too long without level resting areas, ramps become difficult or dangerous for people with mobility problems to use. Wheelchair ramps are limited to a rise of 1:12, but are more comfortable to use with a 1:20 rise. These translate to 12 feet and 20 feet in length for each inch in height.

Short, straight ramps act as beams and may be constructed as wood, steel, or concrete floor systems. Long or curved ramps are usually made of steel or reinforced concrete. Ramp surfaces should be stable, firm, and slip-resistant.

RAMP LANDINGS

The ADA sets requirements for ramp landings. Ramps should have level landings at least 60" (1525 mm) long at each end. A landing should be at least as wide as the widest ramp leading to it.

Where a ramp changes direction, a landing at least 60" by 60" (1525 by 1525 mm) is required. Inserting landings into a long ramp allows it to bend back on itself. This allows the ramp

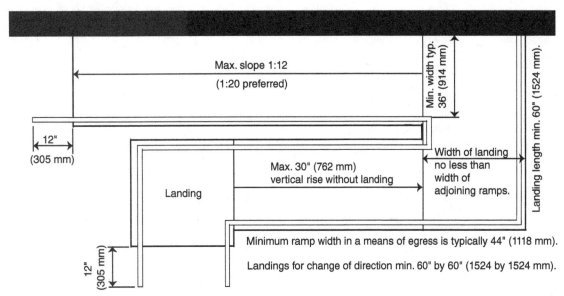

Figure 5.20 Ramp

to be contained in a limited area rather than running throughout the space.

RAMP CURBS, GUARDS, AND HANDRAILS

Curbs, guards, or walls are required to keep people from slipping off a ramp. The 2010 ADA requires a curb or barrier that prevents the passage of a 4" (100 mm) diameter sphere, where any portion of the sphere is within 4" (100 mm) of the finish floor.

Handrails are required along both sides of ramps with a rise of over 6" (152 mm) or a run greater than 72" (1829 mm).

According to the 2010 ADA, the top of the handrail should be between 34" and 38" (864 and 965) above the ramp surface. Some ramps are provided with a second handrail at a lower level to accommodate wheelchair users. Handrails should extend at least 12" (305 mm) horizontally beyond the top and bottom of ramp runs.

We have now explored the basics of floor assemblies, interior walls, and stairs and ramps. In Chapter 6, we look at how windows, skylights, interior glazing, and doors link us to the outdoors and connect interior spaces.

Windows and Doors

Windows and skylights control thermal radiation and offer daylight and views. They also sometimes provide ventilation and affect how doors control entry and fire.

Twentieth-century modernist buildings were often covered in poorly insulated glazing. Rather than designing for the environment, architects and engineers used energy-consuming mechanical heating and cooling to create the interior conditions they desired.

> We can no longer afford the resources to sustain the construction and operation of buildings with poorly designed glazed areas habitable only through the use of active climate control systems. (Walter T. Grondzik and Alison G. Kwok, *Mechanical and Electrical Equipment for Buildings* [12th ed.], Wiley 2015, page 167)

Today, windows, skylights, and doors are designed to work with the building's site conditions to deliver comfortable interior conditions with efficient energy use.

INTRODUCTION

Doors and windows connect the inside to the outside. A building's windows have a huge effect on the appearance and function of interior spaces.

> Windows and doors…are designed to allow controlled penetration of the wall's environmental defenses. A door primarily controls the passage of people but often acts as a valve or filter for the selective passage of air, heat, animals, insects, and light. A window is one of the most fascinating of the building's components. A standard residential window of the most common type allows for the simultaneous and independent control of natural illumination; natural ventilation; view out; view in; passage of insects; passage of water; and passage of heat – radiant, conducted, convected….To architects, the possibilities of a window are limitless. (Edward Allen, *How Buildings Work* [3rd ed.], Oxford University Press 2005, p. 255)

The types, locations, construction, thermal properties, and shading of windows are all concerns of the interior designer. Toplighting with skylights and clerestories bring daylight into building interiors. Door types and thermal performance are also important to interiors; their role in fires is critical.

The **fenestration** (arrangement of windows and doors) of a building greatly influences the amount of heat gain and loss as well as the infiltration and ventilation. The proportion of glass on the exterior affects energy conservation and thermal comfort.

Types of fenestration include windows at eye level and higher **transoms** (small windows above doors or other windows) above them, horizontal skylights, raised **clerestory** windows extending above adjacent rooftops, and **roof monitors** extending above the roof.

History

The first windows were merely holes in a wall, which were later covered with animal hide, cloth, or wood. Early glass windows in Alexandria, Egypt, around 100 CE let in some light but were not transparent. Shutters were later developed to protect and control their openings.

The earliest doors on record are seen in Egyptian tomb paintings, where single or double doors are each made from a single piece of wood. In less arid environments, door panels were framed with stiles and rails to control warping.

WINDOWS

Windows admit clean air and control thermal radiation, air temperature, and humidity. They provide access to view and daylight, and affect visual and acoustic privacy. Windows can control whether animals, and sometimes people, can enter or leave the building. They keep out water and can help control fire. Sunlight coming through a window affects the thermal qualities of surfaces. They sometimes channel communication between inside and outside.

When an architect designs a window for a building, he or she considers the window's orientation, location, size, and proportions. The design also addresses external shading devices, the window's mode of operation, its material and frame, and the type of glass. Interior designers are often involved in the design of shades or shutters as well as types of interior window treatments and their material and color.

Windows can be used to improve energy conservation by admitting solar thermal energy, providing natural ventilation for cooling, and reducing the need for artificial illumination. The proper amount of fenestration is determined by architectural considerations, the ability to control thermal conditions, the first cost of construction versus the long-term energy and life-cycle costs, and the human psychological and physical needs for windows.

Windows do not generally provide for air filtering. Their use can have a significant impact on the functioning of mechanical heating and cooling systems.

Moving or changing the size of a window requires an awareness of structural issues, plumbing equipment such as vent stacks, and structural headers above the window.

Window Selection

Factors to consider when selecting windows include operability, security, privacy, and cost. It is advisable to purchase windows that will last longer, be easier to operate, and make the interior more pleasant and comfortable.

Operable windows open to provide fresh air. They are often composed of many parts, each with its own name. (See Figure 6.1)

Windows may affect the location of equipment and furnishings. The ability to reach above obstructions such as kitchen sinks should be considered. Where security is a concern, select security glass that is difficult to penetrate, plus high-quality lock systems. Consider interior window treatments that are easy to adjust for privacy. Costs can vary a great deal, depending on framing material, size, quality of construction, and energy efficiency.

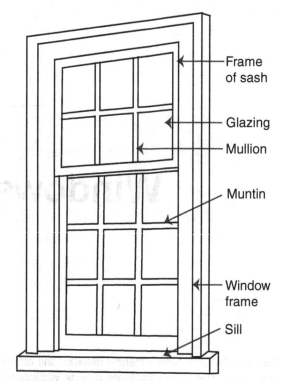

Figure 6.1 Window parts

WINDOW CODES AND STANDARDS

Codes and standards typically prescribe a relationship between floor areas and fenestration area in residences, or between wall area and fenestration area in non-residential buildings. These requirements usually result in small windows, under the assumption that the building relies on electrical energy for lighting, heating, and cooling. Using larger windows for daylighting may require proof of benefit of using on-site, renewable energy resources. Code requirements may vary by climate zone and by jurisdiction.

Glazing materials must be strong enough to resist breaking due to wind loads and thermal stress. Codes require shatter-resistant glazing in certain circumstances. Local building codes specify glazing thickness for various conditions of height and wind speed. Federal Standard 16 CRF 1201 sets breakage performance requirements. Tempered, laminated, or wired glass may satisfy these requirements. Manufacturers recommendations should be followed.

The 2015 IRC considers glazing in or adjacent to doors to be in a hazardous location where the bottom exposed edge of the glazing is less than 50"" (1524 mm) above the floor and within 24" (610 mm) of a closed door or perpendicular to the hinged side of an in-swinging door, with some exceptions. Fixed or operable window glazing is considered to be a hazardous location when the exposed area of an individual pane is greater than 9 square feet (0.836 m^2) with the bottom edge less than 18" (457 mm) and the top edge more than 36" (914 mm) above the floor, where a walking surface is within 36" (914 mm) of the glazing, with some exceptions. Additional requirements apply to glazing in guards and railings, adjacent to stairs and ramps, and in wet areas such as showers and saunas.

WINDOW PERFORMANCE

The National Fenestration Rating Council (NFRC) has developed a standardized rating for window performance. Each window or skylight manufactured in the United States bears an NFRC label certifying that the window has been independently rated. The label carries a brief description of the product and lists U-factor, **solar heat gain coefficient (SHGC)**, and **visible transmittance (VT)**. Energy codes may use NFRC ratings as minimum requirements.

Air leakage (AL) is the amount of heat loss or gain that occurs by infiltration through the cracks and openings in a window assembly. It is measured in cubic feet of air through a square foot of window area. A lower rating is better. The AL does not measure the air leakage between the window assembly and the wall after it is installed. Air leakage is optional on the NFRC label.

WINDOW ORIENTATIONS

In the Northern Hemisphere, windows in internal-load-dominated spaces should open to the north or east. (See Table 6.1) Windows on the west should be kept as small as possible to avoid cooling load.

Window Types

Windows may be fixed or operable. Operable window types include single- and double-hung, casement, hopper, awning, jalousie, and sliding. (See Table 6.2 and Figures 6.2 and 6.3) Bay windows combine fixed and operable panes to project part of the interior space outward.

The type of window used affects natural ventilation. Hopper, awning, or jalousie windows deflect air vertically, up and over people's heads. (See Figures 6.4, 6.5, and 6.6) They also keep out rain while admitting air, a benefit in hot, humid climates.

TABLE 6.1 WINDOW ORIENTATIONS IN TEMPERATE NORTHERN HEMISPHERE

Orientation	Description
North-facing	Lose radiated heat in all seasons, especially winter
East-facing	Gain heat very rapidly in summer when sun enters at very direct angle in mornings
South-facing	Receives low intensity solar heat most of summer day, due to higher angle of sun. In the winter, low sun angle provides sun all day long.
West-facing	Heats up warm building rapidly on summer afternoons, causing overheating, and can result in hot bedrooms at night.

High windows are preferred for horizontal tasks to allow diffuse light to penetrate the space. Low windows are best for vertical tasks, and for distribution of ground-reflected light. Tall, narrow windows provide good light penetration, but can

(a) Double hung   (b) Storm window

Figure 6.2 Windows on Jason Russell house, Arlington, Massachusetts, 1740

TABLE 6.2 TYPES OF WINDOWS

Type	Description	Comments
Single-hung	Only bottom sash moves, may tilt for cleaning	Air leakage may be higher than hinged windows
Double-hung	Two sashes travel vertically in separate tracks or grooves	Maximum 50% ventilation at top and/or bottom; may tilt for cleaning
Casement	Operating sash side hinged, usually swings outward	Open fully for 100% ventilation, can direct breezes
Hopper	Hinged at bottom	Direct draft-free ventilation in, up
Awning	Hinged at top	Draft-free ventilation, some rain protection
Jalousie	Opaque or translucent horizontal louvers pivot in common frame	Glazing strips direct flow of air; difficult to clean; used mostly in mild climates to ventilate with visual privacy
Sliding	Two sashes (one slides) or 3 sashes (middle one fixed)	2 sashes: 50% ventilation; 3 sashes: 66% ventilation

Figure 6.3 Casement windows

Figure 6.4 Jalousie window

Figure 6.5 Awning windows

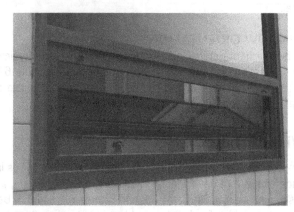

Figure 6.6 Hopper window

produce strong patterns of light and dark. Fixed windows provide view without ventilation, and are often used in interiors. (See Figure 6.7)

OPERABLE WINDOWS

The use of operable windows is coming back after years of fixed windows in large buildings. The open position of a window determines how well it provides natural ventilation. The wind is deflected if it strikes the glass surface. To provide ventilation without drafts, wind needs to be kept away from people. To provide cooling, it needs to flow across the person's body. Windows with multiple positions offer control options.

Casement and pivot windows open the entire window area to airflow. Double-hung, single-hung, and sliding windows open only one-half of the window area. Awning and hopper windows direct air up or down.

Although windows give occupants some control over a source of outdoor air, they generally do not filter air and may interfere with a central HVAC system's attempts to regulate airflow and pressure.

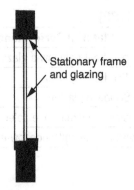

Figure 6.7 Fixed window section
Source: Redrawn from Francis D.K. Ching and Corky Binggeli, *Interior Design Illustrated* (3rd ed.), Wiley 2012, page 180

Noise can be a problem with operable windows. Sealing for thermal purposes also helps improve acoustic properties.

In general, the placement level of most windows for comfortable ventilation is quite low, between 12" and 24" (305 to 610 mm) above the floor for seated or reclining people. This may conflict with desirable heights for views.

Operable windows in a bathroom may compromise privacy or security. A single bathroom window may be inadequate to circulate air adequately to remove moisture. For ventilation, it is best to place the window high on a wall, as warm, moist air rises.

High windows are best for exhausting hot air near the ceiling. High windows require mechanical or automatic operation. Transoms over doors allow ventilation flow throughout the space while maintaining visual privacy at lower levels.

Cross-ventilation requires two openings, an inlet on the windward side of the building, preferably below the middle of the wall, and an outlet on the leeward side. The outlet's position is not critical. The outlet should generally be at least as large as the inlet. Obstructions near the inlets and outlets can significantly reduce wind velocity and its cooling effect.

For more information on ventilation, see Chapter 13, "Indoor Air Quality, Ventilation, and Moisture Control."

Glazing

Glazing refers to the sheets (panes) of glass set into a window. Single glazing does not resist heat flow very well. Double glazing or a separate storm window is required for better thermal resistance (R-value). An even higher R-value can be achieved with a reflective coating or triple glazing.

Double glazing separated by ½" (13 mm) of air space reduces heat loss over that of single glazing by 50 percent. A single glazed window plus a storm window 1" to 4" (25 to 102 mm) away will also produce a 50 percent reduction. Triple glazing generally reduces heat loss about one-third more than double glazing.

Putting an inert (less conductive) gas such as argon or krypton into the air gap between sheets of glazing greatly reduces heat transfer by convection within the air gap. This results in warmer interior surfaces and less condensation and lowers the U-factor.

GLAZING MATERIALS

The selection of glazing materials depends on factors including light transmittance, thermal performance, sound reduction, strength and safety, aesthetic considerations, and life-cycle costs.

Window glass radiates heat to both interior and outdoor surfaces. (See Figure 6.8) Ordinary window glass can transmit over 80 percent of solar infrared radiation (IR) and absorb most of the longer IR from sun-warmed interior surfaces. Most of the absorbed heat is lost by convection to the outside air in cold weather, but the glazing prevents most of the sun-warmed interior air from passing back outside. Insulating glass consists of multiple layers of glass with air spaces between them.

Different glazing materials are often required depending on the orientation of the window. (See Tables 6.3 and 6.4) **Light-to-solar-gain (LSG)** is the ratio between the solar heat gain coefficient (SHGC) and visible transmission (VT). It is used to compare the efficiency of different glass or glazing types in transmitting daylight while blocking heat gains. The higher the number, the more light is transmitted without adding too much heat.

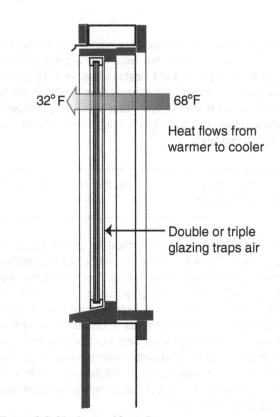

Figure 6.8 Glazing and heat flow
Source: Redrawn from Francis D.K. Ching and Corky Binggeli, *Interior Design Illustrated*, Third Edition, Wiley 2012, page 188

TABLE 6.3 WINDOW ORIENTATION AND GLAZING (NORTHERN HEMISPHERE)

Orientation	Climate	Glazing Material Performance
South	Winter heat gain desired	High solar heat gain low-e glazing
	Winter heat gain not needed	High LSG, low-e glazing
East and west: minimize glazing area	Cold climates	Low LSG low-e glazing
	Hot climates	Low visible transmittance selective low-e glazing
North	All climates	High visible transmittance low-e clear glazing

TABLE 6.4 GLAZING MATERIALS

Glazing Type	Use/Description
Ordinary window glass	Single- and double-glazed windows, doors, skylights. Transmits about 80% of IR solar radiation. Absorbs most longer-wave IR from interior surfaces, keeps heat inside.
Tinted or colored glass	Modifies view. Reduces brightness but not glare. Higher illumination on one side provides privacy on other.
Heat-absorbing glass	Gray or brownish. Absorbs about 50% of solar heat, about half of which is reradiated into building's interior.
Reflective glass	Reflects most solar radiation, reduces brightness. May produce glare and can overheat adjacent buildings.
Shatter-resistant glass	Tempered, laminated, or wired glass, some plastics. May be required by code.
Plastic glazing (acrylic and other)	Translucent, transparent, corrugated, or tinted glazing with transmittance values between 10% and 97% and reflective values from 4% to over 60%. Some may scratch or degrade.

Safety glazing is required for any window that could be mistaken for an open doorway. Any window area greater than 9 square feet (0.84 meters^2) and located within 24" (610 mm) of a doorway, or less than 60" (1524 mm) above the floor must be safety glazed with tempered glass, laminated glass, or plastic. The type and size of glazing allowed in fire-rated walls and corridors is also regulated.

For more information on windows and fire safety see Chapter 18, "Fire Safety Design."

The color of glazing can be critical for certain functions. Artists' studios, showroom windows, and community building lobbies all require high quality visibility between the interior and exterior. Warm-toned bronze or gray glazing can affect the interior and exterior color scheme. Tinted glazing controls glare and excess solar heat gain year round, decreasing solar heat gain in winter as well as summer. Tinting can provide some privacy from the street for occupants, while allowing some view out when the illumination outside is substantially higher than inside during the day. This effect may be reversed at night, putting occupants on display.

Dynamic glazing systems change in response to light, heat, or electricity. (See Table 6.5) Research continues on new products and applications.

GLAZING AND SOUND TRANSMISSION

The sound insulation property of glass depends on its thickness and area. Multiple sheets of heavy glass of differing thicknesses mounted in resilient gaskets and not parallel to each other provide the highest level of sound insulation. The plastic interlayer in laminated glazing dampens sound vibrations between the glass faces on either side.

An open window reduces sound significantly less than a typical cavity wall. The limited acoustic insulation provided by a closed single-glazed window can be increased with a tightly fit sash; heavier glass provides even more improvement. Double or triple glazing improves acoustic performance significantly. Acoustic windows designed specifically to limit sound transmission combine laminated glass with a generous air space and insulation. (See Figure 6.9)

VISIBLE TRANSMITTANCE

High-quality light transmittance is important for spaces such as showroom windows and artists' studios. Commercial building lobbies and restaurants benefit from good visibility between the interior and exterior.

The visible transmittance (VT) of a window is influenced by the glass color (with clear the highest), coatings, and the number of glazings. VT ratings range from 0 to 1; the greater the VT rating, the greater the daylight transmission.

TABLE 6.5 DYNAMIC GLAZING SYSTEMS

Glazing Type	Description
Photochromic glazing	Changes transparency response to UV light using integral compounds or film. Typically used to provide shading.
Thermotropic glazing	Reflective, absorbing, or light scattering types change transparency in response to temperature.
Polymer dispersed liquid crystal (PDLC)	Electricity applied to layer of liquid crystals in laminated assembly aligns (transparent) or scatters (translucent) light. Used primarily for interior privacy control.
Suspended particle device (SPD)	Similar to PDLC glazing. Electrically conductive glass-plastic laminate; can also be used in insulating glass units with low-e glass.
Dispersed particle glazing	Electricity applied to change transparency in range from clear to dark states while preserving view.
Electrochromic glazing	Electrical stimulus used to change VT and SHGC continuously from highly transparent to highly tinted.
Sunlight-activated glazing	Continuously changes tint level, VT, and SHGC based on amount absorbed sunlight. Uses polyvinyl butyral (PVB) film.

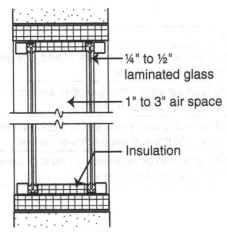

¼" to ½" laminated glass

1" to 3" air space

Insulation

Figure 6.9 Acoustic window

A VT rating is based on the whole window including the frame. Most double- and triple-pane windows have a VT level between 0.30 and 0.70. Some types of glazing coatings reduce the SHGC, which considers the full light spectrum, but not the VT, which considers only visible light (measured as the (LSG). The greater the LSG, the better the window will transmit daylight in hot climates.

Window Frames

Window frames are typically made of wood, vinyl (polyvinyl chloride), fiberglass, or aluminum. (See Table 6.6)

The window's dimensions affect its energy performance. The glass, low-e coating, and gas fill work better at conserving energy than the edge spacer, sash, and frame, so the center is actually more efficient than the edges of the window. True **divided lights** (many small panes, each in its own frame) have a great deal more edge area per window, and are much less efficient.

With double or triple glazing, heat can be lost through edge spacers. High performance windows use edge spacers with thermal breaks, which also keep out moisture to prevent condensation inside the window.

Deep windowsills reflect light indoors, but are a potential source of glare.

CONDENSATION RESISTANCE

The interior surface of an energy efficient window will stay warmer, and is less likely to produce condensation. The NFRC label's U-factor and air leakage ratings aid in selection.

The **condensation resistance (CR)** is a number from 1 to 100; a higher number indicates greater resistance to condensation. Condensation resistance rating is optional on the NFRC label.

Storm Windows and Screens

Old windows can be extremely inefficient. A separate unit—a storm window—can cut conduction and infiltration almost in half. Window screens keep out insects, but affect both ventilation and visibility.

STORM WINDOWS

A **storm window** is a separate sash added to a single-glazed window. A single sash with insulated glazing plus a storm window results in one-third as much heat transmission and half as much infiltration.

There are various types of storm windows. (See Table 6.7) Exterior removable or operable glass or rigid acrylic storm windows are more common and more efficient than interior styles. Storm/screen units are available with low-e coatings on the glass.

TABLE 6.6 WINDOW FRAME TYPES

Frame Type	Description	Performance
Wood frames	Historic material. Remains warm to the touch in winter, stays at room temperature in summer.	Good thermal properties. Moderate insulator. Requires staining or painting unless clad with vinyl or aluminum.
Vinyl (PVC) frames	Hollow or foam-filled; with fiberglass insulate well.	Painting not usually required; maintenance-free. Not a sustainable material.
Fiberglass frames	Fiberglass composite; foam-insulated frames also available	Strong, durable, resists UV. Powder-coat finish for easy care. Insulates like wood.
Aluminum frames	Lightweight. Must have thermal break or conducts heat rapidly.	Maintenance-free. Slowly oxidizes to a dull, pitted appearance. Recyclable.
Combination frames	Mixed materials designed to provide optimal performance.	Example: Vinyl on exterior and wood on interior.
Composite frames	Materials blended together during manufacturing process.	Durability, low maintenance, good insulation properties.

TABLE 6.7 STORM WINDOW TYPES

Type	Description
Plastic film taped to inside of window frame	Simplest, inexpensive, lasts 1 to 3 years. Plastic is heated with blow dryer to shrink tight, can tear.
Interior aluminum frame and 2 sheets of clear glazing film with air layer	Secondary air layer between existing window and storm window. Fasteners screwed into sash or molding.
Exterior aluminum frame	Should be tightly sealed where mounted to window casings, with caulk in all cracks. Do not seal weep holes at bottom.
Exterior wood-framed storm windows	Older storm windows can be repainted and reused. Separate screens are taken up and down yearly.

WINDOW SCREENS

Insect screens decrease airflow by around half. Larger openings may be required for adequate ventilation. Screened-in porches are an effective option.

Pet-proof window screens claim to be resistant to tears and damage caused by household pets. However, a really determined cat can rip through them to escape.

Thermal Transmission

Windows and doors account for about one-third of a home's heat loss, with windows contributing more than doors. Windows should be replaced, or at least undergo extensive repairs, if they contain rotted or damaged wood, cracked glass, missing putty, poorly fitting sashes, or locks that do not work.

Glass conducts heat very efficiently. Windows and skylights are typically the lowest R-value component of the building envelope, allowing infiltration of outdoor air and admitting solar heat. They are much less thermally resistant without some kind of adjustable insulation.

Where there are windows, the temperature inside the building is strongly affected by the exterior temperature. Glazed areas at the perimeter of the building cool adjacent interior air in the winter, with the cooler, denser vertical layer of air along the glass dropping to the floor to create a carpet of cold air. The inside and outside surfaces of a pane of glass are around the same temperature, which is in turn about halfway between the indoor and outdoor temperatures. In walls with a lot of glazing, the interior surface and air temperatures approach the exterior temperature.

In order to conserve energy, building codes and standards prescribe relatively small windows in relationship to residential floor areas and commercial wall areas unless the designer can prove a significant benefit. Large glass areas for daylighting increase heating requirements, but use less electricity for lighting. Less exposed glazing is needed for daylighting in summer than in winter. Increasing insulation in walls or roofs may also justify more glass areas.

SOLAR HEAT GAINS

Solar heat gain through fenestration comes from transmission of heat by conduction due to the difference between indoor and outdoor air temperatures during the day, as well as by solar radiation transmitted through a glazed opening and absorbed within the space. The latter occurs whenever there is a temperature difference, whether the sun is shining or not.

The solar component of building heat gain depends on the orientation of the fenestration, the area of and numbers of layers of glazing, and the shading effects of nearby structures and shading devices. Reflective surfaces such as water, sand, or parking lots to the south, east, or west (Northern Hemisphere) of the building also affect heat gains. Cloudiness is also a factor.

Solar heat gain is a desirable quality for passive solar heating, but is undesirable when you want to prevent overheating in the summer. Energy efficient windows can reduce the cost of the building's HVAC by minimizing the influence of outside temperatures and sunlight. This also reduces maintenance, noise, and condensation problems.

WEATHERSTRIPPING

Weatherstripping all window edges and cracks is a quick and inexpensive way to improve window thermal transmission. Weatherstripping provides a seal against windblown rain and reduces infiltration of air and dust. It also helps block acoustic transmission.

The material chosen should be durable under extended use, non-corrosive, and replaceable. The upper sash of a double-hung window can be permanently caulked if it is not routinely opened for ventilation. Weatherstripping is made of vinyl, metal, felt, foam, silicone, or other materials in many configurations. More than one type of weatherstripping may need to be used in a given application.

WINDOW U-FACTORS AND R-VALUES

Windows and skylights have the lowest R-value (and the highest U-factor) ratings of all the building envelope components. They are chiefly responsible for the infiltration of outdoor air. A U-value is the inverse of an R-value, which indicates the level of insulation, so a low U-value correlates to a high R-value. (See Table 6.8)

The National Fenestration Rating Council (NFRC) was established in 1992 to develop procedures that determine the U-factor of fenestration products accurately. The U-value measures how well a product prevents heat from escaping a building. U-value ratings generally fall between 0.20 and 1.20. The smaller the U-value, the less heat is transmitted. The U-value is particularly important in cold climates.

Ratings involve detailed descriptions of specific window constructions. Windows often exhibit differences in heat flow rates between the center of the glass, the edge of the glass, and the frame. The size of the air gap, type of coating and gas fill, and frame construction all affect the U-factor.

Designers, engineers, and architects can evaluate the energy properties of windows using their U-values. Ratings are based on standard window sizes, and intended to compare windows of the same size.

Refer to Chapter 2, "Designing for the Environment," for more information on U-factors.

TABLE 6.8 COMPARATIVE WINDOW THERMAL RESISTANCE

Glass Type	Subtype	R-Value
Single glass	Without storm window	0.91
	With storm window	2.00
Double-insulating glass	3/16" air space	1.61
	1/4" air space	1.69
	1/2" air space	2.04
	3/4" air space	2.38
Insulating glass 1/2"	With low-e film	3.13
	With suspended film	2.77
	With 2 suspended films	3.85
	With suspended film and low-e	4.05
Triple-insulating glass	1/4" air space	2.56
	1/2" air space	0.29
Addition for tight fitting drapes or shades, or closed blinds		3.23

SOLAR HEAT GAIN COEFFICIENT (SHGC)

The U-factor tells you how much heat will be lost through a given window. The NFRC also provides solar heat gain ratings for windows that look at how much of the sun's heat will pass through into the interior. The solar heat gain coefficient (SHGC) value looks at the performance of the entire glazing unit rather than just at the glass itself.

Solar heat gain is good in the winter when it reduces the load for the building's heating equipment. In the summer, added solar heat increases the cooling load. The SHGC is a number from 0 to 1.0. The higher the SHGC, the more solar energy passes through the window glazing and frame.

A high SHGC (such as 0.9) indicates poor resistance, which is desirable for solar heating applications. A lower SHGC (such as 0.2) indicates good resistance, and would be a good choice where cooling is the issue. The SHGC depends on the type of glass and the number of panes as well as tinting, reflective coatings, and shading by the window or skylight frame.

National Fenestration Rating Council (NFRC) testing does not include related elements such as draperies, building overhangs, and trees. The IRC recommends window SHGC levels based on climate zones.

HIGH-PERFORMANCE WINDOWS

Windows installed prior to the 1980s had the greatest heat flow rate; many of these windows are still in use. However, in today's buildings, the highest rate of heat flow is more likely to be from outside air infiltration, or from deliberate ventilation.

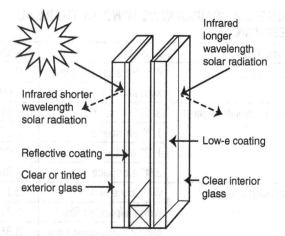

Figure 6.10 Solar heat control window

The development of high-performance **superwindows** combine and refine features such as multiple layers of glazing, multiple low-e coatings, and low-conductivity gases. These advances have produced excellent results, but the drive for the ultimate high-performance window has leveled off. (See Figure 6.10)

The cost of high-performance superwindows has slowed their acceptance for energy-efficient design. Instead, designers are turning to climate-specific solutions that consider the whole building and its environment rather than just the windows.

WINDOW FILMS

Plastic films designed to be glued to the inside face of window glass have the same properties as reflective and absorptive glass. (See Table 6.9) During the winter, films reflect interior heat back inside, reduce drafts from cold glass surfaces, and allow higher relative humidity without condensation. Window films are available that will block up to 99 percent of the sun's UV rays, which are the single largest cause of fading. Films increase the

TABLE 6.9 PLASTIC WINDOW FILMS

Type	Description
Reflective films	Used for tinted one-way mirrors
Darkening films	Produce an effect similar to sunglasses
Silver or gold tinted films	Block slightly more radiant energy than visible light; silver reflects up to 80% of solar radiation
Bronze or smoke films	Block almost as much visible light as radiated heat; intercept more solar radiation than silver films
Selective transmission films	Admit most solar radiation, reflect long IR radiation from warm objects back into room better than ordinary glass; available as sheets for existing windows

shatter-resistance of window glass. They should generally not be used on thermal pane or self-insulating windows, as they can cause the glass to crack with thermal expansion and contraction. For the same reason, tinted films should not be applied to tinted glass or to very large areas of glass. Plastic window films have a relatively fragile surface and limited service life.

During the winter, tinted windows reflect radiated heat back inside the room, and improve the room's operative temperature. They also reduce drafts due to cold glass surfaces.

LOW-EMITTANCE COATINGS

Low-emittance (low-e) coatings consist of thin, transparent coatings of silver or tin oxide that allow the passage of visible light while reflecting IR heat radiation back into the room, reducing the flow of heat through the window. They are applied to one glass surface facing the air gap. The nearly invisible metallic coating blocks radiant heat transfer. It also reduces UV rays to protect carpets and furnishings from fading.

Low-e films are classified as hard-coat or soft-coat. Hard-coat films are durable and less expensive, but less effective. Soft-coat films are more expensive and have better thermal performance.

A building may need different types of low-e coatings on different sides of the building. (See Table 6.10) The south side (Northern Hemisphere) may need low-e and high solar heat gain coatings for passive solar heating, while the less sunny north side may require the lowest U-value windows possible.

Shading and Solar Control

Unshaded glazing can collect unwanted solar heat, resulting in money and energy being spent for mechanical cooling without achieving thermal comfort. Exterior shading blocks the sun's rays from entering the building and heating its interior.

Summer shading of windows, walls, and roofs can be provided by other structures, shading elements integral to the building, or surrounding vegetation. Roof overhangs, screens, and fins are effective. Planting shade trees and installing deep awnings over windows can help. However, trees may grow too large and will eventually die, and so require planning for the future.

SHADING

Shading is beneficial to an entire building, and is crucial for windows. The required period for a given building depends on both the climate and the nature of the building. Sun shading can reject most solar heat gains while aiding in distributing daylight deep into buildings. Shading is often able to cut cooling loads in half.

Vertical windows must be shaded from direct solar radiation and often from diffuse radiation, especially in humid, sunny climates. Reflected solar radiation in sunny climates and in urban areas, especially where highly reflective surfaces predominate, requires shading as well. Additional indoor shading devices or shading within glazing may be needed to control diffuse solar radiation.

TABLE 6.10 LOW-E COATINGS

Coating Type	Description	Use
High-transmission	Coating on inner glazing traps outgoing IR radiation	Cold climates for passive solar heating
Selective-transmission	Coating applied as separate sheets; can be applied to existing windows	Blocks incoming IR, retains heat emitted by objects; high visible light; winter heating, summer cooling
Low-transmission	Coating on outer glazing rejects solar heat gain	May use tinted exterior glazing for lower light transmission

In the Northern Hemisphere, shading is most important for windows that are facing south, east, or west. Northern windows often also need to be shaded from the direct summer sun low in the sky both early in the morning and near sunset. Horizontal overhangs are probably the most effective shading device for south-facing windows in the summer.

East and west windows must be shaded in tropical latitudes. An overhang is necessary to fully shade an east or west window. The overhang must be much longer than on a south window, and should be backed up by interior treatments.

EXTERIOR SHADING DEVICES

Exterior shading of glazing cuts off direct solar rays, allowing only diffuse light through. (See Table 6.11) As a result, it can reduce solar heat gain to the space by up to 80 percent.

Plants used for exterior shading include live vines. Vines can be supported by strong wire, a lath grid or lattice, wire mesh, plastic netting, chain link fencing, or fishnet.

Fixed exterior shading devices are typically the responsibility of the building's architect. Exterior fixed shading can offer benefits of simplicity, low cost, and low maintenance. If fixed sunshades are applied equally to all façades, they may block sun in spring (when it is needed) as well as in autumn (when it is not).

Overhangs decrease light levels near windows for more uniformity throughout the space. Louvered or translucent overhangs block direct sun glare and allow more diffuse illumination to enter the interior.

INTERIOR SHADING DEVICES

Interior shading absorbs around 80 percent of solar energy that comes through the window. Interior shading devices are generally easier to adjust and accumulate less dirt than exterior ones, and do not have to deal with weathering. Interior shades, blinds, and curtains largely absorb solar radiation and convert it to convected heat in interior air, while shading occupants and furniture inside the building. Draperies can become part of a heat trap that radiates heat to the interior in hot weather. Interior window treatments are relatively ineffective in reducing solar heating of air inside a building, but effective in reducing high radiant heat gain and glare from direct sunlight.

TABLE 6.11 EXTERIOR SHADING DEVICES

Device	Comments (Northern Hemisphere)
Horizontal panel overhang	Can trap hot air. Wind and snow load problems. South, east, west façades.
Horizontal louvers in horizontal plane	Allows air movement, small scale, economical. South, east, west façades.
Horizontal louvers in vertical plane	Shorter length required. Restricts view. South, east, west façades.
Vertical panel overhang	Allows air movement, view restricted. South, east, west façades.
Vertical fin	Restricts view if on east or west. North façade.
Exterior insulating operational shutters	Hinged, sliding, folding, or bi-fold configurations. Require mechanical operation for adjustment.
Awnings	Durable, attractive, easy to adjust; manual or automated operation. Can be difficult to maintain.
Light shelves	Reflect sun and sky light toward interior ceiling to increase indirect distribution and depth of penetration. South.
Exterior roller shades	Fabric shades provide security and shade.
Outdoor venetian blinds	Excellent daylight control.

Interior shading devices can provide privacy, glare control, and insulation, and can improve the appearance of interior spaces. (See Table 6.12)

They also help prevent the appearance of a black hole when looking out a window at the dark outside view.

Interior shading devices allow their operation by room occupants. (See Figure 6.11) The exterior surface of interior window treatments should be reflective for maximum effectiveness.

The color of shades and blinds affects their function. Light-colored reflective shades are more effective than dark ones in keeping out unwanted solar heat. Blinds with white or mirrored

TABLE 6.12 INTERIOR SHADING DEVICES

Device	Description
Venetian blinds	Reduce intensity and redistribute light. Block low east/west sun angles. Available with perforations and between sheets of glass.
Insulating shutters	Hinged, sliding, folding, or bi-fold. Require storage area and airtight seal. Can act as rigid window insulation.
Roller shades	Diffuse direct sunlight, eliminate glare, increase uniformity of illumination. Mounting shades to pull up from bottom blocks glare.
Insulating shades	Layers of air- and moisture-tight fabric with internal solar barrier; ultrasonically welded. Need tight seal against wall.
Fiberglass mesh shades	Designed to intercept specific percent of sunshine. Fairly long life spans. Control brightness while leaving view to outside.
Roman shades	Can be made of insulated materials with decorative surface.
Cellular honeycomb shades	Mounted in tracts to move horizontally or vertically on flat or curved surfaces. Motor or manual operation. Improve R-values in winter.
Draperies	Levels of privacy depend on color and tightness of weave. Locate to avoid interference with HVAC equipment.
Insulating draperies	Foam or other insulating backing allows use as thermal barriers. Can be fitted into tracks.
Light shelves	When positioned just above eye level, can help control glare. Increase the depth of daylight penetration zone.

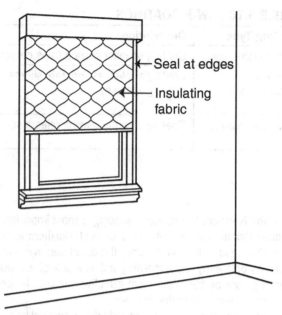

Figure 6.12 Thermal shade

finishes work best for heating, cooling, and daylighting. The side of venetian blinds facing the glazing should be white, to reflect solar radiation back out.

Diffusing shades and curtains can become so bright when illuminated by direct sunlight that they become sources of glare themselves. This problem can be reduced with off-white fabric or by adding opaque drapery.

It is important to check the proposed type, size, and mounting of window treatments to verify that they will not create a problem with the HVAC system.

Both draperies and shutters require storage space when they are not in place across the window. An airtight seal around the edges keeps thermal performance high and prevents condensation from forming. (See Figure 6.12)

TOPLIGHTING AND SKYLIGHTS

Windows at eye level provide views, but also can be sources of intense heat and glare. **Toplighting** involves the use of openings in the roof, while **sidelighting** is provided by vertical windows in walls.

For more information on toplighting and sidelighting, see Chapter 17, "Lighting Systems."

Toplighting

Toplighting comprises skylights, roof monitors, light pipes, and clerestories.

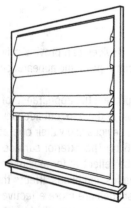

Figure 6.11 Roman shade

Toplighting provides high-quality and high-quantity illumination over a large area. It is used in buildings where view and orientation are supplemented by windows below. It is usually advisable to diffuse toplighting to eliminate bright sources and glare. Toplighting can be diffused by reflecting light off the ceiling, or by using baffles or banners to shield and diffuse light sources.

High windows and toplighting are usually more effective than lower windows, and are most valuable where light but not view is needed. High windows are more secure, free up wall space below them, and distribute illumination more uniformly to all walls and to the interiors of low buildings.

SKYLIGHTS

Skylights allow sunlight to be used for dramatic effects in lobbies, lounges, and other areas without critical visual tasks. When a skylight is placed high in a space, light is able to diffuse before it reaches floor level, preventing glare within the occupants' field of view. Placing the skylight near a wall aids diffusion, making the space appear brighter, larger, and more cheerful.

Traditional skylights are designed like standard windows. Many have fixed glazing, but operable roof windows are also available. Operable skylights or roof windows allow hot air to escape in warm weather. (See Figure 6.13) Building codes require horizontal and vertical clearance from plumbing vents. Skylights are NFRC labeled.

Splayed openings increase the apparent size of skylights, and provide better light distribution while diffusing glare. Adding an opaque bubble to a skylight shields the glass and provides some insulation. High-quality products and high-quality installation are important to prevent problems such as leaks.

Horizontal skylights gain the most solar heat in the summer, when the sun is overhead, and the least in the winter, when the sun angle is lower. They should be avoided in climates that require artificial cooling, unless they can be shaded. Horizontal skylights are most effective for daylighting when the sky is overcast.

Skylights can accommodate accessories, including sun-blocking shades, pleated shades, Venetian blinds, and roller shades. Skylight shades use a track that can be surface mounted or recessed into the opening of the frame. Some styles instead use hook-and-loop attachments where the window shade does not have to be opened and closed frequently.

Skylights that have a smaller area tend to have higher U-factors, indicating higher heat flow. It is advisable to specify the frame with the best thermal performance when selecting smaller skylights.

Skylights are glazed with acrylic or polycarbonate plastic, or with wired, laminated, heat-strengthened, or fully tempered glass. Building codes limit the maximum area of each glazed skylight panel. Codes may also require wire screening below glazing to prevent broken glass injuries when wired glazing, heat-strengthened glass, or fully-tempered glass is used in multiple-layer glazing systems. There are exemptions to these regulations for individual dwelling units.

Skylights can be made with diffusing or translucent materials that reduce contrast, but these should not be used where the sparkle of sunlight is desirable as in an entrance hall.

Good-quality manual or automatic controls for skylights with operable louvers help maintain daylighting illumination at a constant level, and also can block light when needed, such as for audiovisual presentations.

OTHER TOPLIGHTING STRATEGIES

Clerestory windows are high windows above eye level. (See Figure 6.14) Much of their light is reflected off the ceiling. They can be glazed with translucent materials, as view is less important. Light scoops are clerestory windows facing in a single direction, with the opposite side curved to reflect light down into interior space.

Tubular daylight devices (tubular skylights) use a light shaft that provides surface reflection, capped with a clear skylight. (See Figure 6.15) A diffusing lens increases light distribution.

Light pipes are hollow, duct-like light guides that transport daylight, electric light, or fiber-optic lighting in pipes made of prismatic plastic film. They transmit light by total internal reflection. Used as skylights, a roof-mounted plastic dome captures sunlight that passes down a reflective tube that stretches from

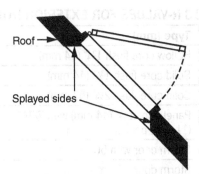

Figure 6.13 Operable skylight
Source: Redrawn from Francis D.K. Ching and Corky Binggeli, *Interior Design Illustrated* (3rd ed.), Wiley 2012, page 181

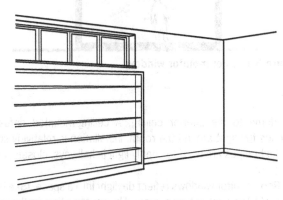

Figure 6.14 Clerestories above bookcase
Source: Redrawn from Francis D.K. Ching and Corky Binggeli, *Interior Design Illustrated* (3rd ed.), Wiley 2012, page 189

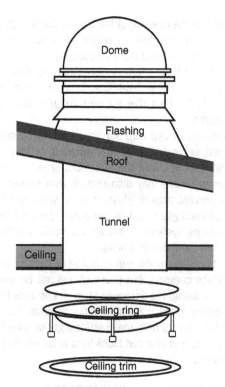

Figure 6.15 Tubular daylight device

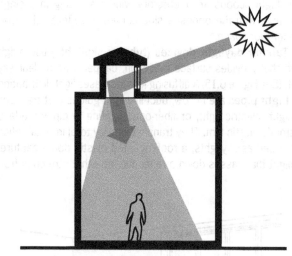

Figure 6.16 Roof monitor window

the dome to the interior ceiling. A ceiling-mounted diffuser spreads the light around the room. Installation is relatively simple. Light pipes are used to supply light in hallways, closets, and other spaces.

Roof monitor windows reflect daylight into a space. (See Figure 6.16) The light enters a scoop-like construction on the roof and bounces off the surfaces of the monitor opening and down into the space. Mirror systems using a periscope-like device

can bring daylight and views underground by reflecting them down through the space. Roof monitor windows usually face in more than one direction and are operable.

DOORS

Doors control the entry of people and animals and direct movement through the building. They provide emergency egress and help control building fires. They help create visual and acoustic privacy. Doors can also provide clean air and control air temperature, humidity, and airflow. They sometimes provide channels of communication and useful surfaces.

Thermal Performance

The National Fenestration Rating Council (NFRC) has established a rating procedure for determining the thermal performance of doors and sidelights, as they have for windows. A permanent label attached to the edge of the door slab lists the certified U-factor. A temporary label also appears on the face of the door. The energy rating is listed as a U-factor, the rate of heat loss; higher numbers mean more heat loss. R-values are also available for exterior doors. (See Table 6.13)

The thermal performance of exterior doors is important. The thin film of air that forms on the inside surface of a closed door resists the passage of heat through the door. When this film is disturbed by forced-air supply registers or return grilles, the door's thermal effectiveness is compromised.

Exterior Doors

A building's exterior doors make a strong statement about the building's function and occupants. Both their interior and exterior appearance should be considered. A sturdy door with break-resistant glass and a high-quality lock system will enhance security.

Doors contribute to building heat loss, and vary in their energy efficiency. Those filled with foam insulation provide better resistance to heat loss than solid wood doors. Doors set in entry

TABLE 6.13 R-VALUES FOR EXTERIOR DOORS

Material	Type (mm)	R-Value
Wood	Hollow core flush 1¾" (44 mm)	2.17
	Solid core flush 1¾" (44 mm)	3.03
	Solid core flush 2¼" (57 mm)	3.70
	Panel door 1¾" (44 mm) with 7/16" (11 mm) panels	1.85
	Storm door with 50% glass	1.25
Metal	Storm door	1.00
	Insulating door 2" (51 mm) with urethane	15.00

vestibules help to keep indoor air from mixing with outdoor air. This helps to control the interior environment, and reduces the amount of unconditioned air that must be heated or cooled.

Larger door openings, especially into residential kitchens, can make it feasible to bring in large equipment.

WEATHERSTRIPPING

Storm doors added to old, uninsulated metal or fiberglass doors may be ineffective. Weatherstripping improves insulation and acoustic separation. It is often supplied and installed by manufacturers of sliding glass doors, glass entrance doors, revolving doors, and overhead doors. Weatherstripping is available made of felt, foam, vinyl, rubber, or metal.

The entire perimeter of the door should be weatherstripped, with a **door sweep** at the bottom of the door. A door sweep is made of aluminum or stainless steel with a brush, and located on the interior bottom of an in-swinging door or the bottom of the exterior side of exterior-swinging door. An aluminum **door shoe** with a vinyl insert seals the space under a door and sheds rain. Thresholds combined with weatherstripping are also available.

Weatherstripping applied to a door bottom or threshold could drag on carpet or break down due to foot traffic. An **automatic door bottom** is a horizontal bar at the bottom of a door that drops automatically when the door is closed to seal the threshold to air and sound.

REVOLVING DOORS

Revolving doors are typically used as entrance doors in large commercial and institutional buildings to provide a continuous weather seal, eliminate drafts, and minimize heating and cooling losses. (See Figure 6.17) They typically consist of three or four leaves that rotate on a central vertical pivot, all enclosed in a cylindrical vestibule.

Revolving doors typically comprise a 6'-6" (1980 mm) diameter revolving door, with a 7'-0" (2135 mm) diameter door used in high traffic areas. Some revolving doors have leaves that automatically fold back in the direction of egress when pressure is applied, providing a passageway on both sides of the door pivot. The 2015 International Building Code (IBC) requires that each revolving door shall have a side-hinged swinging door in the same wall and within 10 feet (3048 mm). Other restrictions also apply.

The revolving door's enclosure may be made of metal, or of tempered, wire, or laminated glass. Door leaves are made of tempered glass with aluminum, stainless steel, or bronze frames.

Interior Doors

A room may have multiple door openings into adjoining spaces; it is not uncommon for a residential kitchen to have up to six openings into a garage, patio, deck, and other spaces. The interior

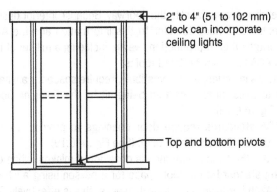

2" to 4" (51 to 102 mm) deck can incorporate ceiling lights

Top and bottom pivots

Elevation

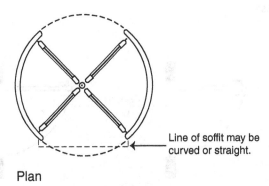

Line of soffit may be curved or straight.

Plan

Figure 6.17 Revolving door

Source: Redrawn from Francis D.K. Ching, *Building Construction Illustrated* (5th ed.), Wiley 2014, page 8.16

designer must consider door swing directions and clearances to avoid conflicts with new or existing elements.

Floor coverings may require specific clearances. Floor heating systems may raise the floor level enough to require trimming the door for clearance.

SOUND CONTROL

The position of doors relative to sources of unwanted sound is an important consideration, especially in residential design and for private offices in commercial spaces. Louvered and undercut doors are useless as sound barriers.

The most important step in soundproofing doors is to seal around the opening. A door in the closed position should exert pressure on gaskets for an airtight seal.

A **sound lock** comprises two doors, preferably with space for a full door swing between them. The doors must be gasketed. All surfaces should be completely covered with an absorbent material and the floor should be carpeted.

ACCESSIBILITY AND DOORS

The Americans with Disability Act (ADA) sets requirements for door openings in spaces that it covers. Minimum maneuvering clearance dimensions up to 60" (1525 mm) vary depending on approach direction, door side, and whether the approach is perpendicular or parallel to the doorway. Because interior design students often need to refer to this information and the ADA encourages their reproduction, we are including a full set of the 2010 ADA figures for related topics.

Doors in series, as in a vestibule, require space for a wheelchair to pass through without being hit by the swinging door. (See Figure 6.18)

ADA standards require door openings to provide a clear width of 32" (815 mm) minimum. (See Figure 6.19)

A 34" (864 mm) clearance is considered minimal when designing standard door floor space for a person using a wheelchair. With hardware and door thickness, this is effectively 30" (762 mm) clear. A swing-away hinge increases the opening from 1" to 1 ½" (25 to 38 mm), which can accommodate the door thickness.

The dimensions of clear floor space required in order for a person using a mobility aid to pull a door open varies based on the type of door and approach. (See Figures 6.20, 6.21, and 6.22)

Door swings that do not interfere with appliances or cabinet doors and drawers are particularly important where people with disabilities use an interior space. Pocket or out-swinging doors are helpful where an aide may need to enter to provide assistance. Hardware is available for pocket doors to make them easier for people in wheelchairs to use.

In older homes, where hallways are sometimes limited to less than 42" (1067 mm), an angled doorway can sometimes help ease circulation problems.

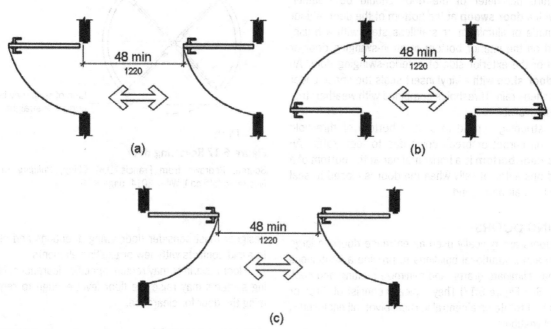

Figure 6.18 ADA doors in series
Source: 2010 ADA Standards for Accessible Design 404.2.6

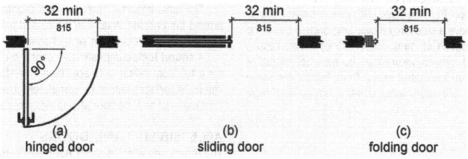

Figure 6.19 ADA clear width of doorways
Source: 2010 ADA Standards for Accessible Design 404.2.3

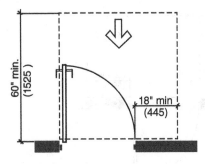

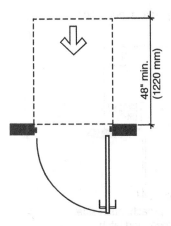

(a) Front approach, pull side

(b) Front approach, push side

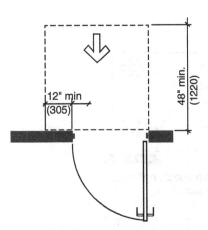

(c) Front approach, push side,
door with both closer and latch

Figure 6.20 ADA front approaches
Source: Redrawn from 2010 ADA Standards for Accessible Design,
404.2.4.1

Doorways to a kitchen in a home with a person with a cognitive impairment may need to be closed with a door that locks to control access. In other cases, doorways may need to be open to maintain visual contact with a resident.

Repetition in the way a door swings, the location of switches and controls, and the order of progression as one enters a room help people with cognitive impairments. Textured or opaque glass can help people with cognitive or visual impairments recognize glass doors and other glazed surfaces.

Door Types

Doors are manufactured in a variety of types. (See Figure 6.23) Doors have wood or metal frames that may be pre-painted, factory-primed for painting, or clad in a variety of materials. Doors may be glazed for visibility and/or have louvers for ventilation. Special doors have fire-resistance ratings, acoustical ratings, or thermal insulation values. Standard interior door widths include 2'-0", 2'-4", 2'-6", 2'8", and 3'-0 (610, 711, 762, 813, and 914 mm).

Glass doors are usually constructed of ½" or ¾" (13 or 19 mm) tempered glass, with fittings to hold pivots and other hardware. Glass doors do not require jamb frames as the door can be butted directly against a wall or partition.

Pocket doors are hung on a track, and slide into a pocket within the wall. They need a pocket free of plumbing, electrical, or HVAC elements. Surface sliding doors (also called barn doors) require wall surface space. Sliding patio doors are similar to large sliding windows. (See Figure 6.24)

HOLLOW METAL DOORS
Hollow metal doors have steel face sheets bonded to a steel channel frame and reinforced with channels. Their cores are made of honeycomb paper, steel-stiffened mineral fiber, or rigid plastic foam. The door face may be seamless or show the panel construction seams.

WOOD DOORS
The frame of a wood hollow core door encases a corrugated fiberboard or wood strip grid core. They are lightweight with little acoustic or thermal value, and are intended primarily for interior use.

Solid core wood doors have cores of bonded lumber blocks, particleboard, or mineral composition. They are used primarily as exterior doors, or wherever increased fire resistance, sound insulation, or dimensional stability is desired.

Wood stile-and-rail doors have a framework of vertical stiles and horizontal rails that hold solid wood or plywood panels, glass lights, or louvers in place. (See Figure 6.25) They are available in a variety of panel designs as well as fully louvered and in French door styles.

DOOR FRAMES
Door frames are standard items. The treatment of the opening and design of the casing trim of a door are opportunities for an interior designer to manipulate the scale and character of a doorway. Casing trim is used to conceal the gap between the door frame and the wall surface. It may be omitted where the wall material can butt up against the door frame neatly.

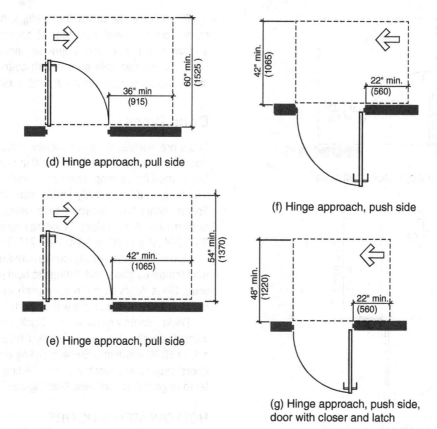

(d) Hinge approach, pull side

(e) Hinge approach, pull side

(f) Hinge approach, push side

(g) Hinge approach, push side, door with closer and latch

Figure 6.21 ADA hinge approaches

Source: Redrawn from 2010 ADA Standards for Accessible Design, 404.2.4.1

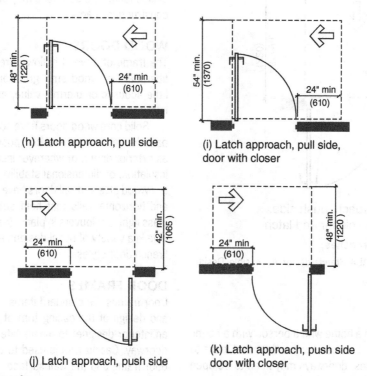

(h) Latch approach, pull side

(i) Latch approach, pull side, door with closer

(j) Latch approach, push side

(k) Latch approach, push side door with closer

Figure 6.22 ADA latch approaches

Source: Redrawn from 2010 ADA Standards for Accessible Design, 404.2.4.1

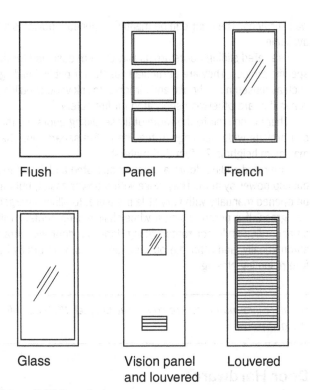

Flush   Panel   French

Glass   Vision panel   Louvered
and louvered

Figure 6.23 Door types

Source: Redrawn from Francis D.K. Ching and Corky Binggeli, *Interior Design Illustrated* (3rd ed.), Wiley 2012, page 191

Hollow metal doors are hung in hollow metal frames. Wood doors use either wood or hollow metal frames. The door opening can be enlarged physically with **sidelights** and a transom above or visually using color and trim work. (See Figure 6.26) Minimizing trim can visually reduce the scale of a doorway or make it appear to be simply a void in a wall. When finished flush with the surrounding wall, it can merge visually with the wall surface.

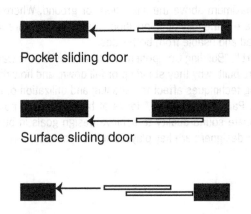

Pocket sliding door

Surface sliding door

Bypass sliding door

Figure 6.24 Pocket, surface, and bypass sliding door plans

Source: Redrawn from Francis D.K. Ching and Corky Binggeli, *Interior Design Illustrated* (3rd ed.), Wiley 2012, page 192

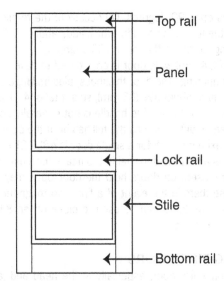

Top rail
Panel
Lock rail
Stile
Bottom rail

Figure 6.25 Wood panel door

Fire Doors

Basic fire door standards indicate that no tool, key, or special knowledge be required to open the door in the direction of egress. No more than 15 pounds (6.8 kg) of pressure be required to release any latch or fully open door, although up to 30 pounds (13.6 kg) pressure may be required to initiate the opening.

FIRE DOOR WIDTHS

Typical building code requirements cover the number of exits that must be provided for fire egress and the width of doors. The width of an egress door is based on the clear opening when

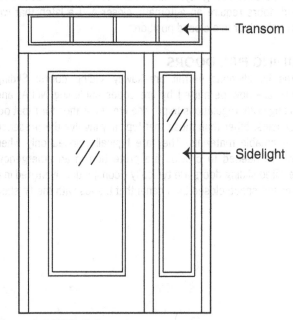

Transom
Sidelight

Figure 6.26 Sidelight and transom

the door is open 90 degrees. On occasion, the requirement is based on the full open position, which adds about 2" (51 mm) to the opening for doors that open 180 degrees.

A 36" (914 mm) wide door is considered to have a 32" (813 mm) clear opening. The door thickness, plus its hinge and hardware, add up to about 3¼" (83 mm), so a total of 4" (102 mm) is often allowed for these. The handle is not generally counted in this allowance, unless it runs the full height of the door.

The maximum width for a single door is 48" (1219 mm), due to concerns about the door being too heavy for a user to open. Automatic closers on doors hold the door open magnetically, and release them in the event of a fire. Two magnetically held 48" (1219 mm) doors can be used to close off an 8-foot (2.4 meter)-wide corridor.

SEALING FIRE DOORS

The seal on a fire door, especially at the head and jambs, is critical to minimize smoke or fire transmission. This seal can be difficult to provide and maintain. Some codes may require the addition of an **intumescent** strip along the head and jambs of a fire door that swells up when heated to seal the gap, making it no longer available as a means of egress.

Closed fire doors that are normally held open continue to offer a means of movement between parts of a building. Fire-rated rolling shutters (overhead doors) are sometimes used to close openings between fire-separated areas in a building, so that no passage through the opening is possible. These shutters can have a passage door or be designed to open manually to allow people to pass through.

FIRE DOOR RATINGS

Most codes require fire stairs to have a 2-hour fire-rated enclosure with 1-½-hour fire-rated doors. Corridors with 1-hour fire-separation ratings typically require ¾-hour rated doors. All rated doors require an automatic closer and a latch that will retain the door in a closed position.

SLIDING FIRE DOORS

Some jurisdictions permit breakaway sliding doors. Sliding doors are now permitted by fire codes (including NFPA) and building code regulations to provide fire separation for most occupancies, other than those that typically involve the presence of flammable materials. They are typically closed only when they are needed to provide fire protection in an emergency. Fire-rated sliding doors are typically inconspicuously stored in a recessed space closed by a panel that blends with the finished

wall surfaces. The door can be manually opened if power is not available.

Fire-rated sliding doors are usually custom designed for the specific project. They are particularly useful for public buildings such as museums, schools, and airports, to link various building zones that are otherwise separated for fire safety.

There is no limit to the width of these sliding doors, so they can provide wide openings in the line of fire separation. Their maximum height is 28 feet (8.5 meters).

Fire-rated sliding doors are power operated and linked to a backup power system. They close with a power assist, but can be opened manually with very little pressure, to allow emergency egress. If a person using a wheelchair makes contact with an accordion-style horizontal sliding door, the door will retract automatically, and after the obstruction clears, wait briefly before resuming closing.

For more information on fire doors, see Chapter 18, "Fire Safety Design."

Door Hardware

Door hardware is selected for function and ease of operation, recessed or surface-mounted installation, and durability. Material, finish, texture, and color are also specified. Security hardware may require electrical wiring.

Hardware for doors includes locksets comprising locks, latches, bolts, a cylinder and stop works, and operating trim. Other hardware items are hinges, closers, panic and fire exit hardware, push and pull bars and plates, and kick plates. Door stops, holders, and bumpers are also included, as are thresholds, weatherstripping, and door tracks and guides.

The ADA requires that door handles, pulls, latches, and locks be easy to grasp with one hand without tight grasping, pinching, or twisting of the wrist. Operable parts of such hardware are required to be 34" (865 mm) minimum and 48" (1220 mm) maximum above the finish floor or ground. Where sliding doors are in the fully open position, operating hardware shall be exposed and usable from both sides.

Part II, "Building Components," has addressed the way buildings are built, why they stand up or fall down, and how different building techniques affect the shaping and utilization of interior space. Part III, "Acoustics," looks at how to control noise and manipulate sound quality to achieve design goals in buildings. Interior designers are key players in this process.

PART

ACOUSTICS

Acoustics is the branch of physics that deals with the production, control, transmission, reception, and effects of sound. **Acoustical design** is the planning, shaping, finishing, and furnishing of an enclosed space to establish an acoustic environment necessary for the distinct hearing of speech or musical sounds. Understanding how we hear sound and how sound interacts with the built environment helps us design spaces that are as acoustically pleasing as they are visually rich.

The quality of the sound within a building depends on many things, some of which originate in the siting and architectural design of the building. Even noises propagated by the building's structure and mechanical systems, as well as those from outside, can be ameliorated by good acoustic design.

Part III, "Acoustics," has two chapters:

Chapter 7, "Acoustic Design Principles," introduces the basic terminology and concepts of sound generation and hearing, and looks at sound sources and paths in general terms.

Chapter 8, "Architectural Acoustics," looks at the acoustical design of buildings, including noise control, airborne and structure-borne sound, sound isolation, and sound transmission between spaces. Acoustical products and applications as well as electronic sound systems are also surveyed.

> The acoustic environment plays an important role in supporting (or disturbing) an overall sense of comfort in many of the spaces we occupy on a daily basis – including both residential and commercial/ institutional spaces...Many design solutions seem to shortchange the acoustical environment. This is partly due to the perceived complexity of architectural acoustics, partly due to lack of coverage of the topic in many architecture programs. Good acoustics is not required by most building codes and is not a key element in the majority of green building rating systems. Nevertheless, providing acceptable acoustical conditions is a

fundamental part of good design practice. (Walter T. Grondzik and Alison G. Kwok, *Mechanical and Electrical Equipment for Buildings, Twelfth Edition*, Wiley 2015, page 1015)

Interior designers are in an excellent position to deal with some of these problems. As designers of office spaces, an awareness of acoustic principles and remedies is essential. Residential design poses its own acoustic problems.

7

Acoustic Design Principles

Our experience of the world is strongly visual, but we are often deeply affected by messages received by our other senses as well. Perhaps the most functionally critical of these is our sense of hearing. Sound in a well-designed space reinforces the function of the space and supports the occupant's experience. A poorly designed acoustic environment hinders both function and enjoyment of the space, and can even damage the health of the user.

INTRODUCTION

> Acoustical design is often neglected or minimized. Although a failure in acoustics is not as fatal as a failure in structures, a failure in acoustics can…be very costly…. Most times a failure in architectural acoustics can be fixed at more modest cost, but it will be an annoyance for the owner and an embarrassment for the architect. (Norbert M. Lechner, *Plumbing, Electricity, Acoustics: Sustainable Design Methods for Architecture*, Wiley 2012, p. 158)

Within a given architectural interior, the acoustical fine-tuning of the space is often the province of the interior designer. Interior designers are concerned with how a space such as an open office or a restaurant functions acoustically. The selection and placement of hard- and soft-surfaced materials and the construction of interior partitions change the way sound is reflected, absorbed, or transmitted.

SOUND BASICS

Sound is essentially a rapid fluctuation in air pressure. It can be defined as a physical wave, as a mechanical vibration, or as a series of pressure variations in an elastic medium. A sound is a range of vibrations to which the human auditory system is specifically sensitive. Vibrations can be transmitted through the air or another elastic medium, including most building construction materials.

Sound is induced through the ear by means of waves of varying air pressure emanating from a vibrating source. In order for sound to exist, there must be a source, a transmission path, and a receiver.

Sound Propagation

Sound waves are mechanical waves, as opposed to the electromagnetic waves of light and electricity. The pressure changes that contain sound information travel in the same direction as the sound wavefront. The energy in a sound wave is capable of moving a great distance, but the medium in which it moves only oscillates in place. In air, it moves forward and backward in the same direction as the longitudinal sound wave.

Amplitude refers to sound pressure, apparent as the distance between a sound wave's maximum compression (its peak) and its maximum rarefaction (its trough). The wave's amplitude is also referred to as its volume, and is perceived as loudness.

Amplitude declines with distance from the source. In an interior space, the sound level from a point source decreases as it reflects off the walls, ceiling, and floor. Eventually, the combined reflections generate a fairly constant sound level.

Within a room in a building, sound waves hit reflecting surfaces. Bouncing off surfaces helps maintain the sound's intensity and audibility at a distance from its source. Reflections also determine how long a sound continues. Areas near walls collect the majority of sound as a result of reflections; these areas are known as **reverberant fields**.

A **free field** is a space free from reflective surfaces or other interferences, such as might occur outdoors. Sound generated by a point source in a free field moves out as a wave front in all directions in an ever-increasing sphere. As it spreads out, the sound intensity diminishes in proportion to the square of the distance from the source. This distribution of sound energy makes it increasingly difficult to hear a sound in open air as the distance from the source is increased.

Sound Waves

Sound energy forms a longitudinal wave comprising the compressions and rarefactions of the air, water, or other material that occur in the direction of travel. Although distortion of the longitudinal wave occurs in the direction of travel, the longitudinal wave is often represented by a sine wave that shows the energy change from positive to negative. (See Figure 7.1)

In theory, sound waves radiate spherically from a point source. In practice, they often originate from sources such as human voice that radiate more strongly in some directions than others. A vibrating object radiates sound waves outward until they hit a surface that either reflects or absorbs them.

WAVELENGTHS AND WAVE FORMS

When sound travels through various media, it is the variation in pressure that travels, not the media themselves. The actual movement of a medium such as air does not have a significant effect on sound transmission.

The distance between the peak of one sound wave and the peak of the next is called its **wavelength**. (See Figure 7.2) This is the distance a sound travels in one cycle.

The form of a sound wave depends on its source. Point sources produce concentric spherical waves. Line sources such as strings on musical instruments produce cylindrical waves. Long vibrating surfaces such as walls produce planar waves.

Wavelengths of audible sounds range from less than an inch (a few mm) for very high pitches to over 50 feet (over 15 m) for very low ones. Their behavior depends on their length and the objects they run into. Acoustic calculations are consequently complex, and will not be dealt with here.

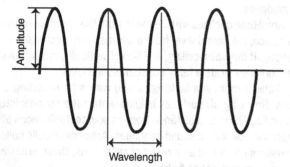

Figure 7.1 Sound wave

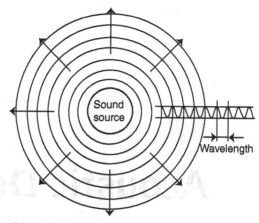

Figure 7.2 Wavelength

Frequency

Whether we perceive a sound as high or low depends on its **frequency**. The peaks in sound waves will pass a stationary point at different rates. A higher pitched sound has peaks that pass at a higher frequency (more frequently), while the peaks of a lower pitched sound pass at a lower frequency (less frequently). The frequency with which these peaks pass a given point is measured as the number of **cycles** completed per second, measured in **Hertz (Hz)**. One Hertz equals one cycle per second, so a wave whose peaks pass at 50 cycles per second has a frequency of 50 Hz.

High-pitched sounds have higher frequencies. High frequency corresponds to a shorter wavelength. Bass notes have lower frequencies; low frequency and long wavelengths go together.

Frequency is an important variable in how a sound is transmitted or absorbed, and must be taken into account in designing the acoustics of a building.

Sound Magnitude

The human ear is sensitive to a very large range of sound power, which is measured in acoustical watts. A source's acoustical magnitude is a measure of the power of sound; its may be termed sound power, sound pressure, or sound intensity. Although there are technical differences between these, we will simply use the term **sound power**.

SOUND POWER

Sound power varies over a very wide range from one source to another. It is expressed in **watts (W) of acoustical power**, the basic unit of acoustical energy. (See Table 7.1)

In theory, sound power is not influenced by the nature of a sound source's surroundings. In practice, acoustic power output depends on the specific environment between the sound source and the listener, and usually varies from location to location within a room.

TABLE 7.1 SOUND POWER EXAMPLES

Sound Source	Acoustical Power
Jet engine	100,000W
Symphony orchestra	10W
Loud radio	0.1W
Normal speech	0.000010W

HEARING

As indicated earlier, in order for sound to exist, there must be a source, a transmission path, and a receiver. The response of an individual to sound involves physiological and psychological reactions, with experience and personal preferences playing a role. The sensitivity of our ears is close to the practical limit for sound reception. The average human ear can withstand the loudest sounds of nature, yet be able to detect the tiny pressures of barely audible sounds. The human hearing range is commonly cited as between 20 and 20,000 Hz, but this varies significantly between individuals and depends on frequency and age.

Human Ear

If you have ever tried to draw someone's ear, you will have noticed that our ears are apparently as individual as our fingerprints. They are small and large, simple and convoluted, smooth and hairy, but all healthy ears have the same parts.

The structures of the ear enable us to collect sound waves, which are then converted into nerve impulses. (See Figure 7.3) The outer ear is a soundgathering funnel. Our outer ear is a more efficient sound-gatherer than the nonexistent external ear in many reptiles and birds, but lacks the collecting and focusing capacity of a cat's or dog's ear. Sound travels from the outer ear through the auditory canal, which is also called the external ear canal, and into the middle ear.

Within the middle ear, sound waves set the eardrum (tympanic membrane) in motion. The middle ear is an air-filled space surrounded by bone and bounded by the eardrum on the outer side and a flexible membrane separating it from the inner ear on the inner side. Its main job is amplification. The vibrations from the eardrum are transmitted by three tiny bones to the inner ear. In the short but intricate journey from eardrum to inner ear, the sound wave is amplified as much as 25 times.

The inner ear is where the sound vibrations are converted to electrical nerve impulses for interpretation by the brain. Rhythmic waves in the inner ear fluid excite a highly delicate organ that is coiled like a snail shell; it's name, cochlea, is from the Latin word for snail. Hair cells at one end respond to sounds at high frequencies, up to 20,000 cycles per second; those at the opposite end respond to low ones, down to 16 cycles per second. The basilar membrane in the cochlea resonates at one end at a frequency of 20 Hz, and at the other end at 20 kHz (kilohertz) establishing the range of frequencies that the human ear can hear.

Vibrations within the ear initiate an electrical impulse that is transmitted to the auditory nerve. These impulses travel to the primary auditory cortex of the brain and ultimately to other brain areas for interpretation as sound.

SENSITIVITY

The human ear tends to be more sensitive to midrange frequencies. The best sensitivity occurs between 500 and 6000 Hz, with maximum sensitivity at 4000 Hz. Generally, a decrease in sensitivity is most noticeable at lower frequencies.

If you are young and your ears are in excellent physical condition, you can hear sounds in the 64 Hz to 23,000 Hz range; you will be most sensitive to frequencies in the 3000 to 4000 Hz range. Very high frequencies may be uncomfortable for young listeners, who may, for example, be very sensitive to the sound of high-speed dental drills.

Our ability to hear upper frequencies decreases with age. By middle age, the typical upper limit is around 10,000 Hz to 12,000 Hz. Upper-range hearing loss is usually more pronounced in men than in women.

Many animals, including dogs, can hear ultrasounds, higher frequencies than humans can hear. For example, beluga whales can hear from 1000 to 123,000 Hz.

Loudness

The way we experience a change in loudness is subjective; it is not related in a linear way to sound power. A sound we perceive as twice as loud as another sound is actually much more than twice as powerful. Perceived loudness depends on sound pressure, the age and health of the listener, the frequency of the sound, and the presence of **masking sound**, the phenomenon when two separate sources of sound are perceived simultaneously and tend to obscure each other.

DECIBELS

The loudness of sounds is measured in a way that relates actual sound intensity to the way humans experience sound, rather than in Pascals (sound pressure) or watts (sound energy). Loudness is measured according to a mathematical logarithmic

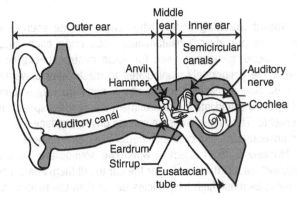

Figure 7.3 Structures of the ear

scale of **decibels (dB)**. A decibel is a unit for expressing the relative pressure or intensity of sounds on a uniform scale from 0 dB for the least perceptible sound to around 130 dB for the threshold of pain. We hear a doubling of sound pressure and intensity not as twice as loud, but as a barely perceptible change.

The decibel scale is also used to represent the energy of sound. The acoustic power generated by a vibrating source can be expressed in watts, but the expression of the full range of human hearing would span 1014 watts. This unwieldy mass of numbers has been condensed into the decibel scale.

The human auditory system will respond to a decibel range from 1 to 130 dB. The decibel scale starts at 0 for the minimum sound intensity or pressure that can be heard. It uses whole numbers, rather than powers of 10 (as logarithms would), so a difference of 10 decibel units represents a doubling (or halving) of loudness.

Our perception of a sound's loudness depends both on the power of the sound and on the distance from the source of the sound to our ear. Every time the sound power is doubled, the actual sound intensity level changes 3 dB. When the distance from the source of the sound is doubled, the sound intensity level changes 6 db. Decibel levels from two sound sources cannot be added mathematically. For example, 60 dB + 60 dB equals 63 dB, not 120 dB. If all this sounds confusing, you may soon get used to relating decibels to sound levels and cease to be bothered by the math. (See Figure 7.4)

The human ear does not perceive all frequencies equally. Because we are not equally sensitive to all frequencies within our audible range, we can hear only certain frequencies at the lowest levels of loudness. Where we are most sensitive, in the range of 3000 to 4000 Hz, we can hear sounds even at −5 dB (technically below the threshold of hearing). The most information in human speech is found between 3000 and 4000 Hz, so we are good at listening for very quiet speech. Our sensitivity drops off at low decibel levels, especially at low frequencies. This is why most stereo amplifiers provide a boost for bass sounds at lower volumes. At the threshold of hearing at 0 dB, we can hear only at 1000 Hz.

The upper limit for loudness is 120 to 130 dB. This level of sound intensity is high enough to produce the sensation of pain in the human ear is called the threshold of pain. At this level, we experience pain in all frequencies.

A-weighted decibels (dBa) express the relative loudness of sounds in air as perceived by the human ear. The A-weighted system reduces the decibel values of sounds at low frequencies to correct for the human ear's lessened sensitivity at low frequencies.

Sound Masking

Studies confirm that poor office acoustics in open plan offices is the number one barrier to the productive use of office space. People are sensitive to sounds that are louder than the background sound, and especially aware of speech that is intelligible over the rest of the sounds in the room.

140 dB
Threshold of pain
Jet engine at 75 feet

120 dB
Jet takeoff at 300 feet
Loudest rock band

105 dB
Tree chipper

95 dB
Popular music group

80 dB
Heavy truck
Average street traffic

70 dB
Conversational speech
maximums

60 dB
Active business office

30 dB
Quiet living room
Empty concert hall

15 dB
Rustle of leaves

0 dB
Threshold of hearing

Figure 7.4 Decibel levels

Sound masking occurs when two separate sources of sound are perceived simultaneously, obscuring each other. When sound masking is used for noise control, background sounds are deliberately manipulated to mask other unwanted sounds. Sound masking introduces a non-intrusive, ambient background sound into the environment that renders speech unintelligible. This helps to ensure speech privacy, reduces stress and absenteeism, and creates a better work environment.

Masking is most effective when two sounds are close in frequency as it is then harder for the ear to tell them apart. Low frequencies mask high frequencies better than the reverse for the same decibel levels.

Background noise that is used for masking unwanted sounds is broadband (containing many frequencies), continuous, and without intelligible information. This helps to cover both lower and higher frequency sounds. Adjustable electronic sound-masking systems can be carefully tailored to combine low-pitched and high-pitched sounds while being generally unnoticeable to people entering the space.

Directivity and Discrimination

Our two ears are separated by our skulls. Our ears and brains are configured to detect the small difference in time it takes sound to reach each ear, allowing us to determine the direction of a sound source. When the distinction is very small, the brain interprets the sound as coming from a point between our ears.

Our ears can pick out specific sounds to which we want to pay attention, but more frequently it combines sounds distinct from each other in frequency and phase as chords in music, for example. Most sounds are actually complex combinations of frequencies. Musical tones combine fundamental frequencies with **harmonics (overtones)**. A trained conductor can pick out one single instrument in a 120-piece orchestra. Amazingly, we have the ability to pick out one voice in background noise much louder than that voice, a phenomenon known as the cocktail party effect.

SOUND SOURCES

Sound may enter a building from outside or be produced within a building's interior. Important building sound sources include speech, music, and noise. Vibrations from mechanical equipment also create acoustic problems.

Speech

The vibration of our vocal cords produces human speech. These vibrations are modified through the throat, nose, and mouth. Most speech is concentrated in the 100 to 600 Hz range. Harmonics outside of this range help give an individual human voice its characteristic sound and specific identity.

Most of the information in speech is carried in the upper frequencies, while most of the acoustic energy is in the lower frequencies. For sounds of equal energy, the human ear is less sensitive to low frequencies than to middle and high frequencies. Higher frequencies carry sound with greater a greater sense of direction, and can be heard around a partial barrier more easily. High frequencies are the most easily absorbed.

Music

Our appreciation of musical sounds results from a combination of physiological and psychological phenomena. Musical sounds are usually of longer duration than speech sounds. Especially in instrumental music, they encompass a much broader range of frequencies and sound pressures than speech does.

Musical instruments often produce very high frequencies in high-pitched overtones. Some large pipe organs produce pitches with frequencies near the extreme lower end of the hearing range.

Musical sound often depends on **resonance**, which occurs when sound is intensified and prolonged by **reverberation**, the persistence of sound within a space caused by multiple reflections after its source has stopped. Sometimes a vibration in one object produces sympathetic vibrations of exactly the same period in a neighboring body.

Noise

Noise is simply defined as any unwanted sound. What constitutes noise is a subjective judgment; one person's noise is another person's music. Children yelling a they run around the yard playing is reassuring and welcome to the parent who is keeping track of their whereabouts, but is a disturbing noise to the neighbor trying to get some sleep before working the night shift.

ANNOYANCE

The amount of annoyance produced by unwanted sound is subjective, psychological, and proportional to the loudness of the noise. The most annoying sounds are high frequency rather than low frequency and intermittent rather than continuous noise. Pure tones are more conspicuous than broadband sounds. When a sound is moving and not locatable rather than from a fixed location, it tends to distract us. Finally, sounds bearing information are harder to ignore than no-sense noise. Speech that is loud enough to be audible but not enough to be intelligible is particularly annoying.

The types of sound that can constitute noise are extremely varied. They may include speech or music and natural sounds like wind and rain. We are surrounded by mechanical and building system noise from motors, compressors, fans, and banging pipes.

HEARING PROTECTION

Prolonged exposure to high noise levels can cause physical damage. Continual exposure to noise levels as low as 75 dB can contribute to headaches, digestive problems, tachycardia, high blood pressure, anxiety, and nervousness. Even lower levels can cause sleeping problems. Most experts consider eight hours a safe upper limit of exposure to levels of 85 dBA. US Occupational Safety and Health Administration (OSHA) industrial regulations set limits for exposure to continuous noise, with limits on duration of louder sounds and requirements for hearing protection. (See Figure 7.5)

Figure 7.5 Hearing protection

Vibration

Vibration is pressure variation perceived through touch. Electrical and mechanical building systems produce both vibration and noise. Vibrations in buildings often have frequencies around 20 Hz, just below the range of human hearing. These vibrations are generally unwanted and can be particularly disturbing to some individuals.

SOUND PATHS

Sound waves travel at different velocities depending on the medium they are traveling through. Sound travels through air at around 1087 feet (331 m) per second at sea level. Sound travels through water more rapidly than through air, at around 4500 feet (1372 m) per second.

Attenuation

Sound energy, like heat energy, can be absorbed or reflected by an object. Sound energy lessens in intensity as it disperses over a wide area. **Attenuation** is the decrease in energy or pressure for each unit area of a sound wave. Attenuation occurs as the distance from the source increases as a result of absorption, scattering, or spreading in three dimensions.

Reflected Sound

When a sound wave strikes a relatively large surface, a portion of the sound energy is reflected (like light from a mirror) and a portion is absorbed. The harder and more rigid a surface the sound wave strikes, the more sound is reflected. **Reflected sound** leaves the surface at an angle equal to the angle at which it strikes it. (See Figure 7.6)

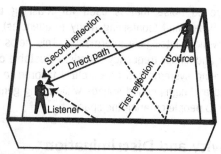

Figure 7.6 Reflected sound

REVERBERATION

Reverberation is the persistence of sound after the source of the sound has ceased, as a result of repeated reflections. The **reverberation time** is the time it takes in a particular space for a sound to drop 60 dB. The reverberation time is directly proportional to the volume of the space, and inversely proportional to the absorption of the surfaces.

The sound in a room is a combination of direct sound from the source and reflected sound from walls and other obstructions. (See Figure 7.7) Our ears sense reverberation as a mixture of previous and more recent sounds. The reverberation time is longer in a room with a larger volume as the distances between reflections are longer. When sound-absorbing materials are added to a space, the reverberation time decreases as sounds are absorbed.

More information about reverberation and reverberation time is found in Chapter 8, "Architectural Acoustics."

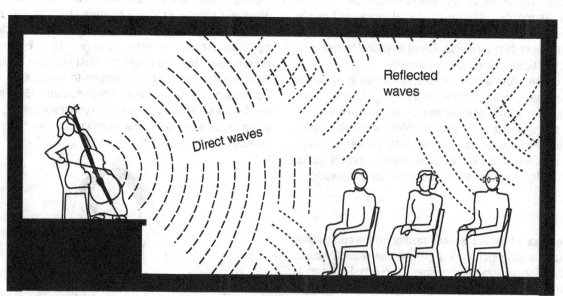

Figure 7.7 Reverberation

When too many reflected sounds are overlaid in an enclosed space, the resulting surface reflections can produce confusion over the source of the sound. If the reflected path is longer than the direct path by about 65 feet (20 m) or more, it may produce an echo. High-frequency sounds closer to the source give the best clues, because they travel in a relatively straight line. Low frequency sounds are difficult to localize, due to the large size of their wavelengths compared to the dimension between our two ears. In addition, low frequencies tend to mix with their own reflections.

An interior designer can control the quality of the reverberating sound by modifying the amounts of absorptive or reflective finishes in a space.

The reverberation time of a room should be appropriate to the use of the space. The reverberation of sounds in lecture halls, theaters, houses of worship, and concert halls sustains and blends sounds, making them much smoother and richer than they would be in open air. Short reverberation times are best for speech, as they allow clarity for consonant sounds. However, some reverberation enriches a speaker's voice, and gives the speaker some sense of how well the voice is carrying to the audience.

Music often benefits from longer reverberation times that extend and blend the sounds of instruments and voices. Music tends to sound dead and brittle with too short a reverberation time, but loses clarity and definition when the reverberation time is too long.

DIFFUSION

Diffusion occurs where sound is reflected from a convex surface. (See Figure 7.8) Convex surfaces scatter sound, reinforcing sound levels in all parts of a room. Diffusion results in the sound level remaining fairly constant throughout the space, a very desirable quality for music performance.

Figure 7.8 Diffusion from a convex surface

DIFFRACTION

Diffraction is the physical process by which sound passes around obstructions and through small openings. When a sound wave strikes an object smaller than or similar in dimension to its wavelength, it is diffracted, and the wave is scattered around the object. Diffraction facilitates the ability to be heard beyond a barrier, and is measured by the amount that airborne sound waves are bent by moving around an obstacle in their path.

When sound reaches the edge of a wall, sound waves will diffract (bend). Long, low frequency waves bend more than shorter high-frequency ones. Although much of a sound wave is blocked by a small opening, the portion that does get through establishes a new wave front at a lower intensity than the original source. (See Figure 7.9) For a small hole, short wavelengths (high frequencies) are attenuated less than long wavelengths (low frequencies). A small hole therefore can block long wavelengths better than short wavelengths.

Attenuation of diffracted sound depends on the frequency, type of source, and dimensions of the barrier. (See Figure 7.10) The best location for a barrier is either very close to the source or close to the receiver. The worst position is halfway between source and listener. A massively thick barrier is only slightly better than a moderately thick one, so there is a practical limit to thickness. Absorptive material on the source side of the barrier will reduce noise reflected back to the source, but will not help the receiver very much.

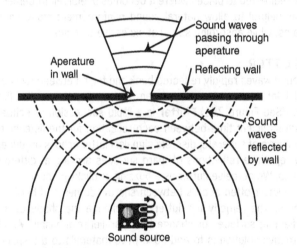

Figure 7.9 Diffraction

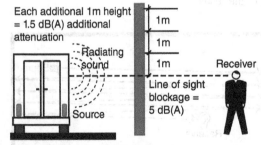

Figure 7.10 Highway noise barrier

Source: Redrawn from US DOT Highway Traffic Noise barrierheight.gif

Natural Sound Reinforcement

Natural (as opposed to electronic) **sound reinforcement** is the amplification of the sound being heard from various reflections as well as directly from the source. Covering the ceilings of meeting rooms, classrooms, and auditoriums completely with sound absorbing material eliminates the potential for useful sound reinforcing reflections off the ceiling, and may result in inadequate sound levels in the rear of the room. You may be able to avoid having to install an electronic sound-reinforcing system by leaving the center of the room as a reflecting surface.

ECHOES

Echoes result when repetitions of a sound are produced by reflection of sound waves from a surface, loud enough and received late enough to be perceived as distinct from the source. Sounds heard twice, with the second hearing arriving 0.07 seconds (70 milliseconds) or more after the initial sound, are likely to be understood as related, but the second sound appears as a separate, distinct sound, an echo. This can be a problem in very large halls. Auditoriums frequently produce echoes between the back wall and the ceiling above the proscenium. Echoes may occur when parallel surfaces are more than 60 feet (18 m) apart.

Echoes can be avoided by careful planning of the room's geometry, or by the selective use of absorptive surfaces. Absorbing the sound energy in echoes wastes energy that could be redirected to places where it becomes useful reinforcement. It is helpful to allow natural sound reinforcement along short paths, while absorbing sound at excessive distances.

FLUTTER

Sound waves rapidly reflected back and forth between two parallel flat or concave surfaces can produce an effect called flutter. (See Figure 7.11) **Flutter** is a rapid succession of echoes with sufficient time between each reflection for the listener to be aware of separate, discrete signals. Flutter echo is produced by repetitive reflections off hard surfaces arriving at different times. We perceive flutter as a buzzing or clicking sound.

Flutter often occurs between shallow domes and hard flat floors. The remedy for flutter is to change the shape of the reflecting surfaces or change their parallel relationship. An alternative solution is to add absorptive materials to the space. Which of these is the best answer depends on the reverberant requirements of the space, the cost of corrections, and the aesthetics of the result.

STANDING WAVES

A **standing wave** is a steady, pure tone between two highly reflective parallel walls. Standing waves operate on the same principle and have the same cause as flutter, but are heard differently. Standing waves are perceived as points of quiet and of maximum sound within a room. Certain frequencies of a voice or music are exaggerated as they bounce back and forth repeatedly between opposite parallel walls. When the walls are exactly one-half wavelength apart, the tone is very loud near the walls and very quiet halfway between them, as the waves cancel each other out in the center of the space.

Standing waves are a serious problem only in rooms that are small with respect to the wavelengths generated in them. To avoid standing waves, the smallest dimension for a room for music should be greater than 30 feet (9 m), and for speech greater than 15 feet (4.6 m).

Standing wave problems in rooms with parallel walls are improved by slightly tilting or skewing two of the walls, or by adding acoustic absorptive material to one of them. Rooms for music rehearsal and broadcast studios often have nonparallel walls, and undulating ceilings can also help. The proportions of the room can minimize the effect, which is especially noticeable for bass frequencies.

RESONANCE

Resonance is the accentuation of a particular frequency, which can be a problem in spaces for music, where it may make one instrument sound louder than the others. Resonance problems can be avoided by using geometrical calculations to design room proportions. Using nonparallel walls and undulating ceilings often helps.

FOCUSING AND CREEP

Focusing occurs when sounds reflected from a concave surface converge at a single point. (See Figure 7.12) The sound is greatly reinforced at the focal point and is less loud elsewhere. Spaces with concave domes, vaults, or walls focus reflected sound into certain areas of rooms. Focusing deprives some listeners of useful sound reflections and causes intense sound spots at other positions.

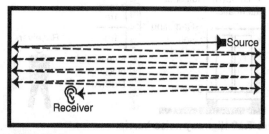

Figure 7.11 Flutter

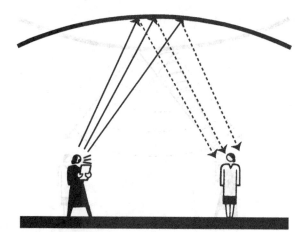

Figure 7.12 Focusing

Figure 7.13 Creep

The reflection of sound along a curved surface from a source near the surface is called **creep**. (See Figure 7.13) The sound can be heard at points along the surface but is inaudible away from the surface. A space with concave surfaces can become a whispering gallery, a room in which two people can stand at two related focal points along a curved surface and hear each other's whispers with startling loudness and clarity, while remaining unheard by other people in the space.

Absorbed Sound

When sound strikes a boundary or any sizeable surface, part of the incident sound energy is absorbed, part is reflected, and part is transmitted. Sound absorption is a key tool in the interior designer's acoustic toolbox.

ABSORPTION

Sound absorption depends on both the area and the absorption characteristics of the material in question. When sound passes through a porous material, some of its energy is converted to heat by friction and is lost as air is pushed through small pores and passageways. Materials do not reflect or absorb sound perfectly; there is always at least an insignificant amount absorbed by any reflecting material, or reflected by an absorbent one.

The effects of materials on sound are different at different frequencies. When selecting materials, it helps to divide them by how they absorb low, medium, and high frequencies.

Greater thicknesses are required to absorb lower frequencies. Soft, porous materials, such as wood, textiles, furnishings, and people, absorb a large part of the energy that strikes them. Deep, porous upholstery absorbs most sound from middle frequencies upward. Smooth, dense, painted concrete or a plastered wall absorbs less than five percent of incident sound, and reflects sound very well. Thin fabric wall coverings absorb only frequencies near or above the top of the audible range of the human ear. Padded carpet and thick drapery absorbs the majority of sound waves in a higher proportion of the audible range.

ABSORPTION COEFFICIENT

The **absorption coefficient** is the ratio of sound energy absorbed to sound energy impinging on the surface of a material. (See Table 7.2) The absorption coefficient depends on the frequency and **angle of incidence** of the sound. It ranges from 0 (total reflection) to 1.00 (total absorption). Absorption coefficients are determined experimentally by manufacturers.

An understanding of basic acoustical principles plays a significant role in helping interior designers create acoustically pleasing spaces. Although part of the absorbed sound energy is dissipated by air movement in the pores of a material, a large portion is usually transmitted through the material. For example, although fiberboard and acoustic tile are both good sound absorbers, neither is a good sound insulator that will prevent sound transmission between spaces. In Chapter 8 we explore how sound is transmitted through buildings and how interior designers can use materials and equipment to control it.

TABLE 7.2 SAMPLE ABSORPTION COEFFICIENTS

Material	Frequencies (Hz)			
	250	**500**	**1000**	**2000**
Acoustical ceiling tile	0.15 to 0.95	0.35 to 0.95	0.45 to 0.99	0.45 to 0.99
Heavy carpet on concrete	0.06	0.14	0.37	0.60
Heavy carpet with pad	0.26	0.48	0.52	0.60
Gypsum board	0.10	0.05	0.04	0.07
Upholstered seating with audience	0.74	0.88	0.96	0.93
Unoccupied upholstered seating	0.66	0.80	0.88	0.82
Heavy drapery	0.35	0.55	0.72	0.70

8

Architectural Acoustics

Architectural acoustics, sometimes referred to as room acoustics or building acoustics, is the branch of acoustics concerned with achieving good quality sound within a building. In Chapter 8 we explore how sound is transmitted through buildings and how interior designers can use materials and equipment to control it.

INTRODUCTION

Architects and engineers design building spaces, structures, and mechanical and electrical systems to meet acoustical needs. The architect is responsible for any decision affecting both acoustics and other architectural requirements. Acoustical consultants can identify possible solutions for satisfactory noise control. The architect uses consultants' advice to consider which acoustical solution can be most successfully integrated with solutions to other building demands.

> Generally, it is the architect's role to recognize a potential noise problem in a proposed building and take steps to solve it. An acoustical defect that appears in the completed building cannot be readily corrected, resulting in inadequate acoustical quality. (Vaughn Bradshaw, *The Building Environment: Active and Passive Control Systems, Third Edition*, Wiley 2006, page 411)

Architectural acoustics address four areas of concern: room acoustics, sound isolation, mechanical equipment, and sound systems. Interior designers are primarily concerned with

room acoustics, including the acoustical environment within a space and the isolation of sound within and between spaces.

History

Awareness of how sound acts in architectural spaces has a long history. In *The Ten Books of Architecture* (Morris Hicky Morgan translation published by Harvard University Press 1914; republished by Dover Publications, Inc., 1960), the Roman architect Vitruvius wrote about how reflections off hard materials could make it difficult to understand speech, due to problems with echoes and reverberation times. He was also concerned with minimizing sound from outside the building.

The history of modern acoustics begins with the efforts of Wallace Clement Sabine to remedy the acoustic problems of the Fogg Art Museum Lecture Hall (built 1895) at Harvard University in Cambridge. In order to study how sound worked in the space, Sabine and his two assistants dragged hundreds of upholstered seat cushions from the nearby Sanders Theater each night and back again in time for morning classes.

From his efforts, Sabine developed reverberation equations and absorption coefficients for many common building materials. He discovered that the reverberation time of a room is directly proportional to the cubic volume of the room, and inversely proportional to the sound absorption provided at the room's boundary surfaces and by the room's furnishings. His equation uses the simple dimensions of the room and absorption coefficients of materials to determine the acoustic effect of the space, offering an easy method for architects to determine favorable room proportions and treatments.

Figure 8.1 International Symbol of Access for Hearing Loss
Source: Redrawn from 2010 ADA Standards for Accessible Design, Figure 703.7.2.4

Acoustic Codes and Standards

Some building codes have recently added limits on noise. City and town regulations or zoning bylaws also set standards, regulations, criteria, and ordinances for noise.

Several organizations set standards for building industry acoustic analysis and test methods. ASTM International has established methods for measuring, analyzing, and quantifying noise. American National Standards Institute (ANSI) sets scientific parameters and criteria used in acoustic analysis. ASHRAE determines sound levels for mechanical systems in buildings.

ASSISTIVE LISTENING SYSTEMS

An assistive listening system uses a hard-wired or wireless system to transmit an audible signal. The 2010 ADA Standards for Accessible Design requires assistive listening systems in assembly areas where audible communication is integral to the use of the space, and specifies the types and placement of systems. (See Figure 8.1) Except in courtrooms, assistive listening systems are not required where audio amplification is not provided. In transient lodging accommodations, such as hotels and motels, a certain number of rooms must be accessible to people with hearing impairments.

Assembly areas with 50 or fewer seats require at least 2 receivers, which must be hearing aid compatible. An additional receiver is required for each 25 seats up to 500 seats, with more for larger quantities of seats. Selecting or specifying an effective assistive listening system for a large or complex venue requires the assistance of a professional sound engineer.

ACOUSTIC DESIGN

The amount of necessary acoustic treatment required can be reduced by limiting the sources of noise. When designing for an existing building, the architect and interior designer must first define the character of the sound problem. For new buildings, they have to imagine what noise sources can be anticipated. All parts of the building and its surfaces are potential paths for sound travel. Noise sources should be placed as far as possible from quiet areas. The internal acoustics of individual rooms must be reviewed. The next step is design of the internal acoustics of specific rooms. Structural precautions that must be taken to reduce noise penetrations are an additional step.

Design Process

The acoustic design of the building should be integrated with other architectural requirements. By carefully planning the building's siting and structure, the architect can reduce noise penetration into the building. The overall building design and function ought to be reviewed in terms of desirable acoustic qualities.

ACOUSTIC CONSULTANTS

For special acoustic issues, an acoustic consultant should be brought into the process at the earliest possible time. Acoustic consultants are most commonly called in for buildings where loud noise is a special problem, or where the quality of interior sound is critical. Music and performance spaces, educational spaces and libraries, and all types of residential structures require good acoustic design. Other commercial, institutional, and industrial buildings also benefit from the expertise of an acoustic designer.

Acoustic consultants play a role in selection of materials and the detail of construction components. They also influence the selection and use of interior surface materials. Their work has direct implications for the interior designer. Acoustic consultants also design and specify sound and communications systems, and detail components for noise and vibration controls in mechanical systems.

ACOUSTIC MODELING

Computer software can predict in advance what the acoustic properties of a space will be. Software can model a performance space with electronic equipment to simulate what music will sound like from any location in the hall. This allows the designers to try out the hall before construction, propose changes that address problems, and then hear the results. Today, interactive acoustic modeling is being developed that allows real-time exploration of the acoustic environment.

Room Acoustics

How sound behaves in a given room depends on the shape, size, and proportions of the room. The amount of sound of various frequencies that are absorbed, reflected, and diffracted from the room's surfaces and contents also determine acoustic effects. The room's shape determines the geometry of the paths along which sound is reflected, and can alter the sound quality, sometimes in unexpected ways.

How much sound energy is absorbed and how much is reflected by a surface has a significant effect on what one hears within a space. Where little sound is absorbed and much is reflected, sounds are mixed together. When steady sounds are mixed together, they reverberate, resulting in a noisier space. Speech becomes less intelligible, but music may sound better in a reverberant space. Where much of the sound energy is absorbed and little is reflected, the room sounds quiet for speech but may sound dead to music.

Attenuation reduces sound energy by separating a sound source from the listener. Attenuation can be increased by enclosing the source to isolate the sound, absorbing the sound with materials that change the sound energy to heat, or canceling sound waves by electronic means.

ACOUSTIC FIELDS

What you hear at any point in a room is a combination of sound that travels from the source directly to your ear and sound reflected from the walls and other obstructions. If the reflections are so large that the sound level becomes uniform throughout the room, you have what is termed a diffuse acoustic field.

Most enclosed spaces have three **acoustic fields**. (See Figure 8.2) The area within one wavelength of the lowest frequency of sound produced in the room is called the near field. Sound produced in a free field attenuates as it spreads out without interruption. The reverberant field is the area closest to large obstructions such as walls, where conditions approach a diffuse acoustic field. In a reverberant field, sound waves are multiplied and interwoven. Applying absorptive materials to the boundaries of a reverberant field decreases the loudness of reverberated sound waves.

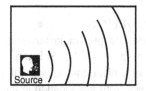

(a) Free field (outdoors)

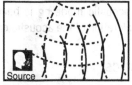

(b) Reverberant field (indoors)

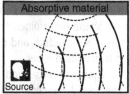

(c) Reverberant field with absorption

Figure 8.2 Acoustic fields

ACOUSTIC REFLECTIONS AND OPENINGS

Ideally, every listener in a lecture hall, theater, or concert hall should hear the speaker or performer with the same degree of loudness and clarity. This is not possible using only direct sound paths from the source to the listener, so the acoustic designer reinforces desirable reflections and attempts to minimize and control undesirable ones to even out the sound in the space. Designers usually only consider the first acoustic reflection, as the second and third times the sound bounces is less noticeable.

When sound reflects off a hard polished surface, the result is termed a **specular** (mirror-like) reflection. For a surface to reflect a sound wave, the reflecting surface must be larger than the sound wave. Acoustic designers sometimes place a reflecting panel above theater seats, sized to a minimum of one wavelength at the lowest frequency they are considering, to bounce the sound from the stage to the audience. The short wavelengths of high frequency sounds are reflected by hard surfaces, and can proceed unaltered through openings with little or no change.

Low-frequency sounds with long wavelengths are not reflected by small surfaces, but may be diffracted (bent) when they pass through openings in walls such as doors or windows. Low-frequency waves are also diffracted by recesses, surface protrusions, and smaller combinations of reflective and absorptive materials. Low-frequency sound can form a circular wave front at an opening, and the opening may then seem to be the source of the sound that spreads from that point.

BUILDING NOISE CONTROL

Noise affects design decisions regarding building siting, space planning, exterior and interior material selection, and natural ventilation (windows open for natural ventilation also let in outside noises). Mechanical equipment continues to be a source of noise, although this may decrease with energy efficient design. Productivity in offices is affected by noise generated as the space between occupants decreases and open office areas with minimal divisions become the norm. Noise is a major complaint of residents in multifamily housing.

There are three basic ways to control noise in a building:

1. Reduction at the source through proper selection and installation of equipment
2. Reduction along the paths of transmission through proper selection of construction materials and construction techniques
3. Reduction at the receiver through acoustic treatment of relevant spaces

Controlling Exterior Noise

As buildings open up for views and ventilation, outside noise comes in. Traffic, construction, industrial plants, and sports facilities all generate noise. On-site noises include children's play areas, refuse collection, and delivery or garage areas. Sound can also be reflected from other buildings.

Solid exterior barriers must be close to either the source or the receiver to be effective, as sound may pass over barriers at in-between locations. Window and door openings can be oriented away from the sources of undesirable noise; multiple glazing of windows also helps.

Interior planning to deal with outside noise sources includes grouping quiet rooms in areas remote from the noise sources as well as clustering noisy areas together and isolating them from clusters of quiet areas.

The interior of the building can be screened from outside noise sources by using mechanical, service, and utility areas as sound buffers. Activities with higher noise levels should be located on the noisier side of the building.

Building materials and construction assemblies designed to reduce transmission of airborne and structure-borne sound help control both exterior and interior noise sources. The walls, floor, and ceiling of the protected room should be heavy and airtight.

Weatherstripping on windows and doors will reduce wind noises and also cut the transmission of outdoor noises into the building, in addition to reducing heat loss. Rain and sleet noises can be reduced with heavier roof and window construction.

Controlling Interior Noise

The noise inside a building comes from the activities of the building's occupants and the operation of building services. As we have already seen, additional sound comes in from outside the building.

EQUIPMENT NOISE

The first step in quieting machine noise is to select quiet equipment and install it away from inhabited parts of the building. Noisy equipment can be mounted with resilient fittings and housed in sound-isolating enclosures.

Laundry machines, mixers, bins, chutes, and other machinery with sheet metal enclosures that vibrate can create a lot of noise. The vibration can be dampened by permanently attaching a layer of foam to the vibrating metal, which converts the noise energy to heat. Adding a heavy limp material to the outside of the foam further reduces the noise. (See Figure 8.3)

Using flexible joints in all pipes and ducts connected to the machine breaks the connection from the vibration source to the building structure. Flexible conduit connections are used for all motors, transformers, and lighting fixtures with magnetic ballasts.

The motors and controls of elevators and escalators are localized sources of noise. If the spaces around them are located judiciously, their noise should not be a major problem.

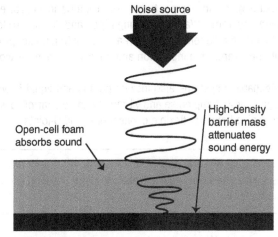

Figure 8.3 Quieting equipment

MECHANICAL SYSTEM NOISE

A building's mechanical equipment has many noise-producing components. The **air handling system** includes fans, compressors, cooling towers, condensers, ductwork, dampers, mixing boxes, induction units, and diffusers, all of which can either generate noise or carry it to other locations.

See Chapter 14, "Heating and Cooling," for more information about mechanical equipment.

Specifying quieter equipment and reducing equipment use with passive design both help reduce noise. Equipment noise level and vibration transmission can be partially controlled by mechanical isolation, shields, baffles, and acoustical liners.

Air turbulence generates noise that increases at sharp bends in ductwork. Lining ducts helps minimize crosstalk between rooms. Gluing damping material on the outside keeps thin metal duct walls from resonating. Separating adjacent ducts as much as possible helps as well. Enclosing ducts behind sound-insulating construction or firmly attaching them to heavy walls can help. Grilles, registers, and diffusers can be selected to minimize noise output.

The building architect works with the mechanical engineer to design heavy construction to enclose rooms with mechanical equipment. Mechanical rooms should be located to avoid equipment sounds intruding into interior occupied spaces or neighboring structures. Equipment spaces should be separated from spaces with critical acoustic requirements by as many barriers as possible. Sensitive spaces include executive offices, conference rooms, sleeping areas, theaters, auditoriums, and worship spaces.

PLUMBING SYSTEM NOISE

The piping for a building's plumbing system can also be a source of noise, including both the normal sounds of water rushing through uninsulated pipes and from **water hammer** (knocking noise in a pipe turned off rapidly) in improperly designed systems. Walls containing pipes and flushing toilets should not be adjacent to quiet areas. To contain noise, pipes should be wrapped in insulation and covered with impervious jackets.

Mechanical systems also include pumps and liquid flowing through piping. Pump noise and vibration can be controlled with resilient pipe hangers, flexible connections, and U-joints.

Interior designers can plan the location for noise-sensitive areas such as bedrooms away from walls with potentially noisy plumbing.

Background Noise

Background noise is any noise other than those sounds that an occupant wants to hear. The degree of noise reduction required depends on the difference between the sound level produced at its source and the level desired at the listener's position. Unnecessary levels of sound reduction add to building costs and can produce an undesirably low background noise level. Consequently, it is important to establish the level of tolerance of the listeners.

Considerations for optimal background noise level include potential hearing damage, interference with speech, and level of annoyance. Critical listening tasks such as music and theatrical performance require special attention. Sudden noises in very quiet surroundings, slight noises that nonetheless disturb sleep, and even low-level conversations in quiet spaces can be sources of annoyance.

RATING BACKGROUND NOISE LEVELS

There are a number of different systems that rate background noise levels. Noise criteria curves rate indoor noise including noise from equipment. Noise criteria (NC) values define appropriate noise levels in decibels (dB) for various spaces. (See Table 8.1) The NC value represents a noise level at certain frequencies. It applies to constant noise that the brain tends to ignore, but not to sudden noises.

TABLE 8.1 RESIDENTIAL AND OFFICE NC RATINGS

Type of Room or Space	Recommended NC Rating Level
Apartment house	25 to 35
Private homes	20 to 30
Hotel/motel guest rooms	25 to 35
Private offices	30 to 35
Open plan offices	35 to 40
Conference rooms	25 to 35

CONTROLLING BACKGROUND NOISE

The first action to take in order to meet appropriate background noise levels is to eliminate or reduce the source of the noise within the space. Next, increase sound absorption within the space and provide good sound isolation by reducing transmission of noises from elsewhere into the space. Other steps include lowering the power of the outside noise source, increasing the separating distance, and/or adding sound absorption in the source space. Modifying the source or transmission paths and relative positions of the source and listener can also help.

SOUND TRANSMISSION

Sound transmission involves the transfer of sound from one part of a building to another. A primary strategy for minimizing sound transmission involves high-quality, airtight construction using acoustical sealant at all edges and joints. Materials should have high mass and be limp rather than stiff, with damping materials on the side away from the noise source. Walls should be full height from the top of one floor slab to the underside of the next for best performance. Double walls with an air space perform better than single ones.

Airborne and Structure-Borne Sound

In practice, all sound transmission involves both airborne and structure-borne sound. **Airborne sound** originates in a space with any sound-producing source. Airborne sound changes to structure-borne sound when the sound waves strike the room boundaries, but is still considered airborne because it originated in the air. **Structure-borne** sound is energy delivered by a source that directly vibrates or hits the structure.

Airborne sound is usually less disturbing than structure-borne sound. The initial energy is usually very small and attenuates rapidly at the room's boundaries.

When airborne sound hits a partition, it can make the partition vibrate, generating sound on the other side. (See Figure 8.4) The sound will not pass through the partition unless an air path exists. If the partition is airtight, then the sound energy may cause the structure itself to become a sound source by vibrating the partition. The partition vibrates mostly

Figure 8.4 Sound passing through partition

in the vertical plane, but also causes some energy to pass into the floor and ceiling, resulting in some structure-borne sound.

AIRBORNE SOUND

Airborne sound changes directions (diffracts) easily. Low-frequency sounds are the most flexible, and can get around barriers. **Sound leaks** can occur at any air passages such as keyholes, cracks around doors or windows, or gaps between walls and floors that form acoustic bridges.

To control airborne sound, locate the source of the noise as far as possible from the listener. Within a room, porous materials absorb sound and inhibit its reflection. Some sound energy is attenuated, but some will pass through a porous material, with thicker materials attenuating more sound. Absorptive materials on the inside surface of a space will reduce sound levels by about 6 dB at most.

STRUCTURE-BORNE SOUND

Structure-borne sound involves energy delivered by a vibrating or impacting source directly contacting the building's structure. (See Figure 8.5) Structure-borne sound generally has a much higher initial energy level and travels much faster than airborne sound. It attenuates less as it travels through the structure, and tends to disturb large sections of the building. With structure-borne sound, the entire structure becomes a network of parallel paths for the sound. Although it radiates very little from a massive structure, it can still be annoying. When it meets a large mass, the mass minimizes the vibration in that direction.

Continuity of the building structure is essential to structural stability, but makes controlling impact noise more difficult. Even when the structure is set in motion by direct contact with vibration or impact, a solid sound path usually terminates by radiating airborne sound. (See Figure 8.6)

Partial solutions fail when sound finds a **flanking path** (the acoustic path of least resistance). Adding mass does not usually block structure-borne sound, especially in buildings with long spans. The floor becomes a diaphragm, improving structure-to-air noise transfer efficiency like a drumhead. Exposed structural ceilings further reduce the attenuation that would occur in a plenum above a suspended ceiling. As most structure-borne sound is carried by floor structures, the sound radiates up and down into the rooms above and below.

DISCONTINUOUS CONSTRUCTION

Discontinuous construction improves both airborne and structure-borne sound transmission problems. (See Table 8.2)

Figure 8.5 Structure-borne sound

Figure 8.6 Vibration and structure-borne sound

TABLE 8.2 DISCONTINUOUS FLOOR CONSTRUCTION

Floor Type	Reduction Compared to Bare Concrete
Thin composition tiles on concrete floor	2 to 5 dB
Cork tile 5/16" (8 mm) thick	Around 10 dB
Thick carpet	More than 20 dB
Wood floor finish on sleepers	6 to 7 dB
Wood floor with resilient strips of mineral wool or fiberglass under battens	Around 12 to 14 dB

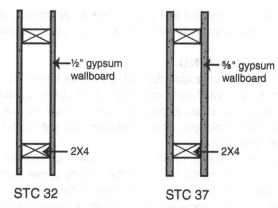

Figure 8.8 Wall construction STC samples

When a wall is divided into two separate layers, vibrations are not easily passed from one layer to the other. Two wall layers generally decrease sound transmission more than a single wall equal to their combined thicknesses.

Layers within walls can be separated with resilient clips or rubber spacers. Layers should preferably be dissimilar, for example, by using different numbers of gypsum wallboard sheets. Insulating the wall cavity also helps.

Measuring Sound Transmission

Sound transmission is measured by transmission loss and by sound transmission class.

TRANSMISSION LOSS (TL)

Transmission loss (TL) is a measure of the performance of a building material or construction assembly in preventing transmission of airborne sound. Transmission loss rates the difference in the sound pressure level of acoustic energy between the incident side and the opposite side of the construction. It is equal to the reduction in sound intensity as it passes through the material or assembly, as derived in controlled laboratory tests. TL is measured using the decibel scale.

A wall's TL indicates its sound-insulating quality. The TL of a wall is related to the wall's physical characteristics, which include **mass** (heaviness and density), **rigidity** (layers and air spaces), absorbency of materials, and method of construction and attachment. (See Figure 8.7) In general, the TL is greater

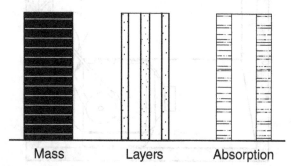

Figure 8.7 Factors affecting transmission loss

for denser, heavier construction, which is more difficult for sound energy to set in motion.

SOUND TRANSMISSION CLASS (STC)

The **sound transmission class (STC)** is a rating of the performance of a building material or construction assembly in preventing transmission of airborne sound. The STC measures the average transmission loss over a wide range of frequencies, adjusted for the sensitivity of the human ear.

STC ratings are typically used for interior walls. (See Figure 8.8) The higher the STC rating, the greater is the wall's sound-isolation value. An open doorway has an STC value of 10. Normal construction has STC ratings from 30 to 60. Special construction is required to achieve an STC rating over 60.

Sound barriers with the same STC can perform quite differently. Increasing mass typically increases the STC rating. Adding acoustic insulation can increase the rating by 6 dB or more. Other effective techniques include resilient mounting and staggered studs.

Without acoustical sealant properly installed, a wall with nominal STC 50 will perform at around 30 STC. Airtight seals are required for acoustic control at doors, pipe and duct penetrations, and all construction joints.

Double glazing and greater glazing thicknesses increase the STC rating for windows and doors. The frame should be caulked, with gasketing all around.

To maximize STC ratings, doors should be either solid core or hollow core filled with a sound-absorbing material. They should not be undercut or have grilles or louvers. A raised sill provides a better seal.

Noise can find a flanking path when partitions reach only up to a suspended ceiling. (See Figure 8.9) Partitions should ideally extend to the structural slab above. When this is not possible, a large amount of sound-absorbing material can be used above the suspended ceiling.

The US **Federal Housing Administration (FHA)** uses STC ratings to specify grades of construction to limit sound transmission. Sound isolation requirements are generally divided into requirements for walls and for floors.

Figure 8.9 Less than full height partitions

SOUND ABSORPTION

A room's ability to absorb sound depends on its size and geometry, the hardness of it surfaces, the frequencies of the sound, and the distance of listeners from the source of the sound. The acoustic treatment of a space starts with reducing the source of the noise as much as possible, followed by control of unwanted sound reflections. Sound will be tempered by how much absorptive material it encounters in the room, and by how much of it can be distinctly heard above any background noise. (See Figure 8.10) **Speech privacy** is another major acoustic concern for the interior designer. Sometimes it is also necessary to decrease or increase reverberation time for sound clarity and quality.

Reducing Acoustic Energy

Noise is reduced within a building by intercepting the sound energy before it reaches our ears. Sound waves that are absorbed are dissipated in immeasurably small flows of heat created by friction between moving molecules of air and the walls of a porous material's pores. Most of this heat can easily be absorbed by the room contents and wallcoverings, and by the structure of the building itself.

ROOM SOUND QUALITY

Both the background noise level and the reverberation time affect the acoustic quality of a space. As noted above, the background noise level is the general noise level within a space, excluding any sound related to the intended use of the space. The reverberation time is the amount of time a sound can be sustained within a space. Reverberation makes a sound fuller, but too long a reverberation time hinders speech intelligibility.

A space with many absorptive surfaces that does not support sound is termed a **dead space**. A **live space** has reflective surfaces to sustain the sound through several reflections before it attenuates to the point of inaudibility. Live spaces aid speech audibility and musical sounds. Too lively a space produces distorted sounds.

Measuring Sound Absorption

Sound absorption is measured by a material's sound absorption coefficient or its noise reduction coefficient.

SOUND ABSORPTION COEFFICIENT (SAC)

The performance of sound absorbing materials is described by the **sound absorption coefficient (SAC)**. (See Table 8.3) The SAC ratings run from 0 to 1, with 1 indicating total sound absorption. The SAC can be used to calculate the actual sound absorption of material in a space by multiplying the SAC by the exposed surface of the materials.

NOISE REDUCTION COEFFICIENT (NRC)

In order to give a useful and general idea of a material's ability to absorb sound at a variety of frequencies, the absorption coefficients at 250 Hz, 500 Hz, 1000 Hz and 2000 Hz are averaged together for the **noise reduction coefficient (NRC)**. (See Table 8.4) The NRC is useful as a single-number criterion for measuring the effectiveness of a porous sound absorber at midrange frequencies. It does not accurately indicate the material's performance at high or low frequencies. Because it is an average, and two materials with the same NRC may perform differently.

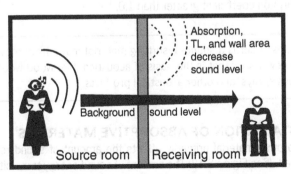

Figure 8.10 Background sound in adjoining space

TABLE 8.3 SAMPLE SOUND ABSORPTION COEFFICIENTS

Material	SAC
Acoustic tiles	0.4 to 0.8
Brickwork, painted	0.01 to 0.02
Cork sheet, ¼" (6 mm)	0.1 to 0.2
Hardwood	0.3
Plaster walls	0.01 to 0.03
Plywood panel, (3 mm)	0.01 to 0.02
Polyurethane foam, flexible	0.95

TABLE 8.4 SAMPLE NOISE REDUCTION COEFFICIENTS

Material	Description	NRC
Brick	Painted	0.00
Concrete	Floor	0.00
Glass	Ordinary window glass	0.15
Gypsum board	½" (13 mm) on 2×4s at 16" (406 mm)	0.05
Plaster	On lath or brick	0.05
Carpet	Heavy, on concrete	0.29
	Heavy, on carpet pad	0.55
Fabric	Light velour, in contact with wall	0.15
	Medium velour draped to half area	0.55
	Heavy velour, draped to half area	0.60
Flooring	Terrazzo	0.00
	Linoleum, rubber, or cork	0.05
	Wood	0.10
Paneling	Plywood 3/8" (10 mm) thick	0.15
	Thin wood, vibrating	0.05
Tile	Marble or glazed	0.00
Acoustic ceiling tiles and panels	5/8" (16 mm) fissured	0.60
	5/8" (16 mm) textured	0.50
	5/8" (16 mm) perforated	0.60

SOUND ABSORPTION STRATEGIES

Adding absorptive materials to a room changes the room's reverberation characteristics. This is helpful in spaces with distributed noise sources, like offices, schools, and restaurants.

For maximum noise reduction, apply sound absorbing materials to cover the ceiling fully. Some spaces will require sound absorption on the walls as well.

ABSORPTIVE MATERIALS

The contents of the space control the noise levels within the space, while the construction of the building controls the transmission of noise between spaces. In a normally constructed room without acoustical treatment, sound waves strike the walls or ceiling, which then transmit a small portion of the sound. The walls or ceiling absorb another small amount, while most of the sound is reflected back into the room.

The amount of transmission to an adjoining space is determined primarily by the mass of the solid, airtight barrier between the spaces, not by the surface treatment. However, the amount of sound that is reflected off the surfaces back into the room is greatly decreased by absorptive materials. When acoustic material is applied to a wall or ceiling, some of the energy in the sound wave is dissipated before the sound reaches the wall, and the portion that is transmitted is reduced slightly. In addition to materials designed specifically for acoustic treatment, furniture, finishes, and even people's bodies can provide sound absorption.

A material's sound absorption depends on its thickness, density, porosity, and resistance to airflow. Paths must extend from one side of the material to the other, so that air will pass through. Sealed pores do not work for sound absorption, and painting may ruin a porous absorber such as an acoustic ceiling tile. A porous, fibrous, thick material that smoke can pass through should make a good sound absorber.

Sound absorbing materials are most effective for high-frequency sounds. Thicker materials absorb more sound, including more low-frequency sound. Soft, fragile sound-absorbing materials are best used out of reach on ceilings rather than on walls. Any decorative or protective coverings for sound absorbent materials must be very thin and open, such as open weave fabric or perforated films or sheets. It is best to distribute sound absorbing materials over a larger area than to concentrate them in one location.

There are three families of absorptive materials: fibrous materials, panel resonators, and volume resonators. Panel and volume resonators are usually used to control specific frequencies.

FIBROUS MATERIALS

Most sound absorbing materials are composed of fibrous or open cell structures that let air pass through tiny passageways, where it is attenuated by friction. (See Table 8.5) The amount of sound absorption depends on a material's thickness, density, porosity, and resistance to airflow. (See Figure 8.11)

Fuzzy materials are good sound absorbers for high- and medium-frequency sounds, but not very effective at low frequencies. Very thick blocks of absorptive material installed at a distance from each other can produce very high edge absorption, especially for high frequencies. These large blocks can have an absorption coefficient greater than 1.0.

Cellulose fiber is a sound-absorbing material made from recycled newspaper that is the basis of acoustical tile, wood wool, fibrous sprays, and other acoustical products.

INSTALLATION OF ABSORPTIVE MATERIALS

Absorptive material primarily affects the amount of sound reflected. (See Figure 8.12) The amount of sound energy transmitted depends mostly on the mass of the solid airtight barrier.

TABLE 8.5 SELECTED FIBROUS MATERIALS

Material	Description	NRC Rating (mm)
Cementitious wood fiber panel	Fibrous surface absorbs sound. Exposed on ceiling or walls to reduce noise and reverberation.	1" (25) plank 0.40; up to 0.65 for 3" (76) thick planks
Acoustical foam, open cell	Air can be blown through foam. Excellent sound absorber if thick. Upholstered seat cushions.	¼" (6) thick 0.25; 2" (51) thick to 0.90
Fiberglass or mineral fiber batts, blankets	Reduce noise and reverberation, depending on facing and thickness. Inside stud walls, or behind fabric or open grill. Ceiling behind perforated pans.	Up to 0.90
Fibrous board	Rigid or semi-rigid boards. Wall or ceiling panels with sound-transparent facings including fabrics.	1" (25) 0.75; 2" (51) around 0.90
Fibrous spray-on insulation	Porous, absorptive, fire-resistant. Performance depends on thickness, application technique.	1" (25) thick coat 0.60 or higher.
Loose acoustical insulation	Blown or dumped in place. Reduces sound transmission through partition.	0.75 to 0.82

Figure 8.11 Acoustical batt insulation

Source: Courtesy of Herb Fremin

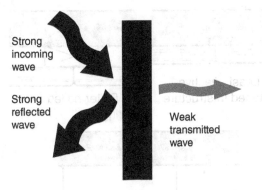

(a) Heavy solid barrier

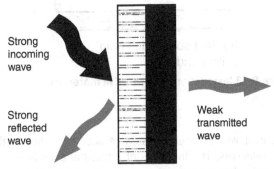

(b) Absorbent material and heavy solid barrier

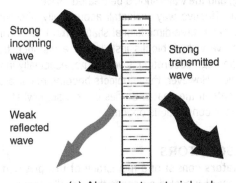

(c) Absorbent material only

Figure 8.12 Sound waves striking barriers

Installation methods are very important in determining the effectiveness of an absorptive material. (See Figure 8.13) As indicated earlier, for effective absorption, air paths must extend from one side of the material to the other, and sealing or painting a material can ruin its ability to absorb sound.

For best results, treat the ceiling, floor, and wall opposite the sound source approximately equally. Treating the ceiling alone may miss highly directive high frequency waves, which may not reach the ceiling until the third reflection off a surface.

Most materials are better at absorbing high frequencies than low ones. A layer of air between the absorptive material and a rigid surface works almost as well in mid-range frequencies as if the same cumulative thickness of absorptive material

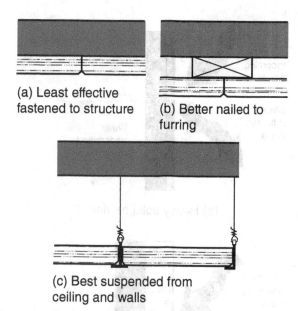

(a) Least effective fastened to structure

(b) Better nailed to furring

(c) Best suspended from ceiling and walls

Figure 8.13 Installation of absorptive materials

were used, which is useful to know because air is cheaper than other materials. The thickness of a material does little to increase its absorbency, except at very low frequencies. To get the best low-frequency absorption, you need a deep air space on the ceiling, and the walls should be treated as well.

The most effective way to install acoustically absorbent material is to hang three-dimensional shapes from the ceiling. When you use very thick blocks installed at a distance from each other, the edge absorption is very large, especially in the high frequencies. However, these objects become major architectural elements in the space. Baffles offer a somewhat less effective but less conspicuous option.

PANEL RESONATORS

Panel resonators consist of a membrane of thin plywood or linoleum in front of a sealed air space that usually contains an absorbent material. The panel is set in motion by the alternating pressure of the sound wave, and the sound energy is converted to heat. Panel resonators are used for efficient low-frequency absorption. They are often used in recording studios.

VOLUME RESONATORS

Volume resonators (also known as cavity or Helmholtz resonators) are usually hollow concrete blocks with open slits that allow sound waves to enter. They create a resonance in hollow constructions whose natural frequencies match that of the sound. Air within the hollow acts as a spring, oscillating at a related frequency. Because a resonating body absorbs energy from the sound waves that excite it, resonating devices can absorb sound energy. Entering sounds reflect off interior surfaces and must reemerge from the small opening to be heard. Volume resonators can be tuned to different frequencies.

Reverberation

Reverberation can be considered as a mixture of previous and more recent sounds. The reverberation time indicates how long a sound will be sustained in a space. It is defined as the time it would take for a sound level to diminish by 60 dB.

Reverberation and reverberation time are introduced in Chapter 7, "Acoustic Design Principles."

Both the physical size of a space and the amount of sound absorption it contains determine the length of time a sound is sustained. The small but appreciable delay between hearing a sound directly from its source and hearing the sound reflected off a room surface contributes to the acoustical ambience of the space.

The reverberation time is dependent on the dimensions and configuration of the space and the amount and placement of sound absorbing materials. In spaces with large wall surfaces, such as those with very high ceilings, sound-absorbing material may be needed on the walls as well as on the ceiling and floor.

Reverberation times can be adjusted by altering the amount of sound absorption in the space, by adjusting the acoustic volume of the space (for example, by combining it with an adjacent space), or electronically with a sound reinforcement system (preferably as a last resort).

SOUND ISOLATION

Sound isolation generally depends on mass, resiliency, and tightness. The more massive a barrier, the less likely sound is to make it vibrate significantly and the less sound will be transmitted. Movement on one side of a resilient barrier is less likely to be transmitted to the other side because an air space allows movement independent of the two sides. The tightness of a barrier's construction will minimize or eliminate openings that could transmit sound.

Mass

Most partitions are built of light, upright framing members with plaster or gypsum wallboard surfaces attached to both sides. This construction does not provide a very good sound barrier. Adding more layers of gypsum wallboard to one or both sides increases the wall's mass and improves acoustic performance. Resilient metal clips improve the sound barrier, and fibrous batts inside the wall help even more.

A thick brick wall makes a good sound barrier between rooms. Concrete blocks are somewhat porous, so more sound passes through; adding a plaster finish helps.

There are practical limits to increasing the transmission loss (TL) by adding mass to a wall. The maximum theoretical increase in transmission loss with increase in mass is 6 dB per doubling of mass; in practice, actual homogenous walls perform even less well.

For information on other roles of massive construction, see Chapter 12, "Principals of Thermal Comfort."

Resiliency

The stiffness of a barrier is determined by its material and the rigidity of its mounting. In a stiff material, the sound energy motion is passed from molecule to molecule, conducting sound very efficiently. The stiffer a barrier, the more it will be set in motion by sound energy.

Less stiffness results in high **internal damping**. The motion of the molecules is not transmitted well, so less stiff materials are good sound insulators. Stiffness transmits the most sound at low frequencies.

Resilient furring channels separate the wall's structure from its surface material. (See Figure 8.14) They make the wall less stiff, inhibiting the transfer of vibrations through to the other side.

The transmission loss of a sound barrier can be improved by constructing it as two separate layers that are not rigidly interconnected. The air cavity between reduces the stiffness of the barrier, and improves its ability to block sound. Filling the void with porous sound-absorbing material increases the transmission loss. Any rigid interconnections decrease performance, and a common stud wall will perform in a manner similar to a single-material wall.

Tightness

A wall construction's tightness affects its sound isolation ability, as any point on an acoustic wavefront is a potential origin of a new wave. This means that any sound that passes through an opening in an acoustic barrier could potentially spread out from the point of entry to fill the entire space.

Compound Barriers

Compound barriers and cavity walls improve transmission loss when the void between the two sides of the wall is filled with porous, sound-absorbent material. This decreases the stiffness of the compound structure, and absorbs sound energy reflecting back and forth between the inside wall surfaces.

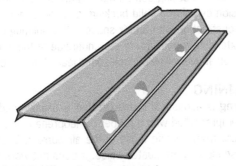

Figure 8.14 Resilient furring channel

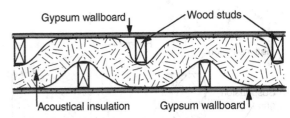

Figure 8.15 Staggered-stud partition plan

Light-gauge steel studs used to frame partitions are lightly resilient, which helps the wall attenuate sound. Heavy-gauge steel studs and wood studs are stiffer and offer less sound attenuation.

When one layer of gypsum wallboard is attached to the framing with resilient metal clips, structure-borne transmission of sound through the partition is reduced substantially.

Staggered-stud partitions are framed with two separate rows of studs arranged in a zigzag fashion and supporting opposite faces of the partition; this produces a wider wall. (See Figure 8.15) This type of wall is often used to reduce sound transmission in recording studios. A fiberglass blanket can be inserted between the rows of studs.

The more layers of rigid but flexibly installed material separated by airspaces that are used, the less sound will make it through an opening in the wall. Special acoustic liners are produced to fit inside the openings in concrete blocks, significantly increasing a CMU wall's acoustic transmission loss.

Gypsum wallboard is not very heavy or thick but provides fairly good sound attenuation. As indicated earlier, the most effective construction detail for blocking sound uses multiple layers of gypsum wallboard with a resilient separation between the two faces of the partition, and with absorptive material in the staggered stud space. The wallboard joints must be perfectly sealed. Gypsum wallboard is highly reflective of higher frequencies and will resonate unless it is attached directly to a solid substrate without an air space, so that it will absorb low frequency sounds.

Floor/Ceiling Assemblies

The most disturbing noise tends to be that which radiates down from the ceiling. A flexibly suspended ceiling with an acoustically absorbent layer suspended in it is effective if there are no flanking paths leading into walls and reradiating into the space below. The insulation is usually 3" to 6" (76 to 152 mm) thick and not packed tightly into the space; it may lie above the ceiling or be attached to the underside of the floor. (See Figure 8.16)

Two or more layers of gypsum wallboard on a metal channel frame suspended from vibration isolation hangers can replace or be added to an existing ceiling. A double-layer gypsum board ceiling can also be installed on resilient channels or clips, with fiberglass insulation in the cavity.

Installing a resilient layer between the structural floor and a hard finish floor treatment like marble, ceramic tile, or wood will help cushion impacts. Resilient products are often installed

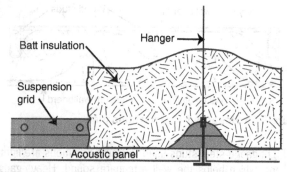

Figure 8.16 Insulation above suspended ceiling

beneath light-weight gypsum concrete or other light-weight leveling materials. Floor underlayment consisting of pre-compressed molded glass fibers is used to control sound transmission of both impact and airborne noise in floor systems. The sound matt is installed between a plywood subfloor and the floor's finish material. It provides enough stiffness to prevent grout cracking in tile floors, while being resilient enough to greatly reduce noise.

For more information on floor/ceiling assemblies, see Chapter 5, "Floor/Ceiling Assemblies, Walls, and Stairs."

Special Acoustic Devices

When the building design calls for the placement of quiet spaces next to, under, or over noisy mechanical equipment rooms, kitchens, or manufacturing spaces, additional measures must be taken to assure that quiet spaces will remain quiet. A room within a room can be created with double partitions, a high mass ceiling, and a floating floor.

Acoustical product manufacturers have developed systems for gypsum wallboard that isolate the partitions from the structure while providing lateral restraint to prevent toppling or collapse. The systems include resilient, load-bearing underlayment, vertical joint isolation material, sway braces, and top wall brackets. Other acoustic devices include air springs, resilient hangers, vibration isolators, resilient mounts, flexible connections, and gaskets.

SOUND TRANSMISSION BETWEEN SPACES

Sound can be transmitted through steel, wood, concrete, masonry, or other rigid construction materials. The sound of a person walking is readily transmitted through a concrete floor slab into the air of the room below. A metal pipe can carry plumbing noise throughout a building. A structural beam can carry the vibrations of a vacuum cleaner to an adjacent room, or the rumble of an electric motor throughout a building.

Sound transmission from one space to another depends on the sound-insulating qualities of the construction between the spaces. Wherever an opening exists—even a keyhole, a slot at the bottom of a door, or a crack between a partition and the ceiling—sound will move from one room to another. To seal construction to sound leaks, weatherstrip cracks around windows and doors and close all other cracks and openings with airtight sealants.

Walls and Partitions

Partitions from the top of the floor slab to the underside of the next floor provide maximum sound isolation. For best results, partitions should be built in as massive and airtight manner as possible.

As the area of a common barrier such as a wall or floor/ceiling between two spaces increases, so does the likelihood that sound will be transmitted.

A basic partition comprising single wood studs 16" (406 mm) on centers, ½" (13 mm) gypsum board on both sides, and an air cavity has an STC of 35. Adding gypsum board, resilient channels, or staggered or double stud construction will increase the STC.

The overall acoustic performance of **composite walls**—those walls with a window, door, vent, or other opening—is strongly affected by the element with the highest sound transmission. The acoustical quality is harmed less if the poor-performing element is much smaller than the better performing parts of the wall. However, even a very small opening seriously degrades the ability to keep sound within the room.

Flanking Paths

Sound will find parallel or flanking paths, acting like an acoustic short circuit. (See Figure 8.17) It is important to avoid locating doors and windows where they will allow shortcuts for sound. The most common flanking path is a plenum with ductwork, registers, and grilles. A plenum will make an excellent intercom unless it is completely lined with sound absorbent material. Even then, low-frequency sound will still get through.

As we have seen, sound can pass through a light, rigid ceiling, then over a partition, and down through the ceiling of an adjacent room. A ceiling can be made more resistive to sound transmission by using a solid backing, a resilient mounting, or airtight construction with an air space with insulating material above. Extending a partition to the underside of the horizontal construction above will help block sound leaks through ceilings.

DUCT LINING

Duct lining is acoustical insulation that is usually made of fiberglass impregnated with a rubber or neoprene compound to avoid fibers from coming loose in the air current. Duct lining absorbs sound and attenuates noise. It does not work as well for low frequencies as for high ones.

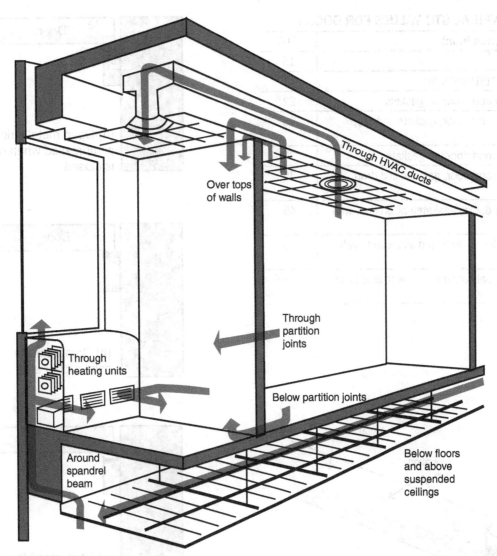

Figure 8.17 Flanking paths

Moisture can condense on linings from cool air moving through the ducts, creating environments for the growth of microorganisms that can then be blown throughout the building. Duct linings should be avoided where the airflow may be contaminated or in health care areas like burn units, unless the air is filtered before entering the room.

SEALING PENETRATIONS

To keep sound from traveling flanking paths, it is important to isolate piping by caulking all penetrations with resilient sealant to avoid air or sound leakage. Envelope perforations such as pipe sleeves, electrical raceways, back-to-back electrical outlets in walls, recessed panelboards, and duct openings must be tightly sealed against sound leaks.

DOORS AND SOUND TRANSMISSION

Doors vary in their ability to block sound. (See Table 8.6) Folding or sliding doors typically provide very low acoustic privacy, although there have been some recent improvements. Hollow core doors do

not provide much acoustical insulation; they may be adequate from a room to a corridor, but probably not to an adjacent room. Heavier doors that close tightly against rubber gaskets and are sealed at the threshold are better. Double doors with an insulated air space between them create a sound trap. Louvered and undercut doors are useless as sound barriers.

For more information on doors, see Chapter 6, "Windows and Doors."

There are many kinds of door seals. All of them block the gap at the doorjamb with a compressible material. An acoustic door sweep is a simple way to block the gap at the bottom. (See Figure 8.18) Other door seals use magnets. (See Figures 8.19 and 8.20)

Two gasketed doors, preferably with enough space between them for a door swing, can be used to create a **sound lock**. (See Figure 8.21) All surfaces in the sound lock are completely

TABLE 8.6 TYPICAL STC VALUES FOR DOORS

Door construction (mm)	STC
Louvered door	15
Any door with 2″ (51) undercut	17
1 ½″ (38) hollow core door no gaskets	22
1 ½″ (38) hollow core door, gaskets and drop closure	25
1 ¾″ (44) solid wood door, no gaskets	30
1 ¾″ (44) solid wood door, gaskets and drop closure	35
Two hollow core doors, gasketed all around, with sound lock	45
Two solid core doors, gasketed all around, with sound lock	55
Special commercial construction, with lead lining and full sealing	45–65

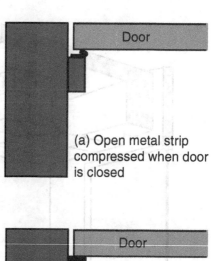

(a) Open metal strip compressed when door is closed

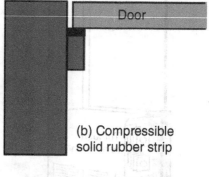

(b) Compressible solid rubber strip

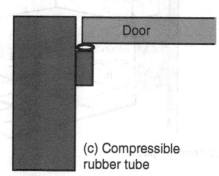

(c) Compressible rubber tube

Figure 8.20 Acoustic door jamb seals

Source: Redrawn from Walter T. Grondzik and Alison G. Kwok, *Mechanical and Electrical Equipment for Buildings, Twelfth Edition,* Wiley 2015, page 1094

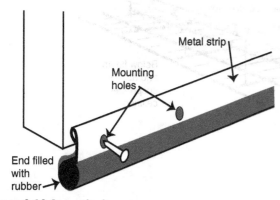

Figure 8.18 Acoustic door sweep

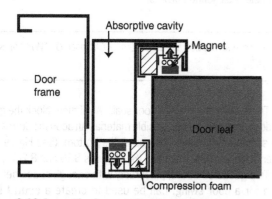

Figure 8.19 Acoustic door seal with magnet

covered with absorbent materials and the floor is carpeted. A sound lock will increase attenuation by a minimum of 10 dB, and by as much as 20 dB at some frequencies.

To avoid doors in residential buildings, including private homes, apartments, dormitories, hotels, and commercial offices from transmitting sound across a corridor, they should not be located directly across from each other. (See Figure 8.22)

Solid-core doors are much superior to hollow-core doors as barriers to sound. (See Figure 8.23) The construction and installation of the frame also make a difference.

Special sound-insulated wood flush doors have their faces separated by a void or a damping compound. They are installed with special stops, gaskets, and thresholds.

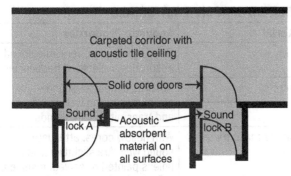

Figure 8.21 Sound locks

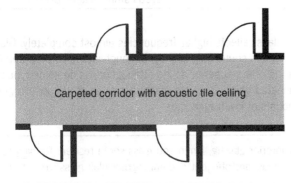

Figure 8.22 Arrangement of doors on corridor

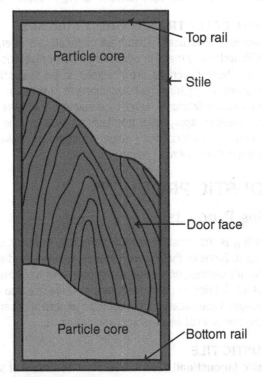

Figure 8.23 Solid-core door

For more information on doors and windows, see Chapter 6, "Windows and Doors."

WINDOWS AND SOUND TRANSMISSION

Exterior walls usually have a high STC, but windows are their weakest part. Sound leaks through cracks in operable windows are more critical for keeping sound out than the type of glazing. Weatherstripping for thermal reasons helps acoustical performance. The manner of opening and the window placement also affect transmission loss.

Plate glass ½″ (13 mm) thick has an STC in the low 30s. Laminated glass of the same thickness may approach an STC of 40.

Impact Noise

Impact noise is often the greatest acoustic problem in buildings with multiple residents. The most common impact noise problem is footfalls. A carpeted or cushioned floor can help, as can a floating floor. Kitchens and bathrooms should be stacked and not located over living rooms or bedrooms. Specify soft acrylic or felt sliders for chairs and other movable furniture.

CONTROLLING IMPACT NOISE

Two basic methods of controlling impact noise include preventing or minimizing the impact, and attenuating the impact noise once it occurs. Cushioning the initial impact that produces a noise will frequently eliminate all but severe problems.

Impact noise on a floor is more serious than on a wall, where the joint between the floor and wall partially attenuates the noise. Floor impacts, on the other hand, introduce noise directly into the building structure. Resilient flooring such as rubber, cork, or linoleum floor tile reduces impact noise transmission. Carpet on a pad works even better.

Install closers on wall cabinetry to decrease impact vibration that reradiates as sound to the adjacent space.

A floating floor separates the impacted floor from the structural floor using a resilient element such as rubber or mineral wool pads or blankets or metal spring sleepers to spread the load evenly over a large area. A floating floor requires airtight construction, especially where partition rests on it.

A suspended ceiling with an acoustically absorbent layer separates the impacted floor slab from the ceiling. The floor above must also be decoupled from walls below it.

IMPACT INSULATION CLASS (IIC)

The **impact insulation class (IIC)** is a rating for floor construction, similar to the STC rating for walls. The IIC rates the ability of a floor system to isolate impact noise. A higher rating is better. The IIC rating is based on tests of actual constructions.

Figure 8.24 Interior brick and concrete wall, Stata Center, Massachusetts Institute of Technology

It is influenced by the weight of the floor system and of the suspended ceiling below, the sound absorption in the cavity between the floor and ceiling, whether the floor is carpeted or not, and the type of building structural system.

Materials and Acoustics

The characteristics of materials affect their acoustic performance. Composite materials such as concrete or organic materials such as wood do not conform to the general rules for homogenous materials made of a single material.

MASSIVE MATERIALS

Acoustically **massive materials** like concrete and brick reflect sound while resisting vibrations that could allow the sound to continue on in adjacent spaces. (See Figure 8.24, Table 8.7)

REFLECTIVE MATERIALS

A smooth, dense wall of painted concrete or plaster absorbs less than 5 percent of the sound striking it, making an almost perfect sound reflector. A gypsum wallboard partition will reflect most of the sound waves that hit it, but allow sound to pass through openings. (See Figure 8.25) Resilient flooring such as cork, rubber, vinyl, or linoleum sheet or tile, reflects sound, although it is acoustically useful to cushion impact noises.

Figure 8.25 Gypsum wallboard partition with openings
Source: Courtesy of Herb Fremin

TABLE 8.7 MASSIVE MATERIALS

Material	Acoustic Properties
Brick	Good attenuation, very little sound absorption, reflects all frequencies.
Concrete	Absorbs almost no sound, attenuates sound well. Will carry and transmit impact sounds.
Concrete masonry units (CMUs)	With hollow cores, attenuate sound well. Are slightly porous unless painted or sealed; if sealed, reflect all frequencies well.
Stone, reconstituted materials, terrazzo	Thick stone walls attenuate sound very well. Marble is very acoustically reflective. Naturally porous stone is less reflective.

Glass reflects higher frequencies almost completely. Glass attenuates sound only slightly, although well-separated double-glazing offers superior sound attenuation, as do some types of laminated glass. Because glass resonates, it will absorb good amounts of low frequencies.

For interior glazing, laminated glass set in resilient framing have more mass and offer better damping than plain glass in rigid frames.

Plywood is relatively ineffective for attenuating sound. Thin plywood furred out from a solid wall is a good absorber of low frequencies. Plywood is quite reflective at high frequencies.

ACOUSTICALLY TRANSPARENT SURFACES

Soft, porous, acoustically absorbent materials are often covered with perforated metal or other materials for protection and stiffness. These coverings are designed to be **acoustically transparent**. Staggering the holes improves absorption.

Open-weave fabric is almost completely transparent to sound. However, applying a thin layer of wallcovering to a sound-reflective material like gypsum wallboard will make little difference in the amount of sound reflected.

ACOUSTIC PRODUCTS

Ceiling Products

The ceiling is the most important surface to treat for sound absorption. Some of the fibrous materials we discussed above are used for ceilings, either openly or covered with acoustically transparent fabrics or perforated panels. There are also products designed specifically for the acoustic treatment of ceilings, the most common of which is acoustic ceiling tile.

ACOUSTIC TILE

Acoustic (acoustical) tiles are excellent absorbers of sound within a room, where they help lower noise levels by absorbing

Figure 8.26 Acoustic ceiling tiles
Source: Courtesy of Herb Fremin

some of the sound energy. Acoustic tiles are usually installed in metal ceiling grids that allow them to be removed for access to the plenum. (See Figure 8.26) To improve resistance to humidity, impact or abrasion, tiles are available factory painted, or with ceramic, plastic, steel or aluminum facing.

Acoustic tile is made of mineral fiber or fiberglass. Mineral fiber tiles have NRC ratings between 0.45 and 0.75. Faced fiberglass tiles are rated up to NRC 0.95 and are often used in open office applications. Acoustic tiles are both light-weight and low density, and can be easily damaged by contact. Consequently, they are not recommended for walls and other surfaces within reach.

Tiles are available in size multiples of 12″ (nominal 305 mm) from 12″ by 12″ (305 by 305 mm) up to 48″ by 96″ (1219 by 2438 mm), and thicknesses of ½″, ⅝″, and ¾″ (13, 16, and 19 mm). The thicker the tile, the better the absorption. Edges may be square, beveled, rabbeted, or tongue-and-groove. Acoustic tiles come in perforated, patterned, textured, or fissured faces. Some tiles are fire-rated, and some are certified by their manufacturer for use in high-humidity areas.

Acoustic tile is usually suspended from a metal grid, but can also be attached to furring strips or glued to solid surfaces. Suspended applications absorb more low-frequency sound than glued-on tiles.

Suspended grids create space for ductwork, electrical conduit, and plumbing lines. They allow lighting fixtures, sprinkler heads, fire-detection devices, and sound systems to be recessed. The grid consists of channels or runners, cross tees, and splines suspended from the overhead floor or roof structure. (See Figure 8.27) The grid may be exposed, recessed, or fully concealed.

In addition to absorbing sound within a room, acoustic tiles may also attenuate sound passing through to adjacent rooms. This can be critical where partitions stop against or just above the ceiling to create a continuous plenum. Tiles for sound attenuation in this use are usually made of mineral fiber with a sealed coating or foil backing.

PERFORATED METAL PANS AND PANELS
Perforated metal pans are backed by acoustic fillers and are used on ceilings and walls to absorb sound. (See Figures 8.28 and 8.29) Perforated metal pans and panels are available as

Figure 8.27 Suspended ceiling grid
Source: Courtesy of Herb Fremin

square and rectangular tiles, rectangular panels and planks, linear grids, vertical baffles, open cells, corridor ceiling panels, and modular island ceilings and curved canopies.

The metal finish is usually baked enamel, available in a variety of colors. Metal panels are easy to keep clean, are highly reflective, and are not combustible. Metal pans will not reduce sound transmission unless they have a solid backing. With the acoustic backing removed, a perforated unit can be used for an air return.

Figure 8.28 Plenum air return with perforated metal panel
Source: Courtesy of Herb Fremin

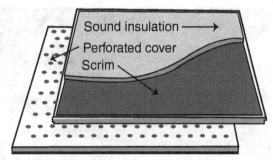

Figure 8.29 Perforated metal pan ceiling pan

Figure 8.30 Linear metal ceiling

Figure 8.31 Wood perforated acoustic panels over sound-absorbing backing

The size and spacing of the perforation—not just the percentage of openness—affect the performance. Depending on the perforation pattern and type and thickness of batt, the NRC of perforated metal pans ranges from 0.50 to 0.95.

LINEAR METAL CEILINGS

Linear metal ceilings consist of narrow anodized aluminum, painted steel, or stainless steel strips. Linear metal ceilings are usually used as part of a modular lighting and air handling system. The panels can be cut for lighting fixtures. (See Figure 8.30) Slots between the strips may be open or closed. The spaces between the strips are designed to accommodate acoustic materials or allow airflow.

SLATS AND GRILLES

Wood or metal slats or grilles in the ceiling are often believed to have acoustic value, but primarily serve to protect the material behind them, which is typically absorbent fiberglass. (See Figure 8.31) The absorption value is maintained if the grilles or slats are small and widely spaced. Increasing the size of the dividers or reducing the space between them will cause high frequencies to be reflected.

ACOUSTICAL FIBER CEILING PANELS

Acoustical ceiling panels or boards of treated wood fibers bonded with an inorganic cement binder are available from 12" by 24" (305 by 610 mm) to 48" by 120" (1219 by 3048 mm). They are 1" to 3" (25 to 76 mm) thick, and receive NRC ratings from 0.40 to 0.70.

Acoustical fiber ceiling panels have high structural strength and are abuse resistant. They have an excellent flame-spread rating. Panels can be used across the full span of corridor ceilings,

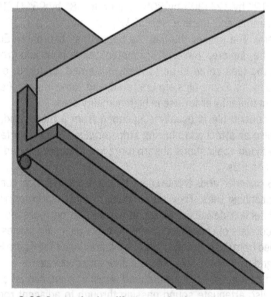

Figure 8.32 Acoustical ceiling panel edge support

or as a long-span finish directly attached to the ceiling. (See Figure 8.32) They are appropriate for wall finishes in school gyms and corridors.

CLOUD PANELS AND ACOUSTIC CANOPIES

Cloud panels perform the same acoustical functions as acoustic ceiling tiles without necessarily interfering with sprinklers

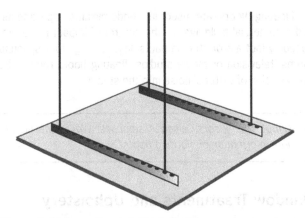

Figure 8.33 Acoustic cloud installation

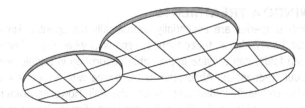

Figure 8.34 Oval acoustic canopies

or lighting. They vary in size, shape, and composition, and can be suspended independently from the ceiling. (See Figure 8.33)

Acoustic canopies are filled with standard acoustic ceiling tile set in a grid. (See Figure 8.34) These can be configured into patterns and suspended over areas that need more sound absorption.

Suspended curved panels can be hung with either concave or convex curves toward the sound source. (See Figure 8.35) Another type of panel is custom-designed stretched open-weave fabric suspended on frames and backed with an acoustic absorbing material.

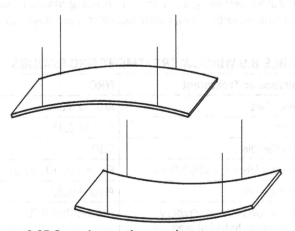

Figure 8.35 Curved acoustic canopies

Wall Panels

Acoustic wall panels have wood or metal backings and mineral fiber or fiberglass substrates. Fabric coverings are usually fire-rated. They are used in offices, conference rooms, auditoriums, theaters, teleconferencing centers, and educational facilities. Perforated metal acoustic panels can also be used on walls as well as ceilings.

Fabric covered panels are available from 1" to 2" (25 to 51 mm) thick. Panels are available from 18" to 48" (457 to 1219 mm) wide, and up to 120" (3048 mm) long. NRC ratings vary from 0.5 for direct-mounted 1" (25 mm) mineral fiber panels to 0.85 for strip-mounted 1½" (38 mm) fiberglass panels.

Manufacturers of acoustic materials market entire systems for finishing and acoustically conditioning the walls of basement spaces. (See Figure 8.36) It is essential to make sure the basement is secure and dry before installing these systems.

Flooring

Floor/ceiling assemblies with relatively low STC ratings may have high IIC ratings, especially if the floor is carpeted. (See Table 8.8) This means that sound may be transmitted through the assembly from an airborne source, yet direct impact on the floor will be muted.

CARPET

Living units above other occupied living units should usually be carpeted to reduce footfall noise. Carpet with padding provides

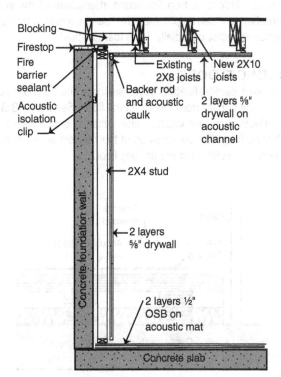

Figure 8.36 Basement acoustic finish

TABLE 8.8 FLOOR FINISH IIC RATINGS

Type	Add resilient material (mm)	IIC rating
Tile	1/16" (1.6) vinyl	0
	1/8" (3) linoleum or rubber tile	3–5
	¼" (6) cork	8–12
Carpet	Low pile, fiber pad	10–14
	Low pile, foam rubber pad	15–21
	High pile, foam rubber pad	21–27

excellent impact isolation. Condominium covenants often require carpets in hallways and foyers and over half of the other living areas to remove most of the objectionable footfall noise.

Carpet is the only floor finish that absorbs sound. Carpets produce a high degree of absorption in the middle to high frequency range. The absorption is proportional to the pile height and density, and increases when installed on a thick pad. Carpet earns an NRC of between 0.20 and 0.55, mainly for high frequencies.

Low-pile carpet on a fiber pad has IIC ratings between 10 and 14. With a foam rubber pad the rating increases to 15 to 21. High-pile carpet with a foam rubber pad earns ratings between 21 and 27.

RESILIENT FLOORING

Resilient tile is almost as sound reflective as concrete. Resilient tile has little effect on the sound attenuation of the floor construction, but can help reduce the sound generated by high-frequency impacts, especially if it is foam backed.

FLOATING FLOORS

As indicated earlier, floating floors separate the floor surface receiving the impact from the structural floor. (See Figure 8.37) Their effectiveness for impact noise depends on the mass of the floating floor, the composition of the resilient support, and the degree of isolation of the floating floor.

Floating floors are used in condominiums, apartments, and commercial buildings for the control of impact noise produced by footfalls or other impacts. In recording studios, sound rooms, television or movie studios, floating floors reduce the transmission of external noise into the studio.

See Chapter 5, "Floors/Ceiling Assemblies, Walls, and Stairs," for more information on floating floors.

Window Treatments and Upholstery

Vertical surfaces can be treated with sound absorptive materials. Window glazing is highly acoustically reflective, and window treatments can help absorb and diffuse sound.

WINDOW TREATMENTS

Most draperies are essentially acoustically transparent. However, heavy, dense, fuzzy fabrics, especially when draped with deep folds, can provide appreciable absorption in middle and upper frequencies. Absorption is increased with an air space between heavily folded drapery and the wall. However, draperies produce little reduction in the passage of noise from room to room through the wall itself. (See Table 8.9)

UPHOLSTERY

Fabric that is not airtight and is stretched over an absorbent material creates an excellent finish that fully preserves the absorption of the underlying material. Deep, porous upholstery absorbs most sounds from mid-range frequencies and upward.

Other finish materials with acoustic properties include acoustical plaster, thin boards and panels, sound blocks, baffles, and hanging panels. (See Table 8.10)

ACOUSTIC APPLICATIONS

Good acoustic design is desirable in all occupied spaces. It is critical for certain types of spaces, including assembly halls, auditoriums, lecture halls, conference rooms, and music rooms.

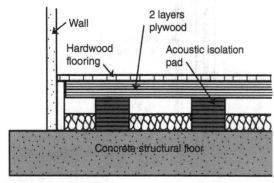

Figure 8.37 Floating floor

TABLE 8.9 WINDOW TREATMENT NRC RATINGS

Surface or Treatment	NRC
Bare glass	Around 0.02
Plaster	Around 0.03
Venetian blinds	0.10
Drapery fabrics, 100% fullness	Between 0.10 and 0.65
Light curtain	Around 0.20
Heavy flow-resistant drapery covering up to half of area	Greater than 0.70

TABLE 8.10 OTHER FINISH MATERIALS ACOUSTIC PROPERTIES

Material	Description
Acoustical plaster	Plaster-type base with embedded fibrous or light aggregate, applied up to 1½" (38 mm) thick. High fire rating. Susceptible to abuse and humidity. Noise absorption varies with composition, thickness, and application technique. Acoustic ratings generally lower than acoustic tiles.
Thin boards and panels	Wood boards or panels attached to furring absorb low frequencies by resonance, can result in a bass deficiency in music rooms.
Sound blocks, baffles, hanging panels	Achieve absorption coefficients greater than 1.0 by exposing multiple absorptive surfaces to impinging sound. Prominent shapes obtrude into space.
	2" (51 mm) fiberglass acoustic baffles for sound absorption in existing high ceiling spaces. Smooth stretched facing. Perforated metal baffles are also available.
	Blocks of fibrous and porous material (space units) made of mineral fibers or fiberglass look like acoustic tile, typically 2" (51 mm) thick. Applied to hard wall and ceiling surfaces; absorb sound efficiently.

Acoustic Criteria

The ratio between sound absorbed and sound reflected affects what is heard within a space. Where little sound is absorbed and much is reflected, intermittent sounds are mixed together. This makes speech less intelligible, but music more pleasant. Steady sounds that accumulate into a reverberant field create a noisy space. Where much sound is absorbed and little reflected, a space is quiet for speech but dead for music.

DIRECT SOUND PATHS

Speech intelligibility depends on relatively weak consonants rather than stronger vowel sounds. In lecture halls, it is important to maintain sound energy by maintaining a direct path. Raising the speaker above the level of the audience helps, as does inclining seating.

Seating in conference rooms and board rooms should be arranged so that all occupants can see one another, allowing direct sound paths between from speakers to listeners. A circular or oval table facilitates this better than a long narrow one.

REFLECTED SOUND

Reflected sound can supplement direct sound. Reflecting surfaces should generally be flat; curved surfaces tend to concentrate sounds in some places and leave others unreinforced.

REVERBERATION TIME

Reverberation is important in lecture halls, theaters, worship spaces, and concert halls where it sustains and blends sounds. Speech has a relatively short reverberation time, but some reverberation enriches the voice's sound and gives the speakers a sense of how their voice is carrying to the audience. From medieval times to the present, liturgical music has often depended on large, echoing interiors for its aesthetic effect.

Reverberation time can be altered by varying absorptive materials with those with different surface finishes to achieve a design that works with a greater variety of frequencies. Fiberglass and similar soft, absorbent materials are very efficient at high frequencies. For low frequencies, a thin panel that conceals an air space lined with absorbent material can perform well.

Offices

Offices continue to become more open, with cubicles replaced by shared worksurfaces and open meeting areas. Exposed ceilings are replacing suspended acoustic tiles. Both of these trends create speech privacy concerns.

SPEECH PRIVACY

Acoustic privacy in an office depends on speech intelligibility, which in turn depends largely on the sound level of background noises. Acoustic privacy requires reduction of noise at the source, along paths of transmission, and at the receiver, plus the use of masking noise where necessary. Providing masking noise in neighboring spaces helps to disguise the information carried by speech and makes it less intrusive.

The amount of speech privacy required within an open office varies. Acoustic consultants identify three levels of speech privacy: normal, confidential, and marginal. (See Table 8.11)

Sound in open offices can travel directly from the source to the listener. It may also be diffracted by objects in its path, or reflected off ceilings or walls. The interior arrangement of the space has a great impact on speech privacy. (See Figure 8.38)

Group spaces according to their speech privacy requirements. Confidential areas should be at the edges of open areas that serve as a buffer zone, with low overall sound levels including any background noise. Speech in perimeter offices with reflective surfaces may bounce out into areas occupied by other workers. High noise production areas should be grouped and placed on the perimeter at a maximum distance from confidential areas.

TABLE 8.11 SPEECH PRIVACY LEVELS

Level	Description
Confidential (good)	Normal voice levels are audible but generally unintelligible. Raised voices may be partially intelligible. 95% of people do not sense sound as intrusive and disturbing, and are able to concentrate on most types of work.
Normal (fair)	Allows normal voice levels from an adjacent space to be heard, but not intelligible without straining to hear. Raised voices generally intelligible.
Marginal (poor)	Around 40% of people find intolerable; decreases productivity. Normal speech in open offices readily understood. Shared workspaces.

MEASURING OFFICE ACOUSTICS

A variety of ratings are used to measure office acoustics. (See Table 8.12) They include absorption coefficients, articulation index (AI), articulation class (AC), ceiling attenuation class (CAC), noise reduction coefficient (NRC), and speech privacy potential (SPP).

In open office environment design, the absorption characteristics of the ceiling are the critical factors in speech privacy. Speech sounds strike the ceiling between 30 and 60 degrees, with most of the sound striking above 45 degrees.

For general sound at a variety of levels, absorption coefficients (AC) are the best indicators. Although the CAC indicates how much sound passes through to the space above the ceiling, this is generally less important, since it can be readily dissipated and absorbed within this space.

OFFICE SPACES

The level of acoustic treatment within the room where the sound originates depends on the loudness of the speech, the effect of the room's sound absorption on the speech level, and the degree of privacy required. The amount of privacy is also affected by the acoustic isolation of the receiving room. This

TABLE 8.12 OFFICE ACOUSTICS MEASUREMENTS

Measurement	Description
Absorption coefficients	Used with materials to measure how well they absorb sound.
Articulation Index (AI)	Relates speech intelligibility, speech intensity, and background sound level at 5 octave-band frequencies. Weighted to emphasize intelligibility. Requires complex computer calculations to derive.
Articulation class (AC)	Derived from open-office cubicle measurements. Indicates absorption at angles of incidence between 45 and 55 degrees. Higher ratings indicate less sound bounces off the ceiling into adjacent cubicles. Ratings range from 170 to 210.
Ceiling Attenuation Class (CAC)	Measures how well a ceiling structure attenuates airborne sound between two closed rooms over the range of speech frequencies.
Noise Reduction Coefficient (NRC)	Measures the average percentage of noise that a material absorbs in the mid-frequency range. Averages absorption at several frequencies. Does not take into account how materials absorb different frequencies.
Speech Privacy Potential (SPP)	US General Services Administration (GSA) rating summarizes background sound level and attenuation between source and listener.

depends on the STC rating of the barrier between the rooms, the noise reduction factor, and the background noise level in the receiving room. The larger the size of the listener's room as compared to the source room, the lower the speech level will be in the receiving room.

Figure 8.38 Sound in open office

Noise carrying information can reduce productivity. Higher levels of noise are annoying for most people. Noise that does not carry information may help reduce annoyance even without completely masking sound.

GENERAL RECOMMENDATIONS

For adequate acoustic absorption, open offices should have a minimum clear ceiling height of 9 feet (2.7 meters), made with highly absorptive materials, and without sound-reflecting surfaces. Above the ceiling, a 3-foot-high plenum with a full layer of sound absorbent blankets further absorbs sound. A carpeted floor reduces noise from footfalls and further increases sound absorption.

Group office areas according to their speech privacy requirements. Be careful about reflections from walls and glass. Make open-area spaces as large as possible to minimize reflection from perimeter walls. Enclose meeting and conference rooms.

ENCLOSED SPACES

For enclosed spaces, the degree of speech privacy depends on the degree of sound insulation provided by barriers between rooms as well as on the ambient sound level in the receiving room. The critical element is the acoustic quality of the airtight barrier between the two spaces. The speech privacy of the source room depends on the loudness of speech, the effect of room absorption on speech, and whether normal or confidential privacy is required.

The isolation rating of the receiving room depends on the STC rating of the barrier, the absorption of the receiving room, and the recommended background noise level. (See Table 8.13)

TABLE 8.13 RECOMMENDED STC FOR ADJACENT PARTITIONS

Wall Location	Walls Adjacent to	STC
Executive office, doctor's suite, confidential privacy	Another office	38–40
	General office area	45–48
	Public corridor	42–45
	Washroom, toilet	47–50
Normal office areas	Other offices, public corridors	38–40
	Washroom, kitchen, dining	40–42
Conference rooms	Corridor, lobby	35–38
	Kitchen, dining, data processing	38–40

OPEN PLAN OFFICE CEILINGS

Sound problems in open plan offices are exacerbated by sound reflecting off ceilings, typically at angles of 45 to 50 degrees. Ceiling materials should be designed to capture sound from almost any angle.

Avoid strongly reflective speech paths such as metal pan air diffusers or flat lighting fixture diffusers. Where these cannot be avoided, highly absorptive vertical baffle strips may be placed on their perimeter to block sound paths.

GLASS SURFACES

Glass walls are becoming increasingly popular in offices. Windows are often located in enclosed offices where confidential discussions routinely take place. Windows and walls that lack absorptive treatment will reflect sound out of the space at an angle. To preserve privacy in these offices, use full-height partitions and fixed glass vision panels, with doors in openings. Heavy drapes can be used to control acoustic reflections. Locate confidential spaces in groups, and buffer them from open office spaces with unoccupied storage areas.

FLOORS IN OPEN OFFICES

Floors in open offices do not affect the overall sound absorption very much. However, cushioned floors do greatly reduce the noise of chair movements and footfalls. For this reason, all floors in open office areas should ideally be carpeted.

MASKING SOUND

Background sound that is close to the frequency of speech reduces the intelligibility of speech. What we hear depends on the level of attention to what we are doing and to the intrusiveness of the outside sound. In a very quiet space with no background noise, any sound is distracting. With a constant ambient sound level in the listener's room, sound transmitted from another room is masked, becoming inaudible or simply less annoying.

Where it is too costly or too difficult to treat a building for a persistent or distracting noise source, low level masking noise may help. Masking sounds are also useful in rooms that are so quiet that heartbeats, respiration, and body movement sounds are annoying as in a bedroom where small noises disturb would-be sleepers.

Masking sound is imperative in open offices. Almost all open-plan office installations use carefully designed electronic masking systems to provide uniform background sound at the proper level and with good tonal characteristics. Equipment should be uniformly placed throughout the space, and set at as low a level as will provide the desired level of speech privacy. Higher sound levels make the masking sounds themselves a source of annoyance. Non-uniform distribution will be noticeable as one moves through the space, and is likely to be annoying.

The masking sound may be white noise, or sound like whooshing sound of air rushing through an opening. Low frequencies are usually emphasized to avoid high frequency hissing.

Figure 8.39 Sound masking equipment

Sound masking units are typically hung above the ceiling where they are out of sight. (See Figure 8.39) The masking sound fills the plenum area and then gently filters down through the ceiling tiles into the office space below. The speakers can be adjusted to the individual acoustical comfort requirement in any given area.

Masking system loudspeakers preferably should not be visible, as they attract interest and eventually become annoying. They can be placed face-up in a plenum to increase dispersion and improve uniformity, but should not be mounted face down in the ceiling. Most ceiling tiles will allow masking sound to penetrate to the office area below.

Music Performance Spaces

The design of spaces for music performance is both an art and a science. For concert halls and other important music spaces, the services of an acoustical consultant are essential. Although the architectural character of a performance space may be worked out well before the interior designer becomes involved in the project, the finishes and details of the hall's interior are critical to its acoustic success.

The design of a space such as a concert hall for good listening conditions starts with developing a room shape that distributes and reinforces sound evenly throughout the audience. Large-volume spaces require direct-path sound reinforcement by reflection. A relatively long reverberation time is needed for music, so the amount of sound reflection and the liveliness of the space matter a great deal. Brilliance of musical tone is primarily a function of high-frequency content, and spaces that are too absorbent will dull musical sounds. A good sound path for musical tone is equivalent to a good visual path, which means that a seat where you have a good view of the performers is likely to also be a good seat acoustically.

Both irregular and convex (outward curving) surfaces are often used in music performance halls. Diffusion is desirable for music performances as it spreads the sound evenly over a wide seating area. Directivity declines if a reinforced signal is excessively delayed by too many reflective surfaces. Sound reflected from convex surfaces is diffuse, producing a constant sound level throughout the space. Concave surfaces are generally avoided in performance halls because they focus sound in some areas and leave others with insufficient sound.

In order to assure that the sound source is loud enough, major room surfaces can be reinforced naturally to direct reflected sound to the audience. Electronic reinforcement systems are used in large rooms or for weak sources.

Auditoriums

Auditoriums must accommodate many activities, including concerts, with varying acoustic requirements. Their acoustic quality depends on the design concept, the budget, and the availability of auditorium staffing to adjust moveable acoustic treatments. Solutions therefore must be either a compromise between the disparate needs, or be adjustable for varied circumstances. The acoustic design of an auditorium involves room acoustics, noise control, and sound system design.

Changing the volume of the space, moving reflective surfaces, or adding or subtracting sound absorbing treatments can alter the acoustic environment of an auditorium. The size of the audience, range of performance activities, and sophistication of the intended audience influence the acoustic design. Once the basic floor area is determined by the size of the audience, the volume of the space is determined by the reverberation requirements.

The ceiling and side walls at the front of the auditorium distribute sound to the audience. They must be close enough to the performers to minimize time delays between direct and reflected sound. The ceiling and side walls also provide diffusion.

To allow adjustments to the acoustics for different events, large areas of tracked sound-absorbing curtains can be installed along the room's boundaries. For movies and lectures without music, permanent sound absorption on the ceiling, rear, and side walls results in a low reverberation time. Curtains can retract into storage pockets when maximum reverberation time is the goal.

A combination of sound-reflecting and sound-absorbing materials should be used. The bodies of the people in the audience constitute a variable sound absorbing material. Fully upholstered seating minimizes the difference between the times when the room is full of people and when it is almost empty, such as during a rehearsal. Upholstery covered with an open weave material is particularly effective.

Auditoriums must frequently accommodate a wide variety of performance events. Adjustable drapes and ceiling panels can help. (See Figure 8.40) Auditoriums and similar spaces need background noise for a lively sound. Absorbent materials should be limited to controlling reverberation time and sound distribution. The surrounding construction must be capable of excluding excessive background noise.

Reflector panels are used on the ceilings of auditoriums, performing arts centers, lecture halls, and churches. They are

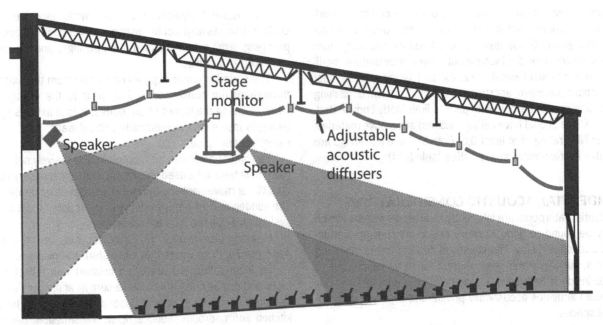

Figure 8.40 Auditorium adjustable sound treatments

designed for large spaces that require improved sound reinforcement and timing of sound reflections to improve listening quality. The sound reflective and diffusive surface of the flat panel is bowed and positioned in the field to the architect's or acoustical consultant's specifications.

Lecture Rooms

In a space used primarily for lectures, it is important to have good sound reflection in the teacher's space at the front of the room. Lecture room boundaries, especially the ceiling, should be shaped for good natural reinforcement of the speaking voice. Applied sound absorbent treatments control reverberation, echo, and flutter.

The rear wall, the perimeter of the ceiling, and the side wall areas between seating and standing height are the most important to treat. Acoustic tile ceilings with hard walls are not an adequate solution for a lecture room.

Schools

Schools have a variety of spaces with a variety of acoustical environments. Noisy corridors with reverberant finishes can project annoying sound intrusions into classrooms.

Design walls for adequate privacy between classroom spaces, and be especially aware of sound leaking through doors between classrooms. Partitions should be full height from floor to ceiling slab or roof construction. Use sound absorbing materials to reduce noise levels. The ceiling tile should have an NRC rating of at least 0.7.

School dining areas are especially noisy. Keep the kitchen and serving areas separate from the eating area so that kitchen noise does not add to the clamor of hundreds of students. The ceiling and walls should have sound-absorbing materials. The ceiling tile should have a minimum NRC of 0.8.

Nothing will quiet the excessive noise in a gymnasium, where noise is expected. The ceiling should be sound absorbing, with an NRC of 0.7. If there is a sound amplification system, the walls should also have sound absorption to prevent echoes.

Public Toilet Rooms

Buildings often have toilet rooms designed to serve occupants and visitors. You should not be able to hear the noises from toilet rooms in adjacent spaces. Good acoustic design helps keep them inconspicuous and pleasant to use.

When designing toilet rooms within service cores, surround them with corridors and mechanical spaces. Avoid seating people adjacent to a wall with plumbing, where they will hear the rush of water through the pipes.

Within the toilet room, a lack of acoustic privacy results in repeated flushing, which is wasteful and noisy. Use a steady masking sound to obscure sounds. A higher-sound-level ventilating system, for example a noisy fan, solves the problem and is less expensive, although its sound can carry through ductwork to other spaces. Music can disguise sounds very well.

Residential Buildings

Understanding how to control noise in a residential building is a critical part of the work of an interior designer. Excessive noise is a major complaint in multifamily housing. Good acoustics are equally important for keeping the peace (and quiet) in single-family residences.

The acoustic design of apartment buildings and other buildings with multiple residents strives to protect privacy and reduce annoyance. Group quiet spaces together and away from noisy activities. Plan the locations of convenience outlets, medicine cabinets, mechanical services, and direct-exhaust duct connections between apartments carefully to avoid flanking paths. Use rugs or carpets with pads to limit footfall noise. Bedroom ceilings should have ceiling-mounted absorptive materials with an NRC rating of at least 0.6. Both STC and IIC ratings are available for residential spaces. (See Table 8.14)

RESIDENTIAL ACOUSTIC CONSIDERATIONS

Residential bathrooms are inherently reverberant spaces (which is why we sound so good when we sing in the shower) that demand acoustic privacy. The sounds of flushing toilets, spraying showers, running water, and whirring fans can be difficult to isolate inside a bathroom. In addition, we prefer to keep our bathroom activities acoustically private and separate from adjacent spaces.

Acoustic privacy can be enhanced by separating the bathroom from bedrooms with intervening closets and hallways. Eliminate cracks around bathroom doors and add resilient bumpers to absorb banging noises. Do not use undercut or louvered doors (which are often provided to improve ventilation). Avoid back-to-back electrical outlets between the bathroom and bedroom. Air grilles should not open into ducts and then out into a bedroom. Do not locate another open window near the bathroom window.

Well-insulated gypsum wallboard walls increase sound control. Use resilient nonhardening caulk around receptacles, plumbing, light fixtures, and other openings, and where wall joins floor and ceiling.

The sound of water in pipes can travel from the bathroom throughout the house and end up next to the dining table. When planning the layout of the house, locate fixtures so that piping is inside less acoustically critical walls. Wrap and resiliently mount pipes.

Equipment noise can be a problem. Excessively noisy bathroom fans tend to be used less often, limiting necessary ventilation. Water movement and motors in jetted tubs generate noise. The sudden rush of water from pressure-assisted toilet flushing systems may be too loud for some users.

Laundry areas tend to be noisy, and may be run at night. Avoid placing them where they will disturb sleeping areas.

Residential kitchens are often finished with reflective surfaces and can be noisy areas. Make sure all appliances are level to reduce vibrations. Placing a closet as a buffer space between kitchen and bedroom reduces noise transmission. Backing a noisy dishwasher up to a toilet can also help. Noisy kitchen ventilation fans are less likely to be used, which can lead to indoor air quality problems.

It is a good idea to let a client listen to the noise produced by a ventilation fan, jetted tub, or washing machine before selecting one.

TABLE 8.14 RESIDENTIAL TRANSMISSION AND IMPACT RATINGS

Rooms	Ceiling Below	STC	IIC	Floor Above	STC	IIC
Bedroom	Bedroom	52	52	Bedroom	52	52
	Living room	54	52	Living room	54	52
	Kitchen	55	50	Kitchen	55	50
	Family room	52	58	Family room	56	48
Living room	Bedroom	54	57	Bedroom	54	52
	Living room	52	52	Living room	52	52
	Kitchen	52	52	Kitchen	52	57
	Family room	54	50	Family room	54	50
Kitchen	Bedroom	55	62	Bedroom	55	50
	Living room	52	57	Living room	52	52
	Kitchen	50	52	Kitchen	50	52
	Family room	52	58	Family room	52	58
Family room	Bedroom	56	62	Bedroom	56	48
	Living room	54	50	Living room	54	60
	Kitchen	52	58	Kitchen	52	52
Bathroom	Bathroom	50	50	Bathroom	50	50
Corridor	Corridor	48	48	Corridor	48	48

ELECTRONIC SOUND SYSTEMS

Although sound reinforcement design may not properly be a part of an interior designer's work, ignoring sound system details creates difficulties that definitely affect the appearance of the project and the smoothness of the construction process. There is a tendency to think of the sound reinforcement system for a smaller project as an add-on, and leave it to the end of the design and construction process. Doing so risks last minute crises when sound wiring, speaker locations, and space and wiring for control equipment has to be retrofit into locations for which it was not planned. Running wires within walls is much easier before the walls are closed in and finished. Wireless sound systems offer a newer solution.

The goal of sound reinforcement is to adjust acoustic problems and to assure that everyone in the listening space can hear well. The ideal sound reinforcement system would give every listener the same loudness, quality, directivity, and intelligibility. Speech should be as clear as if the speaker was 2 or 3 feet (610 to 914 mm) away, with a longer distance acceptable for music. Sound reinforcement systems should be designed to provide adequate sound levels without distortion. Loudspeakers cannot completely correct poor acoustics, but they can improve them.

Specialists in sound system design are sometimes certified members of the National Council of Acoustical Consultants. Others may operate independently, and it is always wise to check into their experience and previous work.

Sound Reinforcement Systems

Sound systems are installed in most spaces where meetings and presentations occur. Electronic sound reinforcement is generally needed when around 100 people gather, although this depends on the space and the speaker.

COMPONENTS

A sound reinforcement system consists of three parts: the input, the amplifier and controls, and the loudspeakers. The input can be a microphone or any of a variety of playback devices. Microphones convert sound waves into electrical signals, which are further amplified, transmitted, and processed as required in the sound system. They may be hand held or on stands, or miniature lavaliere types. Small wireless transmitters are available for any type of microphone.

Signal processing equipment includes equalizers, limiters, electronic delays, feedback suppressors, and distribution amplifiers. The amplifier has controls for volume, tone mixing, input-output selection, and equalization controls for signal shaping. Power amplifiers provide a signal output with sufficient power (voltage and electrical current output) to feed the loudspeakers connected to the system. Amplifiers have controls for volume, tone mixing, and input-output selection.

Wiring for the sound system should be run discretely, preferably within walls. Even wireless controls often have wires running in walls between the amplifiers and loudspeakers. Dedicated presentation spaces may have hardwired podiums for speakers to control equipment.

LOUDSPEAKER SYSTEMS

Loudspeakers convert the electrical signal supplied from the power amplifier into air vibrations that the ear perceives as sound. The design of the sound system should be coordinated early in the design process.

Today's loudspeakers are much smaller than in the past, making it much easier to integrate them into the building. The interior designer should be aware of the size and placement of loudspeakers. Controls should be located where they are accessible but unobtrusive. Signal processing and amplification equipment is often in a rack in a remote location; it needs adequate space and ventilation, and electrical supply of adequate capacity.

Distortion happens when a system changes the shape of the acoustical signal that it receives; some stages of the amplification are overloaded and some frequencies are incorrectly amplified. If the equipment that was purchased and installed was inadequate for the designated use, you will also get distortion from the undersized equipment. If a system was designed for a single person giving a speech and is used for a rock band, the undersized equipment will probably distort the excessive signal. Loudspeakers must be far enough from microphones to avoid feedback.

Sound coloration occurs when a reproduced sound loses its naturalness and acquires and unpleasant, ringing quality. When the system is turned up too loud, it produces **acoustic feedback** or howling. Both of these problems can be corrected, but may require changes or adjustments in the equipment or system design.

Strong speakers will be heard clearly in smaller rooms for about 100 people, but people with weaker voices will require sound reinforcement. The voice level should be a minimum of 25 db above the background noise.

The design and placement of the loudspeaker system must be coordinated with the architectural design. Loudspeakers can be physically integrated in walls, columns, or ceilings; mounted on surfaces; or supported by speaker stands. The speaker placement of conventional distributed loudspeaker systems has to be coordinated with the locations of lights, sprinklers, and air-handling system diffusers.

Depending on the application, loudspeakers are arrayed either in a centralized system or a distributed pattern. Centralized systems are used in large spaces with high ceilings to project sound with strong directionality from a focal point such as a stage or pulpit. Distributed loudspeaker patterns are used in spaces with lower ceilings where the sound is distributed evenly and without a strong sense of source through many smaller speakers as in offices or restaurants.

Sound Systems for Specific Spaces

A familiarity with the equipment required for specific types of spaces will help you anticipate conflicts with the details of the interior design and head off last minute problems.

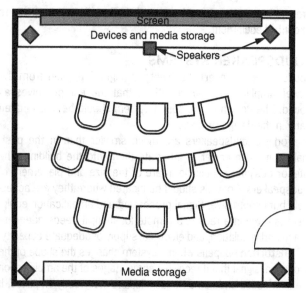

Figure 8.41 Home theater plan

HOME LISTENING ROOMS

The interior designer working with a sound installation company on the design of a home listening room should be aware of several issues. (See Figure 8.41) The goal is to suppress the sound of the room itself, but not to totally deaden it; this requires large areas of sound-absorbing and sound-diffusing treatments on all surfaces. The space will likely experience significant coloration and poor spatial imaging of reproduced sound if it is not extensively and uniformly treated with broadband sound absorbing materials. Paired surfaces should be treated the same. The treatment must be totally symmetrical on the left and right sides of the main listening axis, so that sound transmission paths for left and right playback channels are identical. Surround sound channels have special placement considerations.

OFFICE SPACES

Emerging standards of office design and the huge influx of personal digital devices are significantly changing the need for office sound reinforcement. As indicated above, open offices usually have masking sound systems for speech privacy.

High-rise office towers may include a sound system designed for life-safety announcements and transmission of warning signals. Emergency sound systems are normally separate from other systems and use fire-rated equipment, wiring, and installation materials.

AUDITORIUMS AND LECTURE HALLS

In auditoriums and lecture halls, the controls for sound, projection, and lighting are frequently located within the listening room. Some spaces have a provision for remote controls for projection, volume control of the sound, and dimming of lights at the lecturer's position. It is important for the interior designer to ask whether the facility has its own in-house audiovisual technician (in which case controls are in the back of the room), or whether controls should be in the front for use by the presenter.

In medium to large auditoriums, sound systems should have a sense of directional realism. Where the audience or delegates participate from the floor, microphone arrangements vary from simple to very complex. Assembly halls for multinational conferences have a system for simultaneous translation.

CONCERT HALLS AND THEATERS

The interior designer needs to be aware of the type of music that is being performed in a concert hall. Some musicians travel with their own sound systems, while others use the system installed in the space.

High-quality, full-range sound reinforcement systems are used in concert halls with multiple functions. A central loudspeaker system with high directivity can be suspended in free space or integrated into the architecture. Sound reflectors or canopies may be located above the orchestral platform. Concert halls may also have backstage performance monitoring, intercom, and paging systems.

Theaters may reinforce vocals over orchestral sounds. However, some music fans prefer the sound without amplification, in which case the architectural acoustics of the space become very important.

EXHIBITION HALLS

Exhibition halls use sound reinforcement for announcements and for background music. If the hall includes a platform for stage presentations, it will typically include a central loudspeaker system located above the front of the platform to provide sound reinforcement with directional realism.

HOTELS

Designing for sound reinforcement systems in hotel ballrooms and banquet rooms requires knowing what the room will be used for and designing to user needs. Unobtrusive but accessible controls and proper speaker placement are priorities.

During renovation projects, it is wise to see whether existing equipment mounted on the walls is actually in use. For example, in one hotel, microphone antennas that had not been used for 12 years would have been given special treatment when the wallpaper was installed if an alert technician had not noticed and alerted the designer.

As the interior designer, your awareness of acoustic concerns that affect the health, safety, and enjoyment of the public is an important part of your role as a member of the building design team. Your knowledge of all the systems that make a building work will enhance the quality of your design work, the value of the building to those who use it, and your own enjoyment of your work as an interior designer.

We have now completed our survey of acoustic design. Part IV, "Water and Waste Systems," addresses the elements of water supply and distribution, waste and recycling, and fixtures and appliances.

PART

IV

WATER AND WASTE SYSTEMS

We turn on the tap and out comes fresh, cool, clean water. This is the water we drink, cook, bathe, clean our clothes, wash our cars with, and flush down the toilet. It is easy to think of water as a free, readily available resource. However, fresh water is a limited resource.

Global climate change is producing severe storms and drought. Seawater is becoming more acidic as it absorbs greater amounts of carbon dioxide from the atmosphere, creating problems for sea life. Water withdrawals in coastal areas are allowing saltwater intrusions into aquifers. Water supply crises can result in countries downstream from supplies threatening those upstream over water rights.

> With a finite planetary water supply, pitted against an increasing population, seen in conjunction with an increasing *per capita* consumption of water, we see again (as with fossil fuels) the problem of limited resources versus growing demand. At least in this case, the amount of water is fixed, not diminishing. However, the problem of fair allocation remains… Countries have fought wars over oil; must they also wage war over water? (Walter T. Grondzik and Alison G. Kwok, *Mechanical and Electrical Equipment for Buildings, Twelfth Edition*, Wiley, 2015, page 793)

Building systems control the supply and distribution of rainwater and **groundwater** (water held underground in soil or openings in rocks) relatively lightly. **Potable water** suitable for drinking and wastewater are tightly controlled.

The three chapters in Part IV look at the components of the typical sanitary plumbing system:

Chapter 9 describes the distribution system of piping supplying water to fixtures. Gas supply and distribution in buildings is also briefly included.

Chapter 10 deals with the separate waste piping system, and with other waste and recycling systems.

Chapter 11 examines the fixtures and appliances that consume or use water.

Interior designers and those in the related field of bath and kitchen design are directly involved in the layout and selection of plumbing fixtures and appliances. This necessitates an understanding of water distribution and waste piping systems. Designing for recycling is another important aspect of their work.

Water Supply Systems

A building's plumbing systems are designed to provide clean water and to remove wastes. Plumbing vents introduce clean air. Plumbing systems also aid in suppressing fire, controlling humidity, creating acoustic privacy, and keeping unwanted water out of the building.

Plumbing systems provide water and drainage for sanitation and potable water needs and dispose of precipitation falling on the building through a **stormwater** system. At the building boundary, typical utility connections for water supply, wastewater, and stormwater systems include a potable water supply line with a meter, a sanitary drain line, and a stormwater drain.

INTRODUCTION

Water is a recyclable resource, but not a renewable one. The amount of water on earth is a fixed resource. However, fresh water supplies are limited. Increasing human population and increasing per capita consumption of water in some parts of the world threatens limited resources.

Plumbing systems are primarily designed by mechanical or plumbing consultants. Architects are usually involved in specifying fixtures, and locating stormwater drains and other exterior elements. Interior designers and kitchen and bath designers also select and locate plumbing fixtures.

> While as architects and designers we might not be able to resolve the global water problems, we must make some contribution to their resolution through our own activity, both personally and professionally. Specifically, we can reduce the amount of water used in the buildings we design, adjust the way we access water, and change the way that the water we

do use impacts on the environment. (David Lee Smith, *Environmental Issues for Architecture*, Wiley 2011, pages 401–402)

Environmental efforts are leading to reductions in water consumption, but water resources are threatened in many parts of the world. Water is not always available where and when it is needed. Some parts of the United States—and many parts of the world—are currently experiencing water shortages. Depletion of water from underground **aquifers** (groundwater supplies) is accelerating. Both **hydrological** (dealing with the properties, distribution, and circulation of water) and political causes are responsible for depletion of sources.

History

Water wells dug around 6500 BCE have been found in the Jezreel Valley in Israel. Before 1000 BCE, the Minoan civilization on Crete was using underground terracotta pipes to supply the palace at Knossos.

The Roman Empire had indoor plumbing supplied by covered stone aqueducts and distributed to public fountains, public baths, and private homes. Roman villas included an atrium with a sunken area to catch and store rainwater.

Until the middle of the nineteenth century, London's drinking water came from the Thames River, which was used as an open sewer, or from sewage-polluted wells. Water from any source was scarce for poor people, personal hygiene was almost nonexistent, and epidemics were common.

By the nineteenth century, many large cities built their own aqueducts. In Boston, Massachusetts, in 1829, architect Isaiah Rogers designed the first hotel with indoor plumbing.

Codes and Testing

Plumbing codes protect the health and safety of building occupants from contamination of the water supply, contamination of the air and objectionable odor due to the escape of sewer gas, and chronically blocked drains due to improper size or pitch. Testing and treatment ensure that the water we consume is safe.

CODES AND STANDARDS

The **International Plumbing Code® (IPC)** is issued by the International Code Council. It establishes minimum regulations for plumbing fixtures. Its provisions are intended to protect public health, safety, and welfare without unnecessarily increasing construction costs or restricting the use of new materials, products, or methods of construction. Chapter 6 of the 2015 edition of the IPC covers water supply and distribution.

Local building codes cover almost every aspect of plumbing design and materials. Local health departments and special consultants can assure proper quantity and quality of supply water through water-quality analysis. This process assesses mineral content, **turbidity** (cloudiness), total amount of solids, biological purity, and suitability for intended use.

There are two types of water quality standards established by the United States (US) Environmental Protection Agency (EPA). **Primary Drinking Water Standards** are enforceable by law. They ensure that water is safe to drink or ingest. State and local health departments or environmental agencies and municipal or public water authorities work together to meet primary standards. Primary standards are based on Maximum Contaminant Levels (MCLs) for three classes of pollutants: disease-carrying organisms, toxic chemicals, and radioactive contaminants.

Secondary Drinking Water Standards are voluntary, and ensure that water is functional and aesthetically acceptable for bathing and washing. Secondary standards are based on **Secondary Maximum Contamination Levels (SMCLs)**. They address appearance, taste, odor, residues, and staining, and cover levels of chloride, iron, manganese, and sulfur as well as altered pH.

Secondary Drinking Water Standards are important to residential bathroom water use.

TESTING AND TREATMENT

Most experts recommend annual testing for private water sources such as wells. Annual evaluations should include total **coliform bacteria**, nitrates, pH, and total dissolved solids. Indications of problems include fixture staining or hard water deposits in the kitchen or bathroom. Additional testing may be needed if there are problems with color, cloudiness, discoloration, taste, or odor. Information on testing is available from state or local health departments.

It is advisable to ask clients if they use filters for drinking water, and to evaluate a glass of water for appearance or odor problems.

Most types of water treatment equipment must be maintained regularly to remain safe and effective. Water treatment may include filters, water softeners, iron removal equipment, neutraliziers, distillation units, reverse osmosis units, and disinfection methods. Multiple problems may require a combination of different types of equipment.

Plumbing and Construction Drawings

The design of the plumbing system is communicated on engineering plumbing and architectural construction drawings that show all fixtures and piping on floor plans. In addition, isometric schematic drawings of water supply and drainage show vertical pipes. Interior designers show plumbing fixtures on floor plans.

WATER SOURCES AND USE

In most developed areas of the world, the primary water source is either a lake or a river, which usually requires treatment before being supplied to buildings. A small number of public water systems in the United States use groundwater from deep wells. Locations that are far from public water services rely on private water supplies. These may be wells or springs that supply water directly from the ground, which is usually purer than that from possibly polluted streams or ponds.

Freshwater resources make up only 3 percent of total world water resources, with ocean saltwater accounting for the remaining 97 percent. About 30 percent of the freshwater is groundwater, with nearly 69 percent in icecaps and glaciers. The remaining 0.3 percent is surface water in lakes, swamps, and rivers.

The world's population is expected to continue to grow dramatically, while water resources are fixed. A growing population is related to diminished natural resources and water pollution; the amount of water available per person in the future is likely to continue to decrease.

Water Use

Water is used for building construction, cooling, and cleansing, transportation of wastes, and fire suppression. Water can be an interior or exterior focal point with associations to nourishment, cleansing, and cooling.

In arid regions, water can be used sparingly in tightly controlled channels and small trickling streams. Where water is plentiful, it may be used for landscaping and fountains.

We associate water with cooling. Sunlight playing on the surface of water and the sound of running water both help us feel cooler. Water also has strong cultural and religious significance in Christian, Hindu, Islamic, and Jewish traditions.

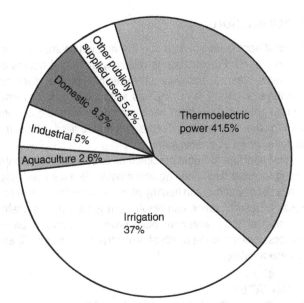

Figure 9.1 US freshwater withdrawals, 2005

Source: EPA US Freshwater Withdrawals (2005) with data from US Geological Service Circular 134: Estimated Use of Water in the United States in 2005, available at http://pubs.usgs.gov/circ/1344/

According to the US **Department of Energy (DOE)**, in 2005, the building sector was estimated to consume 39.6 billion gallons per day (bgd) of water, nearly 10 percent of the total US water use. Residential water use was the third largest category after thermoelectric power generation and irrigation. Commercial buildings used another 10.2 bgd. Population increases are somewhat balanced by efficiency efforts.

In 2005, 86 percent of the US population obtained drinking water from public supplies. (See Figure 9.1) The remaining 14 percent were supplied from private sources. The vast majority of water used in the United States is for the generation of electric power and for irrigation. All remaining uses make up only 18 percent of the total. Water used in homes is less than 1 percent of the total.

BUILDING MATERIALS

A great deal of water is used to create materials for building construction. (See Table 9.1) Plastic uses the most water per ton of material produced; steel also uses substantial amounts. Concrete is 16 percent water, but much more is used for processing and cleaning, and this water becomes acidic and difficult to reuse, although some can be recycled. Very large amounts of water are also used to produce bricks and steel.

TABLE 9.1 WATER USE IN CONSTRUCTION

Material	Water Used for One Ton (907 kg)
Bricks	580 gal (2200 L)
Steel	43,600 gal (165,000 L)
Plastic	348,750 gal (1.32 million L)
Concrete	124 gal (470 L)

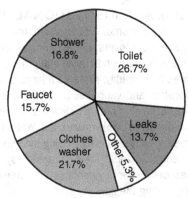

Figure 9.2 Residential water uses

Source: EPA "How Much Water Do We Use?, data source American Water Works Association Research Foundation, "Residential End Uses of Water," 1999

RESIDENTIAL USE

Although electrical production, irrigation, and industry all use substantial amounts of fresh water, residential water use remains a major factor, and one that we, as interior designers, can help to control while maintaining a good quality interior environment. (See Figure 9.2)

About 74 percent of the water used in the home is used in the bathroom, about 21 percent for laundry and cleaning, and about 5 percent in the kitchen. In our homes, toilets and clothes washing make up almost half of all the water used. Showers and faucets are the next greatest users, with leaks responsible for 13.7 percent of water use. Dishwashers and baths make up the remainder, less than 10 percent of the total.

The use of fresh water supplies by toilets should be the primary focus for water conservation in the home. (See Table 9.2) Another area of concern is the nearly 14 percent of water lost to leaks.

The high heat-storage capacity of water allows it to retain comfortable temperatures for bathing. Bathing facilities are usually designed for use in privacy, on a personal scale. Social bathing occurs in swimming pools, bathhouses, and hot tubs. Water is supplied for bathing by spouts, jets, and cascades to create the desired atmosphere.

TABLE 9.2 RESIDENTIAL PER CAPITA WATER USE

Use	Gallons Per Day	Percent
Drinking and cooking	3.0	4.1
Dishwashers	3.5	4.8
Faucets	7.9	10.8
Toilets	21.0	28.7
Showers	20.0	27.3
Baths	1.2	1.6
Clothes washers	15.0	20.5
Other domestic uses	1.6	2.2

COMMERCIAL AND INSTITUTIONAL USE

Almost all the water that small offices and institutional buildings use is for irrigation and toilets. Large buildings use about one-third of their water for air conditioning equipment.

Water stores heat readily, removes a large quantity of heat when it evaporates, and vaporizes at temperatures found on the human skin surface. In hot, dry climates, water surfaces or sprays and **evaporative coolers** are used. Water-filled **cooling towers** are used for large building cooling systems.

Water is central to most fire-suppression systems. Large diameter pipes deliver large quantities of water quickly, with large valves regulating the flow.

See Chapter 14, "Heating and Cooling," for more information on equipment, and Chapter 18," Fire Safety Design," for fire suppression.

Hydrologic Cycle

The process by which water constantly circulates by the sun's power is called the **hydrologic cycle**. (See Figure 9.3) The sun's heat powers the hydrologic cycle by evaporating water into the air and purifying it by distillation. The water vapor condenses as it rises and then precipitates as rain and snow, which clean the air as they fall to earth. Heavier particles fall out of the air by gravity, and the wind (driven by the sun's heat) dilutes and distributes any remaining contaminants when it stirs up the air.

Changing climate conditions affect how much water evaporates from ice, snow, and surface bodies of water. When, where, and how much water precipitates back to earth is also affected by climate, including wind and storm patterns. Warming temperatures influence how much water is stored as ice and snow.

The total amount of water on the earth and in the atmosphere is finite, except perhaps for a small additional amount that may be contributed by comets. The water we use today is the same water that was in Noah's proverbial flood. A quarter of the solar energy reaching the earth is employed in constantly circulating water through evaporation and precipitation via the hydrologic cycle.

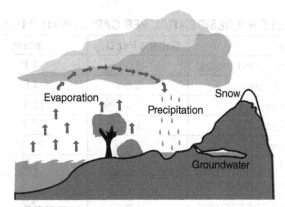

Figure 9.3 Hydrologic cycle

Precipitation

The most accessible sources of water for our use are precipitation and runoff. Precipitation is relatively pure, and occurs as rain, snow, sleet, or hail. Any daily precipitation that does not evaporate or run off is retained as soil moisture. After plants use it to grow, it evaporates back into the atmosphere, or runs down to where all the voids in the ground are filled with water, the zone known as groundwater.

It is important for most precipitation to be absorbed into the ground to avoid flash flooding and erosion. Healthy riverbanks (riparian ecosystems) and healthy plants in watersheds help this process. Desertification can result from ground that is unable to absorb or store water for plant growth. Impervious paved surfaces can lead to the death of urban trees, speed runoff, and increase flooding.

RAINWATER

Historically, rain falling in the countryside ran into creeks, streams, and rivers, and most rivers rarely ran dry. Rainfall was absorbed into the ground, which served as a huge reservoir. The water that accumulated underground emerged as springs and **artesian wells** (wells under pressure in confined aquifers), or in lakes, swamps, and marshes. Most of the water that leaked into the ground cleansed itself in the weeks, months, or years it took to get back to an aquifer.

Streets in towns that developed near rivers sloped to drain into the river, which ran to river basins and the sea. As marshy areas were filled in and buildings were built, paved streets and sidewalks channeled water to storm sewers and pumping stations. The rapid runoff increased the danger of flooding and concentrated pollutants in waterways. Water ran out of the ground into overflowing storm sewers without recharging groundwater levels.

RAINWATER RETENTION

In most of the United States, the rainwater that falls on the roof of a home is of adequate quality and quantity to supply the family's cleansing needs of about 35 gallons (132 L) per day.

For centuries, traditional builders have designed for rainwater retention (**rainwater harvesting**). In the world's drier regions, small cisterns within the home are used to collect rainwater to supplement unreliable public supplies. With the advent of central water and energy supplies in industrial societies, rainwater collection and use became less common, but this may be beginning to change.

In many areas of North America, half of the residential water used is for outdoor purposes. Rainwater can make a major contribution to the irrigation of small lawns and gardens by using a rain barrel below a downspout or a **cistern** (a tank for storing water) located above the level of the garden to collect and store water for later release. **Drip irrigation** emitters that deliver water directly to the root area of a plant are more efficient than lawn sprinklers. The alternative of using recycled or reclaimed water, such as **graywater** (wastewater from lavatories, showers, and baths) or stored rainwater for irrigation is gaining acceptance by North American building codes.

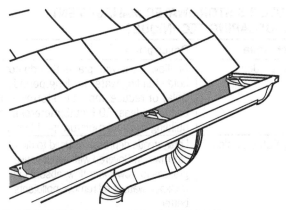

Figure 9.4 Gutter

Xeriscaping is a sustainable landscaping strategy focused on using drought-tolerant, native, and adapted species that require minimal or no water for maintenance.

Rain, snow, and other precipitation spread a supply of relatively pure water over a large area. Rain is generally diffuse, intermittent, and often seasonal. It is most often collected on-site, and is used where other sources are scarce or of poor quality. A system that combines public networks with individual rainwater cisterns offers major environmental benefits. If another supply is available for potable water, around 95 percent of indoor water usage could be provided by rainwater.

Rain affects the design of roofs, overhangs, gutters, and downspouts. **Gutters** are often exposed on the exterior of the building. (See Figure 9.4) Built-in gutters may leak into a building. Rainwater leaders may clog (slowing water flow) or freeze into ice. The downspout is the terminal part for the opening of a rain leader, through which water is discharged onto the ground.

Precipitation can be captured directly in cisterns, or as a more concentrated flow of runoff. Where permitted by code, water collected in a cistern can be used for flushing toilets, washing, or personal hygiene.

Urban rainwater is not commonly used for drinking and cooking. There are sometimes problems with particulates and lead or contaminants on catchment surfaces. Cisterns need to be checked periodically for bacterial growth.

There are no current US guidelines on rainwater harvesting, although several states address the subject. The local health code authority has jurisdiction.

SNOW

Snow—where it occurs—has a great impact on building and site design. Water stored as snow delays runoff after a storm. Snow is a great insulator, both thermally and acoustically. As one of the most reflective natural surfaces, it bounces light into interior spaces; glare can be a problem. Snow can collect on exterior light shelves or restrict the movement of awnings and other exterior sun controls. Exterior building parts can be damaged or impeded by the weight and mass or snow.

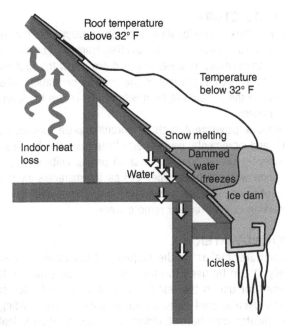

Figure 9.5 Ice dam

As snow collecting on a roof melts due to heat lost through the building envelope, melting snow on exterior overhangs may freeze and build up into ice dams that trap water running off the roof. (See Figure 9.5) If it runs out of room, this water may migrate up under roof shingles and into the building interior. Venting the roof keeps it cooler, and makes it less likely that warmer interior temperatures will melt snow. A vented roof requires adequate thermal insulation below the vent-air passage.

ACID RAIN

Acid rain is precipitation that is unusually acidic. It can harm plants, aquatic animals, buildings, historic monuments, and statues. It is caused by sulfur dioxide and nitrogen oxide emissions reacting with water molecules in the atmosphere. Electrical power plants using coal are a principal source. Acid rain affects the eastern third of the United States, southeastern Canada, and most of eastern Europe as well as the southeastern coast of China and Taiwan.

Surface Water, Groundwater, and Stormwater

Runoff is a concentrated flow of water that is relatively easy to capture in larger quantities. Runoff may contain organic, chemical, or radioactive pollution.

Surface water from precipitation affects the thermal, acoustic, and daylighting conditions of a building site. Evaporative cooling occurs as hot, dry breezes move over the surface of ponds. The pleasant sounds of running water mask other noises. Sparkling, shifting reflections off surface water can bring welcome daylight to a building interior, but may create glare.

SOIL MOISTURE

Soil moisture is precipitation that neither evaporates nor runs off. It is used by growing plants and then transpired by the plant to the atmosphere. As water drains down below the root level of plants, it eventually reaches the zone of saturation, where all voids in the ground are filled with water. This is the level of groundwater.

Porous pavement such as porous asphalt, porous concrete, and incremental paving helps keep rainwater in the soil. Incremental paving consists of small paving units with joints that allow water to pass. Open cell paving materials alternate areas of paving with grass or groundcover plants, and are used for parking in short-term or remote areas.

GROUNDWATER

Groundwater makes up the majority of our available water supply. It can be used to store excess building heat in the summer for use in the building in winter. Groundwater can also be used to cool water stored by solar-heated buildings. Groundwater may harm building foundations when it leaks into spaces below ground.

Groundwater is generally found near the surface where it can be accessed by drilling or as a spring and be recharged by precipitation percolating through the soil. The upper surface of groundwater, the top of the aquifer, is called the **water table**. It is the level of the saturated or water-filled part of the porous subsurface layer, and the level at which water is located below grade.

STORMWATER MANAGEMENT

Stormwater management traditionally involved removing the water from a building and its surrounding property as quickly as possible. Large storm sewers carried stormwater, along with any trash and pollutants, to waterways. Combined stormwater and sewage drainage still used in older cities can overflow in downpours and divert untreated sewage into local water bodies.

Stormwater management today seeks to delay runoff, and to infiltrate the ground to replenish the groundwater water table and landscape. (See Table 9.3) By increasing evaporation from surfaces and plants, stormwater management creates a cooling effect. Water is cleaned before it is returned to streams and lakes.

Site landscaping encourages animal life, reduces cooling needs, and helps clean the air. Plant roots hold soil in place, and their leaves diminish the impact of rain, reducing erosion. Good landscaping reduces the intensity and quantity of stormwater.

Conservation

There are many reasons to conserve water. A population increase means that a limited resource has to go farther. Ancient aquifers that are depleted will take thousands of years to recharge. Some existing water sources continue to be contaminated. The cost of supplying and transporting safe, clean water is increasing.

TABLE 9.3 STORMWATER MANAGEMENT LANDSCAPING TECHNIQUES

Technique	Description
Rain gardens	Shallow areas with plants that do not hold standing water for long periods. Delay or reduce impact of stormwater. Locate at least 10 feet (3 meters) from building to avoid moisture problems.
Vegetated swales	Long, gentle depressions used to direct stormwater. Covered with uncut native grasses to slow water flow, absorb and transpire water, and remove pollutants better.
Infiltration trenches (French drains)	Trenches along drip line from roof, filled with rocks or gravel and often a perforated pipe. Help keep stormwater out of building, slow its absorption.
Dry well	Large hole filled with rocks or gravel, stores stormwater temporarily, allows infiltration into soil.

Most of the potable water in North American buildings is used to carry away organic waste. This use affects everything from the detailed arrangement of bathroom fixtures and interior surfaces to regional plans for very large and complex water and sewage treatment facilities.

Affluence apparently increases water use. Lawns, swimming pools, and golf courses use very large amounts of water.

Increasingly, building designers and others are seeking to reserve high-quality water for high-grade tasks. This involves emphasizing recycling and water conservation.

The EPA estimates that a faucet that loses one drop per second can waste 3000 gallons (11,356 L) annually. Leaking pipes can be prevented by careful installation of high-quality fixtures, fittings, and water-using appliances.

CONSERVATION STRATEGIES

Educating clients and fellow professionals is the first step to water conservation. Interior designers are in an excellent position to advocate the use of efficient plumbing fixtures and water-using appliances and equipment, as well as water restrictors and aerators for faucets and showerheads and sensor-activated faucets on sinks and lavatories. More complex conservation strategies include dual plumbing that separates potable and nonpotable systems, using purple pipes for graywater, using tankless water heaters where appropriate, and recirculating hot water. Leak control also saves water.

See Chapter 11, "Fixtures and Appliances," for more information on water-efficient equipment.

Another way to reduce potable water use is to change the quality of water used. Potable water should be used for all applications involving human consumption or ingestion. Other sources should be sought for landscaping irrigation, fire protection, heating and cooling equipment, toilet flushing, and other uses.

WATER EFFICIENCY STANDARDS

The Energy Policy Act of 1992 set maximum usage for various fixtures, including showerhead restrictions that assume shorter 5-minute showers, 1.6 gallon toilet flushing, and high-efficiency front-loading clothes washers, with no potable water use for irrigation and less leakage.

Per capita water use in the United States is 25 percent lower than in late 1970s. This is primarily due to less agricultural and power generation use as well as awareness of the value of water.

LEED v4 certification requires reducing water consumption by 20 percent from the LEED baseline level. The LEED system also sets standards for appliances.

The US EPA WaterSense program was begun in 2006. It provides a product label that facilitates purchase of high-performing, water-efficient products.

Protecting the Water Supply

Potable water is water that is free of harmful bacteria and safe to drink or use for food preparation. The water carried from the public water supply to individual buildings in **water mains**—large underground pipes—must be potable.

Protecting and conserving our clean water supplies is critical to our health. Until recently, a reliable supply of clean water was not always available, and epidemic diseases continue to be spread through unsanitary water supplies. Water from ponds or streams in built-up areas is unsafe to drink as it may contain biological or chemical pollution.

CONTAMINATED WATER SUPPLIES

For centuries, people knew the benefits of living near a source of fresh water. Unfortunately, it took them a long time to realize that their own wastes and those of their animals could pollute water sources and cause disease and death. Bacteria were unknown to science until discovered in Germany in 1892.

Waterborne bacterial diseases are responsible for an estimated 2 million deaths worldwide per year, mostly children. Diseases include cholera, typhoid fever, dysentery, and Brainerd diarrhea, all of which involve abdominal symptoms and diarrhea. They are caused by bacteria including *Vibrio cholera*, *Campylobacter*, *Salmonella*, *Shigella*, and **Eschericia coli (E coli)**.

These diseases are spread by contaminated surface water sources and large, poorly functioning municipal water distributions. They can be prevented by chlorination and safe water handling. However, water treatment systems are expensive to build and to operate, and have not been able to keep up with growth in population and human migration. Effective alternatives include point-of-use disinfection and safe water storage vessels. Low-cost, locally controlled technologies are becoming more prevalent.

Although earth is largely a water planet, fresh water is not evenly distributed. More than one-third of the people living today do not have access to safe water. In addition to geological and climactic issues, political and economic barriers prevent access to clean water. Some water supplies are contaminated by industry, while others contain naturally occurring pollutants.

Water from wells and mountain reservoirs needs relatively little treatment. River water is sent through sand filters and settling basins, where particles are removed. Additional chemical treatment precipitates iron and lead compounds. Special filters are used for hydrogen sulfide, radon, and other dissolved gases. Finally, chlorine dissolved in water kills harmful microorganisms. The result is an increased supply of clean water to support the development of residential and commercial construction.

WATER DISTRIBUTION

Private water systems serve rural areas and many small communities, with each building having its own water supply. A few locations have their own reliable springs for pure water. However, most private water systems depend on catching rainfall (usually on the roof) or using a well.

Larger communities rely on municipal water supply systems. These systems centralize collection, treatment, and distribution of water.

Well Water

Wells supply water of more reliable quantity and quality than a rainwater system. However, water near the surface may have seeped into the ground from the immediate area, and may be contaminated by sewage, barnyards, outhouses, or garbage dumps nearby. Well water is often hard water and may require a water softener. Most wells require the use of a pump. (See Figure 9.6)

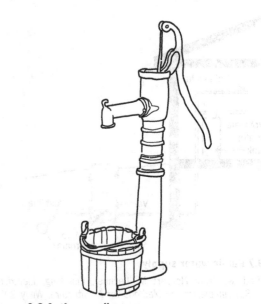

Figure 9.6 Antique well pump

Deep wells are the primary on-site source of privately supplied water. Over 14 percent of US single-family residences have no community water service and rely on wells.

Wells are commonly sunk substantially below the water table to avoid seasonal fluctuations. They may be driven with pipe sections or bored with augers. Jetted wells tap a source close to the surface using a pressure pump.

Deep wells are expensive to drill, but the water deep underground comes from hundreds of miles away, and the long trip filters out most bacteria. Well water sometimes contains dissolved minerals, most of which are harmless, but which may result in hard water conditions.

Well water is usually potable if the source is deep enough. It should be pure, cool, and free of discoloration and odor problems. The local health department can check samples for bacterial and chemical content before use.

Springs and artesian wells are relatively unusual. An artesian well is a well drilled into a pressurized portion of a confined aquifer. Artesian wells do not need a pump to bring water to the surface, but may still need one to distribute the water to fixtures.

Municipal Water Supply Systems

According to US Geological Survey (USGS) 2014 data, about 86 percent of the US population gets water from a public supply system. A centralized public water distribution system includes a **water meter** and shut-off valve installed when a tap is made to the water main. (See Figure 9.7) Remote gauges or sensor units can read a meter located many feet below the surface.

WATER SUPPLY EQUIPMENT

Water mains are large pipes that transport water for a public water system from its source to service connections at buildings. Water mains underneath sidewalks and streets carry water

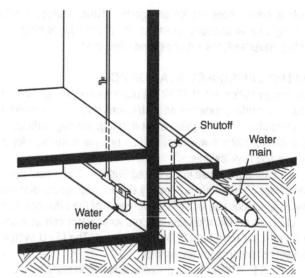

Figure 9.8 Water main and meter

Figure 9.9 Water shutoff cover

from reservoirs to buildings. The water main is connected to each building by a supply line. A service pipe installed by the public water utility runs from the water main to the building, far enough underground to avoid freezing in winter.

Within the building or in a curb box, a water meter measures and records the quantity of water passing through the service pipe and usually also monitors sewage disposal services. (See Figure 9.8) A control valve is located in the curb box to shut off the water supply to the building in an emergency or if the building owner fails to pay the water bill. A shutoff valve within the building also controls the water supply. (See Figure 9.9)

Water Quality

The relationship between epidemic diseases and water quality was not made until the nineteenth century. In 1854, London physician John Snow deduced that local cholera cases originated from a single pump contaminated with sewage from a nearby house.

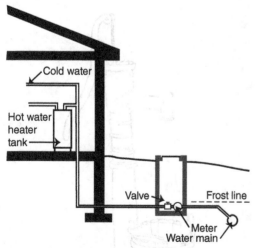

Figure 9.7 Public water supply

Source: Redrawn from Norbert M. Lechner, *Plumbing, Electricity, Acoustics: Sustainable Design Methods for Architecture*, Wiley 2012, page 80

Water quality is regulated by federal, state, and local laws. The 1996 Amendments to the Safe Drinking Water Act, through the EPA, requires water suppliers to publish annual Water Quality Reports that state what water pollutants are present, to what degree, and whether they exceed permissible limits. US Food and Drug Administration (FDA) regulations for bottled water are less stringent than the EPA tap water standards.

Pesticides, cleaning solvents, and seepage from landfills pollute groundwater in some rural areas of the United States. In urban areas, the level of chlorine added to prevent bacterial contamination sometimes results in bad tasting water and deterioration of pipes and plumbing fixtures.

The textile industry uses large quantities of water in fiber production and processing, and fabric finishing, especially dyeing. As an interior designer, you can avoid products whose manufacturing includes highly toxic technologies and seek out those with low environmental impact.

WATER QUALITY CHARACTERISTICS

Communities routinely check on the quality of their municipal water supplies. If a home or business owner is unsure whether their building's supply meets safety standards, a government or private water quality analyst will provide instructions and containers for taking samples, and assess the purity of the water supply. The analyst's report gives numerical values for mineral content, acidity or alkalinity (pH level), contamination, turbidity, total solids, and biological purity, and an opinion on the sample's suitability for its intended use.

Water characteristics are classified as physical, chemical, biological, or radiological. Physical water characteristics derive especially from the surfaces of roofs and water bodies. Chemical characteristics occur when groundwater slowly dissolves minerals in rocks and soils as it moves down from the surface. Chemical problems may be indicated by a tendency for water to stain fixtures and clothing. Water may contain biological organisms that normally do not cause disease, but the presence of which in very large amounts is undesirable. The presence of radioactive chemicals can produce adverse health effects even in very low concentrations.

Hard water is water with an excessive percentage of minerals. It can leave mineral deposits on fixtures and plumbing, which are unsightly, especially on darker materials. Hard water prevents soap from lathering well, and affects some industrial processes. Hard water is not covered by EPA standards. Corrosive materials in water create flakes of scale that, when combined with hair and other debris, clog water pipes.

Water softening treatments can improve the appearance and extend the life of plumbing fixtures. Water softeners are usually installed near the water heater and typically treat all household water. They are used to remove calcium and magnesium. Softening may give the water an undesirable taste.

The softening process increases the sodium content of the water. This may be a problem for people on low-sodium diets. To minimize problems with sodium used for softening water, it is possible to treat only the hot water supply, or to provide a cold water line to the kitchen faucet that bypasses the softener.

WATER QUALITY TREATMENTS

Primary water treatment begins with filtration, followed by disinfection to kill microorganisms in the water. **Secondary water treatment** keeps the level of disinfectant high enough to prevent microorganism regrowth.

Residential water filters include a wide variety of equipment from carafes to whole-house systems. (See Table 9.4) They are becoming simpler and easier to install. Water filtration at the faucet requires space under the sink, with access for filter replacement.

Fiber (mechanical) filters are available as small kitchen faucet filters, or as larger under-sink filters. A bypass on the faucet allows the choice of filtered or unfiltered water, and can extend the life of the filter cartridge.

Distribution within Buildings

Throughout history, a primary concern of architects, builders, and homeowners has been how to keep water out of buildings. It was not until the end of the nineteenth century that supplying water inside a building became common in industrial countries. Indoor plumbing is still not available in many parts of the world today.

Interior designers work with architects, engineers, and contractors to make sure that water is supplied in a way that supports health, safety, comfort, and utility for the client. For indoor plumbing to work safely without spreading bacteria and polluting the fresh water supply, it's necessary to construct two completely separate systems. The first, the **water supply system** delivers clean water

TABLE 9.4 HOUSEHOLD WATER TREATMENT METHODS

Method	Problem Treated
Activated carbon filtration	Taste and odor problems, chlorine residue, organic chemicals, and radon
Chlorination	Coliform bacteria, iron particles, iron bacteria, manganese
Distillation	Metals, inorganic chemicals, other contaminants
Neutralizing filtration	Acidity (low pH)
Particle or fiber filtration	Dissolved solids, iron particles
Oxidizing filtration	Iron particles, manganese
Anion exchange	Nitrate, sulfate, arsenic
Reverse osmosis	Metals, inorganic chemicals, other contaminants
Water softening	Calcium, magnesium, iron particles

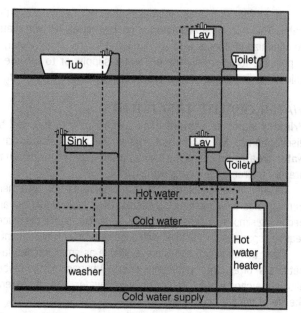

Figure 9.10 Water supply system

In cold climates, the water supply must enter the building below the frost line to prevent freezing. Unheated houses must be entirely free of water that could freeze and burst pipes. House shutoff controls are usually located at the main, at the curb, and within the house. There should be a drain valve at every low point in the system.

LARGE BUILDING WATER SUPPLY

In large buildings with many fixtures, piping is usually concealed except in basements, utility rooms, and points of access to controls. It is often located in **pipe chases**, vertical and horizontal open spaces between walls or ceiling and floor. (See Figure 9.12) Pipe chases often have access doors to allow the pipes to be worked on. The water supply plumbing and the sanitary drainage plumbing must be coordinated with the building's structure and with other building systems.

The weight of the vertical supply pipes and the water they contain is supported at each story and horizontally every 6 to 10 feet (1.8 to 3 m) apart. Adjustable hangars are used to pitch the horizontal waste pipes downward for gravity drainage.

WATER PRESSURE

The water in a community's water mains is under pressure to offset friction and gravity as it flows through the pipes. The

to buildings. (See Figure 9.10) Some of the water entering a building is diverted to the water heater; from there it travels in hot water supply piping to fixtures and appliances that need it. The rest of the water goes through cold water supply piping to appliances and fixtures. The second is a system of drains called the **sanitary waste** or **drain, waste, and vent (DWV) system**, which channels all the waste downward through the building to the sewer below.

For information on the sanitary waste system, see Chapter 10, "Waste and Reuse Systems."

SMALL BUILDING WATER SUPPLY

In small wood-frame buildings, indoor plumbing usually is hidden in floor joist and wall construction spaces. Masonry buildings require spaces that are built out with wood furring strips or metal channels to hide horizontal and vertical plumbing. (See Figure 9.11) It is difficult to run plumbing piping through masonry walls, and not a good idea for those that are subject to freezing.

Figure 9.11 Masonry wall with furring for plumbing

Figure 9.12 Pipe chase

TABLE 9.5 TYPES OF WATER DISTRIBUTION SYSTEMS

Type	Application	Description
Upfeed distribution	Small low buildings with moderate water use	Water main pressure, or pressure from pumped wells
Pumped upfeed distribution	Medium-sized buildings with inadequate water pressure	Pumps provide additional pressure
Hydropneumatic systems	Compressed air maintains water pressure	Forces water into sealed pressurized tanks
Downfeed distribution	Water raised to rooftop storage tanks, drops down to plumbing fixtures	Rooftop tanks may need to be heated to avoid freezing; water available for fire hoses; requires extra structural support

water pressure in public water supplies is usually around 50 to 70 pounds per square inch (345 kilopascals [kPa]). This is also about the maximum achieved by private well systems, and is usually adequate pressure for two- or three-story buildings.

The actual water pressure required for distribution is related to the volume of water supplied, the diameter of the pipes, the length of the pipes, and the height to which the water is raised. For distribution in taller buildings, **upfeed, pumped upfeed, hydropneumatic,** or **downfeed distribution** is used. (See Table 9.5)

Once the water is inside the building, its pressure is changed by the size of the pipes it travels through. Bigger pipes put less pressure on the water flow, while small pipes increase the pressure. If the water rises up high in the building, gravity and friction combine to decrease the pressure. (See Figure 9.13) The

water pressure at individual fixtures within the building may vary between 5 and 30 psi (35 and 204 kPa).

Too much pressure results in splashing; it can be moderated with a flow restrictor on the faucet outlet. Too little produces a slow dribble. Water supply pipes are sized to use up the difference between the service pressure and the pressure required for each fixture.

Installing a pressure-reducing valve where supply pressure is too high can avoid leakage and poor operation of automatic valves in clothes washers and dishwashers.

Within the building, water continues under pressure to supply makeup water for the space-heating boiler, supply and pressurize cold water mains and branches, and supply and pressurize the water system through a hot water heater, hot water storage tank, and water mains, branches, and circulating lines.

SUPPLY PIPES
Pipe systems consist of pipe or tubing and fittings to join pipes to fixtures, valves, water heaters, or other pipes. The overall water demand is established by the number of fixtures, which in turn is based on the types of building occupancies and the number of occupants.

Lead was used for plumbing pipes by the Romans two thousand years ago, and the word *plumbing* is derived from the Latin word for lead, *plumbum*. Indoor plumbing reached a height in the Roman Empire, when public baths combined hot and cold pools, exercise spaces, and even libraries and dining areas. (See Figure 9.14)

Lead pipes were used through the 1950s. The EPA is concerned even today that lead may leach out of old lead pipes and copper pipes joined with lead solder and enter the water supply. Fortunately, lead on the inside surface of a pipe quickly reacts to form a coating that keeps it from leaching out of the pipe. However lead content in water may exceed safe guidelines when the water is highly acidic or is allowed to sit in lead pipes for a long time. Today, plumbing supply pipes are made of copper, red brass, galvanized steel, and plastic.

Figure 9.13 How the weight of water above an opening increases the pressure of that opening

Figure 9.14 Roman bath, Bath, England
Source: Roman Baths and Abbey, Circular Bath, Bath, England, Library of Congress Prints and Photographs Division

Figure 9.15 Plumber opening trap
Source: Bob Giuliani, *Illustrations of Services and Trades*, Dover Publications, Inc., 1995, copyright free

Local codes vary in acceptance of plastic piping. Plastic pipe is widely used in water supply piping, fittings, and draining systems. Most plastic piping is made from thermoplastics that will repeatedly soften when heat is applied. Plastic pipe used for potable water is required to have a seal from the National Sanitation Foundation (NSF). Most plastic pipes are glued together rather than soldered, which makes them more difficult to take apart than metal pipes.

Cross-linked polyethylene, known as PEX, is formed into tubing that is widely replacing the use of copper plumbing. Its cross-linking properties change it from thermoplastic to thermoset, which allows it to remain flexible where not cross-linked and make tight compression joints with itself when heat set without the use of potentially toxic solvent. The use of flexible PEX tubing reduces labor costs as well as the time it takes to get hot water, which in turn reduces water and energy waste.

Threaded connections are used for ferrous pipes and what is termed iron-pipe-size brass. Larger ferrous pipes are often welded or connected with bolted flanges. Copper pipes are soldered after being placed in their final position.

Plumbing pipes must be accessible for cleaning and repairs. (See Figure 9.15) This may require access panels.

The greater the velocity of water flowing through pipes, the greater the resistance and the more noise is produced. High velocity is also likely to produce pipe vibrations, generating more noise. To control noise, the velocity should not be greater than 6 feet per second (fps) for pipes within an occupied area. Supporting pipes resiliently avoids transmitting pipe vibrations to the building's structure. To contain sound, pipes should also be enclosed by at least ½" (13 mm) of gypsum wallboard.

Engineers determine pipe sizes by the rate at which the pipes will transport water when there is the most demand. Pipes in the supply network tend to become smaller as they get farther from the water source and closer to the point of use, since not all of the water has to make the whole trip. The sizes depend on the number and types of fixtures to be served and pressure losses due to friction and vertical travel. Water flowing through a smaller pipe is under greater pressure than the same amount of water in a larger pipe.

The sizing of water pipes depends on the availability of sufficient pressure at fixtures. Local ordinances often state that the flow must be adequate to keep fixtures clean and sanitary. Pipe size is based on the amount of flow in gallons per minute (gpm), the imposed resistance or pressure loss, and water velocity.

The interior designer needs to give the engineer specific information about the number of plumbing fixtures and their requirements as early in the process as possible. Each type of fixture is assigned a number of fixture units. The number of gallons per minute is estimated based on the total number of fixture units for the building, The engineer assumes that not all the fixtures are in use at the same time, so the total demand is not directly proportional to the number of fixture units.

For economy of piping, it is a good idea whenever possible to locate spaces with plumbing needs, such as a kitchen and a bathroom, or two bathrooms, back-to-back.

CONDENSATION AND INSULATION

Supply water piping is insulated to prevent condensation and to reduce heat transfer between the air and pipe contents. All cold-water piping and fittings should be covered with insulation and a tight vapor barrier to prevent condensation. All hot-water piping in recirculating hot water systems should be insulated to conserve energy.

Pipes sweat when moisture in the air condenses on the outsides of cold pipes. This condensation drops off the pipes, wetting and damaging finished surfaces and aiding the growth of mold. Insulation keeps heat from adjacent warm spaces from warming the water in the pipes. When pipes are wrapped in glass fiber ½" to 1" (13 mm to 25 mm) thick with a tight vapor

Figure 9.16 Pipe insulation

retarder on the exterior surface, the moisture in the air cannot get to the cold surface. The vapor retarder prevents heat flow from warmer air as well, keeping the water cooler. Pipe insulation is available preformed to wrap around water pipes, making home improvement easier. (See Figure 9.16)

Hot water pipes are insulated to prevent heat loss. Hot and cold water supply lines often run close to each other. Heat from the hot water can be lost to the adjacent cold water lines. When hot and cold water pipes run parallel to each other, they should be a minimum of 6" (152 mm) apart, to avoid exchanging heat, even when the pipes are insulated.

Storage tanks and water heaters usually are manufactured with integral insulation. However, older devices may have less insulation than desired; these can be retrofitted by adding an insulating blanket.

In very cold climates, water pipes in exterior walls and unheated buildings may freeze and rupture. Avoid locating fixtures along exterior walls for this reason. If water supply pipes must be located in an exterior wall, they should be placed on the warm side (inside) of the wall insulation. A drainage faucet located at a low point will allow the pipes to be drained before being exposed to freezing weather.

Plumbing should not be installed in exterior walls where there is any chance of subfreezing weather. This may require that no fixtures be placed on exterior walls.

BRANCH SUPPLY LINES

A line runs out to each fixture from a branch supply line. (See Figure 9.17) **Roughing-in** is the process of getting all the pipes installed, capped, and pressure tested before actual fixtures are installed. The rough-in dimensions for each plumbing fixture should be verified with the fixture manufacturer so that fixture supports can be built in accurately during the proper phase of construction.

Small-scale piping assemblies normally fit into a 6" (152 mm) interior partition.

Standard ½" (13 mm) supply pipes may be insufficient in bathrooms that include spa tubs, whirlpool tubs, or high-volume

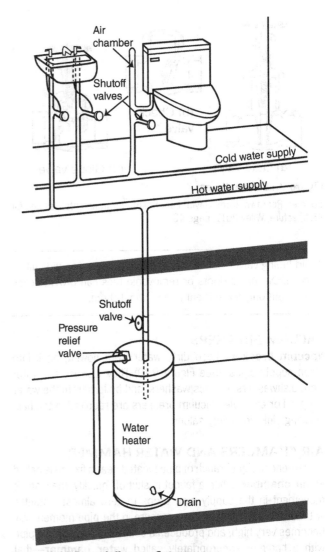

Figure 9.17 Branch lines to fixtures

or multihead showers with large water demands. Larger tubs and whirlpools typically use a ¾" (19 mm) water supply line with a ¾" (19 mm) iron pipe size valve and bath spout.

PLUMBING VALVES

It is a good idea to have a **shut-off valve** to control the flow of water at each vertical pipe (known as a **riser**), with branches for kitchens and baths and at the runouts to individual fixtures. Additional valves may be installed to isolate one or more fixtures from the water supply system for repair and maintenance. A water shut-off valve in or near each bathroom fixture, or where the water supply enters the area, instead of only in basement or outside, allows for emergency shut off.

Valve types include globe, gate, angle, check butterfly, and ball valves. (See Figure 9.18) Mixing valves are required for showers to eliminate temperature change due to variations in line pressure. Mixing valves have a maximum discharge temperature 120 degrees F (49°C).

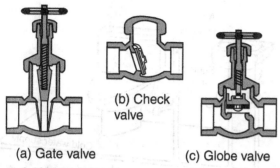

Figure 9.18 Valve types

Source: Redrawn from David Lee Smith, *Environmental Issues for Architecture*, Wiley 2011, page 435

Avoid hiding valves behind panels or doors that are difficult to remove. Disguised doors or removable tiles can provide easy access, but only if the client knows about them.

VACUUM BREAKERS

Vacuum breakers keep dirty water from flowing back into clean supply pipes. (See Figure 9.19) They also isolate water from dishwashers, clothes washers, and boilers from the water supply. For example, vacuum breakers are required in the hair-washing sinks in beauty salons.

AIR CHAMBERS AND WATER HAMMER

A dead-end upright branch of pipe located near a fixture is called an **air chamber**. When a faucet is shut off quickly, the water's movement in the supply pipe drops to zero almost instantly. Without the air chamber, the pressure in the pipe momentarily becomes very high, and produces a sound like banging the pipe with a hammer—appropriately called **water hammer**—that may damage the system. This can happen when due to the operation of an automatic shutoff valve in a clothes washer, or a quick-acting faucet. The air chamber absorbs the shock and prevents water hammer.

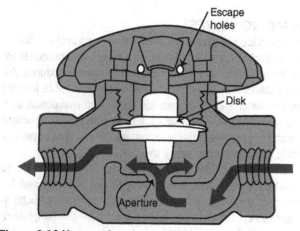

Figure 9.19 Vacuum breaker

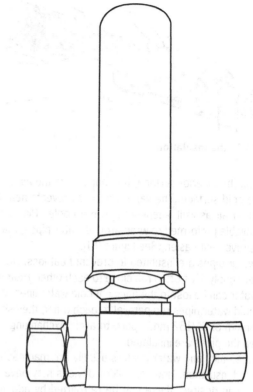

Figure 9.20 Water hammer shock arrestor

Water hammer at its worst sounds like a pipe being hit by a metal hammer, but more often like the thud of a rubber mallet, followed by a rattle of pipes.

Because the air in the pipe could possibly be absorbed over time or harbor bacteria that could contaminate the water supply, many jurisdictions now require anti-knock devices designed specifically as a water hammer arrestor, which use a piston rather than air in a column. (See Figure 9.20) When the water hammer afflicted valve near the shock absorber closes, the initial shock expands the rubber bellows. This displaces the hydraulic fluid, and the shock is absorbed by the inert gas.

Hot Water

The amount of energy used to heat water depends on the building's function and energy conservation efforts. (See Table 9.6) For example, over the past few years hotels have managed to decrease their water efficiency from 40 percent of total energy used to around 33 percent.

The hot water that is used for bathing, clothes washing, washing dishes, and many other things, but not for heating building spaces, is called **domestic hot water (DHW)** or sometimes **building service hot water** (in nonresidential buildings). When a well-insulated building uses very little water for space heating but uses a lot of hot water for other purposes, a single

TABLE 9.6 HOT WATER ENERGY USE

Use	% of Total Energy Use
Hotel	33
Healthcare	27
Educational	22
Residential	15

large hot water heater may supply both. In a large building, a return pipe is usually provided near each fixture to return water to the water heater for rewarming.

HOT WATER USAGE
The efficiency of hand washing dishes in a sink compared to using a dishwasher depends on how the hand washing is done. Older dishwashers used 8 to 15 gallons per load, plus up to 3 times the as much electricity. An efficient dishwasher today uses between 3 and 5 gallons per load. It is possible to do an equivalent amount of dishes by hand with less than 8 gallons, but regular hand-washing can use up to 27 gallons. If you run the washer only when it is full, a newer automatic dishwasher is likely to be more efficient.

A tub bath will use about 13 to 15 gallons of hot water. A short shower requires 6 to 8 gallons, but a 5- to 6-minute shower is equivalent to a tub bath.

Clothes washers vary in the amount of hot water they use. (See Table 9.7)

HOT WATER CONSERVATION
Hot water uses two types of resources: energy to heat water and the water itself. Strategies for saving hot water are in some respects similar to water-saving methods in general. Demand can be reduced by using flow restrictors, aerators, and automatic faucets, and by washing clothes in cold water. Other conservation methods include:

- Insulate hot water pipes and storage tanks.
- Install heat traps that allow water to flow into the tank but prevent hot water outflow from tanks, thus preventing convection currents from moving hot water into pipes.

TABLE 9.7 CLOTHES WASHER HOT WATER USE

Machine Type	Gallons (Liters) per Load
Standard top-loading washer newer models	Under 30 (114)
Front- or top-loading high-efficiency washer	Up to 25 or 30 (95 to 114)
Ultra-high efficiency vertical axis with wash plates	About 15 (57)
Horizontal-axis washers	Less than 15 gal (57)

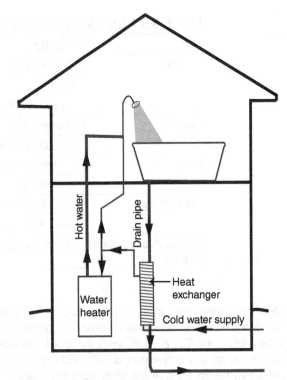

Figure 9.21 Heat exchanger for shower

- Use a timer to turn off electric water heaters when not in use.
- Install a separate meter and heat water at night only for lower electricity costs.
- Use wind or solar energy, with backup heaters for calm or overcast days.
- Increase efficiency with a forced circulating hot water system to continuously supply hot water at each fixture.
- Produce hot water more efficiently with a high-efficiency gas heater.
- Avoid electric resistance heaters, and use an electric heat pump instead.
- For energy recovery, use a heat exchanger to prewarm cold water on way to hot water tank. (See Figure 9.21)

TEMPERATURES
In many buildings, water is heated and stored in large tanks at high temperatures to assure that hot water will always be available in sufficient quantities. Inevitably, some heat is lost during storage and delivery. Sometimes the water tap must be run for an extended period in order to access the water at the desired temperature. In order to save energy and water, wise consumers use water at lower temperatures whenever feasible.

People generally take showers at 105° to 120°F (41° to 49°C), often by blending hot water at 140°F (60°C) with cold water with a mixing valve in the shower. Most people experience temperatures above 110°F (43°C) as uncomfortably hot. Higher water temperatures allow smaller hot water tanks to be installed, since the super-hot water is mixed with cold water before use.

TABLE 9.8 REPRESENTATIVE HOT WATER TEMPERATURES

Use	Activity	Temperature
Lavatory	Hand washing	105°F (40°C)
	Shaving	115°F (45°C)
	Surgical scrubbing	110° F (43°C)
Bathing	Showers, tubs	105° to 120°F (41° to 49°C)
	Therapeutic baths	95°F (35°C)
Laundry	Commercial, institutional	Up to 180°F (82°C)
	Residential	140°F (60°C)
Dishwashing	Commercial spray-type: wash	150°F (65°C) minimum
	final sanitizing rinse	180° to 195°F (82° to 90°C)
General cleaning, food preparation	Typically limited to avoid burns	Not exceeding 140°F (60°C)

Much higher temperatures are used for some commercial purposes, and high temperatures may be required or limited by codes for some applications. The 2015 IPC limits the potable hot water distribution system to a temperature of 140°F (60°C) or less. Local codes may specify the type and efficiency of water heater for remodeling, replacement, or new construction. (See Table 9.8)

Temperatures above 140 degrees F (60°C) can cause serious burns, and promote scalding if the water is hard. However, high temperatures limit the growth of the harmful bacterium *Legionella pneumophila*, which causes **Legionnaire's disease**.

Lower temperatures are less likely to cause burns, but may be inadequate for sanitation. Lower temperature water loses less heat lost in storage and in pipes, saving energy. Smaller heating units are adequate for lower temperatures, but larger storage tanks are needed. Solar or waste heat recovery sources work better with lower temperature water heaters.

Hot Water Heaters

Tank style hot water heaters typically hold 30 to 70 gallons (114 to 265 L); about 70 percent of that amount is usable capacity. Size considerations include the client's flow rate demand, the temperature of the water entering the tank, and the desired output temperature.

Most manufacturers have software to help calculate the size of hot water tank needed.

SOLAR WATER HEATERS

Solar hot water is one of the most common and cost-effective solar applications. Even in less-sunny climates, solar hot water can provide most of the needs for a residence. Solar water heaters manufactured as packages that include collectors, storage tank, and controls are available in many countries worldwide. (See Figure 9.22) Panels are most often located on a roof, but can also be mounted on the ground.

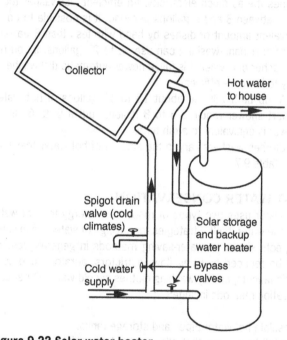

Figure 9.22 Solar water heater

Solar water heating can easily meet most of the summer DHW demand throughout the United States. In places where temperatures fall below 42°F (60°C), solar water heating systems need protection against freezing. Codes in some parts of the United States may require solar water heaters for new construction.

Solar water heaters are designed as either active systems with circulating pumps and controls, or as passive systems. Active solar water heating systems can be located anywhere, but are subject to mechanical breakdowns and require increased maintenance. Panels have tubes for water or a transfer liquid that goes to a heat exchanger. The heated water goes to a storage tank.

Passive solar water heating systems are typically less expensive than active ones. Although they are usually less efficient, they have lower component costs, and are often more reliable, lasting longer than active systems. Passive systems rely on gravity for circulation, so the heavy storage tank must be above the collector, which can create structural problems.

Solar water heaters use either direct or indirect systems. In a direct system, the water to be used in the building circulates through the solar collector. Direct systems are simple, efficient, and do not require a separate fluid loop for heat exchange.

Indirect solar water heating systems use a closed loop containing fluid that circulates through a collector and storage tank. Heat is passed from the fluid to DHW through a heat exchanger. An indirect system allows the use of nonfreezing fluid in the collector loop, and collectors can be operated at low pressure.

Solar energy can be used to heat outdoor swimming pools during the months with most sun. Solar pool heating extends the swimming season by several weeks and can pay for itself within two years.

STORAGE TANK WATER HEATERS

Hot water can be heated at a central location or at the point of use. Central hot water heating systems waste both energy and water. Their primary advantage is their low initial cost. Central locations are most popular in the United States. The rest of the world prefers tankless point of use heaters. This is apparently mostly due to differences in energy costs.

The location of a hot water heater affects how quickly hot water arrives at a bathroom. Adding a second water heater near a residential bathroom may be a good idea, especially where there is a high-demand whirlpool tub or multihead shower.

A storage tank water heater should be energy efficient and sized to the needs of the household. Installing too large a water heater uses more energy than required.

The water enters a storage tank water heater (See Figure 9.23) at the bottom of the tank where it is heated, and leaves at the top. The heat loss through the sides of the tank continues even when no hot water is being used, so storage water heaters keep using energy to maintain water temperature. The tanks usually are insulated to retain heat, but some older models may need more insulation. Local utilities will sometimes insulate hot water tanks for free. High-efficiency water heaters are better insulated and use less energy.

Water heaters are selected based on their energy efficiency, tank storage capacity, and the amount of time it takes for the water temperature to rise a designated amount. Capacity affects how many fixtures can use hot water simultaneously. Recovery rate is the time it takes to heat around half of the demand measured in gallons per hour. The first hour rating (FHR) is based on tank size and recovery rate. Gas heaters have higher recovery rate than electric heaters for the same FHR, so their tanks can be smaller.

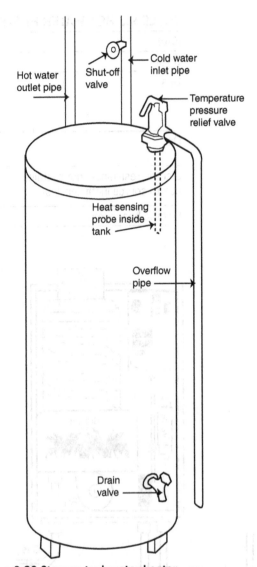

Figure 9.23 Storage tank water heater

The **energy factor (EF)** is used to compare the energy efficiency of hot water heaters. (See Table 9.9) Higher numbers are more efficient. The US Department of Energy (DOE) requires the yellow **Energy Guide Label** on hot water heaters to facilitate comparison.

TANKLESS WATER HEATERS

Small **tankless water heaters** (also called instantaneous, demand, or point-of-use heaters) raise the water temperature very quickly within a heating coil, from which it is immediately sent to the point of use. A gas burner or electrical element heats the water as needed. (See Figure 9.24) These water heaters have no storage tank, and consequently, do not lose heat.

According to the US DOE, homes that use 4 gallons or less of hot water daily can save one-third of the energy they use with tankless heaters. Tankless heaters typically produce a maximum of 5 gpm (gas) or around 2 gpm (electric resistance).

TABLE 9.9 HOT WATER HEATER ENERGY EFFICIENCY

Fuel	Heater Type	Energy Factor (EF)
Solar	Tank	10+
Gas	Tank	0.5 to 0.65
	Tankless	0.7 to 0.85
	Tank-condensing	0.85
Oil	Combined with space heating boiler	0.6
Electric resistance (not sustainable due to source inefficiencies)	Tank	0.75 to 0.95
	Tankless	0.98
Electric heat pump	Tank	1.5 to 2.5

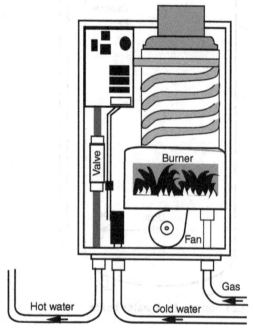

Figure 9.24 Gas tankless water heater

Point-of-use heaters are built into equipment such as dishwashers. An instantaneous hot water dispenser has a separate, smaller faucet for near-boiling water for beverages and instant soups. These dispensers require an extra hole in the surface of a sink or counter, plus a water supply line.

Tankless hot water heaters are also used for remote fixtures. Tankless heaters can be powered by electric resistance, natural gas, or propane; gas heaters produce hot water faster than electric resistance heaters, and are more efficient overall due to electrical generating inefficiencies.

The number of water heaters needed is reduced with back-to-back plumbing design that puts the kitchen, laundry, and a bathroom all in one central location.

GAS WATER HEATERS

Natural gas and propane water heaters typically cost more to buy but are less expensive to operate than electrical water heaters. Gas hot water heaters must be designed to allow gas to escape safely in the rare but serious event of a gas leak.

Higher-efficiency gas condensing water heaters can be exhausted by a horizontal pipe. Combustion gasses pass from a sealed chamber inside the tank through a coiled steel tube that serves as a secondary heat exchanger. Condensing heaters remove so much heat from exhaust gasses that they are cool enough to vent through inexpensive plastic plumbing pipe rather than a stainless steel flue. A fan can blow them horizontally through the outside wall. Condensing heaters are 90 to 96 percent efficient, compared to 60 percent efficiency for an average tank water heater.

Most larger demand water heaters are gas fueled. They require a vent to the outdoors, and sometimes a large diameter gas line. They are typically more expensive to buy than a standard tank gas water heater, but can save money over time.

HOT WATER DISTRIBUTION

Hot water is carried through the building by pipes arranged in distribution trees. When hot water flows through a single hot water distribution tree, it will cool off as it gets farther from the hot water heater. To get hot water at the end of the run, you have to waste the cooled-off water already in the pipes. With a looped hot water distribution tree, the water circulates constantly. There is still some heat loss in the pipes, but less water has to be run at the fixture before it gets hot. Hot water is always available at each tap in one to two seconds.

Hot water is circulated by use of the **thermosiphon** principle. This is the phenomenon where water expands and becomes lighter as it is heated. The warmed water rises to where it is used, then cools and drops back down to the water heater, leaving no cold water standing in pipes. A thermosiphon system comprises a heater with storage tank, piping to the farthest fixture, and piping to return unused cooled water back to heater. In multistory buildings, the increased height produces better thermosiphon circulation.

Forced circulation is used in long buildings that are too low for thermosiphon circulation, and where friction from long pipe runs slows down the flow. The water heater and a pump are turned on as needed to keep water at the desired temperature. It takes five to ten seconds for water to reach full temperature at the fixture. Forced circulation is common in large one-story residential, school, and factory buildings.

A **recirculating hot water pump** delivers hot water instantly at faucets without having to draw off any cold water. The recirculating pump consumes electrical power, and the entire piping system is always filled with hot water, exposing it to greater heat loss. In small buildings such as single-family residences, where fixtures are close together, less energy is wasted in drawn-off water than by continuously recirculating hot water.

Computer controls can save energy in hotels, motels, apartment houses and larger commercial buildings. The computer provides the hottest water temperatures at the busiest hours. When usage is lower, the supply temperature is lowered and more hot water is mixed with less cold water at showers, lavatories, and sinks. Distributing cooler water to the fixture results in less heat lost along the pipes. The computer stores and adjusts a memory of the building's typical daily use patterns.

Hot water pipes expand. This rarely causes problems in smaller buildings such as houses, but can be an appreciable problem in a tall building.

Chilled Water

Most public buildings provide chilled drinking water. Previously, a central chiller with its own piping system was used to distribute the cold water. Today, water is chilled in smaller water coolers at each point of use, providing better quality at less cost. Some systems involve a central water purification system that distributes water throughout an office; water is then chilled at individual water coolers. These systems require more plumbing, but eliminate the need to install heavy water bottles on coolers.

Standard hot and cold bottled water coolers can use more energy than a large refrigerator. Drinking fountains are usually specified with point-of-use water coolers, but are also available without coolers.

The term **water cooler** is used for a drinking fountain with its own self-contained refrigeration equipment. Centralized chiller equipment can be used where a large number of drinking fountains are needed. The central chilled water is pumped continuously, and pipes must be insulated to avoid warming, with a vapor-tight material to avoid condensation.

GAS SUPPLY AND DISTRIBUTION

Natural gas distributed by bamboo pipes was used in Chinese homes for light and heat as early as 400 CE. The first commercial natural gas was produced from coal and used in Britain around 1785 to light houses and streets.

Natural gas is delivered to a building from a nearby pipeline. A natural gas piping connection is installed at the building boundary where required.

When a gas appliance is moved, the gas line should be capped to prevent leaks or a possible explosion. Space for a gas shut-off valve should be located very near a gas device. Existing gas lines can be adjusted or relocated without too much difficulty, as the piping is flexible.

When gas is installed for the first time, it should be done before the floor and wall finishes are completed.

Propane is a byproduct of natural gas processing and petroleum refining. It is provided as either a liquid or a gas (vapor), but a propane system designed for one cannot be used for the other. Propane gas is mainly used for engines, barbeques, portable stoves, and home heating. It is usually supplied from a tank (typically in a backyard) when natural gas is not available. The gas burner at the bottom of the tank is ignited by a standing pilot light or by spark ignition. **Liquefied petroleum gas (LPG)** is a fuel that is in liquid form at or below its very low boiling point of -44°F (-42°C).

Gas equipment generates dangerous carbon monoxide combustion gas. This requires a vent to the outside, typically through the roof. Natural draft ventilation or forced air ventilation with a fan may be used.

In Chapter 9, we have described the water supply system. Chapter 10 deals with the separate waste piping system, and with other waste and recycling systems.

10

Waste and Reuse Systems

When cities began to pave their streets in the nineteenth century, natural streams were enclosed in storm sewers. These pipes channeled water to local rivers. (See Figure 10.1) With the advent of flush toilets, the pipes became **combined sewers** that carried both storm runoff and building wastes to rivers. Eventually, the rivers and associated bodies of water became too polluted for use, and separate **sanitary sewers** and **sewage treatment plants** were built.

Today, residential housing developments slope from lawns at the top to street **storm drains** at the bottom. Once water enters a storm drain, it dumps out in rivers far away from where it started. Huge amounts of storm water also leak into sewer pipes that mix it with sewage and take it even farther away to be processed at water treatment plants. The result is a suburban desert, with lawns that need watering and restricted local water supplies.

Figure 10.1 Storm drain cover, Boston, Massachusetts

INTRODUCTION

Architects and engineers view the use of fresh, potable water to carry waste with concern.

> We expend half our urban water consumption in flushing garbage and excrement from our buildings and then mix this half with the half we have used for washing and other purposes. This is marvelously convenient and keeps our buildings free of disease and odor, but it creates new problems. The water that the municipality just brought to the town at such trouble and expense, sparkling clean and bacteria free, is now *sewage*. It is thoroughly contaminated with odor and potential disease and presents an enormous disposal problem. (Edward Allen, *How Buildings Work* [3rd ed.], Oxford University Press 2005, page 42)

The process of designing to remove human waste from a room within a building is more complex than that required to supply water. Pipes for waste water are required, including large diameter pipes that slope continuously downward from the toilet to the sewer or septic tank. An entire system of vents and traps is needed to deal with sewer gas. In addition, cultural attitudes regarding privacy, personal hygiene, and physical comfort affect the design.

Each building has a sanitary plumbing system that channels all the waste downward through the building to the municipal sewer or a septic tank below. The sanitary system begins at the sink, bathtub, toilet, and shower drains. (See Figure 10.2) The drainage system includes essentially horizontal branches and vertical stacks as well as vents.

The sanitary plumbing system carries wastewater downhill, joining pipes from other drains until it connects with the sewer buried beneath the building. Underground pipes for sewage disposal

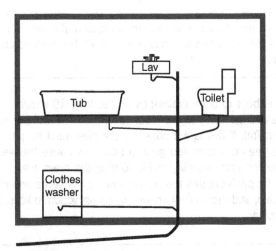

Figure 10.2 Waste drain system

are made out of vitrified clay tile, cast iron, copper, concrete pipe, asbestos cement, or PVC or ABS plastic.

The sanitary system has large pipes to avoid clogs. Since the system is drained by gravity, all pipes must run downhill. The large size of waste pipes, their need to run at a downward angle, and the expense and difficulty of tying new plumbing fixtures into existing waste systems means that the interior designer must be careful in locating toilets.

History

The city of Mohenjo-daro in Pakistan, which existed as part of the Indus Valley Civilization from the twenty-sixth to the nineteenth centuries BCE, had a sophisticated sewage system with drainage channels and street ducts. Major streets were lined with covered drains for wastewater. Water-flushed latrines were found at Mohenjo-daro, and also in Mesopotamian cities dating to 1500 BCE.

An extensive sewage system ran under the Minoan (3000 to 100 BCE) town of Knossos in Crete. Rome's Cloaca Maximus evolved from an open-air canal, and was gradually covered over; it is still in use as a sewer today.

Until the advent of indoor plumbing, wastes were removed from the building daily for recycling or disposal. Urban inhabitants continued to dump sewage and garbage into street gutters until the 1890s. Rural people dumped wastes into lakes, rivers, or manmade holes in the ground called **cesspools** that generated foul smells and created a health hazard.

By the nineteenth century, natural streams came to be enclosed in pipes under paved city streets. Rain ran into storm sewers and then to waterways. When flush toilets were connected to the storm sewers later in the century, the combined storm water and sanitary drainage was channeled to fast-flowing rivers, which kept pollution levels down. Some sewers continued to carry storm water only, and separate sanitary sewers were eventually installed that fed into sewage treatment plants.

Modern plumbing with fresh water supplies and effective sewer systems was not widely available until the late nineteenth century. Older cities still may have a combination of storm sewers, sanitary sewers, and combined sewers, in a complex network that would be difficult and expensive to sort out and reroute.

By the 1950s, most residences in industrialized countries included indoor plumbing. According to the United Nations Human Development Report (2006), it is estimated that 2.6 billion people worldwide lack indoor plumbing.

SANITARY WASTE SYSTEMS

Water has nearly ideal properties for the dissolution and transportation organic waste. However, this is a case of using a high-quality resource for a low-grade purpose.

Human waste can contain disease-causing organisms such as viruses, bacteria, protozoa, and parasitic worms. Fecal coliform bacteria (E. coli) are not themselves pathogenic, but as they indicate exposure to human or animal waste, they signal the likely presence of disease-causing organisms.

Sanitary Piping Elements

Waste pipes are drainpipes that carry only dirty water from sinks, **lavatories** (for washing hands), tubs, showers, **water closets (WC)**, urinals, and floor drains. The sanitary plumbing system also consists of waste, soil, and vent stacks; branch waste and soil lines, and vents; floor drains; cleanouts to remove clogs from pipes; and fresh-air inlets. In order to preserve the gravity flow, large waste pipes must run downhill, and normal atmospheric pressure must be maintained throughout the system at all times.

What we usually call a toilet is referred to in codes as a water closet.

PIPING, FITTINGS, AND ACCESSORIES

Soil pipes carry water and human excrement. Soil and waste piping and venting is made of cast iron, copper, or various types of plastic. Cast iron is durable and corrosion resistant, but difficult to cut. (See Figure 10.3) Plastic waste pipes made of ABS or PVC plastic are lightweight and can be assembled in advance. (See Figure 10.4)

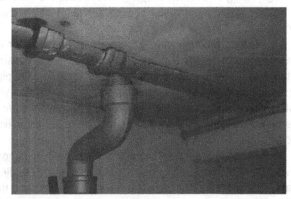

Figure 10.3 Cast iron pipe

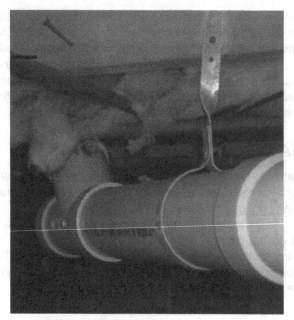

Figure 10.4 Plastic waste pipe

Older buildings may still have old pipes. Drainage lines under 3" (76 mm) in diameter used to be mainly galvanized steel, but were subject to problems from caustic materials. Cast iron, which is very long-lasting but may crack or leak, was used for larger pipes.

Today's 1.6 gallons (6 L) low-flush toilet may not work well with older cast iron waste pipes. The interior surface of the cast iron is rougher than that of plastic pipes, and backups and clogs may result.

DRAINAGE PIPING

Engineers size waste plumbing lines according to their location in the system and the total number and types of fixtures they serve. Waste piping is laid out as direct and straight as possible to prevent deposit of solids and clogging. Bends are minimized in number and angled gently, without right angles.

Waste pipes at least 4" (102 mm) in diameter are used in drainage piping to prevent clogging. **Cleanouts** are installed for access where a horizontal or vertical change of direction occurs, and at intervals in long pipe runs.

The size of drainage piping depends on the number of **drainage fixture units (DFUs)**. The DFU is based on the size of the drainpipe and trap on a particular fixture. Drainage piping is designed with the assumption that not all fixtures will discharge the maximum allowable drainage at the same time. The maximum number of fixtures connected to a drain is limited by code.

At its bottom, the drainage piping is connected to a nearly horizontal pipe. Within the building footprint, this is called the **building drain**. Outside the building, generally 5 feet (1.5 meters) from the building foundation, it becomes the **sewer line**.

High-volume showers may require enlarged drain lines. The required size of a drain line may change if a tub is replaced by a shower.

It is best to pitch drainpipes ¼" per foot (19 mm/m), but pitches range from ⅛" per foot (27 mm/m) to ½" per foot (38 mm/m). If this is not possible, the pipe must be pitched 45 degrees or more. Too great a pitch can cause the waste to flow too fast, inducing solids to plug the drain over time. Too little pitch creates too slow a flow; solids settle and clog the drain, and there is not enough scouring action to keep the drain clean.

Large, sloping drainpipes can gradually drop from a floor through the ceiling below and become a problem for the interior designer.

FIXTURE AND BRANCH DRAINS

Fixture drains extend from the trap of a plumbing fixture to the junction with the waste or soil stack. **Branch drains** connect one or more fixtures to vertical soil or waste stacks. At their bases, drain stacks are connected together by a sloping building drain. The building sewer connects the building drain to the public sewer or to a private treatment facility such as a septic tank.

STACKS

A **stack** is a vertical pipe. Vertical waste pipes are either waste stacks or soil stacks. A **soil stack** is the waste pipe that runs from toilets and urinals to the building drain or building sewer. A **waste stack** is a waste pipe that carries gray liquid wastes from plumbing fixtures other than toilets and urinals. A **drain stack** is a convenient central drain serving a number of fixtures one above the other on different floors. A **stack vent** is a vertical pipe that vents a waste or soil stack.

Codes set minimum diameters for stack and vent pipes in relation to the number of fixtures installed. If additional fixtures are added to an older home, stack and vent pipe sizes may need to be changed.

TRAPS

Originally, the pipe that carried wastewater from a plumbing fixture ran directly to the sewer. Foul-smelling gases from the **anaerobic** (without oxygen) digestion in the sewer could travel back up the pipe and create a health threat indoors.

The **trap** was invented to block the waste pipe near the fixture so that gas could not pass back up into the building. The trap is a U-shaped or S-shaped section of drainpipe that holds wastewater. (See Figure 10.5) The trap forms a seal to

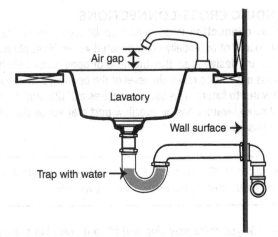

Figure 10.5 Trap

prevent the passage of sewer gas while allowing wastewater or sewage to flow through it. Each time the filled trap is emptied, the wastewater scours the inside of the trap and washes debris away.

Traps are made of steel, cast iron, copper, plastic, or brass. The trap should generally have a height of 2" to 4" (50 to 102 mm). Traps are usually located within 24" (610 mm) of a fixture and should be accessible for cleaning through a bottom opening that is otherwise closed with a plug. On water closets and urinals, they are an integral part of the **vitreous** (glasslike) china fixture, with wall outlets for wall-hung units and floor outlets for other types.

There are a few exceptions to the rule that each fixture should have its own trap. Two laundry trays and a kitchen sink, or three laundry trays, may share a single trap. Three lavatories are permitted on one trap.

When water moving farther downstream in the system pushes along water in front of it at higher pressures, negative pressures can occur behind. The higher pressures could force sewer water through the water in some traps, and lower pressures could siphon water from other traps, allowing sewer gases to get through. (See Figure 10.6) Increasing air pressure by introducing air from a **fixture vent** prevents siphoning.

If the fixture is not used often, the water may evaporate and break the seal of the trap. This sometimes happens in unoccupied buildings and with rarely used floor drains. Adding a special **hose bibb** (threaded faucet connection) provides a source of water directly above the drain as a way to manually refill the drain's trap.

Although traps are necessary, they can collect debris, and are the first place that stoppages occur. Traps in public facilities must be accessible for cleaning out.

Access panels may be required for traps concealed behind walls or floor surfaces.

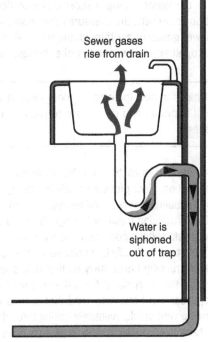

Figure 10.6 Sink without fixture vent

VENTS AND VENT STACKS

The fixture vent fulfills two purposes, providing air pressure to push wastewater down to the sewer, and allowing sewer gases to rise up and out of the building. (See Figure 10.7) Every fixture must have a trap, and every trap must have a vent. Vent pipes

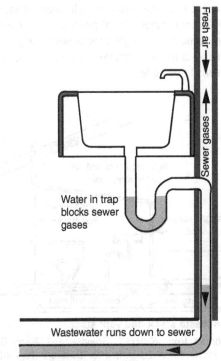

Figure 10.7 Sink with fixture vent

are added to the waste piping a short distance downstream from each trap to prevent the pressures that would allow dirty water and sewer gases to get through the traps. Vent pipes run upward, join together, and eventually poke through the roof.

Adding vent pipes in new locations is more difficult where the roof is several floors up and pipes have to pass through other tenants' spaces.

The vent pipe allows air to enter the waste pipe and break the siphoning action. Vent pipes also release the gases of decomposition, including methane and hydrogen sulfide, to the atmosphere. By introducing fresh air through the drain and sewer lines, air vents help reduce corrosion and slime growth.

The vent must run vertically to a point above the spillover line on a sink before running horizontally so that debris will not collect in the vent if the drain clogs. Once the vent rises above the spillover line, it can run horizontally and then join up with other vents to form the vent stack, eventually exiting through the roof.

A **vent stack** is a system of vertical air vent pipes extended through the roof to admit air and discharge gases from the sanitary piping system. (See Figure 10.8) They are typically located in the attic of single-family residences so that a single vent will penetrate the roof. Vent stacks emit noxious odors and potentially hazardous gases. The vent must extend at least 6" (152 mm) above the roof, and higher where snow accumulates. On a roof that is occupied, the vent must extend around 7 feet (2 m) above the roof to allow for discharge of sewer gas.

More than one fixture can be served by the same vent stack, but the maximum length of drainpipe between the trap and the air vent is limited to a critical distance (the pipe's diameter multiplied by 48). Any number of vent pipes from their respective fixtures can connect to a single central vent stack. There is a limit of 5 feet (1.5 m) for relocation of some fixtures before a new vent needs to be added.

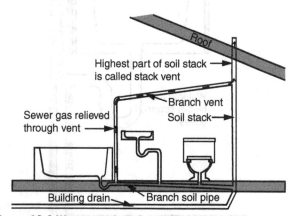

Figure 10.8 Waste stack and vent stack
Source: Redrawn from Walter T. Grondzik and Alison G. Kwok, *Mechanical and Electrical Equipment for Buildings* (12th ed.), Wiley 2015, page 936

AVOIDING CROSS-CONNECTIONS

A **cross-connection** is a connection between supply piping and a source of potentially contaminated water. Most plumbing fixtures are designed so that the level of open water held by a fixture is not able to reach the level of the opening that supplies fresh water to fixture. This usually requires a 1" (25 mm) air gap. A bathroom lavatory has an overflow port that keeps the water below this level.

See Chapter 11, "Fixtures and Appliances," for more information on lavatories, toilets, and other plumbing fixtures.

A tank-type toilet may clog and fill to its rim, but the building's supply water inlet is above both the rim of the bowl and the water level of the tank, avoiding the possibility of a cross-connection.

In public buildings, water closets or urinals have the supply pipe connected directly to the rim, creating the possibility of a cross-connection. A vacuum breaker must be installed in the supply line so that, in the event of a failure of water pressure, air will enter the line and destroy any siphoning action, preventing contaminated water from being sucked into the system. The chrome-plated **flush valve** that we can see on the fixture is the vacuum breaker. Vacuum breakers also prevent siphoning at dishwashers and clothes washers, where pumps can force wastewater into the drain line.

FRESH AIR INLETS

It is important to admit fresh air into the waste plumbing system, to keep the atmospheric pressure normal and avoid vacuums that could suck wastes back up into fixtures. A **fresh air inlet** is a short pipe that connects to the building drain just before it leaves the building.

Fresh air inlets relieve any potential vacuum that might suck all water out of a trap. They lead sewer gas pressure safely out of the building and bring in fresh air. A fresh air inlet provides better ventilation for the plumbing system than only a vent at the top of the vent stacks. Not all codes require a fresh air inlet.

FLOOR DRAINS

Floor drains are located in areas where floors need to be washed down after food preparation and cooking. (See Figure 10.9) Floor drains carry away water from washing floors or drained from heating equipment. They allow floors to be washed or wiped up easily in shower areas, behind bars and other places where water may spill. They are also necessary in mechanical rooms and toilet rooms.

Where floor drains are provided solely for protection against possible basement leaks or to drain condensate from a cooling coil, they may be connected to the storm drain. If they are used to carry away waste from a plumbing fixture such as a clothes washer, they should be connected to the sanitary sewer system.

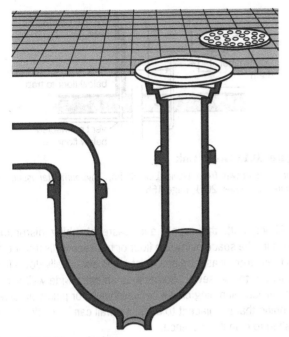

Figure 10.9 Floor drain

All floor drains should have a trap, and when connected to a sanitary sewer, the trap must be vented. The water in their trap seals must be preserved to prevent odors and unsanitary conditions from entering the room. The trap should have a simple way, such as a nearby faucet or hose bibb, to add water to maintain its seal. Using a backflow preventer for a floor drain line that is located in a basement or similar low point is recommended.

SEWAGE EJECTOR PUMPS AND SUMPS
Sewage ejector pumps are used where fixtures are below the level of the sewer. Drainage from the below-grade fixture flows by gravity into a sump pit or other receptacle and is lifted up into the sewer by the pump.

It is best to avoid locating fixtures below sewer level where possible because if the power fails, the equipment shuts down and the sanitary drains do not work. Sewage ejector pumps should be used only as a last resort.

Adding waste drain lines for a basement bathroom may make it difficult to achieve the required ¼" (6 mm) slope to the sewer line. It may be necessary to cut through a concrete slab to add or move pipes below grade.

A **sump** is a tank used to drain groundwater or storm water from basements too deep for gravity draining. Sewage sumps are required whenever drainage, fixtures, or other equipment is installed below the level of public sewers.

Basements may flood due to poor foundation waterproofing, a faulty perimeter drainage system, exceptionally high amounts of rainfall, or a broken water pipe. A sump pump supplies the power to pump out water. An electric pump should have a backup system for electrical outages in storms.

INTERCEPTORS
Interceptors (sometimes called traps) are intended to block undesirable materials before they get into the waste plumbing. They must be readily accessible for periodic servicing.

Among many types of interceptors are ones designed to catch grease, plaster, lubricating oil, glass grindings, and industrial materials. Hair salons and barbershops use interceptors to catch hair. Interceptors are also used in hospitals and clinics.

Grease traps are often required by code in restaurant kitchens and other locations. (See Figure 10.10) Grease rises to the top of the trap, where it is caught in baffles, preventing it from

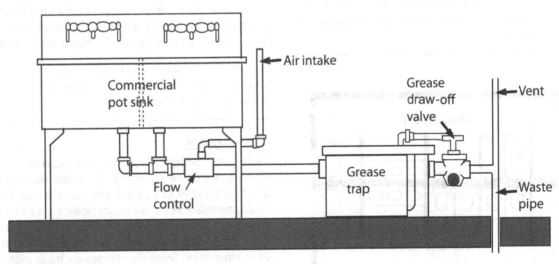

Figure 10.10 Grease trap
Source: Redrawn from Walter T. Grondzik and Alison G. Kwok, *Mechanical and Electrical Equipment for Buildings* (12th ed.), Wiley 2015, page 943

congealing in piping and slowing down the digestion of sewage. Grease is removed through a top cover or drawn of with a valve. Grease traps are often located on a floor adjacent to or below a sink, or recessed in a pit.

CLEANOUTS

All parts of the sanitary piping system from plumbing fixtures to the end of the disposal process should be accessible through cleanouts or other points of access. Cleanouts are distributed throughout the sanitary system between fixtures and the outside sewer connection. They are located a maximum of 50 feet (15.2 m) apart in branch lines and building drains up to 4" (102 mm). On larger lines, they are located a maximum of 100 feet (30.5 meters) apart. Cleanouts are also required at the base of each stack, at every change of direction greater than 45 degrees, and at the point where the building drain leaves the building.

The interior designer should be aware that wherever a cleanout is located, there must be access for maintenance and room to work.

Residential Waste Piping

It is common to arrange bathrooms and kitchens back-to-back. (See Figure 10.11) The piping assembly can then pick up the drainage of fixtures on both sides of the wall.

The plumbing wall behind fixtures should be deep enough to accommodate branch lines, fixture run-outs, and air chambers. The waste piping for a wood or steel frame residence usually fits into a 6" (152 mm) or 8" (203 mm) partition. The piping assembly for a small building typically consists of a 4" (102 mm) soil stack and building drain. The water closet must have a 2" (51 mm) vent.

On bathroom and kitchen remodeling projects, changes may be required to meet current codes.

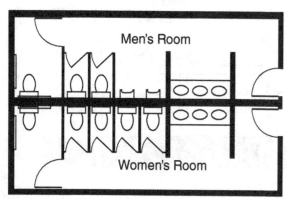

Figure 10.11 Back-to-back plumbing wall

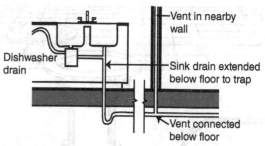

Figure 10.12 Island sink

Source: Redrawn from David Lee Smith, *Environmental Issues for Architecture*, Wiley 2011, page 465

Fitting both the supply and waste plumbing distribution trees into the space below the floor or between walls may be difficult, as larger wastes pipes must slope continually down from the fixture to the sewer. Sometimes an extra-wide wall serves as a vertical plumbing chase between walls for plumbing pipes. Plumbing that is adjacent to a jog in a wall can be enclosed and finished to hide its presence.

ISLAND SINKS

When a sink is located in an island, as in some kitchen designs, there is no place for the vent line to go up. Instead a waste line is run to a sump at another location, which is then provided with a trap and vent. There are various approaches to venting an island sink, but be sure to check local codes as some approaches are not allowed in certain jurisdictions. (See Figure 10.12)

Large Building Waste Piping

In larger buildings, the need for flexibility in space use and the desire to avoid a random partition layout means that plumbing fixtures and pipes must be carefully planned early in the design process. The location of the building service core with its elevators, stairs, and shafts for plumbing, mechanical, and electrical equipment affects the layout of surrounding areas and their access to daylight and views. In high-rise building service cores, plumbing cores grouping fixtures together provide greater flexibility in space planning.

Service cores are introduced in Chapter 4, "Building Forms, Structures, and Elements."

PIPE CHASES

Larger drain piping in large buildings requires the construction of a pipe chase, consisting of a double wall with space for plumbing between its layers. A pipe chase is typically required to accommodate all the necessary piping if more than 2 or 3 fixtures are installed. The minimum width for a pipe chase is about 8" (203 mm). Wall-hung fixtures require chases 18" to 24" (467 to 610 mm) thick. Some pipe chases are made wide enough to accommodate a worker doing repairs. Pipe chases often have holes in the floor for vertical pipes.

HORIZONTAL WASTE PIPING

Where waste piping drops through the floor and crosses below the floor slab to join the branch soil and waste stack, it can be shielded from view by a suspended ceiling. Tubing to accommodate piping has been developed that sits above the structural slab. Casting a lightweight concrete fill over it solves the visibility problem, but raises the floor 5" to 6" (127 to 152 mm). Raising the floor only in the toilet room creates access problems, so the entire floor may be raised. This creates space for electrical conduit and underfloor air distribution as well, but requires considerable planning.

WET COLUMNS

When offices need a single lavatory or complete toilet room away from the central core, pipes can be run horizontally from the core. In order to preserve the slope for waste piping, the farther the toilet room is located from the core, the greater amount of vertical space is taken up by the plumbing.

Integrating piping risers for smaller groupings or for isolated fixtures within the enclosure of structural columns creates a **wet column**. (See Figure 10.13) Wet columns group plumbing pipes away from plumbing cores to serve sinks, private toilets, and other fixtures, and provide an alternative to long horizontal waste

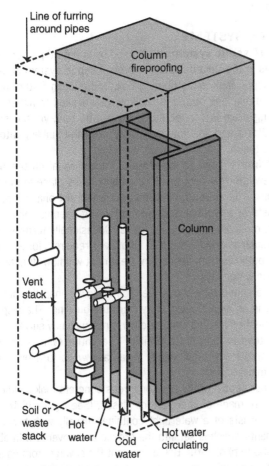

Line of furring around pipes

Column fireproofing

Column

Vent stack

Soil or waste stack

Hot water

Cold water

Hot water circulating

Figure 10.13 Wet column

piping runs. Locating a wet column at a structural column requires coordination with the structural design early in the design process. Individual tenants can tap into these lines without having to connect to more remote plumbing at the core of the building.

When laying out toilet rooms, the designer should consider how the plumbing reaches the fixtures.

TREATING AND RECYCLING WASTEWATER

Fruits, vegetables, grains, milk products, and meats derived from nutrients in the soil are brought into cities, to be later flushed out as sewage. Most cities and towns send their sewage to treatment plants, where the solid matter (sludge) settles out. The remaining liquid is chlorinated to kill bacteria and then dumped into a local waterway.

The sludge is pumped into a treatment tank, where it ferments anaerobically (without oxygen) for several weeks. This kills most of the disease-causing bacteria and precipitates out most minerals. The digested sludge is then chlorinated and pumped into the local waterway.

Waterways are unable to finish the natural cycle by returning the nutrients back to the soil, and end up with increasing amounts of nutrients that promote the fast growth of waterweeds and algae. Eventually plants choke the water and die and decay. Over a few decades, the waterway may become a swamp and then a meadow.

Meanwhile, farmland is gradually drained of nutrients. Farm productivity falls, and the quality of produce declines. Artificial fertilizers are applied to replace the wasted natural fertilizers.

Designers can step into this process when they make decisions about how wastes will be generated and handled by the buildings they design. Sewage treatment is expensive for the community, and becomes a critical issue for building owners where private or on-site sewage treatment is required.

Sewage disposal systems are designed by sanitary engineers and must be approved and inspected by the health department before use. The type and size of private sewage treatment systems depends on the number of fixtures served and the permeability of the soil as determined by a percolation test. Rural building sites are often rejected for lack of suitable sewage disposal.

Water treatment systems range from ones that serve individual residences to ones covering whole regions. Techniques include sand filters with settling basins, aeration, chemical precipitation, chlorine gas, and special filters.

Recycled Water

Water is categorized by its purity. Water grades include potable water, graywater, dark graywater, and blackwater. (See Table 10.1)

TABLE 10.1 WATER GRADES IN BUILDINGS

Grade	Description	Uses
Potable water	Water suitable for drinking; usually has been treated	Most household uses, including flushing toilets
Rainwater	Pure water distilled by hydrologic process; runoff from roofs may contain contaminants; little or no treatment required	Bathing, laundry, toilet flushing, irrigation, evaporative cooling
Graywater	Wastewater from sinks, baths, showers; not from toilets or urinals; likely contains soap, hair, human waste from soiled clothes, grease and food from kitchen wastes	Treatment required for reuse to flush toilets. Filtering required for drip irrigation
Dark graywater	From kitchen sinks, dishwashers, washing diapers	Usually prohibited from reuse
Blackwater	Water containing toilet or urinal waste	Requires high level treatment

Nonpotable rainwater or recycled water must be kept completely separate from the normal potable water supply. It may be possible to use a dual wastewater system that separates **graywater** from **blackwater** sources, allowing water recycling within a building.

GRAYWATER RECYCLING

The 2015 International Plumbing Code (IPC) allows graywater from approved sources that has been disinfected and treated by an approved on-site water reuse treatment system to be used for toilet flushing. Graywater regulations vary from state to state in the United States.

2015 International Residential Code (IRC) details design of graywater recycling systems in section on Sanitary Drainage. The IRC allows systems to collect discharge from tubs, showers, lavatories, clothes washers, and laundry trays for a graywater system. The water can then be used for flushing toilets or urinals or for landscape irrigation. Some local codes now require new residential construction to include connections for graywater plumbing.

Graywater systems were originally developed for single-family homes, and are now also used in apartment buildings, dormitories, hotels, and other buildings with many showers, baths, and clothes washers. Implementing a graywater system requires designing and sizing the system, installing additional plumbing pipes, and providing a collection reservoir or storage tank.

Graywater systems may use pipes as small as 1" (25 mm) in diameter under pressure. The system pipes the water to a short-term storage tank (dosing basin). It is then pumped out in batches to subsurface irrigation or a disposal field. The storage tank must be emptied at least once a day to prevent odors, as anaerobic bacteria take over from aerobic as the oxygen supply is depleted.

A simpler system uses lavatory water to flush an adjacent water closet. The lavatory can be part of the water closet to fit into a tight space. A small hand-washing tank is located directly over the toilet tank, allowing water to flow to the toilet tank for the next flush.

Installation of a graywater system requires careful planning early in the process. Check codes and permit requirements. Expert advice may be needed.

Rural Sewage Treatment

In times past, rural wastes ended up in a cesspool that allowed sewage to seep into the surrounding soil. Cesspools did not remove disease-causing organisms. Within a short time, the surrounding soil became clogged with solids, and the sewage overflowed onto the surface of the ground and backed up into fixtures inside the building.

Private water supply systems usually need to make private arrangements to dispose of sewage without polluting the water source. One of the most common reasons for rejection of a potential rural building site is its lack of suitability for sewage disposal.

SEPTIC SYSTEMS

In 2007, **septic systems** were used by about 20 percent of all homes in the United States. A typical septic system consists of a septic tank, a distribution box, and a leach field of perforated drainpipes buried in shallow, gravel-filled trenches. The building drain is connected to the septic tank via the sewer line. Waste and soil flow by gravity from the building into the septic tank.

Primary sewage treatment for individual on-site building sewage treatment most commonly takes place in the septic tank. The **secondary sewage treatment** phase typically consists of a filtration system comprising seepage pits, drain fields, mounds, and/or sand filters. Occasionally tertiary treatment such as disinfection with chlorine is required, for example, when outflows from secondary treatment will flow directly into surface waterways.

The septic tank should be located lower than the building sewer in an area sloping away from the building. The disposal field must be located under a clear, grassy, sunny surface; tree roots tend to clog the drainfield. The entire system must be separated from water supply pipes and wells to avoid cross-contamination.

Most **septic tanks** are nonporous concrete tanks, although some are made of fiberglass or plastic. (See Figure 10.14) The tank consists of a watertight container, usually with two compartments, which is set into the ground and covered with about 12" (305 mm) of earth. Baffles prevent the sewage from passing quickly from inflow to outflow, giving more time for processing.

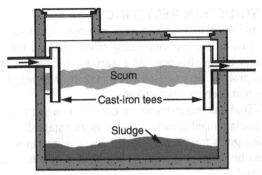

Figure 10.14 Septic tank

Source: Redrawn from Francis D.K. Ching, *Building Construction Illustrated* (5th ed.), Wiley 2014, page 11.29

A tank with a capacity of 1000 to 1500 gallons can serve a family of four.

The interior designer should check the capacity of the septic tank if installing a whirlpool bath, full body-spray shower, or other high-water-use fixtures. A larger or additional septic tank may be needed.

If the building and its occupants practice water conservation, less water and wastes flow through the septic tank, the effluent stays in the tank longer before being flushed out, and emerges cleaner.

The septic tank is intended to hold sewage for a minimum of 30 hours while the sewage decomposes anaerobically, a process that produces methane gas and odor. From the septic tank, the effluent flows to a **drainfield** (variously called disposal, leaching, or absorption fields or beds) for secondary treatment. The drainfield has 4" (102 mm) diameter perforated pipes that allow the effluent to seep into the ground. The effluent in a drainfield becomes oxidized, allowing **aerobic** bacteria that require oxygen to complete waste decomposition and render it harmless.

A septic system has limited capacity and a limited lifetime. It requires regular maintenance and care in use. Most communities have strict regulations requiring soil testing and construction and design techniques for installing septic tanks. As indicated earlier, it is not uncommon for a potential building site to be incapable of accommodating a septic system, making the site unsuitable for construction or expansion of a building. New devices are introduced each year for coping with the problem. Increasingly, buildings separate discarded wash water from toilet wastes.

Nothing that can kill bacteria should ever be flushed down the drain into a septic system. Some systems include a grease trap in the line between the house and the septic tank, which should be cleaned out twice a year.

Trained professionals must clean the tank at regular intervals. Most tanks are cleaned every two to four years. Septic tanks should last around 50 years. Most septic systems eventually fail, usually in the secondary treatment phase.

AEROBIC TREATMENT UNITS

An **aerobic treatment unit (ATU)** is essentially a small sewage treatment plant that may be used to replace septic tanks in troubled systems. By rejuvenating existing drainfields, they can extend the system's life. Air is bubbled through the sewage, facilitating aerobic digestion. After about one day, the effluent moves to the settling chamber where the remaining solids settle and are filtered out. Because aerobic digestion is faster than anaerobic digestion, the tank can be smaller. However, the process is energy intensive and requires more maintenance. The refined effluent is less polluted than that from a septic tank and can be returned to the natural water flow.

Other on-site treatment systems include constructed wetlands, greenhouse ecosystems, and lagoons. (See Table 10.2)

TABLE 10.2 OTHER ON-SITE WATER TREATMENT SYSTEMS

System Type	Description
Constructed wetlands	Break down organic waste and produce nutrients that benefit species doing the work. Can remove organic nutrients and inorganic substances. Can accommodate storm surges and treat the contaminated runoff.
	Free-surface (open) wetlands are lined shallow open basins or channels. Soil supports wetlands vegetation; plants are nourished by flow of effluent from the primary treatment system. Must be designed to avoid human contact with treated effluent and problems with mosquitoes and other insects.
	Subsurface flow wetlands have a layer of soil covering a gravel bed. Conditions encourage growth of aerobic and anerobic microbes and some invertebrates. Reclaimed water is safe for landscape watering, creek habitat restoration, or recharging aquifer. Safer for human contact, also attract birds.
Greenhouse ecosystems	Secondary sewage treatment systems are constructed wetlands moved indoors. Consist of series of tanks, each with its own ecosystem; first a stream, second an indoor marsh. Bacteria, algae, snails, and goldfish help the process. Enclosed system depends on solar energy for photosynthesis and on gravity flow. No final chlorine treatment. System produces about one-quarter the sludge of conventional systems. Pleasant to look at, smell like commercial greenhouses.
Lagoons	Type of secondary treatment for multiple buildings. Require sun, wind, and substantial land area, but simple to maintain, use very little energy.

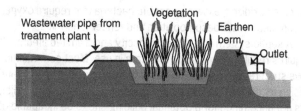

Figure 10.15 Greenhouse ecosystem

On-site wastewater treatment has a significant impact on the design of the building's site. Interiors are also affected, as the system may use special types of plumbing fixtures and may include indoor greenhouse filtration systems. (See Figure 10.15)

Centralized Sewage Treatment Systems

Larger scale sewage treatment plants use aerobic digestion plus chemical treatment and filtration, and can produce effluent suitable for drinking. Clean effluent is pumped into the ground to replenish depleted groundwater. Digested sludge may be dried, bagged, and sold for fertilizer. Some plants spray processed sewage directly on forests or cropland for irrigation or fertilizer. Many municipal sewage treatment plants are now using aerobic digestion and chemical treatment and filtration, producing effluent that is sometimes clean enough to drink.

SOLID WASTE SYSTEMS

According to the EPA, each person in the United States generated 250 million tons of trash in 2011. Less than 35 percent of this was recycled or composted. Many cities in the United States have run out of landfill space, so wastes are taken to large incineration plants and burned, reducing the amount that needs to be buried. Incineration plants must be designed carefully to avoid air pollution.

The design of a building distribution system to bring supplies in and solid waste out requires careful planning. Mechanical equipment for solid waste systems can take up more space than is required for water and waste systems. In addition, handling of solid wastes may present fire safety dangers and could create environmental problems.

As part of the building design team, interior designers are responsible for making sure that the solid wastes generated during construction and building operation are handled, stored, and removed in a safe, efficient, and environmentally sound way. Whether we are designing a place for office recycling or making sure that furniture is reused rather than discarded, we can have a significant impact on how the building affects the larger environment. Many waste substances contain useful energy, and separation and recycling of mingled refuse is becoming a routine task.

Recycling

It is best to reduce, then reuse, recycle, and regenerate. This applies to the construction, operation, and demolition of the building.

CONSTRUCTION RECYCLING

As interior designers, we can work with contractors to ensure that the materials removed during renovation and the waste generated by construction has a second life. Interior designers can prolong the life of interior spaces by designing flexible spaces that accommodate change.

LEED v4 has requirements for construction and demolition waste management planning. They require establishing waste diversion goals for the project and specifying whether materials will be separated or commingled, and how they will be processed.

DEMOLITION RECYCLING

Designers should anticipate the eventual demolition and removal requirements of materials that they specify. **Design for disassembly** is a deliberate design effort to maximize the potential for disassembly rather than demolition, to recover components for reuse and materials for recycling, and to reduce long-term waste generation. Design for disassembly should be an overall building strategy.

Interior designers should select materials that have value for future reuse or recycling.

Demolition by hand salvage produces useful building components and even some architectural gems. Architectural salvage warehouses are a goldmine for interior designers. When checking out a building for a renovation project, consider which elements can be reused in your design or salvaged for another project.

DESIGNING FOR RECYCLING

Convenience has been shown to be a motivating factor for recycling behavior. The kitchen is often a central location for dropping off trash and recyclables from adjacent rooms. Much of the food waste is compostable. Packaging may need to be separated by material for recycling. The majority of paper products can be recycled, while some (napkins, paper towels) can be composted.

Communities vary in their recycling practices. Some allow commingled recyclables, while others require separation. Energy is saved by keeping solid wastes as separate as possible. However, this requires more effort by consumers. Multiple containers for sorting also take up more space.

Small Building Solid Waste Collection

Recycling requires temporary storage space. Community requirements change over time, so flexibility is important. The space should be well ventilated, dry, and easy to clean. An indoor space is easier to access regardless of the weather. Containers should be easy to remove, durable, and washable. A nearby sink is handy for cleaning.

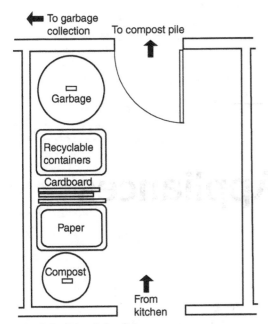

Figure 10.16 Residential solid waste storage

Most of the waste in a home comes from the kitchen. Finding recycling space within a pantry, airlock entry, or a cabinet or closet that opens to the outside makes daily contributions easier, facilitates weekly removal, and simplifies cleaning. A recycling center location between the primary food preparation area and the exit to wherever the trash bins are kept (garage, outdoors) is ideal. Using a smaller bin inside for transfer to a larger bin near the trash may work well.

Kitchens are often hot and humid, and storage for solid wastes needs to be cool, dry, and well ventilated. Ideally, a recycling storage space would open to the kitchen on one side and to the outdoors on the other. (See Figure 10.16)

GARBAGE DISPOSERS AND TRASH COMPACTORS

In-sink garbage disposer units are popular in the United States, where many people consider them convenient and sanitary, but not in the rest of the world. They use from 2 to 4 gallons (7.5 to 15 L) of water per minute plus electrical energy, and create organic sludge that puts an added burden on the septic tank or sewage system. The majority of what goes down the disposal could be composted instead, providing a source of nutritious soil for other growing things.

Check the septic system for adequate capacity before adding a garbage disposer. The added water and waste may require a larger tank. Some communities do not allow garbage disposers.

Trash compactors may reduce the storage space required for small businesses with large quantities of bulky waste. They can be used selectively to compact aluminum for recycling, and

for ferrous metals and box cardboard. Used indiscriminately, trash compactors crush dissimilar items together. They are probably not worth the space they take up for single-family households that recycle.

COMPOSTING

Composting is a controlled process of decomposition of organic material. Naturally occurring soil organisms recycle plant nutrients as they break down the material into humus. The process is complete when dark brown, powdery humus has been produced. Its rich, earthy aroma indicates that the finished compost is full of nutrients essential for the healthy growth of plants.

Compost happens as long as there is air and water to support it. Self-contained composting containers are available commercially. When a compost pile is frequently turned and is quite warm, damp, and well-aired, its odor is minimal.

A food scrap collection container should be conveniently placed in the kitchen to collect each day's compostables. It will need a tight lid and be non-absorbent and easy to clean.

Large Building Solid Waste Collection

Large apartment complexes fence in their garbage can areas to keep out dogs and other pests. This area can be a good place for bins for recycling, and even a compost pile for landscaping. The solid waste storage area needs garbage truck access and noise control, and should be located with concern for wind direction to control odors.

Both the building's occupants and the custodial staff must understand and cooperate with the process for successful recycling in a large building. Office building operations generate large quantities of recyclable paper and box cardboard, along with nonrecyclable but burnable trash, including floor sweepings. Offices also produce food scraps (including coffee grounds) and metals and glass from food containers.

The collection process for recycling in larger buildings has three stages. Waste generation and separation by employees involves depositing paper, recyclables, and compostables in bins near desks, in the lunchroom, and in the copy room. Next, custodians dump the separate bins in a collection cart. The separated waste is deposited in a storage closet with a service sink for bin washing. Finally, paper is shredded and stored for recycling, and recyclable materials are collected by recycling trucks at the ground floor service entrance. Compostable materials are stored or sent to a roof garden compost pile.

The storage area should be supplied with cool, dry, fresh air. Compactors and shredders are noisy and generate heat, and must be vibration-isolated from the floor. A sprinkler fire protection system may be required, and a disinfecting spray may be necessary. Access to a floor drain and water for washing is a good idea.

Interior designers who understand water supply and conservation, wastewater treatment options, and solid waste recycling are well prepared to design sustainable interiors. Chapter 11 looks at how interior designers contribute to the selection, installation, and use of plumbing fixtures and appliances.

Fixtures and Appliances

Some parts of buildings, such as sinks, bathtubs, cooking ranges, and dishwashers, were considered separate items in the past. They are now less portable and more commonly viewed as fixed parts of the building.

INTRODUCTION

Fixtures and appliances are integrated into plumbing, electrical, and HVAC systems. Their design has significant implications for acoustical privacy and indoor air quality. Bath and kitchen spaces are particularly involved.

> Fixture spacing and clearances are important for safe and comfortable movement within a bathroom space.... The layout of bathrooms and other restroom facilities should also take into account the space for and locations of accessories such as towel bars, mirrors, and medicine cabinets [as well as the] number of plumbing walls required and the location of stacks, vents, and horizontal runs. (Francis D.K. Ching, *Building Construction Illustrated* [5th ed.], Wiley 2014, page 9.26)

Selection of plumbing fixtures is usually a joint decision of the client, architect, mechanical engineer, and interior designer. The work of interior designers requires an understanding of plumbing requirements and universal design issues.

History of Bathrooms

Indoor bathrooms were not common in homes until around 1875, but their history goes back thousands of years. Around 1700 BCE, Minoan buildings at Knossos, Crete, included bathrooms with water supplied through clay pipes, and toilets comprising bench seating with openings over a drainage channel. Water was poured into the channel to carry away waste to a sewer.

Privies in Medieval European castles used small stone chambers (closets) with toilet seats. The closets projected from the tops of walls, designed to drop excrement to the base of the wall or the moat.

Hygiene has been a religious imperative for Hindus since around 3000 BC, when many homes in India had private bathroom facilities. At Mohenjo-daro in today's Pakistan, the Great Bath was designed to hold water, perhaps for ritual bathing or religious ceremonies.

In the Minoan palace at Knossos on Crete, the Queen's Room held washing basins and a painted terracotta bathtub. The Minoans also had the first known flushing toilet, screened off with partitions and flushed with rainwater or water carried from a cistern through conduits built into the wall.

In Western Europe from medieval times on, wastes from chamber pots were supposed to be collected early in the morning by night soil men, who carted them to large public cesspools, but many people avoided the cost of this service by throwing waste into the streets.

By the seventeenth century, plumbing technology reappeared in parts of Europe, but indoor bathrooms did not. When Versailles was constructed in France, it included a system of cascading outdoor water fountains, but did not include plumbing for toilets and bathrooms for the royal family, 1000 nobles, and 4000 attendants who lived there.

Britain in the eighteenth century had no residential or public sanitation. Cholera decimated London in the 1830s, and officials began a campaign for sanitation in homes, workplaces, public streets, and parks. Throughout the rest of the nineteenth century, British engineers led the western world in public and private plumbing innovations.

Figure 11.1 Victorian commodes

Before the installation of indoor toilets, the options included outhouses (cold in winter, smelly in summer) and chamber pots (indiscrete). Chamber pots in bedrooms were sometimes enclosed in "night tables" or "end stools" that hid their function under a lid. (See Figure 11.1)

PLUMBING FIXTURES

Water closets, showers, and bathroom faucets use almost three-quarters of the water consumed in a typical home. All plumbing fixtures are supplied with clean water and discharge contaminated fluids. (See Figure 11.2)

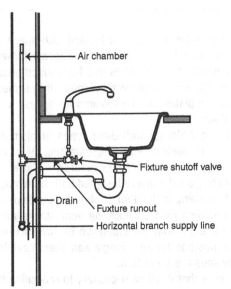

Figure 11.2 Water fixture plumbing

Source: Redrawn from Francis D.K. Ching, *Building Construction Illustrated* (5th ed.), Wiley 2014, page 11.24

WaterSense® plumbing fixtures are independently tested and certified in a partnership between EPA and manufacturers, retailers, and distributors, and utilities to provide water-efficient products. Most products available are residential bathroom fixtures, including toilets, lavatory faucets and aerators, and showerheads. In general, products meeting WaterSense standards are 20 percent more efficient than comparable products.

General Code Requirements

The 2015 International Plumbing Code (IPC) specifies the number of required plumbing fixtures for a building. (See Tables 11.1 and 11.2) It also specifies code-approved materials and installation requirements. Equipment efficiencies are required to be in accordance with the International Energy Conservation Code. The IPC requires that plumbing fixtures be constructed of approved materials, with smooth, imperious surfaces, free from defects, and concealed fouling surfaces.

The IPC does not apply to single-family residences, but it is advisable to follow requirements for minimum clearances between plumbing fixtures. The International Residential Code (IRC) has the same requirements for fixture materials as the IPC. There are requirements for showers, lavatories, water closets, bidets, bathtubs and whirlpool baths, sinks, laundry tubs, food disposers, dishwashers, and clothes washers.

Remodeling an older bathroom may require upgrades to comply with safety and health codes, such as the use of a vent fan or operable window, safety glass in certain locations, or scald prevention shower control valves.

TABLE 11.1 MINIMUM NUMBER OF SELECTED ASSEMBLY PLUMBING FACILITIES

Occupancy	Water closets (1 per) Men	Water closets (1 per) Women	Lavatories (1 per)	Drinking Fountains (1 per)	Others
Theaters	125	65	200	500	1 service sink
Nightclubs	40	40	75	500	
Restaurants	75	75	200	500	
Halls, museums	125	65	200	500	
Churches	150	75	200	1000	

TABLE 11.2 MINIMUM NUMBER OF SELECTED RESIDENTIAL PLUMBING FACILITIES

Occupancy	Water closet	Lavatory	Bathtub, Shower	Others
Hotels, motels	1 per sleeping unit	1 per sleeping unit	1 per sleeping unit	1 service sink
Apartment house	1 per dwelling unit	1 per dwelling unit	1 per dwelling unit	1 kitchen sink per dwelling unit, 1 clothes washer connection per 20 dwelling units
1- and 2-family dwellings: per dwelling unit	1 per dwelling unit	1 per dwelling unit	1 per dwelling unit	1 kitchen sink per dwelling unit, 1 clothes washer connection per 20 dwelling units

ACCESSIBILITY CODES, LAWS, AND STANDARDS

The 2010 ADA Standards for Accessible Design cover plumbing elements and facilities in new construction and alterations. There are many topics covered, including toilet and bathing rooms and compartments, urinals, lavatories and sinks, bathtubs, showers, washing machines and clothes dryers, saunas and steam rooms, and drinking fountains.

Fair Housing Act Accessibility Guidelines provide technical guidance for compliance with accessibility requirements of the Fair Housing Amendments Act of 1988. They cover newly constructed multifamily buildings (minimum 4 dwellings) for occupancy on/after March 13, 1991.

The International Building Code (IBC) references the ICC/ANSI A117.1-2009 Accessible and Usable Buildings and Facilities, a nationally recognized standard of technical requirements for compliance with accessibility requirements of the IBC and other state and local codes.

The ADA does not apply to private residences, but many designers incorporate the principles of universal design to accommodate present or future needs of their clients. Structural reinforcement for future grab bars and wall-mounted water closets may be required, and are a good idea anyway.

Bathroom Fixtures

Plumbing fixtures must be made of nonporous material. Enameled cast iron, vitreous china, stainless steel, copper, and brass are commonly used. Plumbing fixtures typically have rounded interior corners and are smooth, hard, and capable of years of rugged use.

High water demand fixtures such as large tubs and multiple-head showers sometimes exceed available hot water supply, volume, or pressure.

Bathroom fixtures should be located with space for easy cleaning around and within the fixture and access for repair and part replacement. Access panels may be required behind tubs, showers, and lavatories in the walls of rooms they serve. Trenches with access plates may be required for access to pipes in concrete floors.

In existing buildings with older types of plumbing, it is necessary to evaluate how existing plumbing affects plans to change fixtures and whether new fixtures will fit in old locations. Installing a toilet in a redesigned bathroom may require changes to framing in the floor so that the waste pipe can be installed between joists. If existing vent stacks and drains cannot be used, it may be expensive to move fixtures. It is easier to reposition fixtures along a wall where flexible plumbing supply lines are permitted.

Remember that it will be necessary to reposition the drain as well. For renovation work, select new fixtures with the same drain location where possible. For example, showers come with the drain opening at the center or at a side.

When renovating a bathroom, you may need to deal with mold inside the floor area where water has been trapped.

Water Closets

As indicated in Chapter 10, what most of us call a toilet is technically called a water closet (WC). Toilets are not usually designed to facilitate proper washing while eliminating. A toilet seat that provides a cleansing spray is available from several manufacturers for use on existing toilets. Toilets are available without a separate toilet seat, with a warmer for the seat, and with warm water within the toilet for washing. A **urinal** is a bowl or other receptacle usually attached to a wall in a public toilet, into which men may urinate. **A bidet** is a low oval basin used for washing the genital and anal area. Bidets are popular in many parts of the world, but are less often used in the United States.

The dimensions of water closets vary, and include compact, accessible, and children's models as well as standard designs. (See Table 11.3) The best height for a person transferring from a wheelchair is one matching the wheelchair; heights vary, but average about 18″ (457 mm).

Each water closet or urinal should be provided with a means to ensure privacy. Exceptions include single-occupant toilet rooms, as well as daycare and childcare facilities. Privacy partitions for both toilets and urinals should cover from a maximum of 12″ (305 mm) above the floor to a minimum height of 5 feet (1.5 m).

A direct view into a public toilet room should be cut off by an entry vestibule or by the door swing. This is also preferred, but not as critical, for single-occupancy facilities.

HISTORY OF INDOOR TOILETS

In 1596, Queen Elizabeth had an indoor toilet installed by Sir John Harrington. As Harrington's toilet was connected directly to the cesspool, with only a loose trapdoor in between, the queen complained about cesspool odors.

In 1775, Alexander Cummings patented the first effective WC with a controlled water supply. The design included a standing water trap in the sewer pipe, plus standing water in the bowl, along with a vent tube. A backward curve in the soil pipe directly underneath the toilet bowl retained water and cut off the smell from below. Because the valve opened after use to let the water flow out, it became contaminated relatively quickly.

In 1900, George (J.G.) Jennings invented the basic WC design still used today. It relies on a water trap to seal the pipe and maintain the water level in the bowl, and siphoning action to extract water and waste from the bowl.

Thomas Crapper developed the flushing mechanism that is essentially the one still used today. His Water Waste Preventer controlled the water supply to flush the tank while filling the drained closet bowl. The release of accumulated water from the tank into the closet began the siphoning action.

Toilet bowls could never be leak-proof and free of contamination until all the metal and moving parts were eliminated. In 1885, English potter Thomas Twyford succeeded in building the first one-piece earthenware toilet that stood on its own pedestal base. Today, water closets, urinals, and bidets are made of vitreous china.

Indoor plumbing resulted in the use of washout (washdown) water closets that used a lot of water. Water was collected from rain or pumped up from below to a water supply in the attic, then flowed down to the home's bathroom on the second floor. The indoor toilet did not become a common fixture in North American homes until the early twentieth century.

WATER CLOSET CODE REQUIREMENTS

The 2015 IPC requires that water closet bowls for public or employee toilet facilities be of the elongated type. Seats provided for public or employee toilet facilities are required to be the hinged open-front type.

The Energy Policy Act that went into effect in 1994 mandates the use of low flow toilets in new construction and remodeling. It made 1.6 gallons (6 L) per flush a mandatory federal maximum for new toilets. Some states or local jurisdictions may have adopted even lower requirements.

Codes set specific clearances on each side and in front of the toilet bowl. A minimum of 30″ (762 mm) clear space in front of the toilet is recommended, with more for larger persons

TABLE 11.3 WATER CLOSET, URINAL, AND BIDET DIMENSIONS

Fixture	Width (mm)	Depth (mm)	Height (mm)
Water closet	Standard 20′ to 24″ (510 to 610)	Round front bowl 27″ (686) average	Standard 14″ to 19″ (356 to 483) to seat; around 30″ (762) with tank
	Compact 22″ (559)		Compact toilet 26″ (660)
	Accessible around 20″ (510)	Elongated bowl around 29″ (737)	Accessible bowl height 17″ to 19″ (432 to 483)
Urinal	18″ (455)	12″ to 24″ (305 to 610)	24″ (610) rim height
Bidet	14″ (355)	30″ (760)	14″ (355)

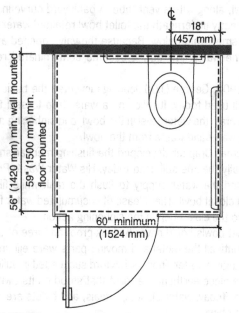

Figure 11.3 Wheelchair accessible toilet compartment
Source: Redrawn from 2010 ADA Standards Figure 604.8.1.1 Size of
Wheelchair Accessible Toilet Compartment

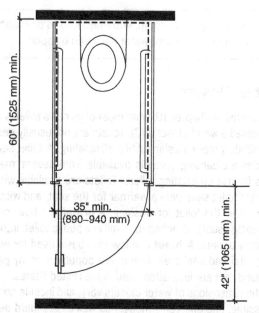

Figure 11.4 Ambulatory accessible toilet compartment
Source: Redrawn from 2010 ADA Standards Figure 604.8.2 Ambulatory
Accessible Toilet Compartment

or people who need personal assistance. The International
Residential Code (IRC) allows a minimum of 21" (533 mm) front
clearance, which makes room for a person's legs but not enough
to manage clothing. For a side wheelchair approach and transfer, at least 30" (762 mm) is required to the side of the toilet.
Wall-hung toilets offer better clear floor space for maneuvering
and maintenance.

A wheelchair can negotiate a 90-degree turn in a 36" (914
mm) corridor, but requires 60" (1524 mm) to completely pivot
around. The ADA requires wheelchair accessible toilet compartments to be 60" (1525 mm) wide minimum measured perpendicular to the side wall, and 56" (1420 mm) deep minimum for wall
hung water closets, and 59" (1500 mm) deep minimum for floor
mounted water closets measured perpendicular to the rear wall.
(See Figure 11.3) The ADA does not allow toilet compartment
doors to swing into the minimum required compartment area.
There are requirements for approach and toe clearances as
well. There are also requirements for children's use.

An **ambulatory accessible toilet** is designed for someone who can walk with some support; it is not intended as a replacement for a wheelchair accessible toilet. (See Figure 11.4)
The ADA allows ambulatory accessible toilet compartments to
be a minimum of 60" (1525 mm) deep and between 35" and 37"
(890 and 940 mm) wide. Both side walls must have grab bars.

GRAB BARS
Standard towel bars are not strong enough to hold a falling person, and the hardware holding them to the wall is even less
strong. Grab bars should be installed with blocking to be able to
support 250 pounds (11.34 kg).

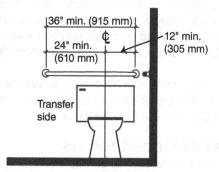

Figure 11.5 ADA rear wall grab bar at water closet
Source: Redrawn from 2010 ADA Standards Figure 604.5.2 Rear Wall
Grab Bar at Water Closets

The ADA sets requirements for grab bars at water closets.
Requirements include cross section dimensions, position, location,
and coordination with toilet paper dispensers. (See Figure 11.5)

The ADA does not require grab bars in a residence. However, reinforcement around the toilet area for grab bar installation allows bars to be placed according to the user's needs,
including their method of transfer. Recommended placement
for residential grab bars is on the rear and side wall closest to
the toilet. The side grab bar should be a minimum of 42" (1067
mm) long, running from 12" to 42" (305 to 1067 mm) from the
rear wall. The rear bar should be a minimum of 24" (610 mm)
long, centered on the toilet. It is preferable to make the rear bar
36" (914 mm) long where possible, with the extra length on the
transfer side of the toilet. Where the side wall is too short for a
42" (1067 mm) bar, a fold-down grab bar off the back wall or a
seat with integrated bars or hand holds might be used.

Traditionally, most grab bars have been placed horizontally. Vertical bars are used for reaching controls or stepping to enter. Angled bars let the wrist be bent in an easier position, and aid in rising to step out of a tub.

WATER CLOSET OPERATION

When a water closet is flushed, water either comes from a flush tank or flows through a flush valve and washes the closet bowl. This cleans the bowl and raises the water level. As the water rises, the draining passageway eventually fills and creates a siphon, draining water and other contents from the bowl. Siphoning is reinforced by ejection of some water into the drain. When the water and other contents are removed, the siphon is broken, but the water level continues to be raised in the bowl.

The vast majority of toilets in public restrooms are tankless. A tankless toilet receives its water directly from a pressurized supply line. It may be floor or wall mounted. Wall mounted water closets are easier to clean and more vandal resistant, but require a substantial fixture carrier support and higher water pressure. In private homes that lack adequate water pressure, tankless toilets add pumps or other technologies to increase flushing power.

Flush tank water closets are typically used in homes and small buildings. Their small water pipes slowly fill tanks that may be separately mounted on the wall or integral to the WC.

In a pressure-assisted toilet, water is compressed in a vessel inside the tank, and forced into the bowl quickly, for a clean flush with fewer clogs. (See Figure 11.6) The process may make more noise than other types of WCs. A pressure-assisted toilet installs in the same space as a conventional toilet, and requires 20 psi (138 kPa) of water pressure, which is typical in residential housing. They are used in homes, hotels, dormitories, and light commercial applications, and are available in accessible models. More and more states are mandating the use of pressure technology in commercial structures, primarily to prevent blockages.

A toilet bowl is designed to clean itself with each flush. (See Figure 11.7) A portion of the water flows out around the top rim, swirling to wash down the sides of the bowl.

Figure 11.6 Pressure-assisted flushing system

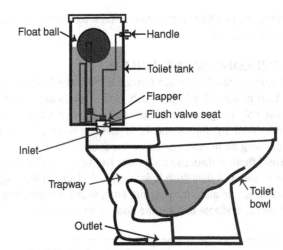

Figure 11.7 Toilet bowl and tank

You can flush a **touchless toilet** by simply holding your had over the tank. This is helpful to avoid picking up or leaving behind germs.

Water closets have large traps that are forced to siphon rapidly during the flushing process and are refilled with fresh water to retain the seal. The WC must be vented nearby to prevent accidental siphoning between flushes.

Most toilets bolt to the floor about 12" (305 mm) from the finished back wall surface to the center of the toilet flange, but this may vary. It is important that the plumber know the dimensional requirements of the model of toilet selected.

Most residential toilets are either one-piece, with a combined tank and bowl, or two-piece, with the tank bolted to the top of the toilet bowl. Two-piece toilets are generally less expensive and easier to carry upstairs. Without a seam, one-piece toilets are easier to clean. Wall-mounted residential toilets are also available that are even easier to clean, but are more expensive to install.

Conventional toilets do not provide for personal cleansing. Manufacturers offer toilet seats that provide for cleansing by building a source for clean water into the toilet. However, this may eliminate the possibility of using recycled water in the toilet. Other added features include heated seats, automatic flushing, remote controls, integral washing systems, and an LED light source within the fixture for night visits.

Water closet tanks are available with insulating liners of plastic foam inside to raise the tank's outside surface temperature above the dew point and prevent condensation.

An alternative type of toilet is made that consists of a toilet or urinal with no plumbing connections that reduces waste to a small volume of ash. It requires connection to electric power and a 4" (102 mm) diameter vent to the outside.

Toilets in many parts of the world that do not include a seat are known as squat toilets, Turkish toilets, or Asian toilets. They are designed to either sit or squat; some are designed to wash, others to wipe.

WATER-CONSERVING TOILETS

US Environmental Protection Agency (EPA) WaterSense toilets use 1.28 gallons (4.8 L) per flush. Some toilets are available that use only 1.0 gallon (3.8 L) per flush. These low figures are accomplished by improvements in existing flushing systems and by using air pressure to power the flush.

Dual flush toilets use buttons or handles to flush different levels of water. A dual flush toilet uses only 1 gallon for a light flush that is adequate most of the time, with a second, larger flush option available to clear the bowl when required.

When remodeling a bathroom, it is important to make sure that the water pressure is adequate for a new, more efficient toilet.

Another water-saving option is a central compressed air system that amplifies the water supply system pressure. A small compresser with an air tank can operate up to three toilets.

Automatic flushing controls add to the toilet's accessibility, keep toilets clean, and may reduce water use. They work by radiant heat from body pressure or by reflecting a light off the user and back to the control. Touch screen controls are also available.

COMPOSTING TOILETS

Composting toilets are probably the most ecologically sound method for disposing of human waste. (See Figure 11.8) Composting toilets convert human waste into organic compost and usable soil. A fan (which can be solar powered) continually draws air into the chamber and vents it through the roof. A composting toilet can operate very efficiently without odor problems if use is not excessive.

Figure 11.8 Composting toilet

Composting toilets use no water and keep wastes out of the wastewater system. They are used where water is scarce and/or sewage disposal is a problem. They are often used in federal and state parks, and are increasingly used in residential and institutional buildings.

No water inlet, sewer connection, or chemicals are needed. Solid waste is broken down and the residue is sent to a collection tray at the unit's bottom. The resulting compost needs to be removed about once a year. Although the aerobic digestion of waste is typically odor-free, ventilation is important to reduce odors and facilitate evaporation of excess moisture. The composting process creates around 1 cubic foot (0.03 cubic meter) of waste per person per year in the form of earth (compost), which can be used to fertilize nonedible plants.

EJECTOR TOILETS

Ejector toilets are used where the toilet is below the level of the sewer connection. The typical ejector toilet has a 5" to 6" (125 to 150 mm) high polyethylene pedestal for mounting directly on or recessing level with the floor. Where there is adequate ceiling height, a raised floor may be used.

A set of impellors and a sewage ejector pump inside the unit process waste and push it to the main sewer line. The toilet ejector tank can also drain nearby tubs, showers, or lavatories. In some models, the pump, vent, and pipes are located at a distance behind the toilet. This allows a wall to be built between the toilet and the equipment for a neater installation.

Sewage ejector pumps are introduced in Chapter 10, "Waste and Reuse Systems."

Urinals

Urinals reduce contamination from water closet seats and require only 18" (457 mm) of width along the wall. Urinals are not required by code in every occupancy type. They are usually substituted for one or more of the required WCs for men. The wall-hung type stays cleaner than the stall type, but tends to be too high for young boys and for men in wheelchairs; one or more in a group may be positioned lower. Stall-type urinals provide greater accessibility for a broader range of persons.

The 2010 ADA requires that urinals (where provided) be stall-type or wall-hung fixtures with the rim 17" (430 mm) above the finish floor. They are required to be at least 13½" (343 mm) deep measured from the outer face to the back of the fixture. Clear front space 30" wide by 48" (760 by 1220 mm) deep is required for a wheelchair front approach.

The height of the urinal's front lip off the floor should be 19½" (495 mm) for boys and 24" (610 mm) for men. Custom installation should locate the lip 3" (76 mm) below the man's pants inseam.

The standard wall-hung urinal may not meet accessibility mounting-height requirements. Newer styles should be checked for accessibility.

The recommended distance from the centerline of a urinal to a toilet, wall, or other obstacle is 18" (457 mm), with a minimum distance of 15" (381 mm). There should be a minimum 3" (76 mm) clearance from the edge of the urinal to a side wall. The recommended clearance in front of a urinal is 30" (762 mm), although a minimum of 21" (533 mm) is allowed.

A protective durable surface material should be installed at least 12" (305 mm) on either side of a urinal. Durable flooring needs to be used below and in front of the urinal.

Urinals should be separated by walls or partitions extending at least 18" (457 mm) off the wall or 6" (152 mm) beyond the outermost front of lip of the urinal, whichever is greater. Privacy partitions should cover from a maximum of 12" (305 mm) above the floor to a minimum height of 5 feet (1.5 m).

Older urinals used up to 5 gallons (19 L) per flush. In the United States, urinals are now required to use 1.0 gallon per flush (3.8 L) or less. High-efficiency urinals use a pint of water per flush.

Until waterless urinals became popular, a flush valve was often used, or a separate flush tank was hung on the wall above. **Waterless urinals** use no water, and are installed in public restrooms and large assembly venues. (See Figure 11.9) Some states are making waterless urinals mandatory. Residential waterless urinals may be a good idea in homes occupied mainly by men and boys. (See Figure 11.10)

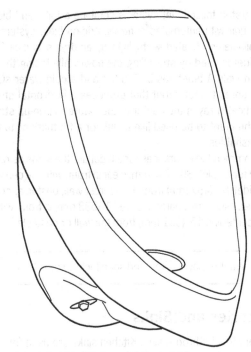

Figure 11.10 Residential waterless urinal

Bidets

A bidet looks like a toilet but works like a sink. (See Figure 11.11) A bidet may be installed to provide cleansing for the pelvic area. The need for and benefit of a bidet tends to increase as a person ages, as they are especially helpful for adults who have difficulty cleaning themselves.

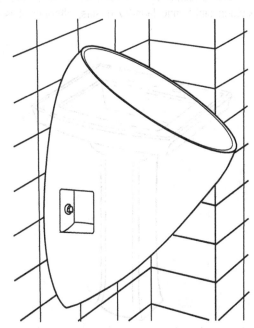

Figure 11.9 Waterless urinal

Figure 11.11 Bidet

Integrated toilet/bidet systems combine a toilet and bidet in a single unit with automated controls. Add-on bidet systems are also available with heated seats, lighting, air dryers, and air filters.

A bidet is used by straddling the bowl while facing the controls and wall. A bidet has both a hot and a cold water supply. It has a spray faucet spout that produces a horizontal stream, or a vertical spray in the center of the fixture. A pop-up stopper allows the bidet to be used like a sink for a footbath or to wash hand-washables.

Recommended clearances for a bidet are the same as recommended for a toilet: 30" (762 mm) clear in front, with the centerline of the bidet 18" (457 mm) from the nearest wall, obstacle, or adjacent toilet. Minimum clearances are 21" (533 mm) in front, with the bidet's centerline 15" (381 mm) from the wall or obstacle.

It is important to have towels and soap located next to the bidet.

Lavatories and Sinks

A lavatory is a bathroom sink. Kitchen sinks are used for washing dishes and preparing food. Service (slop) sinks are used for filling buckets, cleaning mops, and general cleaning.

LAVATORIES

Few lavatory designs seem to consider the way our bodies work and the way we wash. Most lavatories are designed as collection bowls for water, but we tend to use them for washing our hands, faces, and teeth quickly with running water. Historically, some washstands were combined with commodes. (See Figure 11.12)

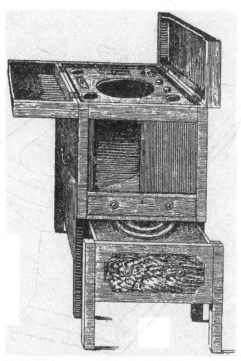

Figure 11.12 Antique washstand with slide-out commode

The sink and adjacent counter area are often difficult to keep clean and dry. For cleanliness and durability, lavatories must be made of hard, smooth, scrubbable materials like porcelain, stainless steel, or solid surfacing materials.

Lavatories are made in a wide variety of forms and sizes. The distance from the centerline of the lavatory to the sidewall or other tall obstacle should be at least 20" (508 mm). The minimum distance from the centerline of the lavatory to a wall required by the IPC is 15" (381 mm). According to the IRC, the minimum distance between a wall and the edge of a freestanding or wall-hung lavatory is 4" (102 mm). It is recommended that the distance between the centerlines of two lavatories should be at least 36" (914 mm).

The height of a lavatory or vanity can vary from 32" to 43" (813 to 1092 mm) to fit the needs of the user. Factory-standard vanities are 36" (914 mm). Adjustable height lavatories require a flexible plumbing line. Wall-mounted lavatories and those on wall-mounted counters can be located at the heights desired; some clients may prefer to have two lavatories at different heights. Vanities are usually 21" (533 mm) deep to allow the user to move close to the mirror.

Pedestal lavatories and above-counter vessels bowls have specific requirements for drainpipes or wall-mounted faucets. Pedestal lavatories come in a variety of heights. (See Figure 11.13) A platform base in a material to match the floor may be needed to achieve the height desired.

Vessel lavatories may be several inches in height. (See Figure 11.14) They can be set on top of or cut into a countertop. Concrete and ceramic lavatories are also available. (See Figures 11.15 and 11.16) Lavatories designed for installation in a corner may have special requirements.

SINKS

The term **sink** is technically reserved for service sinks, utility sinks, kitchen sinks, and laundry basins, although it is often

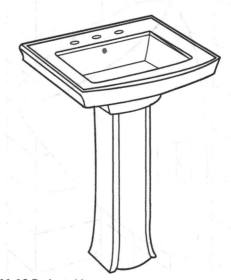

Figure 11.13 Pedestal lavatory

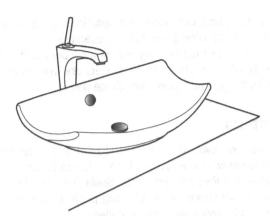

Figure 11.14 Vessel lavatory

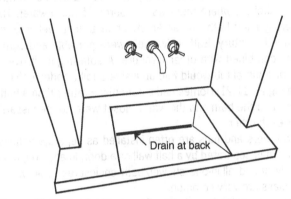

Figure 11.15 Concrete ramp lavatory

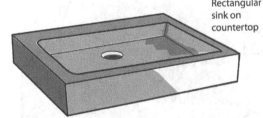

Figure 11.16 Ceramic lavatories

used for lavatories as well. Kitchen sinks are usually made of enameled cast iron, enameled steel, or stainless steel. Utility sinks are made of vitreous china, enameled cast iron, or enameled steel.

Building codes require sinks in some locations, and local health departments may set additional requirements. Kitchen or bar sinks in break rooms and utility sinks for building maintenance

Figure 11.17 Kitchen sink with drainboard

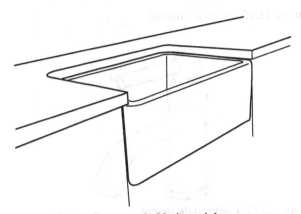

Figure 11.18 Farmhouse style kitchen sink

are often installed for convenience even when not required by code. (See Figure 11.17) The ADA sets standards for accessible kitchen sinks, including wheelchair clearances.

Services sinks are located in janitor's rooms for filling buckets, cleaning mops, and other maintenance tasks. Wash fountains are communal hand-washing facilities sometimes found in industrial facilities.

Farmhouse style sinks are wide and deep, with a large face set into the countertop and cabinet below. (See Figure 11.18) Both recycled and newly manufactured of types of this style are available in enameled porcelain, soapstone, and copper.

FAUCETS

Because of the design of most faucets, we usually have to bend at the waist and splash water upward to wash our face. Most lavatory faucets are difficult to drink from and almost impossible to use for hair washing.

Look for faucet designs that are washerless, drip-free, and splash free, and made of noncorrosive materials. Touch-on and touchless faucets are available. Public restroom lavatories should have self-closing faucets that save water and water heating energy. ADA compliant faucets come in variety of spout heights, and feature single lever, easy-to-grab models, wing handles, and 4" or 5" (102 or 127 mm) blade handle designs.

Faucets are manufactured as 4" (102 mm) **centerset** faucets that fit counters and sinks with three holes using a (102 mm) deckplate, or as **center hole faucets** that fit counters and sinks with a single hole. Centerset faucets are most often found on smaller vanity tops and sinks. (See Figure 11.19) Center hole faucets are found on a wide range of vanity tops and sinks. (See Figure 11.20)

Figure 11.19 Centerset faucet

Figure 11.20 Center hole faucet

Federal requirements limit residential bathroom and kitchen faucets to a maximum flow rate of 2.2 gallons (8.3 L) per minute, and nonresidential public restroom faucets to 0.5 gallon (1.9 L) per minute. Screwing an aerator onto the end of the faucet mixes air with the water to produce a fuller flow. A faucet meeting WaterSense specifications has a 1.5 gallon (5.7 L) per minute flow at 60 psi (414 kPA) water pressure.

A faucet should be designed so that the spray stays within the lavatory and does not spray on the user, counter, or floor.

The length of the spout should be proportional to the size of the lavatory. A larger bowl generates less overspray.

Deck-mounted faucets may require a deeper counter and a longer spout neck. Wall-mounted faucets are more difficult to install as the plumbing goes through the wall.

Bathtubs

Bathtubs have evolved from a barrel filled with water from a bucket to the whirlpool baths of today. Modern bathing is done in private on a very personal scale. Social bathing is limited to recreation, not cleansing, in swimming pools, bathhouses, and hot tubs with spouts, jets, and cascades.

In seventeenth- and eighteenth-century Europe, bathing in a middle-class home was usually done in a portable bathtub in the kitchen, where there was a source of warm water. The Saturday night bath was an American institution well into the twentieth century. Bathing vessels were portable and sometimes combined with other furniture. A sofa might sit over a tub, or a metal tub would fold up inside a tall wooden cabinet. (See Figure 11.21) Homes had a bath place rather than a bathroom, and the bath and the water closet were not necessarily near each other.

Showers and tubs are often installed as separate entities, sometimes separated by a half wall or a door. In addition, moderately priced all-in-one shower/bath enclosures in acrylic or fiberglass are very common.

We use bathtubs primarily for whole-body cleansing, and also for relaxing and soaking muscles. We follow a sequence of wetting our bodies, soaping ourselves, and scrubbing—all of which can be done well with standing water. Then we rinse,

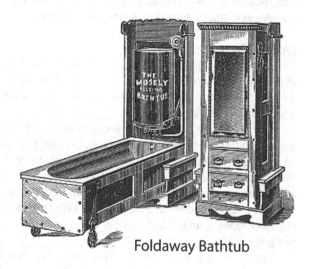

Foldaway Bathtub

Clawfoot bathtub with shower ring

Figure 11.21 Antique bathtubs

preferably in running water. Tubs work well in the wetting through scrubbing phase, but leave us trying to rinse soap off while sitting in soapy, dirty water. This is particularly difficult when washing one's hair.

There are alternatives to our typical showers and tubs. Traditional Japanese baths have two phases. You wet, soap, and scrub yourself on a little stool over a drain, rinse with warm water from a small bucket, then (freshly cleansed) you soak in a warm tub. An updated version uses a whirlpool hot tub for the soak. Locate the hot tub in a small bathhouse with a secluded view, and you approach heaven.

BATHTUB ACCESSIBILITY AND SAFETY

Tubs are often uncomfortable and dangerous for people to get into and out of. Turning water on and off and adjusting the temperature requires reaching. Shaving legs is easier and safer sitting on a bench or ledge. Tubs should be well lit and have easily cleaned but nonslip floors.

People in wheelchairs have to transfer onto a tub seat by sliding across from the chair. This can be an awkward maneu-

ver, and allowing enough access is important. A roll-in shower is easier to use.

Where a bathtub is required to be accessible, the ADA specifies the clear floor space in front of the tub, a secure seat within the tub, the location of controls and grab bars, the type of tub enclosure, and fixed/hand-held convertible shower sprays. (See Figure 11.22)

For safety's sake, all tubs should have integral braced grab bars horizontally and vertically at appropriate heights, with no unsafe towel or soap dishes that look like grab bars. (See Figures 11.23 and 11.24) Manufacturers offer very stylish grab bars that avoid an institutional look. A vertical grab bar at the control end wall of a tub is helpful. Freestanding tubs can use J-shaped supports intended for pools.

The safest way to design a tub is with the tub floor at the same level as the bathroom floor, with the deck or top of tub at around 18″ (457 mm). Even a single step at the tub makes balancing more difficult. Installing the tub so that its top is at floor level requires the user to either step down to tub level or sit on the floor to get in. Sunken tubs also present a greater risk

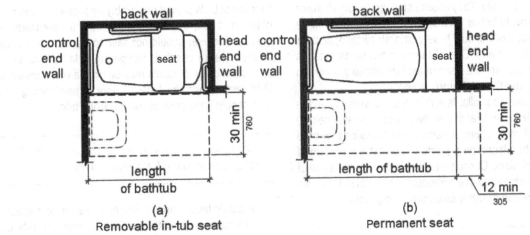

(a)
Removable in-tub seat

(b)
Permanent seat

Figure 11.22 2010 ADA bathtub clearances
Source: Adapted from 2010 ADA Standards Figure 607.2

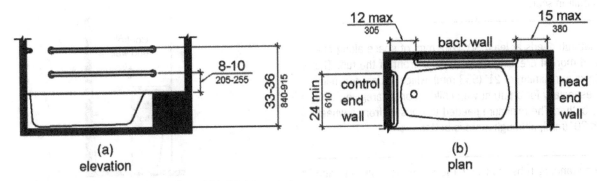

(a)
elevation

(b)
plan

Figure 11.23 ADA grab bars for bathtubs with permanent seats
Source: 2010 ADA Standards Figure 607.4.1

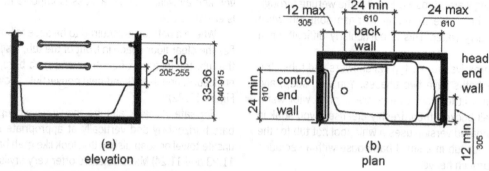

Figure 11.24 ADA grab bars for bathtubs with removable seats
Source: 2010 ADA Standards Figure 607.4.2

of tripping and falling in. Steps are not advised, but where they exist, they should have a grab bar or handrail and an alternate, nonstep way to enter the tub.

A transfer seat can be placed at the head of a standard tub. A minimum seat depth of 15" to 16" (381 to 406 mm) is recommended, with a height of 17" to 19" (432 to 483 mm), but the user's size and weight should be considered.

Tubs are available with integral or fold-away seats. Removable seats must be stable. They require storage space when not in use. Seats provide better leverage if the user's feet can be put under the seat. They should have a smooth surface; slots or openings improve hygiene. One of the best tub seats extends from outside the tub into the head of the tub, allowing a person to maneuver outside the tub before sliding in.

Accessible tubs are available with doors that swing open before the tub is filled. The seat should be at a comfortable height, typically 18" (457 mm). There are also tubs with sides that drop down; they can be placed on a platform so that the bottom of the tub is at seat height. Consult a professional regarding use of a lift for people with more severe impairments. Lifts may involve reinforcing the ceiling or other structural components.

Examine the specific way that a client can approach and enter a tub with doors or drop-down sides. Check the strength and clear space required to open the door, how much time it takes to fill and empty the tub, and the threshold height into the tub or to a built-in seat.

A bathtub needs at least 30" (762 mm) of space along one side, and more if dressing takes place in front of the tub. The IRC allows a minimum of 21" (533 mm), which is tight for many users, especially for a parent with children or a caregiver helping the bather. The minimum needed to transfer from a wheelchair is 30" (mm), although more is better.

For free-standing tubs, consider the side(s) for entering and exiting, and allow space for passage and clearances.

Building codes require that a window with its bottom edge less than 60" (1524 mm) above the finished floor must be tempered glass when located next to a bathtub. Tempered glass must also be used for glazing the door or enclosure for a tub or shower.

BATHTUB CONTROLS

The ADA sets standards for bathtub controls. (See Figure 11.25) Faucet controls should preferably be located within 6" (152 mm) of the front wall, where they will be easy to reach before entering the tub. The user should not have to lean across the tub to turn on the water or check its temperature. In an enclosed combination tub/shower, controls may be on the wall a maximum of 33" (838 mm) above the finished floor. Free-standing or platform tubs should have controls on the front side.

The location of bathtub controls should not conflict with the transfer area required for wheelchair users.

A single-lever shower faucet is easier to manipulate than round handles, and both temperature and flow rate can be adjusted with a single motion. Scald-proof thermostatically controlled, pressure-balanced, or combination valves must be used for tub/shower combinations.

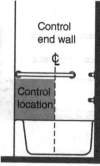

Figure 11.25 ADA bathtub controls location elevation
Source: Redrawn from 2010 ADA Standards Figure 607.5

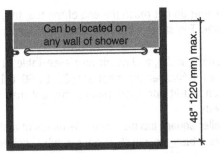

Figure 11.26 Elevation of ADA roll-in shower wall control locations

Source: Redrawn from 2010 ADA Standards Figure 608.5.2

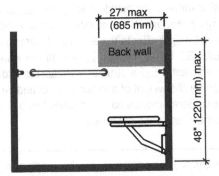

Figure 11.27 ADA roll-in shower with seat control locations

Source: Redrawn from 2010 ADA Standards Figure 608.5.2

A hand spray and 60" (1524 mm) hose allows a caregiver to assist a bathtub user. A hand-held spray aids in hair washing, and is easy to use for a seated bather. A trickle valve allows easy control. The hose length can be increased to 72" (1829 mm) or the length needed to reach bather.

The ADA sets standards for shower control locations. (See Figures 11.26 and 11.27) If installed on a sliding bar to double as a showerhead, the lowest position should not be greater than 48" (1219 mm) and well within the reach of the seat in the tub. The highest position for the showerhead should be 72" to 78" (1829 to 1981 mm).

Old bathtubs with a supply spigot in the tub my have ineffective or missing overflow limits, creating a potential backflow problem. Newer tub faucets are usually on the wall above the top of the tub to prevent this problem. A properly designed hand shower including antisiphon device is a good idea to include.

Bathtubs are made of a wide variety of materials. (See Table 11.4) Durability and sanitation are important concerns.

BATHTUB STYLES

Bathtubs come in a variety of styles, including corner and recessed alcove baths, among others. (See Figures 11.28 and 11.29) They are available either left- or right-handed, depending on which end has the drain hole. The design of the tub should ideally support the back, with a contoured surface and braces for the feet. Tubs can accommodate different leg lengths. If

TABLE 11.4 BATHTUB MATERIALS

Material	Description
Porcelain on steel (POS) or enameled steel	Resists acid, corrosion, abrasion. Flameproof, colorfast, sanitary, durable. If chipped, can rust. Can be noisy.
POS composite	Light-weight, better heat retention. If chipped, can rust.
Acrylic reinforced with fiberglass	Low cost, light-weight, repairable. Easy to clean. Keeps water warm. May scratch or discolor. Do not use abrasives.
Fiberglass-reinforced plastic (FRP)	With gel coating. Low cost, light-weight, easy to install. Less durable than acrylic; quality varies. Do not use abrasive cleansers.
Cast iron with enamel	Very durable. Keeps water warm. Extremely heavy. Difficult to repair. Sand-blasted non-skid bottom very difficult to clean.
Cultured marble (limestone, resin, gelcoat)	Tough, durable. Can repair slight chips. Moderate cost. Surface may scratch. Very brittle, thermal shock can crack.
Stone	Very heavy, requires structural support. Marble not recommended; very porous.
Solid surfacing materials	Warm to touch, retain water temperature. More damage resistant, repairable than cast iron and acrylic. White exterior can be painted.
Wood	Most woods will warp, crack, or rot. Teak will dry out and split if not used regularly. May not meet codes. Unique designs, expensive.
Ceramic tile	Durable, great variety of patterns, shapes, sizes. Water-resistant grout. Requires skilled installation. Tile cleans well, grout may not.

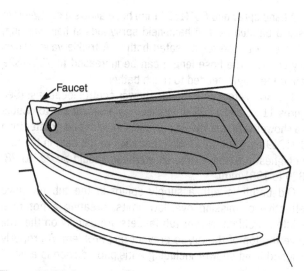

Figure 11.28 Corner bath

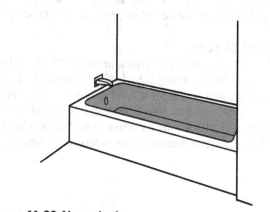

Figure 11.29 Alcove bath

the user's feet do not reach the end of the tub, they may slide under water. Too short a tub may leave bent knees out of the water.

Tub length, width, and depth vary. (See Table 11.5) A standard tub or tub/shower combination is 32" by 60" (813 by 1524 mm), which will fit into most bathrooms, but may not fit the user's needs.

A smaller square tub may meet the needs of a shorter person. Soaking tubs are usually deeper. A tub specified for bathing children should be smaller and shallower than one for soaking a larger person.

Whirlpool tubs are filled with heated water each time they are used. Different jetted actions move the water in the tub, some forcing water out at certain locations while others create a rotating pattern. Air bubblers lining the bottom create a soft massaging movement. Jetted tubs must be filled to above the jets, so jet height affects the amount of water used. Small whirlpool tubs 32" by 60" (813 by 1524 mm) use water and energy more efficiently. Extra structural support is added in new construction for the weight of additional water and people in an oversized tub. Some models have the pump located separately for noise control and ease of service.

Carefully evaluate the floor structure when adding an oversized tub. The floor may have to be stripped down to the joists to verify their size and spacing.

Two-person tubs are often jetted, and may be used only sporadically. Tubs for two people sitting side by side are 42" (1067 mm) wide; 36" (914 mm) suffices for sitting opposite.

TABLE 11.5 BATHTUB STYLES AND SIZES

Style	Description	Sizes (mm)
Recessed (3-wall alcove)	Unfinished on 3 wall sides, with decorative front	60" to 72" (1524 to 1829) by 32" to 42" (813 to 1069) by 16" to 24" (406 to 610)
Corner	Usually triangular with unfinished wall sides and finished front	Triangle 60" by 60" by 21" (1524 by 1524 by 533), other sizes available
Drop-in	Undermounted in platform cutout or platform built from floor	Sizes similar to recessed tubs; installed overlapping deck or undermounted
Free-standing	Traditional claw feet or contemporary, finished surfaces	60" to 75" (1524 to 1905) by 29" to 44" (737 to 1118) by 18" to 27" (457 to 686)
Whirlpool	Motorized circulation jets; available all styles, usually platform installed	Lengths 48", 60", 65" (1219, 1524, 1651). Up to 75" (1905) round. Corner tubs 60" by 60" (1524). Maximum practical size 72" by 42" (1829 by 1069)
Spa tub (hot tub)	Water remains in tub and is reused; thermostat controlled water stays heated	Must be kept tightly covered and insulated with cover and around the bottom and sides
Soaking tub	Can be designed to support back or legs; Japanese soaking tub is smaller and deeper, with seat	Length up to 75" (1981), 24" (635) depth. Requires large hot water tank, extra structural support

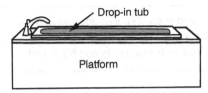

Figure 11.30 Drop-in bath

The deck of a drop-in or undermount tub can serve as a transfer seat, but must be designed to support a minimum of 250 lbs. (113 kg). (See Figure 11.30) If undermounted, the surface deck should overlap the tub flange to eliminate a permeable seam, and should slope slightly toward the tub.

Air tubs have a champagne bubble-type effect, while river jets simulate the undulating motion of white water river flow. Underwater lights, vanity mirrors, and wall-mounted CD/stereo systems with remote control are other luxurious options. Some tubs have built-in handrails and seats, while others have integrated shower or steam towers. Tubs are available with integral skirts for easy installation and removable panels for access.

Clients may request a big, two-person whirlpool tub, but often they do not use it as much as they think they will. People conscious of water use do not want to fill up a 300-gallon (1136 L) tub.

Thermoformed acrylic tub liners that can be installed over existing cast iron or steel tubs are a fast and economical way to upgrade a bathroom. However, existing moisture problems or imperfect installation can result in water collecting between liner and tub where mold can grow.

Showers

Modern showers evolved from military barracks and gymnasiums for men. The first showers used a hand pump to move water up a pipe and over a portable or outdoor tub.

Showers are seen as a quick, no-nonsense way to clean the whole body. They waste a lot of fresh running water while we soap and scrub, but do an excellent job rinsing skin and hair. With luck, you get a nice invigorating massage on your back, but a real soak is impossible. If you drop the soap, you may slip and fall retrieving it, so grab bars and an integral seat are good ideas.

Shower floors should have textured surfaces for slip resistance. It is also important that the flooring near the shower be slip resistant.

SHOWER CODES AND SAFETY

Where there is more than one shower in a public facility, the ADA requires at least one to be accessible. Accessible showers have requirements for sizes, seats, grab bars, controls, curb heights, shower enclosures, and shower spray units.

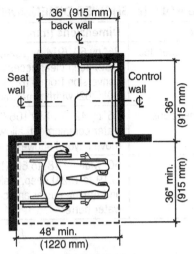

Note: inside finished dimensions measured at the center points of opposing sides

Figure 11.31 ADA transfer-type shower minimum dimensions
Source: Redrawn from 2010 ADA Standards for Accessible Design Figure 608.2.1

With enough adjacent clear floor space, most standing or seated users are able to use the shower on their own. How a bather with disabilities will enter the shower is an important design issue, particularly if a person is using a wheelchair. There are two types of accessible showers, transfer showers and roll-in showers.

Transfer showers are for the bather who can physically transfer from a wheelchair to a shower seat. (See Figure 11.31) The seat and grab bars must be positioned to facilitate that entry. The minimum finished dimension for a transfer is 36" by 36" (914 by 914 mm), with a full 36" (914 mm) opening. An L-shaped grab bar is required on the control wall and half of the back wall. A fold-up seat should be located on the wall opposite the controls. A transfer shower needs clear floor space at least 48" (1219 mm) along the opening and extending beyond the opening on the seat wall, by 36" (914 mm) deep.

A roll-in shower allows a person to remain in their wheelchair while showering, and is easier for most people to use, including children and people with balance issues. (See Table 11.6 and Figure 11.32) Clear floor space is required at least 60" (1524 mm) long next to open face and at least 30" (762 mm) wide. The threshold cannot be more than ¼" (25 mm) high to permit roll in, and the shower floor must be sloped to contain the water. A 60" (1524 mm) wheelchair-turning circle allows the user to access the shower seat or controls.

The IRC requires only 24" (610 mm) in front of the shower, but this is rather tight; a floor clearance 30" (762 mm) minimum is better. A dressing circle 42" to 48" (1067 to 1219 mm) in diameter allows space for drying off and changing clothing.

The area outside the shower can become a wet area by extending a waterproof membrane and sloping the floor gently toward the drain. A second drain outside the shower or a trench-style drain is helpful; many products are available to help. (See Figure 11.33)

TABLE 11.6 ROLL-IN SHOWER DIMENSIONS

Shower Type	Dimensions (mm)
Minimum recommended dimensions	36" by 60" (914 by 1524) with 30" (762) deep clear access. Allows conversion from a traditional bath. 30" (762) width is permitted.
Preferred dimensions	36" by 42" (914 by 1067) facilitates water containment, allows user to move beyond shower spray.
Ideal dimensions	60" wide by 48" to 60" (1524 by 1219 to 1524) deep. Makes access and turning easier and contains water better.

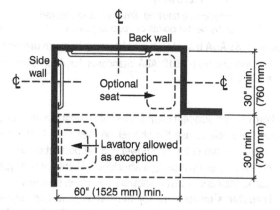

Inside finished dimensions measured at the center points of opposing sides

Figure 11.32 Standard ADA roll-in shower

Source: Redrawn from 2010 ADA Standards for Accessible Design Figure 608.2.2

Figure 11.33 Shower drains

A glass or other inflexible door can be problematic for a roll-in shower user. Where a shower is not more than 42" (1067 mm) deep, a door opening 36" (914 mm) is required to manage the turn at the entry. A deeper shower can work with a minimum 32" (813 mm) door.

Generously sized shower spaces address universal design principles. They can accommodate more than one person, and can be also be used for bathing pets and other tasks.

Grab bars are required at the shower's back and sides that support a minimum of 250 pounds (113 kg). Walls can be reinforced by installing 2 × 4 or 2 × 8 blocking horizontally between framing joists. If using a complete plywood surround, make sure it is watertight and protect it with a waterproof membrane. Placing reinforcing throughout shower walls allows clients to add supports as needed. A vertical bar at the shower entry provides support when getting in and out of shower. Grab bars should have slip-resistant surfaces. Towel bars, soap holders, and hand-held spray bars that are designed to function as grab bars are available.

Grab bars cannot extend more than 1½" (38 mm) from the wall to prevent getting a hand or arm caught between bar and wall. Controls should be installed above the grab bar.

The IRC requires a shower surround to be at least 80" (2032 mm) high. Waterproof material must extend a minimum of 3" (76 mm) above the showerhead rough-in, which is typically at 78" (1981 mm). The IRC requires coverage by waterproof wall materials to at least 72" (1823 mm) above the finished floor.

Showers may be required by code in assembly occupancies such as gyms and health clubs, and in manufacturing plants, warehouses, foundries, and other buildings where employees are exposed to excessive heat or skin contamination. The codes specify the type of shower pan and drain required.

Different kinds of shower seats are available, including adjustable, fold-up and stationary seats. Regardless of type, the seat must be installed where it will allow a seated bather to reach the showerhead, control valves, and soap caddie. A folding seat maximizes the open floor space in a roll-in shower. In a privately owned single-family home, it may be possible to match the specific client's needs and preferences.

Built-in seats are recommended to be between minimum 15" and maximum 16" (381 to 406 mm) deep with finish material in place, and 17" to 19" (432 to 483 mm) high. The seat should slope gently toward the shower base at not more than ¼" per 12" (6 mm per 35 mm) to avoid collecting water.

SHOWER ENCLOSURES

No-threshold (curbless) showers improve universal design access. They must be designed so that water will not splash out the opening. Considerations include the size of the shower, slope of the floor, and drain location; the location of a hand-held showerhead and whether the water spray can be controlled; and including the flooring outside of the shower in a wet area. Where a door is used, consider maneuverability around the door swing. A long, weighted shower curtain helps to control the water in a curbless shower. A trough-style drain at the back collects escaping water.

The recommended size for a one-person shower is 36" by 36" (914 by 914 mm); this is acceptable for transfer from a mobility aid. The IRC requires a minimum of only 30" by 30" (762 by

762 mm), which tends to be too tight for most adults. Angled showers require that a disc at least 30" (762 mm) in diameter fit into the shower floor, but larger is better. Large open showers accommodate more than one person.

Shower enclosures are usually enameled steel, stainless steel, ceramic tile, fiberglass, or acrylic. Other options include marble and other stones, glass block, and solid surfacing materials.

Shower pans are typically made of terrazzo or enameled steel and are available in solid surfacing materials as well. Barrier-free shower pans are available. A shower pan that converts a standard 60" (1524 mm) tub to a shower without moving the plumbing can improve safety. In this process, the old tub is removed and replaced with a slip-resistant shower pan. An acrylic wall surround can cover up old tile and unsightly construction work.

Frames for shower doors come in a variety of finishes. Glass panel anti-derailing mechanisms add to safety. Open, walk-in styles of showers with no doors are another option, as long as water is controlled.

A shower door should be at least 32" (813 mm) wide. The shower door should either slide or open outward toward the bathroom. This provides more clear space, and allows a person to get in to assist someone who has fallen inside the shower.

A shower door may drip water onto the floor, a consideration when specifying flooring types and materials.

Heavy glass frameless enclosures that can be joined with clear silicone are available up to ½" (13 mm) thick, although ⅜" (10 mm) is usually adequate. Body spray jets pounding at a frameless door will inevitably leak, so pointing them against a solid wall is a better option. A vinyl gasket can deter leaks, but may defeat the visual effect of the frameless glass, and is unlikely to be effective for very long. Totally frameless enclosures always lose a certain degree of water, and glass doors generally will not keep steam in and do not retain heat as well as framed doors. Complete water tightness may encourage mildew growth, so a vented transom above the door may be necessary.

The minimum finished interior size for all prefabricated shower stalls is 300 square inches (0.2 square meter), with a minimum 30" (762 mm) interior dimension. Smaller prefabricated two-wall units may be 35¼" (895 mm) square and from 71¼" to 83" (1810 to 2108 mm) high. Larger three-panel types range up to 40" by 60" (1016 by 1524 mm). Accessible units are available measuring 30" by 60" (762 by 1524 mm) minimum, with all necessary fixtures including floor, grab bars, seats, faucets, and showerheads. Pre-plumbed, all-in-one shower enclosures that include a steam generator are also available.

One-piece prefabricated showers are available. Multipiece units assembled on-site may solve problems with fitting through doorways.

SHOWERHEADS AND CONTROLS

Showerheads are available in two basic types. Regular stationary showerheads are hands-free but permit only limited aim adjustments. Hand-held types are attached to a flexible hose that can be clipped onto a wall-mounted hanger, swivel, or bar for hands-free use.

Hand-held heads may help to save water and energy by directing water only where it is needed, and by reducing the distance between the showerhead and the body, which encourages use of cooler water temperatures. Both standard and hand-held showerheads are available with adjustable sprays; those with rings outside the head are easier to adjust than those with controls in the head's center. A shutoff at the head reduces water to a trickle and saves water. An adjustable height showerhead on a vertical bar should not obstruct the use of grab bars.

To avoid problems, check whether the showerhead's slidebar is strong enough to act as a support or grab bar.

The IRC requires shower control valves to be either pressure balanced, have thermostatic mixing, or be a combination the two to prevent scalding due to changes in water pressure.

Standard showerheads use 2.5 gallons (9.5 L) per minute. Codes may limit showerhead flow. A WaterSense showerhead is limited to 2.0 gallons (7.6 L) per minute maximum. Shower water use can be decreased to less than 5 minutes by not running water when soaping up.

Low water pressure heads are designed to produce a satisfying stream at pressures below 80 pounds per square inch (psi). Low-flow functions are designed into the showerhead or retrofit with flow restrictors. Most bathers apparently do not notice the difference.

Manufacturers design low-flow showerheads that feel luxurious or more invigorating but use less water, by using wider, pulsing, and massage sprays. However, these showerheads may use more water to accommodate additional showering time and greater flow rates.

Multihead showers should have individual controls on each fitting to save water. This is especially important when a two-person shower is used by one person.

A fixed showerhead is typically roughed-in at 72" to 78" (1823 to 1981 mm) above the floor. The rough-in height can be designed to accommodate the user.

When helping children bathe, you should be able to reach the controls from the outside without wetting your arm, and to manipulate controls from inside without seeing them. Systems are available that allow the sprays to be moved to accommodate people of different sizes, and some systems come with programmable showerheads. Showers for two people should have two showerheads, each with separate controls.

Although it is convenient for a plumber to line control valves up vertically under a showerhead, this may make it more difficult for the user to reach while standing outside of the shower spray. A location 6" (152 mm) from the outside of the fixture is accessible. The best reach height is between 38" and 48" (965 and 1219 mm) above the finished floor. Transfer-type showers should have controls, showerhead, and handheld spray on a control wall within 15" (381 mm) of the centerline of the seat.

An alternative to locating all controls within a seated person's reach is to place the control for the overhead shower near the entry point, with a second control or diverter near a handheld spray located near the seat.

Steam Rooms and Saunas

A **steam room** is an enclosed space that creates a high-humidity environment with large amounts of high-temperature steam. On the other hand, a sauna warms and relaxes the body with dry heat created by pouring water on hot rocks, with only about 15 percent humidity.

STEAM ROOMS

An average steam bath consumes less than one gallon of water. Steam generators are usually located in a cabinet adjacent to the shower enclosure, but may be located up to 20 feet (6 m) away. Look for equipment with minimal temperature variations, an even flow of steam, quiet operation, and steam heads that are cool enough to touch. Plumbing and electrical connections are similar to those of a common residential water heater. Controls can be mounted inside or outside the steam room.

Steam rooms can be custom designed or installed as a prefabricated package. (See Figure 11.34) In either case, the steam generator must find a home nearby but out of the way. Prefabricated modular acrylic steam rooms are available in a variety of sizes that can comfortably fit from two to eleven people. They include seating and low voltage lighting.

Steam shower door openings need to be sealed to the ceiling with a transom or fixed panel. Allow a minimum of 4 square feet (0.37 square meter) per person, with 6 square feet (0.56 square meter) preferred. Doors must swing out and should not lock. Steam units need a drainpipe for the steam generator. The generator may be located under the bench, in a closet or vanity, or in an attic or basement. The steam shower also needs a waterproof ceiling, sloped or curved, to allow drainage. The assembly may need an access panel for servicing the mechanical equipment.

SAUNAS AND INFRARED HEAT

The traditional **sauna** itself is part of a larger process. After a short shower, the sauna user spends 5 to 15 minutes sitting or lying on a wooden bench in the insulated sauna room. This is followed with another shower or a visit to a pool or plunge bath, then a few minutes of rest. Then it is back to the sauna for about 20 minutes, followed by a 20-minute rest and then a final shower.

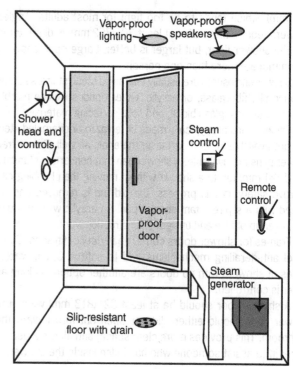

Figure 11.34 Steam room

Modular saunas combine wood and glass in sizes from 4 feet (1.2 m) square to 12 feet (3.7 m) square. There are even portable and personal saunas that can be assembled in minutes. Heating units are made of rust resistant materials and hold rocks in direct contact with the heating elements. Models are available in cedar, redwood, hemlock, and aspen.

Infrared (IR) heat is used as a warm-up for athletes, physical therapy, and massages. An **IR heat therapy room** uses IR heat at a lower temperature than a sauna, with normal room humidity; this takes less time to heat up. Pre-built units have some features similar to saunas, including cedar or alder lining, cedar benches with a backrest, doors, controls, and lighting. Multiple IR heaters along wall direct heat on each user.

RESIDENTIAL BATHROOM DESIGN

The design of bathrooms and public restrooms involves not only the plumbing system, but also the mechanical and electrical systems. There are special space-planning considerations—including acoustics—in bathroom design that have an impact on the plumbing layout.

The minimum code requirements for a residence include one kitchen sink, one water closet, one lavatory, one bathtub or shower unit, and one washing machine hookup. In a duplex, both units may share a single washing machine hookup. Each water closet and bathtub or shower must be installed in a room offering privacy. Some jurisdictions require additional plumbing fixtures based on the number of bedrooms. Many homes have more than one bathroom.

Bathroom Design History

The first indoor rooms designed specifically as bathrooms were spare bedrooms converted for wealthy clients. By the middle of the nineteenth century, fine homes were built with a separate bathroom.

The earliest bathrooms were often finished in wood, with a wooden toilet tank and seat, and a wood trimmed tub. Some were heavily draped, elaborately wallpapered, and carpeted. Finishes included marble, glass, and glazed tile.

In the late nineteenth and early twentieth century, hot and cold running water was still considered a luxury. Bathrooms in wealthy homes may have included a sitz bath in which only the buttocks and feet were submerged, foot bath, bidet, pedestal lavatory, siphon-action WC, enameled tub, and shower bath with a receptor.

By the early twentieth century, middle class homes tended to have stark, simple, hygienic bathrooms with plaster walls and hardwood floors. Pipes were often left exposed. Bathrooms were frequently designed for three fixtures: a toilet, a lavatory, and a bathtub or shower.

Bathroom Planning

On residential projects, the interior designer or architect helps the client with the selection of bathroom fixtures. The interior designer is often the key contact with the client, representing their preferences and providing specification information to the architect and engineer. Kitchen and bath designers, who may work for businesses selling fixtures, often help owners select residential fixtures on renovation projects.

For remodeling projects, verify as-built plans for accuracy. Measure and locate existing equipment and fixtures.

Bathroom floors need to be stable and even, especially around the toilet to avoid wobbling that could break its seal. Floor tiles, tubs, and shower pans should also be level. Avoid future damage by using materials that water cannot penetrate, and by caulking joints. Seal floors so moisture does not reach subflooring and joists, especially around toilets, tubs, and showers.

Some showers or tub/shower combinations have doors that are hinged directly to the wall rather than to the shower frame. The extra-heavy plate glass that is often used with frameless glass doors needs additional studs inside the wall for support. Some bathroom equipment and fixtures, such as full-body spray showers or wall-mounted toilets, specify a minimum 2×6 stud wall to accommodate valves and pipes.

INSPECTIONS

Several inspections by the local building inspector are required during the construction process, to assure that the plumbing is properly installed. As indicated in Chapter 9, roughing-in is the process of getting all the pipes installed, capped, and pressure-tested for leaks before the actual fixtures are installed. The interior designer should check at this point to make sure the plumbing for the fixtures is in the correct location and at the correct height. The critical dimension when installing a toilet is the rough-in distance from the wall to the center of the floor drain. The first inspection usually takes place after roughing-in the plumbing.

The contractor must schedule the inspector for a prompt inspection, as work in this area cannot continue until it passes inspection. The building inspector returns for a final inspection after the pipes are enclosed in the walls and the plumbing fixtures are installed.

RESIDENTIAL BATHROOM TYPES

There are a variety of types of residential bathrooms designed for use by residents and guests. (See Table 11.7)

Universal design provides for use by a variety of visitors. Locating a first floor bathroom near a bedroom or adaptable room helps

TABLE 11.7 RESIDENTIAL BATHROOM TYPES

Type	Description
Basic three-fixture bathroom	Lavatory, toilet, and tub and/or shower for single user. Minimum of 35 square feet (3.25 m²); master baths may be much larger.
Compartmentalized bathroom	Used by more than one person. Lavatory in hallway, bedroom, or alcove, with toilet and bath in separate nearby space. Toilet may be separate with own lavatory. Single door limits access if guests and family share.
Guest bath	Lavatory, toilet, and shower stall (not full bathtub). Minimum of 30 square feet (3 m²).
Half bathroom or powder room	Lavatory and toilet. Powder room may be under stairs or near mudroom entrance. 25 square feet (2.3 m²).
Hall bathroom	Lavatory, toilet, and bathtub and/or shower serving secondary bedrooms.
Shared bathroom	Between two bedrooms. May have lavatory and toilet room on each side, tub and/or shower in middle. Extra space for 2 doorways and circulation.
Bathroom suite	One or more lavatories, toilet, tub, shower, possibly bidet, vanity, dressing area. Usually located adjacent to master or guest bedroom.
Bathroom spa	Bathroom suite plus whirlpool or jetted tub, soaking tub, spa tub, sauna, and/or steam bath. Individualized to client.
Children's bathroom	Located in hall or connected to child's bedroom. Consider ages, needs. Lower mirror, adjustable showerhead, stepstools, and toilet training seat.

avoid the need to use stairs. It is wise to plan a small powder room that can be used by any guest. Maintain audio and visual privacy and avoid opening the powder room directly to a social area.

DESIGNING PUBLIC TOILET ROOMS

On many projects, the interior designer allocates the space for the public restrooms and places the fixtures. Toilet rooms in public facilities are often allotted minimal space and have to be designed with ingenuity to accommodate the required number of fixtures. The location of public restrooms should be central without being a focal point of the design.

Usually a licensed engineer designs the building's plumbing system. On small projects like adding a break room or a small toilet facility, an engineer may not be involved, and licensed contractors will work directly off the interior designer's drawings or supply their own plumbing drawings. The design of public restrooms also involves coordination with the building's mechanical system. The type of air distribution system, ceiling height, location of supply diffusers and return grills on ceilings, walls or floor, and the number and locations of thermostats and HVAC zones influence the interior design.

Interior designers must be aware of the specific numbers and types of plumbing fixtures required by codes for public buildings. The IPC helps interior designers determine the minimum number and types of fixtures required for particular occupancy classifications. Plumbing code requirements also include privacy and finish requirements and minimum clearances.

The entrances to public restrooms must strike a balance between accessibility and privacy, so that they are easy to find and enter, but preserve the privacy of users. Frequently men's and women's toilet rooms are located next to each other, with both entries visible but visually separate. This avoids splitting families up across a public space, is convenient for those waiting, makes finding the restrooms easier, and saves plumbing costs. The area just outside the restroom should be designed to allow people to wait for their friends, but should avoid closed-off or dark areas where troublemakers could loiter.

Restrooms with multiple water closets must have toilet stalls made of impervious materials, with minimum clearance dimensions and durable privacy locks. Urinals have partial screens but do not require doors. Generally, lavatories are located closer to the door than toilets.

Toilet Room Accessibility

The ADA requires that all restrooms it covers to be fully accessible to the public with adequate door width and turning space for a wheelchair. A single toilet facility is usually required to be accessible or at least adaptable to use by a person with disabilities. A door is not allowed to impinge on the fixture clearance space, but can swing into a turn circle. The ADA also regulates accessories such as mirrors, medicine cabinets, controls, dispensers, receptacles, disposal units, air hand dryers, and vending machines. Where non-accessible toilets already exist, it may be possible to add a single accessible unisex toilet rather than one per sex.

Generally, washroom accessories must be mounted so that the part that the user operates is between 38" and 48" (970 mm and 1220 mm) from the floor. The bottom of the reflective surface of a mirror must not be more than 40" (1020 mm) above the floor. Grab bars must be 33" to 36" (840 mm to 910 mm) above the floor.

Wheelchair access generally requires a 5 foot (1520 mm) diameter turn circle. The turn circle should be drawn on the floor plan to show compliance. When this is not possible, a T-shaped space is usually permitted. Special requirements pertain to toilet rooms serving children aged three to twelve. States may have different or additional requirements, so be sure to check for the latest applicable accessibility codes.

The conventional height of a toilet seat is 15" (381 mm). The recommended height for people with disabilities is 17" to 19" (430 to 480 mm). Doors on accessible stalls should generally swing out, not in, with specific amounts of room on the push and pull sides of doors.

The ADA requires a minimum of one lavatory per floor to be accessible, but it is not usually difficult to make them all usable by everyone. An accessible lavatory has specific amounts of clear floor space leading to it, space underneath for knees and toes, covered hot water and drain pipes, and lever or automatic faucets. The ADA requires clearance at water closets without obstructions or overlaps.

Some occupancies with limited square footage and minimal numbers of occupants, such as small offices, retail stores, restaurants, laundries, and beauty shops, are permitted to have one facility with a single water closet and lavatory for both men and women. These facilities must be unisex and fully accessible.

In larger buildings, fixtures may be grouped together on a floor if maximum travel distances are within the limits established by code. Employee facilities can be either separate or included in the public customer facilities. It is common to share employee and public facilities in nightclubs, places of public assembly, and mercantile buildings.

Fixtures should be located back-to-back and one above the other wherever possible for economical installation. This allows piping space to be conserved and permits greater flexibility in the relocation of other partitions during remodeling. Wherever possible, locate all fixtures in a room along the same wall.

Some types of occupancies present special plumbing design challenges. Plumbing fixtures for schools should be chosen for durability and ease of maintenance. Resilient materials like stainless steel, chrome-plated cast brass, precast stone or terrazzo, or high impact fiberglass are appropriate choices. Controls must be designed to withstand abuse, and fixtures must be securely tied into the building's structure with concealed mounting hardware designed to resist exceptional forces.

Drinking Fountains

Access to fresh drinking water is necessary for good health; for some people, frequent drinks of water are an essential health

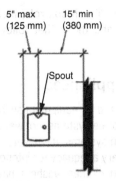

5" max (125 mm)   15" min (380 mm)

Spout

Figure 11.35 ADA drinking fountain spout location
Source: Redrawn from 2010 ADA Standards for Accessible Design Figure 602.5

need. Drinking fountains are not permitted in toilet rooms or in the vestibules to toilet rooms, but are often located in the corridor outside.

The ADA requires that one drinking fountain per floor be accessible. (See Figure 11.35) If there is only one fountain on a floor, it must have water spigots at wheelchair and standard heights. Accessible fountains have controls on the front or side for easy operation, and require clear floor space for maneuvering a wheelchair. Cantilevered models require space for a front approach and minimum knee space. Free-standing models require floor space for a parallel approach.

Drinking fountains are available with quick-disconnect cartridge filter systems that remove lead, chlorine, and sediment from the water, and remove *Cryptosporidium* and *Giardia* as well. Safety bubblers flex on impact to prevent mouth injury. To address the desire to reuse water bottles, some drinking fountains are now made for water bottle fill-ups.

APPLIANCES AND EQUIPMENT

Although such appliances as dishwashers and clothes washers are not usually considered to be plumbing fixtures, we are including them here as an aid to interior designers, who frequently assist clients in selecting them, and who locate them on their plans.

For more information on appliances, see Chapter 16, "Electrical Distribution."

Domestic hot water (DHW) heating uses a great amount of energy in residences, hotels, schools, restaurants, prisons, and other buildings. Solar hot water systems should be used wherever possible.

Residential Kitchens

The location of kitchen appliances has a great impact on the layout of a residential kitchen. Some residential appliances are plumbed for hot and cold water and some for gas. Locate the kitchen near another room with plumbing if possible.

KITCHEN HISTORY

Thirteenth-century Flanders (now Belgium) was the location of the first known kitchen separate from the main living space and fireplace hearth. The kitchen featured tables on trestles for food preparation. Horizontal boards above tables for were used for storage.

In a North American colonial kitchen, the room with the hearth was often the only room with a heat source, and was often also used for bathing. In wealthy households, the kitchen was used only by servants, and often located on a lower level or in a separate building.

The early nineteenth-century kitchen had a dry sink, until it was eventually possible to hand-pump water in the kitchen. Nineteenth-century Victorian kitchens had a free-standing range, sink, and table. Gas stoves were becoming available, but many cooks continued to prefer wood or coal burning stoves.

The Beecher Kitchen was designed in 1869 by Catherine Beecher and her sister Harriet Beecher Stowe. Their book, *The American Women's Home*, included an improved stove designed for multiple tasks and a plumbed sink.

Cities promoted cleanliness and sanitation in the late nineteenth century. The twentieth century saw standardization, improved appliances, and less help in the home. The Great Depression in the 1930s resulted in more efficient, smaller, and less expensive kitchens. Wiring of homes led to electric built-in kitchen appliances. University researchers studying efficiency, accuracy, safety, and usage of appliances developed kitchen planning concepts, including the work triangle, as well as U-shaped, L-shaped, corridor, and one-wall kitchens. The kitchen was becoming a place for family interaction, and opening up to dining and living areas.

WORK CENTERS AND AISLES

Twentieth-century researchers identified three primary **work centers**: sink, refrigeration, and cooking. Various secondary centers such as baking and salad preparation were also identified. Today, kitchens often include serving and dining areas, plus communications, laundry, and office work areas.

Access to appliances requires coordination with circulation. Work aisles for a single cook should be 42" (1067 mm). You need 48" (1219 mm) of space for a cook in front of an open dishwasher or oven door. A work aisle 60" to 66" (1524 to 1676 mm) allows room for a second person to pass, and makes a 360-degree wheelchair turn possible.

Circulation should be planned so that traffic does not interfere with work aisles. Minimum circulation for one person is 36" (914 mm). If circulation turns a corner, one leg should be 48" (1219 mm), which will allow a person using a mobility aid to turn. Where two people will frequently pass, allow 48" to 60" (1219 to 1524 mm) circulation.

KITCHEN UNIVERSAL DESIGN AND ACCESSIBILITY

People with sensory, cognitive, and physical issues are at increased risk of injury from hot surfaces of kitchen appliances and cooking items. Easy-to-understand appliances and controls use color cuing, blinking lights, beeps, and icons. Black glass surfaces make comprehension more difficult. Intelligent appliance controls such as programmed favorites and smart options help.

For people with decreased strength, stamina, and balance, minimum passage clearances, compact work triangles, and continuous counters can provide support in a small kitchen. A person using a mobility aid needs more generous spaces that avoid sharp turns.

Generally, wheelchair users prefer a perpendicular approach to a parallel approach. Avoid putting a sink in a corner, except when there is open knee space at least 30" (762 mm) wide for a seated user, and preferably 36" (914 mm) wide to be used as a leg of a T-turn.

For wheelchairs, a turn circle is preferred, but a T-turn may be easier to plan in a kitchen. Where paths are at right angles, at least one should be 42" (1067 mm) for a wheelchair to turn. Check the measurements required for a specific client.

The ability to sit at a sink or prep area is often welcome. Open knee space under the counter adds flexibility, and can also be used to house a waste bin, chair, or step stool.

The height of the counter can be designed to fit the user, typically 27" to 34" (686 to 864 mm); the landing space on each side should be the same height. A sink depth maximum 6½" (165 mm) works best.

A garbage disposer uses knee space under a sink. A compact model may be able to be offset to the side to leave more open space.

Cooking surface and ventilation controls can be offset toward the room for easy operation, at a height appropriate for the cook, typically between 15" and 44" (381 and 1118

mm) above the floor. Touch controls, including tactile cues, textured surfaces, and those that follow burner configurations help.

Residential Appliances

Water supply is required for equipment such as sinks, dishwashers, steam ovens, coffee systems, hot water dispensers, refrigerator water filtration systems, icemakers, or pot filler faucets. Evaluate water delivery adequacy for intended appliances.

Dishwashers and clothes washers have relatively simple plumbing requirements. Both dishwashers and clothes washers use vacuum breakers to prevent clean and dirty water from mixing. Be sure to leave adequate space for access, especially in front of front-loading machines.

For remodeling projects, check whether existing fixtures can be removed and reinstalled or relocated. Check whether new fixtures can be installed in old locations with old plumbing lines, or if new lines and fittings are needed. Where its use is permitted by codes, flexible PEX water supply plumbing makes it much easier to reposition fixtures.

The EPA estimates that a faucet that loses one drop of water per second can waste 3000 gallons (11,356 L) of water in a year. Select high quality fittings, fixtures, and appliances that are easy to maintain and less likely to develop leaks.

Kitchen appliances vary in width, and dimensions for specified equipment should be verified. However, preliminary appliance widths are helpful to use in early stages of kitchen planning. (See Table 11.8)

TABLE 11.8 PRELIMINARY KITCHEN APPLIANCE WIDTHS

Appliance	Type	Range of Widths (mm)
Kitchen sink	Single bowl	30" to 33" (762 to 1524)
	Double bowl	Up to 48" (1219)
	Triple bowl	Up to 60" (1524)
Dishwasher	Standard	24" (610)
	Compact or portable	18" (457)
	Single drawer	24" (610)
Range	Standard electric	30" (762)
	Standard gas	36" (914)
	Smaller sizes	12" to 24" (305 to 610)
Refrigerator	Side-by-side	30" to 36" (762 to 915)
	French door bottom freezer	29" to 35¾" (737 to 908)
	Top mounted freezer	23¼" to 35¾" (591 to 908)
	Bottom freezer	29" to 35¾" (737 to 908)
	Compact	14" to 24" (356 to 610)

KITCHEN SINKS AND FAUCETS

Kitchen sinks may have one, two, or three integral, self-rimming, or undermounted bowls. Materials include enameled cast iron, composite granite or quartz, solid surfacing, or stainless steel.

The kitchen sink should be located in a central, accessible spot near major cooking and refrigerator storage areas. Kitchen sinks have traditionally been located with a window view, but this may be less important when a dishwasher is used. There should be a landing area (counter) at least 6" (152 mm) deep and totaling between 28" and 48" (711 and 1219 mm) wide, located adjacent to the sink.

Kitchen sinks are equipped with water shutoff valves. Consider locating a central shutoff in or near kitchen to control multiple appliances in the entire area.

An auxiliary sink is useful where there is more than one cook. When close to the serving or dining area, a sink plus a dishwasher can serve as a cleanup area. With a knee space, a sink at a lower height can be used by a seated person. An auxiliary sink should have landing areas of at least 18" (457 mm) on one side and 3" (76 mm) on the other.

Faucets with one or two handles or gooseneck sprays are available. The easier it is to turn a faucet on and off, the less water is wasted. Touch control operates with a light touch anywhere on a large area of the faucet. A leaning bar faucet controller below the edge of a sink counter is another option.

DISHWASHERS

Dishwashers are available in a variety of types. Features include adjustable shelves, electronic or hidden controls, flatware trays, multiple racks, special cycles, and stem storage. Look for speedier cleaning cycles, quieter running, and power-scrubbing features for heavily soiled items. The interior finish may be plastic or stainless steel.

Conventional built-in dishwashers are permanently installed underneath a counter and connected to a hot water supply, drain, and electricity. Compact dishwashers are available for small kitchens. Portable dishwashers on wheels and countertop dishwashers get water from the kitchen faucet. Single or double dishwashing drawers slide out.

For information on water use by dishwashers, see Chapter 9, "Water Supply Systems."

The dishwasher should be placed within 36" (914 mm) of the cleanup sink. Although dishes do not need to be rinsed first, many users prefer to do so before putting them in the dishwasher. The counter above a built-in dishwasher makes a convenient landing area. Storage for dishes, glassware, and flatware should be nearby.

The open dishwasher door can become an obstacle to work flow. A 30" by 48" (762 by 1219 mm) clear floor space should be provided adjacent to dishwasher door.

Newer dishwashers with improved insulation are quieter. Specify a quiet model, especially for people with hearing loss.

COOKTOPS AND RANGES

Surface cooking appliances should be located adjacent to or across from the sink center, with a clear and uninterrupted path. Due to the risk of scalding, burns, or fires, do not place surface cooking equipment below an operable window, where it is unsafe to reach over hot pots. A minimum 15" (381 mm) landing area should be provided on one side of the cooking surface, and minimum 12" (305 mm) on the other, so that pot handles can be turned away from the traffic flow.

Follow manufacturer's clearances for the distance between the cooking surface and the wall. Wall materials should be fire-retardant and easy to clean. Large professional or cast iron ranges may require structural reinforcement for floors to support their weight.

Locating controls at the back of the range or cooktop should be avoided to keep from reaching over hot surfaces. However, locating them on the front may tempt small children. Controls on the top at one side improve access for most users.

Range types include free-standing, drop-in, slide-in, integrated, and professional style ranges. They may use electric, natural gas, or propane for fuel.

Cooking appliances designed for commercial use should not be used in a residential kitchen. Their high heat levels and difficult venting arrangements make them dangerous. **Professional style ranges** look like professional equipment and are high performance, but are typically permitted under residential code requirements.

A range with an oven results in a low oven location and the door becomes an obstacle, especially for short or seated cooks. Some ranges have a smaller oven just under the cooktop. Models are available with two equal-sized ovens.

Induction cooktops with an automatic temperature sensor and shut off reduce the risk of fires and burns for most people. A mapping feature activates the cooking surface according to the size of the pan used. The surface itself gets warm but not hot; heat only occurs where it is reflected back from a pan.

KITCHEN VENTILATION SYSTEMS

The IRC requires that range hoods discharge to the outdoors through a duct that does not terminate in an attic or crawl space or inside the building. There is an exception for properly installed listed and labeled ductless range hoods where mechanical or natural ventilation is otherwise provided.

The ventilation systems should be matched with the features of the cooking appliance, following manufacturer's specifications. Range hood are typically located over the cooking surface, ideally extending past it by at least 3" (76 mm) on both sides. They should be placed a minimum of 24" (610 mm) above the cooking surface; this minimum height could place the bottom of the hood at 60" (1524 mm), which may be too low for a tall person. The hood should be made of fire-resistant, nonflammable materials.

See Chapter 13, "Indoor Air Quality, Ventilation, and Moisture Control," for more information on residential ventilation systems.

A **proximity ventilation system** may be part of the cooking appliance or within the counter next to or behind it. These locations allow the space above the appliance to remain open. A cabinet or other flammable object placed above the cooktop with proximity ventilation should be a minimum of 30" (762 mm) above the cooking surface; a minimum of 24" (610 mm) is adequate for a protected fire-resistant surface.

Proximity or downdraft ventilation or a shallow, retractable overhead hood with eased or radius edges reduce the risk of head collision by tall people and those with visual disabilities. In-line and remote motors can be used to control ventilation noise levels. Electronic sensor controls are available for turning the exhaust on or off. Remote switching for ventilation is available.

OVENS

Placing an oven with a range in the primary cooking center allows easy observation and saves space in a small kitchen. An oven located below a cooking surface is at an inconvenient height for many people. The door of an oven should not open into a traffic path. Locating a single oven with its bottom at 30" to 36" (762 to 914 mm) above the finished floor makes it easier to transfer food to a counter at a similar height. For double ovens, the lower oven will be at a height similar to a range oven. Some larger kitchens accommodate two separate ovens at comfortable heights. Controls should be no higher than 48" (1219 mm), but high enough to reduce bending.

A 15" (381 mm) landing area should be placed on either side of the oven. If located across from the oven, the landing area should be within 48" (1219 mm) of the front of the oven, without crossing any major traffic path.

Ovens are available in a great variety of sizes, shapes, functions, and door designs. They may have conventional knobs or electronic controls. Technology-related intelligence is rapidly developing.

Built-in ovens can be single or double, with wall or under-counter installation. A conventional oven relies primarily on radiation from the oven walls. A **convection oven** uses fans to circulate air around food, cooking more evenly in less time at lower temperatures. A steam oven has a water reservoir that begins the baking process for faster cooking, lower fat content, and more vitamin retention. Combined convection and steam ovens are available. Speed cooking combines a microwave with a convection oven, allowing it to be used as a separate oven or a microwave. (See Table 11.9)

MICROWAVE OVENS

More than 90 prevent of US households use microwaves for food preparation, heating frozen or leftover foods, or as a major cooking appliance.

Microwaves that are intended primarily for heating leftovers or defrosting frozen food can be located next to the refrigerator.

TABLE 11.9 TYPES OF MICROWAVE APPLIANCES

Type	Description
Countertop	Free-standing, on wall cabinet shelf, or base cabinet with open space around it to dissipate heat.
Over-the-range	Vents in front remove heat and particles from cooking surface.
Over-the-counter	Mounted on wall with wall cabinets adjacent. Do not block vents on bottom.
Built-in	In base or tall cabinet. Trim kit helps vent heat out front.
Drawers	Built-in units under counter.

A location between the sink and refrigerator works well for these tasks plus some food preparation. Microwaves used for major cooking tasks can be located between the sink and a cooking appliance surface to combine a cooking area and prep area.

The ideal height for installation of a microwave is debatable. The location of a microwave should be tailored to its intended use and to the user's dimensional requirements.

Placing a microwave over a range saves counter space, but may be too high for some users, and may have an inadequate ventilation system. When locating a microwave with a wall oven in a cabinet, consider the cook's height.

A 15" (381 mm) landing area should be provided above, below, or adjacent to the microwave. Door handles are usually on the right, so this is the best side for the landing area.

REFRIGERATORS AND FREEZERS

Refrigeration units are always on, so they use significant amounts of energy. Selection should be based on volume, access requirements, and the parameters of the space. Consider how noisy the refrigerator is as well.

Refrigerators are accessed frequently for storage of fresh and frozen foods. They are often placed at the end of the kitchen work area. The refrigerator door should be able to open more than 90 degrees for storage and cleaning.

Refrigerator types vary in door and freezer location. (See Table 11.10) Refrigerator installation can be free-standing, boxed or built-in, integrated with cabinetry, under-counter, with decorative panels, or professional style. Older homes may not be able to support the weight of an oversized refrigerator and are likely to require structural reinforcement.

Features available with refrigerators include adjustable shelves, humidity controlled compartments, icemakers, door ice dispensers, mini-doors, and temperature controlled compartments. Water dispensers inside or outside can filter water; they are convenient, but lower the energy efficiency rating. Internet-connected refrigerators aid home care and monitoring, and reduce maintenance issues.

There should be a prep area 36" (914 mm) wide next to the refrigerator. Locate a landing area a minimum of 15" (381 mm) wide on the door-handle side so the door is not in the way when open.

TABLE 11.10 REFRIGERATOR TYPES

Type	Description
Side-by-side	Smaller doors, with storage at any reach level.
Top mount freezer	More convenient and economical for some users.
Bottom mount freezer	Freezer at seated height.
French door refrigerator/freezers	Smaller doors, plus freezer drawer at bottom.
Column refrigerator	Full height compartments sized for specific use.
Drawer refrigerator	May be difficult to lift items from drawer. Shallower, higher drawer above another drawer is available.
Modular units	All refrigerator, all freezer, mixed refrigerator and freezer drawers

Stand-alone freezers are sometimes located in the basement. A convertible freezer/refrigerator provides frost-free operation and a cooling system that allow conversion from a freezer to a refrigerator with the push of a button to provide extra refrigerator space.

Laundry Areas

The process of collecting, sorting, washing, drying, folding, and distributing laundry requires careful planning if it is to be done efficiently and comfortably. A laundry sink is useful for diluting laundry products, pre-rinsing stains, and hand washing or soaking soiled items. An extra deep utility-type sink may not be required; a small sink may work quite well. A gooseneck or pull-out faucet will allow bulky items to fit under it. Touch or automatic controls are helpful.

Locating a laundry area in or near the bathroom saves time and the effort of carrying dirty clothes. With plumbing nearby, installation is easier and less expensive. A clothes washer can also be located in or near a kitchen.

Space should be allowed in the laundry area for moving, turning, bending, and twisting while moving the laundry. There needs to be space for a laundry basket or cart. A clear space 30" (762 mm) wide and 42" (1067 mm) deep is recommended in front of a washer or stacked washer/dryer. Clearances can overlap is appliances are placed at right angles or across from each other. Increase clearances for clients using mobility aids such as a wheelchair or cane.

Remember to check the door swing clearance for front-loading kitchen and laundry appliances.

LAUNDRY EQUIPMENT INFRASTRUCTURE

Installing laundry equipment above a finished basement or on an upper floor can create vibration problems from their spinning

drums. Depending on the joist spacing, noise can travel to other parts of the home. Some manufacturers recommend extra floor support for weight and vibration. An alternative to installing additional joists is to use a thicker subfloor; however, this may affect door clearances.

A floor drain to handle overflow from the washer requires a recessed or sloped floor and connection to the building drain and waste system. An alternative may be to add a clothes washer overflow tray.

CLOTHES WASHERS AND DRYERS

A clothes washer requires water supply and drain piping and an electrical connection. A floor drain is also a good idea. A dryer needs an electrical supply, and for gas equipment, a gas connection and exhaust ventilation. Some of these can be difficult to retrofit in a renovation. A vacuum breaker prevents siphonage at clothes washers, where pumps can force wastewater into the drain line.

Manufacturer product specifications typically require vinyl, rubber, or other moisture-resistant flooring under laundry equipment. Utility service requirements for washers and dryers are specified by the manufacturer and may be controlled by local building codes.

A full-sized load for a standard clothes washing machine uses around 40 gallons (152 L) or less. **High-efficiency (HE) washing machines** are designed to save energy, but may have longer cycle times. High-efficiency front- or top-loading washing machines use from 15 to 30 gallons (57 to 114 L). Adjustable water levels allow less water to be used with smaller loads.

Clothes washers and dryers vary in size. A typical North American model of washer or dryer is 27" to 29" (686 to 737 mm) wide, and 25" to 32" (635 to 813 mm) deep. Most are around 36" (914 mm) high, but some are up to 45" (1143 mm) high. European models tend to be smaller. Taller machines may make it difficult for shorter people to retrieve laundry.

Both front-loading and top-loading washers and dryers are available. Allow adequate clearance in front of washers and dryers for laundry baskets and users. (See Figure 11.36)

Door swing direction affects moving laundry from washer to dryer. Some top-loading washers have doors hinged at the

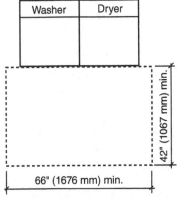

Figure 11.36 Laundry equipment clearances

back. Front-loading appliances with front controls are best for access within the universal reach range.

Lint in exhaust air can create problems. Adding lint filters requires regular maintenance, without which lint can clog the exhaust vent and create a fire hazard.

Not all washers and dryers are designed to be stacked. Check the models' specifications to make sure that stacking is an option. Careful planning of water connections, shut offs, and dryer venting is required for stacked laundry equipment. Smaller stacked washer and dryer combinations with integrated controls and reduced capacity are good for a second laundry area or for a smaller household. Combination laundry machines combine a washer and dryer in a single unit.

Washers are connected to plumbing, and dryers have both plumbing and ventilation requirements. (See Table 11.11)

TABLE 11.11 TYPICAL LAUNDRY PLUMBING REQUIREMENTS

Equipment	Plumbing and Ventilation
Clothes Washer	Check distance from water supply to washer, and water pressure required. Requires vented drain.
Electric Clothes Dryer	Exterior ventilation for dryer exhaust outlet. Distance from dryer to outside depends on number of elbows. Cold water supply if has steam feature.
Gas Clothes Dryer	Natural gas or LP connection. Exterior ventilation for dryer exhaust outlet. Cold water supply if has steam feature.

Check manufacturer specifications for required clearances on sides, front, back, and top for ventilation and equipment connections for laundry equipment located in a closet, cabinet, or under a counter. Doors must have adequate ventilation screening or be louvered.

Compressed Air

Compressed air is used in some toilet flushing systems. It is also used to power portable tools, clamping devices, and paint sprayers. An electric-powered compressor furnishes compressed air that is supplied through pipelines for use in workshops and factories. Air-powered tools tend to be cheaper, lighter, and more rugged than electrical tools.

This concludes our look at water and waste systems. In Part V, "Heating, Cooling, and Ventilation Systems," we consider how buildings keep us comfortably warm or cool, and how that affects their design and energy use.

PART

V

HEATING, COOLING, AND VENTILATION SYSTEMS

Heat is a form of energy that flows from a point at one temperature to another point at a lower temperature. When we are able to give off heat and moisture at a rate that maintains a stable, normal body temperature, we achieve a state of **thermal comfort**, the result of a balance between the body and its environment.

Architects and engineers use heating and cooling systems to provide clean air and control thermal radiation, air temperature, humidity, and airflow. These systems also help control fire and influence the thermal qualities of surfaces and acoustic privacy. Engineers are directly concerned with measureable environmental factors that affect thermal comfort, such as air temperature, surface temperature, air motion and humidity. We will be looking at these factors in Part V of this book.

Chapter 12, "Principles of Thermal Comfort," describes the conditions that comprise thermal comfort and the principles of heat transfer. It also addresses the thermal capacity and resistance of materials, and looks at the mechanical engineering design process.

Chapter 13, "Indoor Air Quality, Ventilation, and Moisture Control," looks at the requirements for indoor air and the role of ventilation and infiltration. Humidity and moisture control issues are also discussed.

Chapter 14, "Heating and Cooling," addresses the basics of heating, cooling, and HVAC systems and equipment.

The siting, orientation, and construction assemblies of a building should minimize heat loss to the outside in cold weather and minimize heat gain in hot weather. Any excessive heat loss or gain must be balanced by passive energy systems or by mechanical heating and cooling systems in order to maintain conditions of thermal comfort for the occupants of a building. While heating and cooling to control the air temperature of a space is perhaps the most basic and necessary function of a mechanical system, attention should be paid to the other three factors that affect human comfort—relative humidity, mean radiant temperature, and air motion. (Francis D.K. Ching, *Building Construction Illustrated, Fifth Edition*, Wiley -2014, page 11.06)

12

Principles of Thermal Comfort

Thermal comfort occurs when body temperatures are held within narrow ranges, skin moisture is low, and the body's effort of temperature regulation is minimized. Thermal comfort is affected by personal factors such as our body's metabolism and the clothing we wear. We can choose to move to a more comfortable place or consume warm or cold foods to control our thermal comfort.

"Factors affecting human comfort include air temperature, relative humidity, mean radiant temperature, air motion, air purity, sound, vibration, and light. Of these, the first four are of primary importance in determining thermal comfort." (Francis D.K. Ching, *Building Construction Illustrated*, Fifth Edition, Wiley 2014, page 11.04)

Interior designers are skilled at manipulating the psychological factors affecting thermal comfort. By working with color, texture, sound, light, movement, and scent, interior designers create spaces that support the thermal, spatial, and other sensory comfort of users. Their understanding of heating and cooling systems encourages good communication with architects and designers and aids in the Leadership in Energy and Environmental Design (LEED) certification process.

INTRODUCTION

Thermal comfort is the thermal transfer from our bodies to the surrounding environment at the same rate that we generate excess heat without undue stress. A building's heating system is not designed to create an overall net flow of heat into the human body, but rather adjusts the thermal characteristics of the indoor environment to reduce the rate of heat loss from the body to a comfortable level. In hot weather, the building's cooling system increases the body's rate of heat loss.

A building's interior thermally interacts with its surroundings through heat transfer through the building envelope. Heat transfer is also the mechanism by which mechanical heating and cooling equipment delivers thermal comfort.

Heat transfer through the building envelope is introduced in Chapter 2, "Designing for the Environment."

Building designers analyze the heat flow characteristics of the building environment to enable them to control heat gain and loss by the construction, orientation, and use of building materials. They compare the desired indoor temperature with average climate conditions to estimate the building envelope's ability to control thermal transfer and regulate interior conditions for thermal comfort.

History

When energy was cheap and plentiful prior to the 1970s, building designers lowered first costs of building construction by minimizing roof and wall insulation and using heating, ventilating, and air conditioning (HVAC) mechanical equipment to make the building comfortable. Increasing fuel costs resulted in energy becoming one of the largest expenses in a building's operating budget; this led to increased spending on construction to avoid future energy costs. Buildings that cannot be retrofitted for energy efficiency are destined to have shorter lives.

Today, energy conservation is a primary issue in building design and operation. By integrating an awareness of building heat-flow characteristics, the designer can conceptualize an energy-efficient building from the start of the design process.

This leads to less reliance on mechanical systems, resulting in lower costs, less maintenance, and less energy expense.

Better isolation of the building interior from adverse outside conditions leads to better control over occupant comfort. Although some energy conservation strategies are at the expense of comfort, building design today strives for a balance among initial costs, energy costs, and comfort.

THERMAL COMFORT

Factors affecting satisfaction with the thermal environment involve complex and subjective responses to many interacting variables. Engineers measure environmental factors including air temperature, radiant temperature, air motion, and humidity. Factors that affect a building occupant's thermal comfort include metabolism, clothing, location, and consumption of warm or cold foods. In addition, psychological factors such as color, texture, sound, light, movement, and aroma play a part.

Interior designers are particularly capable of addressing the psychological factors affecting thermal comfort, which are admittedly difficult to quantify.

ASHRAE's Standard 55-2010—*Thermal Environmental Conditions for Human Occupancy* defines the range of indoor thermal environmental conditions acceptable to a majority of occupants. The standard accommodates an increasing variety of design solutions intended both to provide comfort and to respect the need to design sustainable buildings.

Thermal comfort indices are very important for effective control of the thermal environment using no more energy and equipment than necessary. Common recommended conditions include new effective temperature, dry-bulb temperature, relative humidity, and air velocity. (See Table 12.1)

New effective temperature is an indicator of discomfort or dissatisfaction with the thermal environment. It is experimentally determined from dry-bulb temperature, humidity, radiant conditions, and air movement at specific conditions. The **dry bulb temperature** of a gas or mixture of gases is the temperature taken with a dry bulb that is shielded from radiant exchange,

such as the familiar wall thermometer. It may be the most important determinant of comfort. Relative humidity (introduced in Chapter 1) is the amount of water vapor present in air expressed as a percentage of the amount needed for saturation at the same temperature. Mean radiant temperature (MRT) (introduced in Chapter 2) evaluates the way an interior space and its furnishings radiate and emit heat to a human body in a given location. To achieve the maximum range of activity in which people feel comfortable, engineers minimize dry-bulb temperature and relative humidity (RH), while compensating with an MRT sufficient to maintain comfort.

See Chapter 2, "Designing for the Environment," for more information on relative humidity and mean radiant temperature (MRT).

Comfortable room air temperatures in a residence can range from 68° to 75°F (20° to 24°C) (See Table 12.2) People engaged in physical work need lower effective temperature than sedentary people. Spaces for very active, ill, or nude persons may require considerably different conditions for comfort.

Designing for Thermal Comfort

Buildings commonly deal with thermal radiation in two distinct parts of the electromagnetic spectrum. Solar radiation is emitted at many thousands of degrees temperature and composed of relatively short wavelengths. Thermal radiation coming from most terrestrial sources such as sun-warmed floors, warm building surfaces, or human skin is emitted at much lower temperatures and much longer wavelengths.

THERMAL RADIATION

Thermal radiation can be manipulated by controlling sun penetration, using thermal insulation inside walls and ceilings, insulating windows with multiple layers of glass or window treatments, and heating large floor or ceiling surfaces. Thermal radiation can also be increased by heating small surfaces

TABLE 12.1 COMMON DESIGN CONDITIONS FOR COMFORT

Design Condition	Recommended Level
New effective temperature (ET)	75°F (24°C)
Dry-bulb temperature	Equal to mean radiant temperature (MRT)
Relative humidity (RH)	40% (20% to 60% range)
Air velocity	Less than 40 fpm (02.m/s)*

*fpm = feet per minute, m/s = meters per second

TABLE 12.2 SAMPLE COMFORTABLE ROOM AIR TEMPERATURES

Type of Space	Summer	Winter
Residential	74° to 78°F (23° to 26°C)	68° to 72°F (20° to 22°C)
Bathrooms, showers	75° to 80°F (24° to 27°C)	70° to 75°F (21° to 24°C)
Restaurants	72° to 78°F (22° to 26°C)	68° to 70°F (20° to 21°C)
Retail stores	74° to 80°F (23° to 27°C)	65° to 68°F (18° to 20°C)

such as electric filaments, ceramic tiles, metal stoves, or fireplaces to high temperatures.

Thermal discomfort may exist if one part of the body is warm and another is cold. Noticeably uneven radiation from hot and cold surfaces, temperature stratification in the air, wide disparities between air temperature and MRT, chilly drafts, contact with warmer or cooler floor surfaces, or other factors can generate local discomfort.

Direct sunlight from large windows or skylights necessitates lower room air temperature along with shading and window treatments.

TEMPERATURE SCALES AND DESIGNATIONS

Heat energy is a result of the vibratory motion of molecules. Temperature is an indication of the intensity of heat energy. It is designated in degrees Fahrenheit (F), Celsius (C), and Kelvin (K). (See Table 12.3)

Absolute zero represents no thermal energy and no molecular motion. Above absolute zero there is some thermal energy in proportion to the temperature. In the Kelvin scale, 0° equals absolute zero.

Engineers use the MRT or operative temperature to help determine the amount of supplementary heating or cooling needed in a space. (See Table 12.4)

TABLE 12.3 TEMPERATURE SCALES

Scale	Description
Fahrenheit (F)	Scale with freezing at 32° and boiling at 212°; used in US
Celsius (C)	Scale based on 0° to 100°; used in most countries
Kelvin (K)	Uses absolute zero; not referred to as degrees; measure of color temperature

TABLE 12.4 TEMPERATURE DESIGNATIONS

Designation	Description
New effective temperature	Experimentally determined from dry-bulb temperature, humidity, radiant conditions, and air movement at specific conditions.
Operative temperature	Average of the dry-bulb temperature and the MRT.
Dry-bulb (DB) temperature	Ambient air temperature as measured by a standard thermometer or similar device.
Wet-bulb (WB) temperature	Expresses the humidity of air. Measured by thermometer with wetted bulb rotated rapidly in air to cause evaporation of its moisture.

MEAN RADIANT TEMPERATURE (MRT)

Mean radiant temperature (MRT) control is critical to achieving thermal comfort because it measures how the human body receives radiant heat from or loses heat by radiation to the surrounding surfaces. The MRT is the weighted average of temperatures of all surfaces in a direct line of sight of the body. Because infrared (IR) radiation spreads out, moving closer to a radiant source increases MRT warmth. The MRT is the sum of the temperatures of the surrounding walls, floor, and ceiling of a room, weighted according to the way that the IR radiation spreads out.

The MRT varies with the location where it is measured within a space, and calculations are complex. It is lowered by moving closer to a cold surface, and tends to stabilize near room air temperature. Space-defining interior elements tend to be at the interior air temperature unless they are located along the building perimeter, when they will have a surface temperature between the interior air temperature and the exterior air temperature. MRT is also affected by large glass areas, the degree of insulation, hot lights, and other factors. Window shading and better-insulated windows can greatly affect MRT. If the MRT is 10°F (5°C) hotter or colder than comfortable room air conditions, an occupant will tend to feel uncomfortable.

With a high MRT in winter, air temperatures can be somewhat lower. In summer, a building with high thermal capacity is likely to have cool interior surface temperatures, allowing comfort at higher air temperatures.

MRT is not a sufficient measure of thermal comfort in itself. There must be a balance of temperatures of surfaces to which body is exposed to prevent excessively rapid gains or losses from any one area of body. Conditions also must be balanced among heat gains by convection, conduction, and evaporation.

HUMIDITY

Humidity is the amount of water vapor in a given space. Relative humidity is commonly used in weather forecasts. Along with absolute humidity, specific humidity, degree of saturation, and percentage humidity, it is used by engineers as well. (See Table 12.5)

For more information on humidity, see Chapter 13, "Indoor Air Quality, Ventilation, and Moisture Control."

AIR MOVEMENT

Air movement significantly affects body heat transfer by convection and evaporation. It results from natural and forced convection along with the occupants' body movements. The faster the motion, the greater the rate of heat flow by both convection and evaporation.

Natural convection over the body surface results in continual dissipation of body heat at ambient temperatures within acceptable limits. At higher ambient temperatures, velocity must be increased with fans or other means.

TABLE 12.5 HUMIDITY DESIGNATIONS

Designation	Description
Relative humidity (RH)	Amount of water vapor in an air sample at a particular temperature and pressure, as a percentage of the maximum that the sample could contain at saturation.
Absolute humidity	Density of water vapor per unit volume of air, expressed in units of pounds of water or cubic feet of dry air.
Specific humidity (humidity ratio)	Weight of water vapor per unit weight of dry air, given as grains per pound or pound per pound (kg/kg).
Degree of saturation	Amount of water present in air relative to maximum amount it can hold at a given temperature without causing condensation.
Percentage humidity	Degree of saturation multiplied by 100; a lower percentage indicates lower moisture content.

Air motion also disperses body odors and air contaminants. Insufficient air motion results in **stratification** (varying air temperatures from floor to ceiling) and stuffiness. Air motion that is too rapid results in drafts. (See Table 12.6)

Air motion is measured in feet per minute (fpm) or meters per second (m/s).

Limits to air motion depend on a combination of overall room temperature, humidity, and MRT, and the temperature and humidity of the moving air stream. The cooler the moving air stream is, relative to the room air temperature, the less

TABLE 12.6 OCCUPANT REACTIONS TO AIR MOTION

Air Velocity	Comments
0 to 10 fpm (0 to 0.05 m/s)	Stagnation is evident
10 to 50 fpm (0.05 to 0.25 m/s)	Air outlet devices usually designed for 50 fpm occupied areas
50 to 100 fpm (0.25 to 0.51 m/s)	Motion evident, may be comfortable depending on moving air temperature and room conditions
100 to 200 fpm (0.51 to 1.02 m/s)	Aware of air movement, may be acceptable if intermittent and air temperature and room conditions acceptable
200 fpm and above (over 1.02)	Blowing papers and hair, other annoyances; about 2 mph

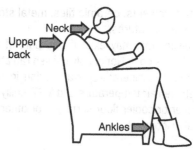

Figure 12.1 Draft sensitivity

velocity it should have. The neck, upper back, and ankles are most sensitive to drafts, especially when the entering air is 3°F (1.5°C) or more below the normal room temperature. (See Figure 12.1) Every 15 fpm increase in air movement above the velocity of 30 fpm is sensed by the body as a 1-degree drop in temperature.

A draft is excessive air motion that produces localized cooling of a building. The perception of a draft depends on air speed, activity, and clothing. Air conditioning systems are often drafty. Fans have a lower dissatisfaction rate.

Concerns about drafts affect air distribution as well as velocity. They affect placement of outside air openings for natural ventilation, radiation devices, and air registers.

For more information on air movement, see Chapter 13, "Indoor Air Quality, Ventilation, and Moisture Control."

THERMAL ENERGY

Temperature measures the degree of heat intensity. A unit of **thermal energy** is the amount of energy required to cause a specific change in temperature of a specific quantity of a material. The same quantity of thermal energy added to different quantities of the same material will result in different temperatures (thermal intensities.)

Thermal quantity is measured in British thermal units (BTUs), calories, or Calories (kilocalories). **Thermal intensity** is measured as temperature in degrees Fahrenheit (F) or Celsius (C).

The British thermal unit (BTU) is used for some purposes as the unit of heat energy in North America. Most of the world measures heat in calories.

THERMAL PROPERTIES OF MATERIALS

Materials can be classified by their thermal properties. **Insulators** retard the flow of heat and are good thermal barriers. **Conductors** encourage heat flow and make good thermal storage materials.

Thermal conductors and insulators are often used together. A wall may have a thermally massive inner layer with high conductivity for thermal storage, an outer layer with the same properties for durability and weathering, and highly insulating, low-mass material in between.

Comfort Range

Thermal comfort can be described as the state of mind that is satisfied with the thermal environment. The range of comfort varies with individual metabolism, physical activity, and the body's ability to adjust to a wider or narrower range of ambient conditions. Conditions affecting comfort include light, air, thermal comfort, acoustic comfort, and hygiene. Sometimes they are perceived by an absence of discomfort, such as temperatures that are not too hot or too cold, air that is not odorous or stale, or the absence of uncomfortable drafts or humidity.

Interior designers have a direct impact on many of the factors affecting comfort, including indoor air quality, acoustics, and lighting.

Historically, our tolerance for a range of indoor temperatures that changes with the seasons has become more limited. Until the 1920s, most people in the United States preferred indoor temperatures around 68°F (20°C) in winter, and tolerated higher temperatures during the summer. They would save the cost of expensive energy in winter by wearing warmer clothes. Low energy costs between 1920 and 1970 led to a preference for year-round indoor temperatures in the range of 72° to 78°F (22° to 26°C). We are now reevaluating our needs in light of energy savings.

In general, the comfort range that 80 percent of occupants would find thermally acceptable varies from 68°F (20°C) in winter to 78°F (25°C) in summer. This large range is partly due to the extra clothing worn in winter.

Our body's internal heating system slows down when we are less active, and we expect the building's heating system to make up the difference. Air movement and drafts, the thermal properties of the surfaces we touch, and relative humidity also affects our comfort.

METABOLISM

The complex physical and chemical processes involved in the maintenance of life are called our **metabolism**. People metabolize (oxidize) food, converting it into electrochemical energy used for growth, regeneration, and operation of the body's organs. Only about 20 percent of all the potential energy stored in our food is available for useful work. The remaining 80 percent is heat produced as a byproduct of conversion.

The rate at which our bodies generate heat is called our **metabolic rate**. Excess heat is carried away by perspiration evaporating off our skin. Thinner people, who have less insulation, stay cooler than fatter people. A very active person generates heat at a rate more than 8 times that of a reclining per-

TABLE 12.7 BODY HEAT PRODUCTION

Activity	BTU per hour (Watts)
Sleeping	340 BTUH (100)
Sedentary work	680 BTUH (200)
Walking	1020 BTUH (300)
Jogging	2720 BTUH (800)

son. This heat production can be measured in **BTU per hour (BTUH)**, or in watts. (See Table 12.7)

When the body's heat loss increases and its internal temperature begins to drop, the metabolism increases its efficiency to stabilize the temperature, even without additional mental or physical activity. As a result, all of the additional energy that is metabolized is converted to heat.

Our skin senses the rapidity with which objects conduct heat to or from the body better than it does actual temperatures. Our fingertips, nose, and elbows are more sensitive to heat and cold than other areas. Because fingertips readily detect the rate at which heat is being conducted away from our bodies, steel feels colder than wood at the same temperature as it conducts heat away from our fingers more quickly. (See Figure 12.2)

The hypothalamus in our brain constantly registers the temperature of our blood, and seems to be stimulated by minute changes in blood temperature anywhere in our body. Skin sensors also signal the brain with the level of heat gain or loss at our skin. The body controls the blood flow to our skin by constricting or dilating blood vessels, or by stimulating our sweat glands to increase evaporation.

Our body skin temperature is around 90°F (32°C) under normal unstressed conditions. In winter, with less blood near the surface, the skin becomes an insulator, and skin temperature is much lower than in summer. Increasing blood flow to the skin at the body's perimeter (vasodilation) usually means that more heat is lost to the surrounding environment. Decreased blood flow to the skin (vasoconstriction) induces the temperature at the body's extremities to drop, reducing the body's heat loss.

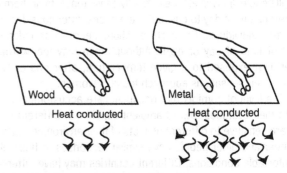

Figure 12.2 Touch and heat conduction

Source: Redrawn from Francis D.K. Ching and Corky Binggeli, *Interior Design Illustrated, Third Edition*, Wiley 2012, page 99

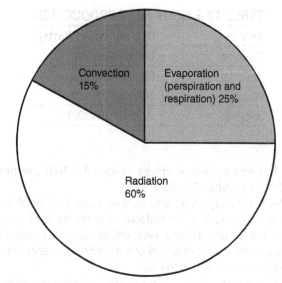

Figure 12.3 Body heat loss

Three conditions that contribute to loss of body heat are air temperature, relative humidity, and air movement. Heat is lost by convection, evaporation, and radiation. (See Figure 12.3)

Deep body temperature remains relatively constant at 98.6°F (37°C), but skin temperature may vary from 40° to 105°F (4° to 41°C), depending on surrounding temperature, humidity, and air velocity. When the ambient temperature in a space reaches 98.6° (37°C), no heat loss can occur by conduction, convection, or radiation. On the other hand, heat loss by evaporation works best at higher temperatures.

The heat given off by a building's occupants—their metabolic heat—affects temperatures in buildings. The heat of the people in crowded auditoriums, full classrooms, and busy stores warms these spaces. Small residential buildings usually gain less heat internally than larger office buildings housing many people and much equipment, and may turn on their heating systems sooner than larger buildings as outdoor temperatures fall.

INDIVIDUAL DIFFERENCES

Thermal comfort standards make no provision for individual differences, and there are no conditions that will produce comfort for all people. Individuals are normally consistent in their thermal preference from day to day, but preferences differ considerably between individuals. Comfort conditions seem to be independent of time of day or night. Although men may feel warmer when first exposed to a given temperature, after one or two hours, their sensations approach those of women.

The level of comfort that occupants are accustomed to affects their attitudes toward ambient conditions. Different people are sensitive to different aspects of their environment. Both younger and older people have similar responses to thermally comfortable conditions. Different countries may have different comfort standards, due to different climate extremes and the relative economics of providing and operating heating and cooling systems.

The differences among individuals tend to vary more than differences among countries. Different conventions of dress for men and women or style preferences among individuals can produce greatly different comfort environments.

The range of individual issues results in building designers trying to satisfy the majority of occupants and minimizing the number of people who will be dissatisfied. Interiors can be designed with a variety of conditions within one space, allowing people to move to the area in which they are most comfortable. Sunny windows and cozy fires offer many degrees of adjustment as a person moves closer or farther away. Providing localized user controls is another approach.

Short-term acclimatization is the period of time required to readjust to a conditioned indoor environment. A space conditioned for long-term comfort may be too cool for a person arriving from a hot outdoor environment. In short-term occupancy spaces such as stores or public lobbies, it is preferable to maintain a relatively warm, dry climate so that the perspiration rate changes only a small amount. For periods over one hour, it is recommended to reduce inside-to-outside temperature difference at least 5°F (2.8°C).

CLOTHING

Most clothing contains small air pockets separated from each other. The insulating properties of clothing modify body heat loss and comfort. These properties are quantified in the unit of thermal resistance called the **clo**, a numerical representation of a clothing ensemble's thermal resistance. A heavy two-piece business suit has insulating value around 1 clo, and a pair of shorts around 0.05 clo.

During the heating season, the average surface temperature of an adult indoors wearing comfortable clothing is approximately 80°F (27°C). Most people lose sensible heat at a rate that makes them feel comfortable with a surrounding temperature around 70°F (21°C). Adding 1 clo of insulation allows a reduction of air temperature of around 13°F (7.2°C) with no change in the thermal sensation.

At lower temperatures, comfort requires a fairly uniform level of clothing insulation over the entire body. For a person performing a sedentary occupation for more than one hour, the operative temperature should not be less than 65°F (18°C).

Conduction heat loss or gain occurs through contact of body with physical objects (surfaces or furnishings); clothing insulates the body from this contact.

THERMAL CONDUCTIVITY

Thermal conductivity is the ability of a material to transmit heat by conduction. Materials that conduct heat rapidly may feel cooler to the touch than they really are, when in fact they are warmer than air temperature. The high thermal conductivity of these materials gives designers a tool for creating spaces that feel cooler than their surface temperatures indicate that they are.

PRINCIPLES OF HEAT TRANSFER

As we noted earlier, heat energy is a result of the vibratory motion of molecules. One way to visualize how heat moves from one place to another is to think of it as the energy of molecules bouncing around. The bouncing of molecules causes nearby, less active molecules to start moving around too. The motion that is transferred from one bunch of molecules to another also transfers heat from the more excited group of molecules to the less excited group. A cold area is just an area with quieter molecules, and therefore, with less heat energy. A warm area is one with livelier molecules.

As long as there is a temperature difference between two areas, heat always flows from a region of higher temperature to a region of lower temperature, which means that it flows from an area of active movement to one of less movement. This tendency will decrease the temperature and the amount of activity in the area with higher temperature, and increase temperature and activity in the area with the lower temperature. When there is no difference left, both areas reach a state of **thermal equilibrium**, and the molecules bounce around equally.

Note that heat flows from a higher temperature to a lower temperature, not necessarily from a source of more heat (quantity) to one of less heat. The greater the difference between the temperatures of the two things, the faster heat is transferred from one to another. In other words, the rate at which the amount of molecular activity is decreased in the more active area and increased in the less active area is related to the amount of temperature difference between the two areas. Some other factors are also involved, such as conditions surrounding the path of heat flow and the resistance to heat flow of anything between the two areas.

Scottish mathematical physicist James Clerk Maxwell (1831–1879) illustrated heat transfer with a drawing of a spoon in a teacup: (See Figure 12.4)

For instance, if we put a silver spoon into a cup of hot tea, the part of the spoon in the tea soon becomes heated, while the part just out of the tea is comparatively cool. On

account of this inequality of temperature, heat immediately begins to flow along the metal from A to B. The heat first warms B a little, and so makes B warmer than C, and then the heat flows on from B to C, and in this way the very end of the spoon will in course of time become warm to the touch. The essential requisite to the conduction of heat is, that in every part of its course the heat must pass from hotter to colder parts of the body. (James Clerk Maxwell, *The Theory of Heat*, Dover Publications 2001, page 12. Unabridged republication of ninth edition, published by Longmans, Green and Co., London and New York, in 1888)

Heat energy is transferred in three ways, by radiation, conduction, and convection. We investigate each of these, along with evaporation, in upcoming sections.

Thermal Energy Transfer

The control of a building's thermal environment involves establishing conditions that allow occupants to lose excess body heat at the rate that they generate it. Heat loss or gain through the building envelope involves all three modes of heat transfer: conduction, convection, and radiation. Evaporation provides a fourth way in which thermal energy can be exchanged. (See Table 12.8)

For more information on thermal energy transfer through the building envelope, see Chapter 2, "Designing for the Environment," where conduction, convection, radiation, and evaporation are introduced.

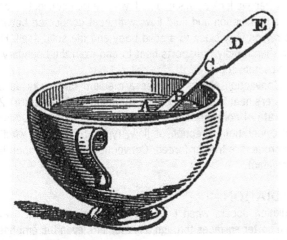

Figure 12.4 Maxwell's Teacup

Source: James Clerk Maxwell, *The Theory of Heat*, Longmans, Green and Co. 1888, unabridged republication by Dover Publications 2001, page 12

TABLE 12.8 HEAT TRANSFER

Process	Heat Flow Mechanism	Factors Affecting Transfer
Conduction	Heat transferred directly from molecule to molecule, within or between materials	Proximity of molecules (material density), surface temperature
Convection	Exchange between a fluid (typically air) and a solid	Motion of the fluid due to heating or cooling; air motion, temperature, humidity
Radiation	Heat flow via electromagnetic waves from hotter surfaces to detached, colder ones	Includes transfer across empty space and potentially great distances; surface temperature, orientation to the body
Evaporation	Moisture flow carries latent heat of vaporization away from wet surfaces	Flow through building envelope and via air leakage; humidity, air motion, air temperature

CONDUCTION

As introduced earlier, conduction is the flow of heat through a solid material, as opposed to radiation, which takes place through a transparent gas or a vacuum. (See Figure 12.5) Molecules vibrating at a faster rate (at a higher temperature) bump into molecules vibrating at a slower rate (lower temperature) and transfer energy directly to them. The molecules themselves do not travel to the other object; just their energy does. For example, when a hot pan comes in contact with our skin, the heat from the pan flows into our skin. Conduction is responsible for only a small amount of the heat loss from our bodies. When the object we touch is cold, like an iced drink in a cold glass, the heat flows from our skin into the glass. Conduction can occur within a single material, where there is a temperature difference across the material. The rate of transfer is directly proportional to temperature difference, surface area, and material **conductance**.

The ability of materials to conduct heat differs considerably. Poorer conductors are insulators, such as wood, plastics, gases, and ceramics. Air is a gas with spread-out molecules, so it minimizes conduction. For a material of given conductivity, conductance decreases as thickness increases.

Surface conductance (film conductance) is the transfer of heat from air to a surface or from a surface to air. Surface conductance of building materials is a function of both the properties of convection and radiation, and is dependant on color, smoothness, temperature, area, position of the surface, and temperature and velocity of surface air.

Thermal resistance is the reciprocal of conductance, and measures the insulating quality of a material. Resistance increases in direct proportion to the thickness of a material.

Figure 12.5 Conduction

Generally, materials that are good electrical conductors are good thermal conductors. Glass is an exception. Although it is a reasonably good thermal conductor, it is often used as an electrical isolator.

CONVECTION

Convection is similar to conduction in that heat leaves an object as it comes in contact with something else—in this case a moving stream of a fluid (liquid or gas) rather than another object. Our skin may be warmed or cooled by convection when we run warm or cold water or air over our skin. The amount of convection depends upon how rough the surface is, its orientation to the stream of fluid, the direction of the stream's flow, the type of fluid in the stream, and whether the flow is free or is forced. When there is a large difference between the air temperature and the skin temperature, plus more air or water movement, more heat will be transmitted by convection.

Convection occurs when the addition of heat to a liquid or gas (but not within a rigid solid) increases molecular activity, normally resulting in expansion of the substance and reducing its density. Liquid or gas particles are not held in a rigid position, so less dense molecules are able to rise. They are then replaced by other, more dense molecules. As more thermal energy is added, the more dense molecules heat up, expand, become less dense, and are themselves displaced. With convection, thermal energy is carried through the material by the movement of molecules, a process called fluid flow. Within a narrow air space where the friction of surrounding surfaces inhibits air movement, convection is limited.

Natural (free) convection currents that occur without wind or a fan tend to create layers that are at different temperatures (stratification). **Forced convection** occurs when a fluid (gas or liquid) is circulated between hotter and cooler areas by wind, fans, or pumps. Natural convection is not usually as powerful as forced airflow.

When we look at this in detail, we see that convection combines conduction and fluid flow, with heat conducted between the region very close to a solid body and the solid itself. Fluid flow then rapidly transports heat to and from the boundary region by convection.

Convection always involves a medium that picks up and transfers heat from one location to the other. (See Figure 12.6) The rate of convection is a function of the surface roughness and orientation, direction of flow, type of fluid, and whether the process is free or forced. Convection is often temperature dependant.

RADIATION

Radiation occurs when heat flows in electromagnetic waves from hotter surfaces through any medium, even the emptiness of outer space, to detached colder ones. It has a close relationship to the transfer of light. Radiant energy cannot go around corners or be affected by air motion.

Figure 12.6 Convection in a heated fluid

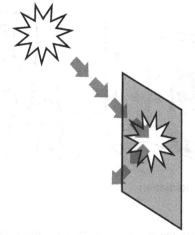

Figure 12.7 Sunlight reflected at angle of incidence

Figure 12.8 Black surface absorbs sunlight

Internal energy or molecular vibrations set up by the electromagnetic waves that emanate from the warm object carry energy to all bodies within a direct line of sight. The waves excite molecules of the receiving bodies, which increases their internal energy. Examples of thermal radiation include the energy of the sun's rays or of hot coals in a fire.

Radiation involves the release or reception of energy in the form of photons. The emission of radiation occurs when molecular activity at the surface of a substance releases a photon.

Radiation that is emitted from a substance is related to its ability to emit radiation (**emissivity**), which is usually equal to its ability to absorb radiation (**absorptivity**). The emissivity and absorptivity of a substance will depend on the wavelength of the radiation.

All objects and surfaces in a room radiate thermal radiation, but the only net transfer of heat occurs from warm bodies to cooler ones. Two objects at the same temperature will radiate to each other, but no net transfer will take place. Radiant heat transfer is influenced by the exposed surface areas of the warmer and cooler objects. An object can simultaneously gain radiation from hotter objects and lose radiation to cooler objects.

There are four possible interactions of radiation with materials—transmittance, absorptance, reflectance, and emittance—and more than one can occur at a time. (See Table 12.9) The type

of interaction depends on the material and on the wavelength of the radiation. The rate of radiant transfer depends on the temperature differential, the thermal absorptivity of the surfaces, and the distance between the surfaces.

Solar radiation reflects off glass at its angle of incidence. (See Figure 12.7). At the other extreme, a black surface will absorb most of the sunlight striking it, raising its temperature. (See Figure 12.8)

Radiation characteristics of a material are determined by temperature, emissivity, absorptivity, reflectivity, and transmissivity. Most materials have high **emittance**, but polished metal surfaces have low emittance. Materials that reflect visible light also reflect radiant heat.

Glass is mostly transparent to short-wave radiation and opaque to long-wave radiation, reflecting ultraviolet (UV) rays. Glass absorbs most long-wave radiation, then reradiates it inward or outward from the glass, so some long-wave radiation is

TABLE 12.9 INTERACTIONS OF RADIATION WITH MATERIALS

Interaction	Description
Transmittance	Radiation passes through material and continues on its way.
Absorptance	Radiation is converted into sensible heat within material.
Reflectance	Radiation is reflected off material's surface and diffracted from its path.
Emittance	Radiation is given off by materials' surface, reducing the sensible heat content of object.

Figure 12.9 Radiation

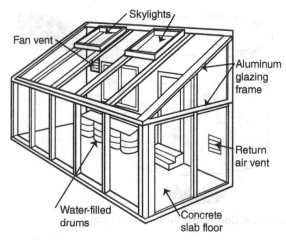

Figure 12.11 Sunspace

blocked (absorbed and reradiated) by the glass. Most plastics used for glazing act like glass, except polyethylene, which is transparent to IR radiation.

Draperies that block the line of sight can also block radiant heat from going out a cold window.

Air is a poor absorber of radiant heat; nearly all of our bodies' radiant exchanges are with solid surfaces to which we are exposed. The human body gains or loses heat by radiant heat according to the difference between body surface (bare skin and clothing) temperature and the MRT of the surrounding surfaces. You will feel the radiated heat from a fireplace if you are sitting in a big chair facing the fire, but if you are behind the chair, the heat will be blocked. (See Figure 12.9)

As we indicated earlier, heat will continue to move from a warmer area to a cooler one until both objects reach the same temperature. (See Figure 12.10) The greater the difference in temperature between the two objects, the faster the heat transfer will take place.

Buildings get heat in the shorter infrared (IR) wavelengths directly from the sun. Buildings also receive thermal radiation from sun-warmed earth and floors, warm building surfaces, and contact with human skin, which is emitted at much lower temperatures and at longer IR wavelengths.

As introduced in Chapter 2, the solar heat entering a sunspace warms its concrete slab floor and a row of water-filled drums, both of which are capable of storing large amounts of heat and slowly releasing it at a later time. (See Figure 12.11)

Emittance is important for cooling. In desert climates, hot days are followed by cool nights. The heat stored by the earth or by a building's mass during the day is emitted into the cooler night sky. (See Figure 12.12)

Air Temperature and Air Motion

Air motion may be caused by natural convection, be mechanically forced, or be a result of the body movements of the space's occupants. The natural convection of air over our bodies dissipates body heat without added air movement. When temperatures rise, we must increase air movement to maintain thermal comfort. Insufficient air movement is perceived as stuffiness, with the air stratifying with cooler air near the floor and warmer air at the ceiling.

A noticeable amount of air movement across the body when there is perspiration on the skin is experienced as a pleasant cooling breeze. When surrounding surfaces and room air

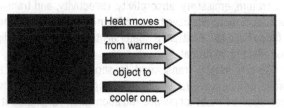

Figure 12.10 Heat transfer between objects

Figure 12.12 Emittance

temperatures are 3°F (1.7°C) or more below the normal room temperature, we experience that same air movement as a chilly draft. When the moving air stream is relatively cooler than the room air temperature, its velocity should be less than the speed of the other air in the room to avoid producing a draft. Air motion is especially helpful for cooling by evaporation in hot, humid weather.

Air passing over the skin surface affects transference of sensible heat to and from the body, in addition to evaporating moisture. The faster the rate of air movement, the larger the temperature difference between the body and the surrounding air. The larger the body's surface area, the greater the rate of heat transfer.

When the air temperature is lower than the skin temperature (also taking clothing into account), the body loses heat to the air; when warmer, it gains heat from the air. Convection is increasingly effective at dissipating heat as air temperature decreases and air movement increases.

Water Vapor and Heat Transfer

Heat loss by conduction, convection, and radiation decreases with increasing air temperature. On the other hand, evaporative heat loss increases with increasing air temperature. When people work under conditions of high temperature and extremely high humidity, both sensible heat loss and evaporation of moisture from their skins are reduced. Consequently, the rate of evaporation must be increased by blowing air rapidly over their bodies.

EVAPORATION

Evaporation is a process that results from the three types of heat transfer (radiation, conduction, and convection). Evaporation is vaporization that occurs below the normal boiling point of water. When a liquid evaporates, it removes a large quantity of sensible heat from the surface it is leaving. Sensible heat is measurable by a thermometer. It is a form of energy that flows whenever there is a temperature difference; it is apparent as the internal energy of atomic vibration within all materials. For example, when we sweat, and the moisture evaporates, we feel cooler as some of the heat leaves our body.

Latent heat and sensible heat are introduced in Chapter 2, "Designing for the Environment."

Water evaporates when the internal pressure of the liquid is greater than the vapor pressure (not the atmospheric pressure) of the air. (See Figure 12.13) The process still requires the addition of latent heat.

Our bodies contain both sensible heat (most of the body's heat) and latent heat (given off when perspiration evaporates). **Evaporative cooling** takes place when moisture evaporates and the sensible heat of the liquid is converted into the latent

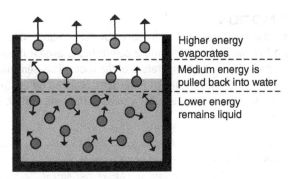

Figure 12.13 Evaporation

heat in the vapor. We lose the water and its heat from our bodies, and we feel cooler. Adding humidity to a room will decrease evaporative cooling, and is a useful technique for healthcare facilities and spaces for older people, where occupants may feel cold even in a warm room.

Evaporation results in body cooling only, and does not produce heat. The body's evaporative rate is determined by the evaporative potential of the air. This depends primarily on the velocity of air motion, and also on the relative humidity. Our skin typically sweats only at moderate to high temperatures. The heat required to vaporize perspiration from the skin is drawn from the body.

The evaporation of moisture from respiratory passages and lungs is constant. Exhaled air is generally saturated (100 percent relative humidity). Even at rest, our body requires around 100 BTUH (30W) of heat to evaporate this moisture from our lungs into the air we inhale. Because it takes a significant amount of heat to convert liquid water into water vapor, the evaporative loss from our lungs and skin is an important part of disposing of body heat.

SENSIBLE HEAT AND LATENT HEAT

Sensible heat is all about the motion of molecules; latent heat describes the structure of the molecules themselves. The **latent heat of fusion** is the heat needed to melt a solid object into a liquid. The **latent heat of vaporization** is the heat required to change a liquid into a gas. When a gas liquefies (condenses) or a liquid solidifies, it releases its latent heat. For example, when water vapor condenses, it gives off latent heat. The same thing happens when liquid water freezes into ice. The ice is colder than the water was because it gave off its latent heat to its surroundings. Sensible plus latent heat flows added together equal the total heat flow.

Every material has a property called **specific heat**. Specific heat identifies how much a material's temperature changes due to a given input of sensible heat.

A large amount of the heat of vaporization is drawn from the skin when sweat evaporates as sensible heat in skin is turned into latent heat of water vapor. As water evaporates, the air eventually becomes saturated, which inhibits further evaporation. Either air motion is required to remove some of the humid air, or very dry air is required to make evaporative cooling effective.

AIR MOTION

Air motion increases heat loss caused by evaporation, which is why a fan can make us feel more comfortable, even if it does not actually lower the temperature of the room. This sensation produces apparently higher or lower temperatures by controlling air moisture without actually changing the temperature of the space.

For information on humidity and air motion, see Chapter 13, "Indoor Air Quality, Ventilation, and Moisture Control."

THERMAL CAPACITY AND RESISTANCE

Thermal capacity is the ability of a material to store heat, and is roughly proportional to a material's mass or its weight. A large quantity of dense material (such as a large rock) can hold a large quantity of heat. Light, fluffy materials and small pieces of material can hold only small quantities of heat. Thermal capacity is measured as the amount of heat required to raise the temperature of a unit (by volume or weight) of the material by one degree. (See Figure 12.14)

Thermal capacity is introduced in Chapter 2, "Designing for the Environment," in relation to building envelope design.

A material with a high thermal capacity slows down the transfer of heat from one side of a wall to the other. In general, heavier materials have higher heat capacity. Water is an exception; it has medium weight, but has a higher thermal capacity than any other common material at ordinary air temperatures. Consequently, the heat from the sun retained by a large body of water during the day will only gradually be lost to the air during the cooler night. This is why, once a lake or ocean warms up, it will stay warm even after the air temperature cools off.

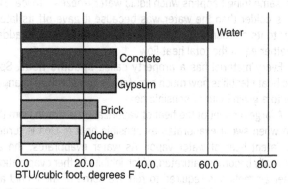

Figure 12.14 Thermal capacity of materials

Thermal Mass

Masses of high thermal capacity materials heat up more slowly and release heat over a longer time. For example, a cast iron frying pan takes a while to heat up, but releases a nice even heat to the cooking food, and stays warm even when off the burner.

Thermal resistance is a measure of a material's resistance to heat flow by conduction, convection, and radiation. Materials with high thermal capacity have low thermal resistance. When heat is applied on one side of the material, it moves fairly quickly to the cooler side until a stable condition is reached, when the process slows down.

Brick, earth, stone, plaster, metals, and concrete all have high thermal capacity. Fabrics typically have low thermal capacity. Thin partitions of low thermal capacity materials heat and cool rapidly, so the temperature fluctuates dramatically. For example, a tin shack can get very hot in the sun and very cold at night. Insulating materials have low thermal capacity since they are not designed to hold heat. It is the air that is incorporated in the air spaces between their thin fibers that has high thermal capacity.

Massive constructions of materials with high thermal capacity heat up slowly, store heat, and release it slowly. For example, a brick or stone fireplace evens out the otherwise rapid heat rise and fall of temperatures as the fire flares and dies. Masses of masonry or water can store heat from solar collectors to be released at night or on cloudy days.

Thermal mass and thermal lag are introduced in Chapter 2, "Designing for the Environment."

THERMAL LAG

The operative temperature of a room may be derived from radiant energy stored in thermal mass, allowing the room's air temperature to even out over time. This thermal lag can help moderate changes and is useful in passive solar design, but may also mean that the room will not heat back up to the desired level fast enough to suit its occupants. Heating or cooling the room's air temperature more than the usual amount can compensate for this slow change in temperature during warm-up or cool-down periods.

Only after a wall warms up substantially does heat begin to exit on the other side. High-capacity materials have greater time lag than low-capacity materials. There is no thermal lag under steady-state conditions when temperature across a material remains constant for a long period of time.

High thermal mass materials can be an integral part of the building envelope, or may be incorporated into the furnishings of the space. For maximum benefit, they must be within the insulated part of the building. A building's envelope with a large amount of mass may delay the transmission of heat to the interior for several hours or even for days; the greater the mass, the

Figure 12.15 Mud building in Sahara desert, Agadez, Niger

Figure 12.16 Amazon indigenous building, Peru

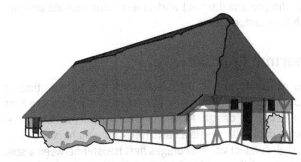

Figure 12.17 Danish vernacular house

Figure 12.18 A-frame ski lodge

longer the delay. Where thermal mass is used inappropriately, excessively high temperatures or cooling loads may result on sunny days, or insufficient storage may occur overnight.

The choice of whether or not to use high quantities of thermal storage mass depends on the climate, site, interior design conditions, and operating pattern of the building. High thermal mass is appropriate when outdoor temperatures swing widely above and below the desired interior temperature. Low thermal mass may be a better choice when the outside temperature remains consistently above or below the desired temperature.

Heavy mud or stone buildings with high thermal mass, such as adobe structures, work well in hot desert climates with extreme changes in temperature from day to night. (See Figure 12.15) The hot daytime outdoor air heats the exterior face of the wall, and migrates slowly through the wall or roof toward the interior. Before much of the heat gets to the interior, the sun sets and the air cools off outside. The radiation of heat from the ground outside to the sky cools the outdoor air below the warmer temperature of the building exterior, and the warm building surfaces are then cooled by convection and radiation. The result is a building interior that is cooler than its surroundings by day, and warmer by night.

In a hot, damp climate with high night temperatures, a building with low thermal capacity work best. The building envelope reflects away solar heat and reacts quickly to cooling breezes and brief reductions in air temperatures. By elevating the building above the ground on wooden poles to catch breezes, using light thatch for the roof, and making the walls from open screens of wood or reeds, the cooling breeze keeps heat from being retained in the building. (See Figure 12.16)

Buildings in cold climates typically have both high thermal capacity and high thermal resistance. Thick walls and insulating roofs, with only a few small openings, help. (See Figure 12.17)

In a cold climate, a building that is occupied only occasionally (like a ski lodge) should have low thermal capacity and high thermal resistance. (See Figure 12.18) This will help the building to warm up quickly and cool quickly after occupancy, with no stored heat wasted on an empty interior. A well-insulated frame coupled with a wood-paneled interior is a good combination.

Where the weather is very cold with high winds, the earth can shelter the building and moderate temperature changes. (See Figure 12.19) The high thermal capacity of soil ensures that basement walls and walls banked with earth stay fairly constant in temperature, usually around 50°F (13° to 15°C) year-round. Earthbound walls are not exposed to extreme air temperatures in cold weather. They should be insulated to thermal resistance values similar to the aboveground portions of the building. Burying horizontal sheets of foamed plastic insulation just below the soil's surface can minimize frost penetration into the ground adjacent to the building.

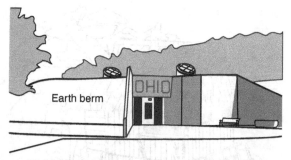

Figure 12.19 Earth-sheltered highway rest area, Ohio

A massive building can act as a heat sink, losing heat by convection to cool night air and by radiation to the cold night sky. This process does not work in very humid regions with high nighttime temperatures.

Thermal Conductivity

Thermal conductivity is the rate at which heat will flow through a homogenous solid. It is an important property in passive heating and cooling design. A material's conductivity is determined by empirical tests, and is a basic rating for a material.

High conductivity encourages heat transfer between a solid material and the air. Good conductors tend to be dense and durable, and to diffuse heat readily. (See Figure 12.20)

Homogenous materials have uniform properties with the same conductivity throughout. Their total conductance or resistance is determined on the basis of their thickness.

Many building materials, such as plywood and gypsum board, come in standard thicknesses. The heat flow of others, such as glass blocks or concrete masonry units, may vary through their thickness. The heat transfer characteristics of composite materials are typically given in terms of overall conductance or resistance.

Figure 12.20 Heat is conducted more rapidly through metal pan than through muffin

Thermal Resistance

Measurements of resistance are very helpful when comparing insulating materials. Thermal resistance indicates how effective any material is as an insulator. A good insulating material resists the conduction of heat. The bigger the difference between the temperature inside and outside a building is, the faster the building gains or loses heat. Designing a building's walls, roofs, and floors for the maximum amount of thermal resistance results in the best body comfort and the most energy conservation.

For building materials, thermal resistance is largely a function of the number and size of air spaces they contain. (See Table 12.10) For example, a 1" (25 mm) thickness of wood has the same thermal resistance as 12" (305 mm) of concrete when the temperature across the material remains constant for a long period of time (disregarding thermal lag).

If you keep air from moving by trapping it in a loose tangle of glass or mineral fibers, you create materials with very high thermal resistance. The fibers themselves have poor resistance to heat flow, but create resistance to air movement, and thereby trap the air for use as insulation. When the air is disturbed, this insulating property drops to about a quarter of its value. If air circulates within a wall, a convective flow is created, which transfers heat from warmer to cooler surfaces pretty quickly.

Thermal Feel

Earlier we looked at how our bodies perceive the movement of heat from warmer to cooler objects, and how our senses are influenced by the rapidity with which objects conduct heat to and from our body rather than by the actual temperatures of objects. Steel feels colder than wood at the same temperature, as heat is conducted away from our fingers more quickly by steel than by wood. Smooth surfaces make better contact than highly textured ones, resulting in better conduction of heat and a cooler feeling. These sensations are very useful to interior designers, who can specify materials that suggest warmth or coolness regardless of their actual temperatures.

Interior designers can use an awareness of how we perceive the temperature of different materials to select appropriate materials for projects. A wood edge on a tabletop will be perceived to be warmer than a brass edge. Some of the materials we like close to our skin—wood, carpeting, upholstery, bedding, some plastics—feel warm to the touch, regardless of their actual

TABLE 12.10 THERMAL RESISTANCE OF MATERIALS

Material	Thermal Resistance
Metals	Very low
Masonry	Moderately low
Wood	Moderately high
Glass	Low
Air	Best resistor commonly found in buildings

temperature. We perceive materials to be warm that are low in thermal capacity and high in thermal resistance, and that are quickly warmed at a thin layer near their surfaces by heat from our bodies. Materials that feel cold against the body, like metal, stone, plaster, concrete, and brick are high in thermal capacity and low in thermal resistance. They draw heat quickly and for extended periods of time from our body because of the relatively larger bulk of cooler material.

MECHANICAL ENGINEERING DESIGN PROCESS

The decisions made by the designers of a building's HVAC system are crucial in determining thermal comfort, the quality of the indoor air, and the efficiency of energy use by the building. Air exchange rates affect the amount of energy used to heat or cool fresh air, and the energy lost when used air is exhausted. American Society of Heating, Refrigerating, and Air Conditioning Engineers (ASHRAE) requirements for ventilation include minimum rates for replacing previously circulated air in the building with fresh air.

Energy efficiency can be increased by improving building insulation, lighting design, and the efficiency of HVAC and other building equipment. The architect and engineers usually make the decisions on what systems to employ, but the responsibility for finding appropriate solutions depends on creativity and integrated efforts of the entire design team, in which the interior designer should play a significant role.

As an interior designer, you may rarely need to refer to the mechanical codes, but you should be familiar with some of the general requirements and terms, especially those affecting energy conservation requirements. In buildings where there is a minimum of mechanical work the mechanical engineer or contractor may work directly off the interior designer's drawings. For example, the interior design drawings may be the source for information in a renovation project where a few supply diffusers or return grilles are being added to an existing system. In any event, you need to coordinate your preliminary design with the mechanical engineer or contractor to make sure you leave enough room for clearances around HVAC equipment.

The mechanical engineer, like the interior designer, is trying to achieve an environment where people are comfortable, and to meet the requirements of applicable codes. By calculating how much heating or cooling is needed to achieve comfort, the engineer develops design strategies that affect both the architecture and the mechanical systems of the building, such as the optimal size of windows, or the relative amounts of insulation or thermal mass.

The engineer will size the HVAC system components for the most extreme conditions the building is likely to experience, calculate the amount of energy used for normal conditions in a typical season, and adjust the design to reduce long-term energy use. This involves considering the number of people using the building both seasonally and hourly, and the amount of heat gained or lost from the outside environment. The materials, areas, and rates of heat flow through the building's envelope affect this calculation. Engineers also determine the amount of fresh air needed and may suggest window locations and other design elements that minimize the heat gain within the building.

Phases of Design Process

The phases of the engineering design process are similar to those of architects and interior designers, with preliminary design, design development, design finalization and specification, and construction phases.

PRELIMINARY DESIGN

During the preliminary design phase, the engineer considers comfort requirements and climate characteristics and lists the schedule of activities that will take place in the space along with the conditions required for comfort. The process involves analysis of the site's energy resources and strategies to design with the climate. The engineer considers building form alternatives with the architect. He or she reviews alternatives for both passive and active building systems and the sizes of one or more alternatives using general design guidelines.

In smaller buildings, the architect may do the system design. For larger, more complex buildings, the mechanical engineers will work as a team with architects, landscape architects, and interior designers. The team approach helps to assess the value of a variety of design alternatives arising from different perspectives. When mutual goals are agreed upon early in the design process, this team approach can lead to creative innovations.

DESIGN DEVELOPMENT

During the design development phase, one alternative is usually chosen as presenting the best combination of aesthetic, social, and technical solutions for the building's program. The engineer and architect list the range of acceptable air and surface temperatures, air motions, relative humidities, lighting levels, and background noise levels for each activity to take place in the building, and develop a schedule of operations for each activity. The engineer then determines the thermal comfort zones for the building, establishes the **thermal load** (the amount of heat gained or lost) for the worst winter and summer conditions and for average conditions for each zone, and may estimate the building's annual energy consumption. The engineer then selects and sizes the HVAC systems, identifies its components, and locates the equipment within the building.

DESIGN FINALIZATION AND CONSTRUCTION

The process of design finalization involves the designer of the HVAC system verifying the load on each component and the component's ability to meet this load. Then the final drawings and specifications are completed. During construction, the engineer may visit the site to assure that work is proceeding according to design and to deal with unanticipated site conditions.

Thermal Comfort Zones

The engineer establishes each of the **thermal comfort zones** with its own set of functional, scheduling, and orientation concerns that determine when and how much heating, cooling, or ventilation is needed. (See Figure 12.21) Functional factors depend on activity levels or users' heat tolerances, and include the need for daylight and the affect of each function on the air quality of the others. Scheduling affects the need for electric lighting and the heating and cooling of unoccupied spaces. Orientation considerations include the degree of exposure to daylight and direct sun, and to wind, especially for perimeter spaces. Cooling for interior spaces is also considered.

The way zones for heating and cooling are set up by the mechanical engineer has implications for the architecture and interior design of the space. Zones may occupy horizontal areas of a single floor, or may be vertically connected between floors. The function of a space affects both its vertical and horizontal zoning. Some functions may tolerate higher temperatures than others. Some functions require daylight, which may add heat to the space, while others are better off away from the building's perimeter. In some areas, such as laboratories, air quality and isolation is a major concern. The input of the interior designer can be an important component in making sure that the client's needs are met.

In multistory buildings, interior spaces on intermediate floors—those spaces not at the building's perimeter or on the top or ground floors—may be so well shielded from the building's exterior that they may not need additional heat, and can be served only by cooling. The amount of electrically generated heat, plus that produced by human activity and other heat-generating sources, may outweigh the cooling effect of the amount of outdoor air supplied by minimal ventilation, even in winter weather. In the summer, most of the interior cooling loads are generated inside the building. The perimeter areas of the building are the most weather sensitive.

Heating and Cooling Loads

Heating and cooling loads are the amounts of energy required to make up for heat loss and heat gain in the building. Buildings may change from losing heat to gaining heat in a short time. Different parts of a building may be gaining or losing heat at the same time. (See Figure 12.22) The rate of flow of hot or cold air coming into the building from ventilation and infiltration influences the amount of heating or cooling load. It is also dependent on the difference in temperature and humidity between the inside and outside air. The amount of outside air coming in is expressed in cubic feet per minute (cfm) or liters per second (L/s).

HEAT LOSS AND HEATING LOADS

The **heating load** is the hourly rate of net heat loss in an enclosed space, expressed in BTU per hour (BTUH). It is used for selecting a heating unit or system.

A heating load is created when a building loses heat through the building envelope. Convection, radiation, or conduction of heat through the building's exterior walls, windows, and roof assemblies and the floors of unheated spaces are the main sources of heat loss in cold weather. Cold outside air entering a building through ventilation, such as an open window, or as a result of infiltration, as when air leaks through cracks in the building envelope, also add to the heating load. This heat loss places a heating load on the building's mechanical system, which must make up heat in spaces that lose it.

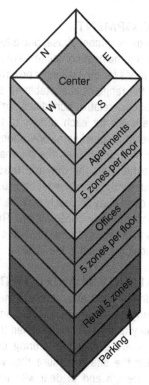

Figure 12.21 HVAC horizontal and vertical zoning

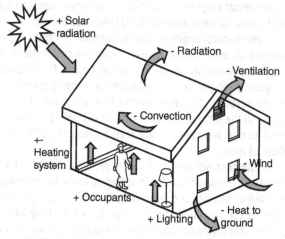

Figure 12.22 Heat losses (-) and heat gains (+) in a residence

For more information on ventilation and infiltration, see Chapter 13, "Indoor Air Quality, Ventilation, and Moisture Control."

HEAT GAINS AND COOLING LOADS

Buildings gain heat from occupants and their activities. **Cooling loads** are defined as the hourly rate of heat gain in an enclosed space, and are also expressed in BTUH. Cooling loads are used as the basis for selecting an air conditioning unit or a cooling system.

Cooling loads represent the energy needed to offset the heat gained through the building envelope in hot weather or from hot air entering by infiltration or ventilation. People's body heat, showering, cooking, lighting, and appliances and equipment use also add to cooling loads.

The heat generated by lighting is often the greatest part of the total cooling load in a building. Most electric lighting converts electrical power into light plus heat. All the electrical power that enters a lighting fixture (including that delivered initially as light) ends up as heat in the space. Energy efficient lighting sources make a big difference in the amount of heat generated by the lighting system.

For information on energy efficient lighting sources, see Chapter 17, "Lighting Systems."

Electric, gas, or steam appliances and equipment in restaurants, hospitals, laboratories, and commercial spaces such as beauty salons and restaurants release heat to interiors. Hoods over kitchen appliances that exhaust air may reduce heat gain, but the exhausted air must be replaced with outdoor air, which may need to be cooled. Steam or hot water pipes that run through air-conditioned spaces and hot water tanks within spaces contribute to the cooling load.

In warm or hot weather, buildings gain heat by convection, radiation, and conduction through the exterior walls and window and roof assemblies. The amount of heat gain varies with the time of day, the orientation of the affected building parts to the sun, the exposure to the wind, and the amount of time it takes for the heat to reach the interior of the building (thermal lag). The heat gain from sun shining on windows varies with the orientation to the sun and the ways the windows are shaded.

For more information on heat gain through the building envelope, see Chapter 2, "Designing for the Environment."

In hot weather, warm make-up air enters when spaces are ventilated to remove odors or pollutants. The use of a **dehumidifier** to lower relative humidity in a space adds to heat gain, due to the latent heat released into the space when moist air is condensed and the heat produced by running the dehumidifier's compressor.

Heat gain calculations tend to be more complex than those for heat loss, as they take into account the sun's position and internal loads. They include both sensible heat gain loads directly added to a building by conduction, convection, and/or radiation as well as latent heat associated with moisture added to the space.

Most buildings need some additional heat to warm up interior space after an unoccupied cold night. Many commercial buildings need little additional heat when only moderately insulated. Some traditional buildings have virtually no insulation. Residential spaces with fewer internal heat sources such as people, lighting, and equipment may consequently require internal heating when a commercial space would not.

Chapter 12 has described the basics of thermal comfort and the principles of heat transfer. We have also looked at the thermal capacity and resistance of materials and the mechanical engineering design process. In Chapter 13, we explore indoor air quality, ventilation, and infiltration. The role of humidity and moisture control issues is also addressed.

13

Indoor Air Quality, Ventilation, and Moisture Control

It is estimated that approximately 90 percent of our time is spent indoors. The interior environment is permeated with new chemicals that may produce a wide array of potential air pollutants. These pollutants are emitted by synthetic products permanently installed in buildings, by equipment used indoors, and by cleaning products. Control of **indoor air quality (IAQ)** depends on limiting pollution sources and on proper ventilation. Moisture control is also an important part of keeping interior spaces healthy.

INTRODUCTION

Human respiration in confined indoor spaces results in air losing some of its oxygen and gaining carbon dioxide. Bacteria and viruses tend to accumulate, along with odors from sweating, smoking, toilet functions, cooking, and industrial processes.

All buildings need to bring in outdoor air for health reasons. Because we use materials that give off toxic components, indoor-air quality (IAQ) has become an important issue. Small buildings, such as residences, have traditionally relied on infiltration to supply the needed fresh air, while large buildings have relied on a designed ventilation system. Because energy-efficient buildings have a tight envelope, all buildings now need a carefully designed ventilation system, and in winter preheating this fresh air will save a great deal of energy. (Norbert Lechner, *Heating, Cooling, Lighting* [3rd ed.], Wiley 2009, page 203)

Rapid rates of air replacement are necessary in damp or overheated spaces or where heat and odor are generated, such as commercial kitchens or locker rooms. Overly high air replacement rates can blow objects about the room. Lower rates of ventilation are acceptable for most residential occupancies, offices with few people, warehouses, and light manufacturing plants.

Buildings also hold pleasant odors, such as baking bread or flowers. Too rapid ventilation can spoil our enjoyment of these desirable scents.

ASHRAE Standards and LEED

ASHRAE Standard 62.1-2013—*Ventilation for Acceptable Indoor Air Quality* and ANSI/ASHRAE Standard 62.2-2013—*Ventilation and Acceptable Indoor Air Quality in Low-Rise Residential Buildings* are the recognized standards for ventilation system design and acceptable IAQ. ASHRAE's *Indoor Air Quality Guide: Best Practices for Design, Construction and Commissioning* is designed for architects, design engineers, contractors, commissioning agents, and all other professionals concerned with IAQ.

LEED v4 standards go beyond ASHRAE Standard 62 minimums. Required minimum indoor air quality performance covers both ventilation and monitoring as well as environmental tobacco smoke control. Points are given for enhanced IAQ strategies for mechanically and naturally ventilated spaces, low-emitting materials, a construction IAQ management plan, and indoor air quality assessment.

INDOOR AIR QUALITY

ASHRAE Standard 62.1.2013 defines acceptable **indoor air quality (IAQ)** as "air in which there are no known contaminants at harmful concentrations as determined by cognizant authorities and with which a substantial majority (80% or more) of the people exposed do not express dissatisfaction." This definition depends more on a subjective response to comfort and health as well as on quantifiable data. It reflects the complexity of identifying contaminants and their sources in indoor air; it also speaks to the difficulty of diagnosing illnesses caused by contaminants in indoor air.

As our buildings become more tightly controlled environments, the quality of indoor air and its effects on our health becomes an increasingly critical issue. Three major reasons for poor indoor air quality in office buildings are the presence of indoor air pollution sources; poorly designed, maintained, or operated ventilation systems; and unanticipated or poorly planned uses of the building.

According to the US Environmental Protection Agency (EPA), in 2014, there were over 84,000 chemical substances manufactured or processed in the United States, most of which have not been tested individually or in combination for their effects on human health. Materials used in building, furnishing, and maintaining a building potentially contain many toxic chemicals, some of which may become airborne.

Engineers consider four approaches when confronting indoor air problems. These include choosing materials and equipment that limit pollution at its source, isolating unavoidable sources of pollution, providing adequate fresh and filtered recirculated air, and maintaining a clean building and equipment. These efforts are seen as preferable to increasing airflow rates and energy consumption.

Flushing the building with air following construction and after every unoccupied weekend or holiday period to remove pollutants from finishes and furnishings is another option. It is recommended that any material assembly deemed likely to contribute one-third of a projected target for allowable concentrations and used in large quantities be tested before acceptance. Testing may shorten or eliminate the need for flush-out period.

Interior designers are key players in the renovation of buildings to new uses and to accommodate new ways of working. They play a significant role by specifying materials that do not contribute to indoor air pollution.

Research has shown that improving IAQ can have significant impact on health and productivity, making it cost effective.[1]

Illnesses Related to Buildings

Efforts to conserve heat by constructing tight buildings in the 1970s resulted in indoor air quality and sick building problems. As a result, building codes sought to strike a balance between energy efficiency and air quality. Today, careful material selection and ventilation make it possible to build tight buildings with both high energy efficiency and good indoor air quality.

Sick building syndrome (SBS) is a term that has been falling out of favor as a clinical diagnosis as there are typically no objective findings, but the term is still used to designate the set of symptoms in the context of workplace health.

BUILDING RELATED ILLNESS (BRI)

Building related illness (BRI) describes a variety of recognized disease entities including allergic rhinitis, asthma, hypersensitivity pneumonitis, Legionanaires' disease, and humidifier fever. It is characterized by objective clinical findings related to specific exposures in the indoor environment. BRI can be caused by microorganisms, including many species of bacteria and fungi.

Interior designers have a responsibility to make sure that building owners, managers and maintenance staff understand the original IAQ design elements and principles in order to ensure benefits in the future. Maintaining and monitoring the heating, ventilating, and air conditioning (HVAC) system reduces sick building risk and assures that problems are identified and corrected at minimum expense. The interior designer will benefit by visiting the space a few months after the construction is finished, and again a few years later, to learn how the process could be improved for future projects. The feedback of an environmental consultant is very helpful as well.

ALLERGIES AND MULTIPLE CHEMICAL SENSITIVITY

Interior designers are often being called on to help people with allergies or other physical sensitivities in the design of healthy, nonpolluting homes. Businesses too are becoming more aware of the cost of sick employees, and more concerned about the health of the indoor environment. Some interior designers have made environmentally sensitive and healthy design a specialty. As our homes and workspaces are exposed to increasing levels of more exotic chemicals, it becomes ever more important that designers have the knowledge and skills necessary to create safe indoor environments.

The causes, symptoms, and treatment of allergies have been recognized by health professionals as an immune system problem. Allergies to mold, household dust, and dust mite allergens are quite common. Interior designers are often asked to consider allergies when designing residential interiors.

Multiple chemical sensitivity (MCS) is a controversial illness. There is no standard medical definition, diagnosis, or cure. The American Medical Association and the Centers for Disease Control do not recognize it, and scientists and doctors argue about whether it is a real or an imagined disorder. Disbelievers claim that those who complain of chemical sensitivities often have a history of other problems, including depression,

---

1. P. Wargocki, D. P. Wyon, J. Sundell, G. Clausen, and P. O.Fanger, "The Effects of Outdoor Air Supply Rate in an Office on Perceived Air Quality, Sick Building Syndrome (SBS) Symptoms and Productivity." *Indoor Air* 10:222–236. doi: 10.1034/j.1600-0668.2000.010004222.x.

which has led many to label MCS a psychological problem. In the United States, the Social Security Administration and HUD now recognize MCS as a disability and as a chronic condition in which people develop increased sensitivities to synthetic chemicals or irritants.

Individuals with heightened sensitivity to materials in the built environment are often forced to make changes in the way they live and work. Interior designers who are sensitive to these issues can help to create safe, healthy interior spaces.

Sources of Pollution

Air normally contains oxygen and small amounts of carbon dioxide, along with various particulate materials. Concentration of people in confined spaces requires the removal of carbon dioxide given off by respiration and its replacement with oxygen.

Air contaminants can be particulate or gaseous, organic or inorganic, visible or invisible, and toxic or harmless. Pollutants can be inhaled, absorbed, or ingested. Certain particularly dangerous pollutants, including asbestos, radon, and pesticides, must be excluded from the building. Other sources of pollution include odors, **volatile organic compounds (VOCs)**, and a wide variety of chemicals, particulates, and biological contaminants. (See Table 13.1)

ODORS

Odors are an indicator of IAQ problems. Foul-smelling air is not necessarily unhealthy, but can cause nausea, headache, and loss of appetite. Our noses are more sensitive than most detection equipment available today. We can detect many harmful

substances by their odors, so that they can be eliminated before they reach dangerous levels.

Odors are most strongly perceived on initial encounter, so visitors to the building are often the most likely to notice them. As our environment is freed of multiple odors, we become more sensitive to the remaining ones. Warning odors such as that of leaking natural gas should not be masked.

Sources of odors are often complex. Odors in buildings come from outdoor air intakes, body odors, grooming products, tobacco smoke, air conditioning coils, food, copy machines, cleaning products, vinyl, paint, upholstery, rugs, draperies, and other furnishings. Materials such as cotton, wool, rayon, and softwoods absorb odors readily and give them off later at varying rates. Bathroom exhaust fans should exhaust directly to the outside. (See Figure 13.1)

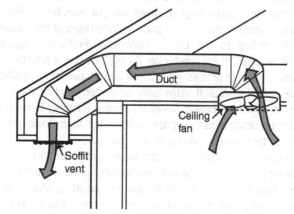

Figure 13.1 Bathroom ventilation

TABLE 13.1 COMMON INDOOR AIR POLLUTANTS

Pollutant	Sources	Effects	Control Strategies
Carbon dioxide	Human respiration	Stuffiness, discomfort at high concentrations	Used as indicator for adequate ventilation rate
Carbon monoxide	Incomplete combustion in furnace, stove, fireplace, motor vehicles	Headache, dizziness, sleepiness, muscle weakness, can be lethal	Sealed combustion burners, exhaust flues, adequate combustion air
Polynuclear aromatic hydrocarbons	Smoking, burnt food, combustion of wood or coal	Irritants and carcinogens	Prohibit smoking, use clean fuels properly, exhaust burning food
Ozone	Laser printers, electronic air cleaners, photocopiers	Bronchial inflammation, shortness of breath, asthma attacks	Remove at source or exhaust, maintain electronic air cleaners
Volatile organic compounds (VOCs)	Particle board, laminated panels, adhesives, fabric treatments, paints	Irritated eyes, nose, skin; headaches, dizzy, shortness of breath	Use alternative materials, seal particle board, ventilate
Fungus	Grows on damp surfaces, walls, ceilings	Very allergenic, irritates eyes, nose, skin, lungs	Keep surfaces clean and dry; borax treatments
Dust mites	Carpet, bedding, fabrics	Very allergenic	Vacuum, barrier cloths
Bacteria (e.g. Legionella)	Standing warm water, hot tubs, drain pans	Severe, potentially fatal respiratory illness	Prevent standing water, clean and treat hot tubs
Radon gas	Natural radioactivity in soil	Increased lifetime lung cancer risk	Seal foundation and floor drains, ventilate

Lower temperatures and relative humidity tend to decrease odors. Odors can be removed by ventilation with clean outside air, or by air washing or scrubbing. They can sometimes be controlled by filtering with electronic or activated charcoal filters.

Odor masking modifies the perceived odor quality to make it more acceptable. It is rarely beneficial to introduce a masking odor. **Ozone generators** are devices that intentionally produce ozone. They primarily reduce the sensitivity of occupants' sense of smell rather than actually reducing the odor.

Beta-cyclodextrin is a chemical used in consumer products that remove odors. It is a molecule that is formed from an enzymatic conversion of starch (usually from corn). It partially dissolves the odor so that it cannot bind to your odor receptors and you cannot smell it. However, these products do have their own scents.

VOLATILE ORGANIC COMPOUNDS

Volatile organic compounds (VOCs) are chemical compounds that tend to evaporate at room temperature and normal atmospheric pressure (and are thus volatile), and that contain one or more carbon atoms (and so are organic compounds). They are invisible fumes or vapors. Some VOCs have sharp odors, while others are detectable only by sensitive equipment.

Awareness of IAQ problems with VOCs has led to the introduction of alternative low-voc products, including low VOC latex paints, water-based varnishes, and formaldehyde-free wood products.

VOCs enter the air when the surfaces of solid materials evaporate or **offgas** at room temperatures. Some products will offgas VOCs for a limited period of time during which the space must be ventilated, and then revert to a safe state.

Most of the more serious effects are the result of VOC exposures at levels higher than those normally expected indoors. However, some common situations are likely to cause mild to serious health effects from VOC exposure. These include installation of large volumes of new furniture or wall partitions, dry cleaning of large volumes of draperies or upholstered furniture, large-scale cleaning, painting, or installation of wall or floor coverings, or higher air temperatures.

Both building maintenance workers and other building occupants are exposed to higher levels of VOCs when the HVAC system is shut down at night or during weekends. This can result in VOCs not being adequately cleared from the building, and the accumulation is circulated through the building when the system is turned on again.

The period immediately following the finishing of the building's interior is critical for VOC exposure. Aging materials before installation may help release some of the VOCs outside of the space. Flush-out periods increase ventilation to exhaust VOCs from the building.

Some building materials may act as sponges for VOCs, absorbing them for later release. Carpeting, ceiling tiles, and free-standing partitions with high surface areas absorb VOCs. Rougher surfaces and lower ventilation rates increase absorption. VOC emissions can be managed by limiting sources, providing proper ventilation, and controlling the relative humidity of the air.

Interior design finishes that do not produce or retain dust and designs that limit open shelving or dust collecting areas help control VOC retention over longer periods of time. Durable materials, such as hardwoods, ceramics, masonry, metals, glass, baked enamels, and hard plastics, are generally low in VOC emissions. Fibers like cotton, wool, acetate, and rayon have low VOCs, but their dyes and treatments may release toxic chemicals.

Re-release of VOCs can be controlled by increasing outside air ventilation during and following installation of finishes and furnishings. Fresh air should be introduced constantly and exhausted directly to the outside rather than through the HVAC system. HVAC systems in newly occupied buildings can be operated at the lowest acceptable temperatures to slow VOC emissions.

BIOLOGICAL CONTAMINANTS

Biological contaminants cause allergic reactions and can result in infections as well as noninfectious diseases. Most of us do not even want to think about bacteria, fungi, viruses, algae, insect parts, and dust in the air we breathe. These microorganisms release **bioaerosols**, which include tiny spores from molds and other fungi that float through the air and irritate skin and mucous membranes. Yet these contaminants are all less common than human skin scales. We shed skin cells constantly from our skin and in our breath, and our environment is littered with our dead cells, which created dust and provide food for dust mites.

Office buildings provide an exceptionally favorable environment of high humidity and standing water. The number of bacteria able to reproduce in an office environment is frequently in the range of 1000 colony-forming units per cubic meter. It is often difficult to test for biological contaminants, as many of the specific test substances are not widely available, and the symptoms are varied and similar to those from other causes. Biological contamination is often the result of inadequate preventive maintenance. Whenever an area is flooded, cleanup must be thorough and prompt.

Biological contaminants need four things to grow in a building. These include a source such as outdoor air, occupants, pets, or houseplants; water; nutrients; and temperatures between 40° and 100°F (4° and 38°C).

Fungi are plantlike organisms that lack the chlorophyll needed for photosynthesis. They include molds, mildew, and yeasts. Fungi live on decomposed organic matter or living hosts, and reproduce by spores. Dry spores can become airborne and can result in allergic and toxic reactions.

Molds are a type of fungi that grow where there is moisture or a relative humidity over 70 percent, often from water damage due to leaks in pipes or floods or from condensation on walls and ceilings. They produce chemicals that are irritants to most people; some people are allergic to specific species of molds. Cellulosic materials support mold growth, including paper, wood, textiles, insulation, carpet, wallpaper, and drywall (gypsum wallboard).

Indoor mold prevention includes eliminating water sources by repairing leaks, preventing condensation, venting moisture-producing equipment or appliances to the outdoors, and maintaining interior relative humidity below 50 percent (ideally between 30 and 50 percent). Air conditioning systems should be adequately drained to prevent standing water. Any damp or wet spots should be cleaned up and dried within 48 hours. Water should drain away from the building foundation. Building systems should be routinely inspected and maintained.

Mildew is a fungus that appears as a thin layer of black spots on a surface. Drywall with a fiberglass nonwoven mat facing does not feed mold. Noncellulose adhesives that do not support mildew growth are available for use with wallcoverings and hardwood flooring.

Dust mites settle out of the air and usually live at floor level, only becoming airborne when the dust is disturbed. Dust mite allergens include enzymes in their feces, saliva, and body parts. Reactions include nasal inflammation, asthma, itching, inflammation, and rash. Dust mites require at least 60 percent relative humidity to survive, so maintaining a relative humidity of 30 to 50 percent will help deter them.

Legionella pneumophila is a bacterium that rarely causes disease except under certain indoor conditions and in susceptible hosts. Legionnaire's disease is a progressive form of pneumonia that infects only about 5 percent of those who inhale droplets of water with the bacteria; about 5 percent of those infected die of the disease. Improper maintenance of HVAC systems is responsible.

Sometimes ultraviolet (UV) radiation is used for control of biological contaminants. Filters are rarely effective, and evaporative humidifier filters may harbor bacteria that eat cellulose and thrive in the warm, wet environment. Air conditioning coils can hold skin cells, lint, paper fibers, and water, a perfect environment for mold and bacteria. Building air quality specialists tell horror stories of mold-covered mechanical systems that are supplying the air for an entire building.

Synthetic carpets containing large amounts of dust make excellent mold environments, especially after water damage. Thoroughly clean and dry water-damaged carpets and building materials within 24 hours if possible, or removal and replacement may be required.

Installing and using exhaust fans that are vented to the outdoors in kitchens and bathrooms and venting clothes dryers outdoors can reduce moisture and cut down on the growth of biological contaminants. Ventilate the attic and crawl spaces to prevent moisture build-up. If using cool mist or ultrasonic humidifiers, clean appliances according to manufacturer's instructions and refill with fresh water daily.

Keeping the building clean limits exposure to house dust mites, pollens, animal dander, and other allergy-causing agents.

As an interior designer, avoid specifying room furnishings that accumulate dust, especially if they cannot be washed in hot water. Using vacuums with **high efficiency particulate arrestance (HEPA) filters** may also help. Do not finish a basement below ground level unless all water leaks are patched, with outdoor ventilation and adequate heat to prevent condensation provided. Operate a dehumidifier in the basement if needed to keep relative humidity levels between 30 and 50 percent.

Antimicrobial finishes are designed to protect materials and products from biological contaminants. Mold can still grow if it finds food source on the surface, so good maintenance and moisture control are still required. There are concerns about the safety of products classified as pesticides, and regarding overuse of antimicrobials.

Indoor Air Quality Equipment

Once the sources of IAQ problems have been removed or isolated wherever possible, increased ventilation and improved air filtration are usually the next most practical measures. Increasing ventilation for improved air quality must strike a balance with energy conservation. Energy conservation efforts have resulted in reduced air circulation rates in many central air-handling systems. Fewer fans use less power, but distribution is poorer, and the air mix within individual spaces suffers. Individual space air-filtering equipment provides a higher circulation rate and a proper air mix.

Air cleaning equipment is incorporated in the HVAC system of a home to filter air before it is returned through ducts throughout house. Sometimes portable, tabletop, or larger console air cleaners are used individual rooms to control particulates such as dust, pollen, or tobacco smoke. Typically, a fan takes air through filtering medium, and then blows the air back into the room or through ducts. Equipment capacity should be matched to the size of the room.

AIR CLEANERS AND FILTERS

Air filters protect the HVAC equipment and its components and the furnishings and décor of occupied spaces, and protect the general well being of residents. (See Table 13.2) They reduce housekeeping and building maintenance and reduce equipment fire hazards. Local air-filtering equipment should have a high circulation rate and proper mixing of new and old air. Each unit has its own fan, and operates with or without a central HVAC fan. Outdoor **makeup air** that replaces exhausted air should be filtered to keep coils and fans free of dirt.

Electrostatic air filters filter and ionize air to collect dirt, then filter it again. (See Figure 13.2)

ANSI/ASHRAE *Standard 52.2-2012—Method of Testing General Ventilation Air-Cleaning Devices for Removal Efficiency of Particle Size* sets the test procedure for evaluating performance of air cleaning devices as a function of particle size. This standard addresses the device's ability to remove particles from the air, total dust-holding capacity, and resistance to airflow.

TABLE 13.2 AIR FILTERS

Type	Description
Panel filters	Used with HVAC equipment to protect fans from large lint and dust particles. Usually fiberglass, and thrown out when dirty.
Pleated media air cleaners	Thick pleated filter paper within a frame. Trap large and small particles. Require regular maintenance, can damage equipment if blocked. Replace often or they increase energy consumption.
High-efficiency particulate arrestance (HEPA) filters	Highest efficiency filter. Able to filter out microscopic particles, including bacteria and pollen.
Self-charging electrostatic filters	Add static electric charge to particles in air, then pass air across metal plates holding opposite charge; dust particles held on plates.
Air washers	Used to control humidity and bacterial growth. Moisture can become a problem if not well maintained.
Activated carbon filters	Often used with other filters to adsorb odors and gases and neutralize smoke.
Hybrid filters	May combine mechanical filters plus electrostatic precipitator, or ion generator in integrated system or as single device.
Chemisorption filters	Active material attracts and bonds polluting gas molecules onto its surface. Absorbing material requires regular replacement.
Adsorption filters	Use activated-charcoal or chemically-impregnated porous pellets to remove small percent of specific gaseous contaminants from air.

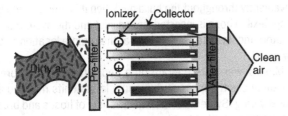

Figure 13.2 Electrostatic air filter

There are many types of air cleaners, which vary in use and effectuveness. (See Table 13.3) Self-contained portable room air cleaners must sometimes be used to obtain sufficiently high levels of filtration effectiveness. (See Figure 13.3)

The effectiveness of each type of air cleaner filter depends on the size of the particle to be filtered. The best air cleaners have a dust arrestance rating above 80 percent. A lower initial air resistance rating is less likely to decrease the heating and cooling system's efficiency. Some types emit ozone, a lung irritant.

TABLE 13.3 AIR CLEANER TYPES

Type	Description
Portable room-sized air cleaners	Residential use for continuous, localized air cleaning. Sized for particular room; can be moved from room to room.
In-duct air cleaning units	Installed in residential unducted return air grilles or ducted air plenums of central HVAC system. Recirculate building air through unit. HVAC fan must always be on for air cleaning.
Negative ion generators	Electronically charges particles to remove them from air. Personal air purifiers can reduce airborne particles. Require frequent maintenance and cleaning. Produce ozone.
Electronic air cleaners	Filter out very small particles. Effective with tobacco smoke. Produce ozone.
Electrostatic precipitators	Charge airborne particles and collect them. Precipitating cell is reusable with maintenance. Capture particles on collector plates; some generate ozone.
Air washers	Used to control humidity and bacterial growth in large ventilation system. If not well maintained, moisture in air washer can add to pollution.
Ozone generators	Primarily act to reduce sense of smell. According to EPA, at concentrations that do not exceed public health standards, ozone has little potential to remove indoor air contaminants.
Ion generators	Charged ions emitted to air often produce dirty spots on nearby surfaces by forcing impurities to cling to surface. Emit ozone.
Ultraviolet (UV) light	Destroys germs, viruses, bacteria, fungi. Installed within HVAC systems or directly in kitchens, sickrooms, overcrowded dwellings. In some personal air purifiers. Mounted high in room, shielded to protect eyes, skin.

Figure 13.3 Room air cleaner

Flat or panel filters usually contain a fibrous medium that can be dry or coated with a sticky substance such as oil so that particles adhere to it. (See Figure 13.4) Less-expensive lower efficiency filters that employ woven fiberglass strands to catch particles restrict airflow less, so smaller fans and less energy are needed. The typical, low-efficiency furnace filter in many residential HVAC systems is a flat filter, ½" to 1" (13 to 25 mm) thick, that is efficient in collecting large particles, but removes only between 10 and 60 percent of total particles and lets most smaller, respirable-sized particles through.

HEPA filters are generally made from a single sheet of water-repellent glass fiber similar to blotter paper that is pleated to provide more surface area with which to catch particles. (See Figure 13.5) To qualify as a HEPA, the filter must allow no more than three particles out of 10,000 to penetrate the filtration media, a minimum particle removal efficiency of 99.97 percent, including smaller respirable particles. Similar HEPA-type filters with less efficient filter paper may have 55 percent efficiencies.

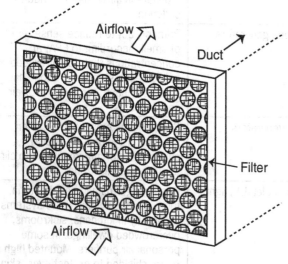

Figure 13.4 Flat or panel filter

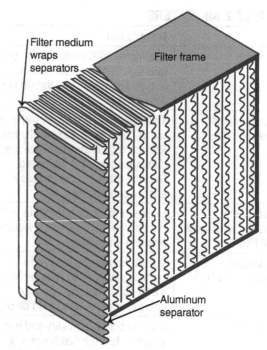

Figure 13.5 HEPA filter

These filters, which are still very good when compared to conventional panel type and even pleated filters, have higher airflow, lower efficiency, and lower cost than their original version.

CENTRAL CLEANING SYSTEMS

Central cleaning systems have been used in homes and commercial buildings for years. They are commonly found in commercial office buildings and restaurants. They are essentially built-in vacuum cleaners with powerful motors. As such, they can be used to trap dirt and dust inside the power unit equipment and away from rooms where people live and work. They can be vented outdoors, decreasing exposure for people with dust allergies. The power unit is usually installed in a utility room, basement or garage. Tubing running under the floor or in the attic connects through the walls to unobtrusive inlets placed conveniently throughout the building. When it is time to vacuum, a long flexible hose is inserted into an inlet and the system turns on automatically. The noise is kept at the remote location of the power unit.

Most power units operate on a dedicated 15 amp normal residential electrical circuit, but some larger units may require heavier wiring. Systems come with a variety of hoses and brushes. Installation is simplest in new construction.

Plants and Indoor Air Quality

Indoor plants can help reduce stress, improve the aesthetics of a space, and increase humidity in overly dry indoor air. Research by retired NASA scientist Dr. B.C. Wolverton measured the rate of air cleaning of various plants with a variety of pollutants, and showed that microbes associated with the root systems consume toxins from the air. Mixing activated carbon with the growing media greatly increased the amount of pollutants removed from the air.

Overly damp planter soil conditions may promote the growth of biological contaminants. In restaurants and other places with water and food, planters can provide a home for cockroaches and other unwanted forms of life.

Controls for Indoor Air Quality

Today's control systems monitor for carbon dioxide along with possible leakage of fuels such as propane, butane, and natural gas, and inadequate levels of oxygen. Alarms activate equipment automatically when levels are too high. The system can regulate ventilation heat exchangers, especially in unoccupied spaces, to prevent buildup of VOCs from finishes and furnishings, thereby avoiding the need for flushing of the room's air after a weekend. Units are available as stand-alone alarms or set up to activate equipment. The unit's size is about that of a programmable thermostat, and the mounting height depends on the gas being monitored.

INFILTRATION AND VENTILATION

Outdoor air enters a building by infiltration and/or ventilation. Ventilation is the process of intentionally bringing fresh air into a building. Infiltration happens when fresh air accidentally enters through openings or cracks in the building.

Ventilation and infiltration were introduced in Chapter 2, "Designing for the Environment."

Infiltration

Wind creates local areas of high pressure on the windward side of a building, and low pressure on the leeward side. Fresh air infiltrates a building on the windward side through cracks and seams. On the opposite side of the building, where pressure is lower, stale indoors air leaks back outside.

Anywhere that walls and ceilings or floors meet or where openings pierce the building's exterior presents an opportunity for air to infiltrate. The most common sources of air leaks are where plumbing, wiring, or a chimney penetrates through an insulated floor or ceiling, or on top of the building's foundation wall.

Air can leak where the tops of interior partition walls intersect with the attic space and through recessed lights and fans in insulated ceilings. Missing plaster allows air to pass through a wall, as do electrical outlets and switches on exterior walls. Outdoor air infiltrates through cracks and crevices around doors and windows and through open doors and windows. Window, door, and baseboard trims can leak air through the joints they

cover, as can ceiling soffits above bathtubs and cabinets. Air can also leak at low walls along the exterior in finished attics, especially at access doors, and at built-in cabinets.

Gaps less than ¼" (4 mm) wide can be sealed with caulk, which is available in a variety of types and colors for different materials. Backer rod or crack filler is used for larger or deep cracks, which are then sealed with caulk.

Infiltration is caused by the **stack effect** that occurs when warm air moves upward in a building, combined with the force of wind. The amount of infiltration depends on the area of an opening and the difference in pressure across the opening. The stack effect is a factor in buildings of more than five stories, over about 100 feet (30.5 m) tall.

It can be difficult to estimate infiltration accurately, due to the wide variety in type, construction quality, shape, and location of buildings. The variety of types of heating systems add to the difficulty.

Infiltration is calculated by either the crack length method or the air change method. The crack length method is more precise, but requires specific information about building dimensions and construction details. The air change method uses tables listing the number of air changes per hour in rooms with various exposures to determine the rates of air leakage. The actual infiltration performance of a constructed building can be measured with a blower-door test.

The blower-door test is introduced in Chapter 2, "Designing for the Environment."

Very leaky spaces have two to three air changes or more per hour. Even when doors and windows are weather-stripped and construction seams are sealed airtight, about one-half to one air change per hour will occur, but this may be useful for the minimum air replacement needed in a small building.

Infiltration can be minimized by weatherstripping doors and windows, providing a continuous air barrier around the building perimeter, and sealing construction seams airtight. Weatherstripping materials generally have a lifespan of less than ten years, and need to be replaced before they wear out. Vestibules with two doors in series can cut infiltration by up to 60 percent in buildings where doors are opened frequently. Revolving doors can cut infiltration by 98 percent.

Ventilation

Ventilation involves bringing in fresh air to provide oxygen for people and to help carry away of carbon dioxide and body odors. Ventilation is always needed whatever the climate. With the advent of modern HVAC systems, fresh air is delivered via the heating and cooling system. Both ventilation and passive cooling consider window positions together with requirements of human occupants.

Passive cooling replaces heated indoor air with cooler outdoor air. Cooling breezes are only available at specific times and in specific places. Much greater amounts of airflow is required for passive cooling than for control of air quality.

Before the invention of mechanical ventilation, the high ceilings common in buildings created a large volume of indoor air that diluted odors and carbon dioxide produced by occupants. Fresh air was provided by infiltration, which along with operable windows created a steady exchange of air with the outdoors.

For more information on windows and passive cooling, see Chapter 6, "Windows and Doors," and Chapter 14, "Heating and Cooling."

Very airtight buildings can starve fireplaces and gas heating appliances for air. This can allow odors to build up, and can contribute to IAQ problems. Eventually, oxygen for breathing can be in short supply. Consequently, buildings with very little infiltration require ventilation.

CODES AND VENTILATION

A minimum amount of fresh air is important for interior air quality control, including removal of odors and pollutants. Many building codes regulate the minimum outdoor airflow rate based on either the number of people in the building or its floor area. Ventilation requirements are generally based on ASHRAE Standard 62.1-2013—*Ventilation for Acceptable Indoor Air Quality* or ASHRAE Standard 62.2-2013—*Ventilation and Acceptable Indoor Air Quality in Low-Rise Residential Buildings.*

VENTILATION SYSTEMS

The basic components of a building ventilation system begin with an air source of acceptable temperature, moisture content, and cleanliness. A force is required to move air through the building's inhabited spaces, with a means to control its volume, velocity, and direction of airflow. Finally, the system requires a means of recycling or disposing of contaminated air.

The ventilation rate actually needed depends on the effectiveness of the ventilation system's design, performance, and location of supply (inlet) outlets and return outlets. The inlet locations determine the velocity and air flow pattern; the outlet locations have little effect on these. It is generally a good practice to have both inlets and outlets the same size.

Rates for both infiltration and ventilation are expressed in cubic feet per minute (cfm) or liters per second (L/s).

Natural Ventilation

Natural ventilation moves a source of fresh air at an appropriate temperature and humidity through a building without fans. Wind or convection moves air from higher to lower pressure areas through windows, doors, or openings provided for the

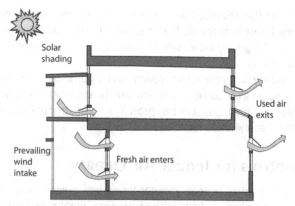

Figure 13.6 Natural ventilation

purpose, or though nonpowered ventilators. (See Figure 13.6) Mechanical controls adjust the volume, speed, and direction of the airflow. Contaminated air is either cleaned and reused or exhausted from the building.

Using natural ventilation helps keep a building cool in hot weather and supplies fresh air without resorting to energy-dependent machines. However, in cold climates energy loss through buildings that leak warm air can offset the benefits of natural cooling. Careful building design can maximize the benefits of natural ventilation while avoiding energy waste.

Opportunities for natural ventilation may be limited by outside noise. Almost all methods of blocking noise will also slow down breezes.

WIND VENTILATION

Wind-powered ventilation is most efficient if there are windows on at least two sides of a room, preferably opposite each other. Where only a single wall abuts the outdoors, a casement window can help create a pressure differential that induces airflow.

Factors influencing airflow through buildings include pressure distribution around the building; the direction of air entering windows; the size, location, and details of windows; and interior partitioning details. The design of open floor plans can maximize airflow by minimizing the use of full height partitions.

Depending on the leakage openings in the building exterior, the wind can affect pressure relationships within and between rooms. The building should be designed to take advantage of warm season prevailing winds when it is sited and when the interior is laid out.

Pressure variations occur when air flows over and around a structure. (See Figure 13.7) Wind on the windward side produces positive pressures. When deflected by a physical obstacle, higher velocities result as the volume of air passes through the reduced area. As air flows around building edges, it creates negative pressure, but as it continues, it can again establish positive pressure. As it passes by the structure, negative pressure

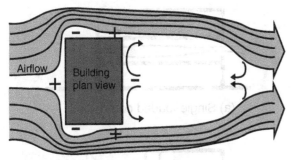

Figure 13.7 Airflow pressures around building

Source: Redrawn from David Lee Smith, *Environmental Issues for Architecture*, Wiley 2011, page 195

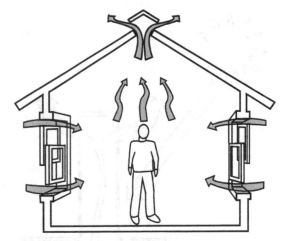

Figure 13.8 Convective ventilation

is formed on the leeward side of the structure, and reversing eddies of airflow are likely to occur.

Cross ventilation is driven by wind through windows. Narrow building plans with large ventilation openings on either side are also compatible with daylighting. There is a trend for office workspaces to be near operable windows, and LEED offers credits for daylighting and views. A supply of tempered fresh air is needed even in cold weather.

Ventilation openings must have both an inlet at a point of positive exterior pressure, and an outlet at a point of negative pressure. Where there is only a single exposure, cross ventilation will probably depend on spaces connecting with ventilated space. Where a single window is used for both in and out, the window itself must form pressure differential. Double-hung windows with both operable tops and bottoms can do this. Regular casements have only a single opening, but architectural projected casements with hinged sides create two openings.

Where there is a smaller inlet and a larger outlet, a good pressure differential is created, and the inlet air is at a higher velocity. More velocity aids cooling, but may blow papers and other objects around.

CONVECTIVE VENTILATION

In **convective ventilation**, differences in the density of warmer and cooler air create the differences in pressure that move the air. (See Figure 13.8) Convective ventilation uses the stack effect referred to earlier, due to warm air buoyancy and differential air pressures. Heated air becomes less dense and more buoyant and tends to rise, displacing colder air downward. The warm air inside the building rises and exits near the building's top. Cool air infiltrates at lower levels.

The stack effect works best when the intakes are as low as possible and the height of the stack is as great as possible. The stack effect is not noticeable in buildings less than five stories or about 100 feet (30.5 m) tall. Fire protection codes restrict air interaction between floors of high-rises, reducing or eliminating the stack effect.

To depend on convective forces alone for natural ventilation, you need relatively large openings. With operable widows, pressures vary. Insect screens cut down on the amount of airflow.

Systems using only convective forces are not usually as strong as those depending on the wind. In cold weather, fans can be run in reverse to push warm air back down into the building.

The stack effect produces differences in atmospheric pressure at the top and bottom of a vertical shaft. It can be controlled if the building is sealed so that it not exposed to variations in atmospheric pressure. This is one of the reasons revolving doors were developed. Limiting the number of floors served by same air distribution helps minimize problem.

COMFORT VENTILATION

Comfort ventilation is the technique of using air motion across skin to promote thermal comfort. It is used in lightweight construction without air conditioning in hot and humid climates. Comfort ventilation still requires some insulation to keep sun on roof and walls from overheating interior surfaces. Comfort ventilation is rarely completely passive. Window fans or whole-house fans are usually needed to supplement wind.

Insulation is required and thermal mass is helpful for mostly air-conditioned buildings, even in humid climates. The mass slows temperature changes, allowing the air conditioning to be turned off during peak electrical demand times.

For comfort ventilation, operable window area should be around 20 percent of the floor area, with windows split roughly evenly between windward and leeward walls. Windows need to remain open during rain to deal with humidity, so roof overhangs are needed to keep rain out. Windows are closed when it is much hotter outdoors than inside, with ceiling fans used to circulate cooler indoor air.

CHIMNEYS AND FLUES

When not in use, a large quantity of air rises through a fireplace chimney and is replaced by infiltrating outdoor air. Closing the fireplace damper reduces, but does not eliminate, this problem as dampers rarely fit tightly. Air loss actually increases when the fireplace is in use, inducing more infiltration throughout the building, while the heat produced is localized near the fireplace.

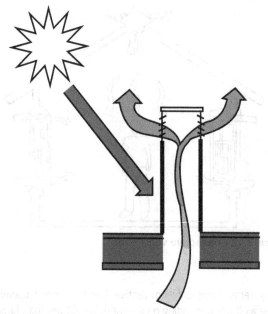

Figure 13.9 Solar chimney

Modern fireplaces with integral combustion air intakes and glass doors significantly reduce infiltration. For more on fireplaces, see Chapter 14, "Heating and Cooling."

Flues for heating equipment can also result in loss of building air. Direct-fired warm-air furnaces installed within the space they heat require an air supply for combustion. If combustion air is not brought inside through a closed duct, conditioned air will be drawn from the space and exhausted through the flue.

A **solar chimney** increases the stack effect without heating the indoors by moving air vertically through a building on calm, sunny days. (See Figure 13.9) It induces a draft to create additional updrafts that pull the breeze through the building.

DOOR AND WINDOW VENTILATION

In residences, ventilation is tied to the quantity of exterior windows and the amount of natural ventilation they supply. If the bathroom does not have a window, it is required to have a fan with a duct leading directly to the exterior. A percentage of the windows must be operable for ventilation and emergency egress.

Open plans are best for ventilation. Partitions increase resistance to air flow and decrease total ventilation. Within a single apartment or tenant area, cross ventilation can be achieved by leaving doors open in partitions between rooms.

Doors should not be relied upon for essential building ventilation unless they are equipped with a holder set at the desired angle. An ordinary door cannot control the amount of air that flows past it.

In commercial buildings, cross ventilation is almost never possible in a double-loaded corridor plan, although transoms above doors may allow some cross ventilation. An open single-loaded corridor permits full cross ventilation. (See Figure 13.10)

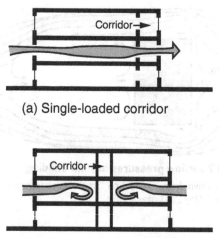

(a) Single-loaded corridor

(b) Double-loaded corridor

Figure 13.10 Sections showing corridors and cross ventilation
Source: Redrawn from Norbert Lechner, *Heating, Cooling, Lighting*, Wiley 2009, page 278

Attic and Roof Ventilation

Thermal buoyancy—the rising of warm air—is a major cause of air leakage from a building's living space to the attic and then out through the roof. Solar heated air also comes in through the roof and attic.

ATTIC VENTILATION

Ventilating an attic reduces temperature swings. It makes the building more comfortable during hot weather and reduces the cost of mechanical air conditioning.

Adding ventilation without sealing air leaks into the attic can increase the amount of air leaking from the house, wasting valuable heat. (See Figure 13.11) Installing rigid insulation in

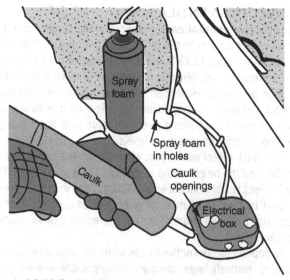

Figure 13.11 Sealing attic air leaks

the eaves (the projecting overhang at the lower edge of a roof) reduces heat loss in that area. Another option is to change the framing detail to one that leaves more room between the top plate and the rafters.

Air leaking out of air handlers and ducts, and heat leaving the system by conduction can be major factors contributing to heat loss in cold weather. Placing HVAC equipment and ductwork in attics will waste leaking air and should be avoided. If there is no alternative, all ducts should be sealed tightly and run close to the ceiling, buried in loose fill insulation to the equivalent R-value of the attic insulation.

ROOF WINDOWS

Roof windows, also called operable or venting skylights, can create the same updraft throughout the house as an old-fashioned cupola. When shaded to keep direct sunlight out, they are one of the best natural ventilating devices available. Roof windows allow moisture to escape from kitchens, baths, laundry rooms, and pool enclosures. They are available with remote controls and rain sensors.

Roof windows are introduced in Chapter 6, "Windows and Doors."

ROOF VENTILATORS

Roof ventilators also increase natural ventilation. (See Figure 13.12) Passive roof ventilators create suction when wind blows across the top of a stack, pulling air up and out of the building. Roof ventilators require control dampers to change the size of the opening as necessary. With high enough winds, roof ventilators that are large enough and located high enough can ventilate habitable spaces.

The term "roof monitor" is used for a type of roof ventilator, and also for a raised structure running along the ridge of a roof. See Chapter 6, "Windows and Doors," for more information.

Figure 13.12 Roof ventilators

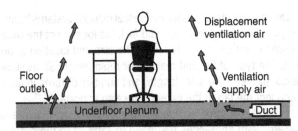

Figure 13.13 Displacement ventilation

Source: Redrawn from David Lee Smith, *Environmental Issues for Architecture*, Wiley 2011, page 372

Mechanical Ventilation

Displacement ventilation may be required for most high-rise structures with central air distribution, depending on their design and engineering. (See Figure 13.13) Displacement ventilation introduces fresh cool air at floor level, and exhausts polluted warm air through the ceiling. Vents must be kept clear of obstructions. Openings should close tightly to prevent unwanted infiltration.

PREHEATING VENTILATION AIR

Heating or cooling of makeup air can be used to heat building air. A south-facing (Northern Hemisphere) solar wall with a collector can be used as a winter preheating device.

One residential system combines a wall with an exhaust air heat pump. The house is under negative pressure, with forced exhaust air. Heat is supplied from exhaust air heat pumped for space heating or heating hot water. Fresh air is drawn through the specially constructed walls. As a result, a slow, steady stream of cold air warmed by insulation enters the building.

AIR SUPPLY RATES

The 2013 version of ASHRAE Standard 62.2 includes a new ventilation formula that requires ventilation of high-performance homes at a higher rate than in the past, 7.5 cfm per person, plus 3 cfm per 100 square feet. The new formula is controversial, with some experts arguing that it is too high, and others that the old standard gave too much leeway for infiltration. Local codes may call for greater levels of ventilation.

Especially high rates of air replacement are needed in buildings housing heat and odor producing activities. Restaurant kitchens, gym locker rooms, bars, and auditoriums require extra ventilation. Lower rates are permissible for residences, lightly occupied offices, warehouses, and light manufacturing plants.

Residential Ventilation Systems

Exhaust systems remove air, heat, moisture, odors, combustion pollutants, and grease to the outside. Residential kitchen ventilation systems are usually located near a cooking surface. They control odors and pollutants with local exhaust and positive building pressure.

Design considerations for residential exhaust systems include the type of cooking appliance and fuel, the location of the range or cooktop within the kitchen, and the size and location of any hood. The type of fan and makeup or replacement air available are considered. The size, length, and number of turns in ducts to connect the fan to the exterior are also important.

In tightly constructed houses, negative pressure can create problems with appliances that need to exhaust to the outside, such as furnaces and water heaters. **Backdrafting** can occur, pulling **carbon monoxide**, excess moisture, and radon into the home. Backdrafting can be prevented by opening a window when operating exhaust fans (including passive fresh air intake vents), or by using whole-house mechanical ventilation that balances airflow.

Kitchen appliances often require ventilation. (See Table 13.4) **Whole-house ventilation** systems typically include an exhaust vent in the kitchen. There are two common types of residential kitchen ventilation systems, updraft and downdraft. Ceiling or wall-mounted exhaust fan are generally considered less effective.

Ventilation is more critical with gas appliances, due to carbon monoxide and other combustion products. Gas also produces water vapor, so there is more moisture to exhaust.

A residence with gas appliances should have a carbon monoxide detector. This may be required by code.

Range hoods are often mounted above residential kitchen ranges. (See Figures 13.14 and 13.15) Slide-out ventilation hoods are mounted below wall cabinets, and can be vented or unvented. Some manufacturers offer hoods with dishwasher safe grease filters.

Residential range hoods are available in a wide variety of styles and materials, including stainless steel and glass. Some models extract air almost noiselessly. Innovative self-cleaning features and lighting fixtures are included with some styles. Where hoods are installed without ducts, heavy-duty charcoal filters are advertised for the removal of smoke and odors.

TABLE 13.4 KITCHEN APPLIANCE VENTILATION

Appliance	Venting
Range oven	Typically vents through/near burner on cooktop, delivers moisture, odors near ventilating system
Built-in or wall oven	Vents to front of appliance into room air
Professional-style range	Needs larger capacity ventilation system; follow manufacturers recommendations and local codes
Microwave oven	Vents to front, side, or back, often not near ventilating system
Dishwasher	Vents warm, moist air

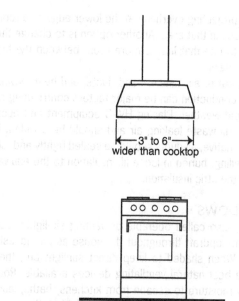

Figure 13.14 Range hood front view

Source: Redrawn from Kathleen Parrott, Julia Beamish, JoAnn Emmel, and Mary Jo Peterson, *Kitchen Planning* (2nd ed.), National Kitchen and Bath Association, Wiley 2013, page 280

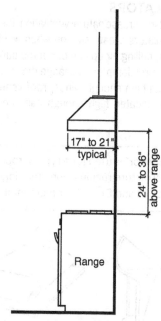

Figure 13.15 Range hood side view

Source: Redrawn from Kathleen Parrott, Julia Beamish, JoAnn Emmel, and Mary Jo Peterson, *Kitchen Planning* (2nd ed.), National Kitchen and Bath Association, Wiley 2013, page 280

See Chapter 11, "Fixtures and Appliances," for more information on residential ventilation systems.

UPDRAFT VENTILATION SYSTEMS

Updraft ventilation systems have a fan mounted above the cooktop or range, usually with a hood. They may either have a ducted exhaust system or a ductless recirculating one. Hoods are available in many styles, and may be a focal point or retractable models that are barely noticeable.

The hood should be at least the size of the cooking surface, but preferably 3" to 6" (76 to 152 mm) larger in all directions, although this may be difficult to achieve. (See Table 13.5) Greater interior height increases efficiency. A larger hood can be mounted higher on the wall than a smaller hood, with comparable efficiency. Higher mounting avoids bumping heads and provides better visibility.

Combination microwave ovens and updraft ventilation systems can be installed over a cooking surface. They are available with either recirculating or exhaust ventilation systems. The flat microwave bottom traps fewer pollutants than a canopy-shaped hood, and also tends to be shallower (typically 12" to 13" [305 to 350 mm]), providing less coverage of the cooking surface. The combination also tends to have limited fan size, and may not be appropriate over larger ranges or cooktops. Maximum ventilating efficiency and safety is achieved when installed with at least 24" (610 mm) clearance above the cooking surface. With a 36" (914 mm) counter height, this typically brings the microwave oven to 60" (1524 mm) above the floor, which may be too high for some people. As a result, combination microwave oven/updraft ventilation systems are often installed only 15" to 18" (381 to 457 mm) above the cooktop or range. It is important to follow the manufacturer's recommendations.

DOWNDRAFT VENTILATION SYSTEMS

Downdraft or proximity ventilation systems are installed in the cooktop or adjacent to a cooking surface where they capture pollutants near the source. (See Figure 13.16) Most are exhaust systems. Downdraft ventilation systems can be effective alternative for grilling, frying, and other cooking from shallow pots and pans. They require larger fan, as they operate without a hood, and work against tendency of warm air to rise. They do require space for ductwork. Retractable downdraft vents behind cooktop burners have washable grease filters.

Recirculating systems provide filtering but not ventilation. They are generally only used when it is not possible to install ductwork for the exhaust system. A recirculating system may

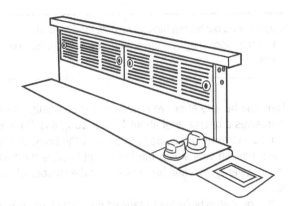

Figure 13.16 Kitchen downdraft ventilation system

have only a simple grease filter screen, or include activated carbon filter to remove odors. The system does not remove combustion pollutants such as carbon monoxide or water vapor. Recirculating systems are less expensive and easier to install, but less effective.

The 2015 International Residential Code sets minimum residential exhaust rates for kitchens and bathrooms in cubic feet per minute (cfm). (One cfm is equal to 0.0004719 cubic meters per second.) Kitchens are required to exhaust at the rate of 100 cfm or 25 cfm continuously. Bathrooms and toilet rooms must have a mechanical exhaust capacity of 50 cfm intermittently or 20 cfm continuously.

REDUCING ENERGY CONSUMPTION

There are several ways to reduce energy consumption while providing adequate ventilation. Ventilation against a wall is more efficient than at an open island or peninsula that requires larger fans and/or hoods due to cross-drafts. Heat recovery from exhaust air can be used to preheat makeup air from outdoors. Computerized controls can track occupancy and provide only the ventilation necessary for the current number of people in the building.

Fans

Mechanical ventilation options include unit ventilator fans on the outside wall of each room to circulate room air and replace part of it with outdoor air. Window or through-wall air conditioning units can also be run as fans. A central heating and cooling system with coils of hot or chilled water will temper the air in room ventilation units. Fixed location fans can provide a reliable, positive airflow to an interior space.

Any time that air is exhausted from a building, makeup air must be supplied. This can be done in a limited manner by infiltration through the building's envelope. Opening windows and doors can also provide a supply of fresh air. Where mechanical equipment exhausts a large volume of air, makeup air is introduced through vents in the building envelope, and directed to the equipment through ducts.

TABLE 13.5 RESIDENTIAL HOOD DIMENSIONS

Direction	Typical Dimensions (mm)
Width	24" to 54" (610 to 1372)
Depth	17" to 21" (432 to 533
Bottom of hood	24" to 36" (610 to 914) above cooking surface

Makeup air for a bathroom fan can be provided by undercutting the bathroom door slightly. Louvered doors may create privacy concerns.

Fans can be very effective in cooling small buildings. A person perceives a decrease of about 1°F per 15 fpm (1°C for every 1 m/s) increase in the speed of air past the body. The air motion produced varies with the fan's height above the floor, the fan's power, speed and blade size, and the number of fans in the space.

Fans are used in buildings to exhaust hot, humid, or polluted air, to bring in outdoor air to cool people (comfort ventilation), or to cool the building at night (night-flush cooling), and to circulate indoor air when it is cooler than outside air.

Types of fans include stand-alone fans, fans packaged as a product (for example, a bathroom exhaust fan), fans in ductwork, and those that are part of larger equipment. Built-in fans in residential kitchens and baths dump air directly or through a short ductwork run to the outdoors. When air is replaced by leaking from outdoors through the house, it may result in heating or cooling losses.

One or more unit ventilating fans may be located on the outside wall of each room. Window or through-wall air conditioner or room ventilating units work this way.

Fans are located in the ceiling or another high location for warm moist rising air. Their intake should be close to the source of the air and/or moisture. In a bathroom, this is usually near a toilet, bathtub, and/or shower. Vapor-proof or moisture-proof fans can be located in a shower or directly over a jetted tub.

CEILING FANS

Ceiling fans are run at higher speeds in summer to increase comfort through increased air motion. They can be run at a slow speed to destratify warm air at the ceiling in winter.

Ceiling fans are preferable to other locations if a ceiling is high enough. Ceiling hugger fans are available for standard height ceilings, but may annoy tall people. A lighting fixture below the fan can lower the clearance even more. Putting a recessed lighting fixture above fan blades produces an undesirable flickering effect. Newer ceiling fans with LED lighting resolve this problem. (See Figure 13.17)

TABLE 13.6 CEILING FAN SIZES

Room Size in square feet	Fan Size (mm)
100 (9 m²)	36" (914)
150 (3.9 m²)	42" (1067)
225 (21 m²)	48" (1219)
375 (35 m²)	52" (1321)
Over 400 (37 m²)	Two fans

Ceiling fans are available in a variety of sizes to accommodate various room sizes. (See Table 13.6) Ceiling fans with aerodynamically curved fan blades are more efficient, and can be run at lower speeds, saving energy. Remote controls and temperature sensors encourage using the fan only when it will improve conditions in the room. Ceiling mounted fans to be located in an insulated ceiling should be specified for this type of installation.

ATTIC, WINDOW, AND WHOLE-HOUSE FANS

An attic fan reduces attic temperature and condensation damage. Window fans should be located on the downwind side of a house, facing outward. When in use, a window in each room and interior doors should be opened.

A whole-house motor-driven fan pulls stale air from living areas of the house and exhausts it through attic vents. (See Figure 13.18) The large fan is mounted in the ceiling of the top floor hallway, and draws air in through open windows and doors.

Without an adequate exhaust fan, the building may not have enough air for combustion equipment such as furnaces and stovetop barbeques, and fumes may not be exhausted properly. Equipment that demands a large amount of exhaust should have another fan supplying make-up air running at the same time.

The visual impact of the fan depends on its grille color and style, the choice of material, and its relationship to other features such as ceiling lights and wall finishes.

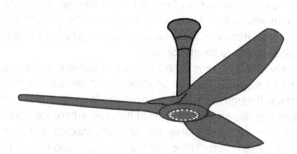

Figure 13.17 LED ceiling fan

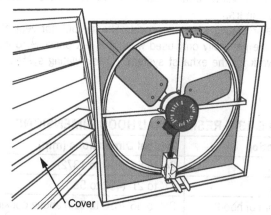

Cover

Figure 13.18 Whole-house fan

WHOLE-HOUSE VENTILATION SYSTEMS

The International Energy Conservation Code sets requirements for whole-house mechanical ventilation system fan efficiency. It requires that the building be provided with ventilation that meets the requirements of the International Residential Code (IRC) or International Mechanical Code as applicable, or with other means of ventilation.

A whole-house fan system includes fan, ductwork, controls, and installation. Noise level are rated in sones, with less than 1.0 to 1.5 sones generally considered quiet enough to be background noise. Larger fans and those with longer, more complex ducts are noisier.

Minimizing the length of the duct run and the number of elbows or bends from the interior air intake to the exterior air exhaust increases the efficiency of the fan system.

There are different types of whole-house systems, and the design may need to be matched to the local climate and whether the building is heating or cooling dominated.

Exhaust ventilation systems work by depressurizing the building. (See Figure 13.19) The system exhausts air from the house while make-up air infiltrates through leaks in the building shell and through intentional passive vents. Exhaust ventilation systems are most appropriate for cold climates. In climates with warm humid summers, **depressurization** can draw moist air into building wall cavities, where it may condense and cause moisture damage. Depressurization can cause problems with combustion appliances, including a gas furnace or water heater, and combustion products could spill back into a home.

Supply ventilation systems use a fan to pressurize a home, forcing outside air into the building while air leaks out of the building through holes in the shell, bath, and range fan ducts, and intentional vents (if any exist). (See Figure 13.20)

Balanced ventilation systems, if properly designed and installed, neither pressurize nor depressurize the home. (See Figure 13.21) Rather, they introduce and exhaust approximately equal quantities of fresh outside air and polluted inside air.

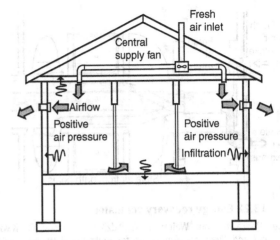

Figure 13.20 Supply ventilation system

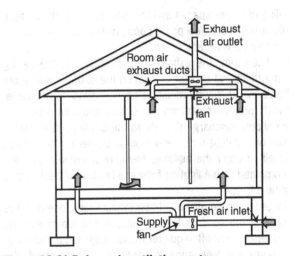

Figure 13.21 Balanced ventilation system

A whole-house ventilation system uses a **heat recovery ventilation system (HRV)** or **energy recovery ventilator (ERV)**; both have similar operation and both are used in all types of climates. Fans exhaust and bring in air, with exhaust fans where moisture and pollutants are likely to be generated (for example, in bathrooms). Fresh air intakes are centrally located but away from main living areas, often in an entryway or closet. Exhaust and intake air goes through a heat exchanger for ventilation with energy conservation.

Energy recovery ventilators can exhaust air from two bathrooms in adjacent apartments. (See Figure 13.22) They are controlled by individual switches in each bathroom. A heat pump adjacent to the ERV warms fresh makeup air.

BATHROOM EXHAUST FANS

Materials in bathrooms are often cooler than the air, producing condensation. Drywall and textiles are absorbent and stay damp. Warm, moist air tends to move through bathroom walls

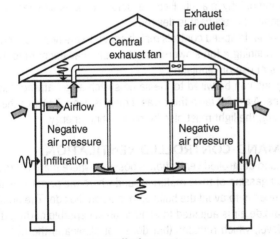

Figure 13.19 Exhaust ventilation system

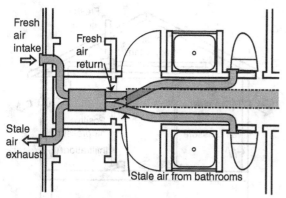

Figure 13.22 Energy recovery ventilator

Source: Redrawn from Walter T. Grondzik and Alison G. Kwok, *Mechanical and Electrical Equipment for Buildings* (12th ed.), Wiley 2015, page 153. Design by Harry James Boody.

and ceiling to cooler spaces and the exterior. When the air temperature reaches the dew point, condensation can form on surfaces and inside walls and attics.

In a bathroom, the exhaust fan should be in the ceiling above the toilet and shower or high on the exterior wall opposite the door. Exhausting moisture accumulation deters bacterial growth, promotes sanitary conditions, and controls odors. The fan should discharge directly to the outside, at a point a minimum of 36" (914 mm) away from any opening that allows outside air to enter the building. Residential exhaust fans are often combined with a lighting fixture, a fan-forced heater, or a radiant heat lamp.

A large bathroom may need two exhaust fans, one over the toilet or near the floor, and a second over the tub or shower. Two smaller fans are often quieter than a single large one. Larger bathrooms may have additional fixtures, such as jetted tubs or both a shower and a tub, that produce more moisture, requiring additional ventilation.

Fan models are available with a high-efficiency centrifugal blower that provides virtually silent performance, and a lighted switch that indicates when the fan is on. Highly energy-efficient motors use about a third of the electricity of standard versions. Some designs allow easy installation in new construction as well as retrofit applications. Some models activate automatically to remove excess humidity. Lighting fixtures and even nightlights are included in some designs.

Toilets are available that are direct-vented to control odor and vapor spray from flushing. Their exhaust venting capacity should be considered in planning total bathroom ventilation. However, they are unlikely to control moisture from other bathroom sources adequately.

Residences with spa areas set up for hydrotherapy or steam treatments generate a great deal of moisture in air, which needs to be ventilated after the treatment. Indoor spas or hot tubs where water remains in the tub, and where chemicals are used to treat the water, require special care. Strong odors and candle soot also may need to be removed by ventilation.

PUBLIC TOILET ROOM EXHAUST

Public toilet room plumbing facilities must be coordinated with the ventilation system to keep odors away from other building spaces while providing fresh air. The toilet room should be downstream of the airflow from other spaces. The air from toilet rooms should not be vented into other spaces, but exhausted outdoors. By keeping slightly lower air pressure in the toilet rooms than in adjacent spaces, air will flow into the toilet room from the other spaces, containing toilet room odors. This is accomplished by supplying more air to surrounding spaces than is returned. Exhaust vents should be located close to and above toilets.

LOCALIZED EXHAUST SYSTEMS

In open offices with few walls but with copying machines, the designer can erect a barrier around a contamination-producing machine and provide mechanical ventilation to ventilate the task area immediately.

Hoods can be built over points where contamination originates. Commercial kitchen hoods collect grease, moisture, and heat at ranges and steam tables. Sometimes outside air is introduced at or near the exhaust hood with minimal conditioning, and then quickly exhausted, saving heating and cooling energy.

Most buildings are designed to have positive air pressure as compared to the outdoors, so that unconditioned air does not enter through openings in the building envelope. Corridors should be supplied with fresh air, and residential units including apartments, condominiums, hotels, motels, hospitals, and nursing homes, should have exhausts.

Multistory buildings have chases for exhaust ducts through successive floors, which can double up with plumbing in apartments, hotels, and hospitals. Kitchen exhausts must remain separate, due to the risk of fires. In major laboratory buildings, many exhaust stacks may be seen rising high above the roof.

FAN CONTROLS

Sensor controls turn fans on and off based on humidity, providing excellent moisture control. Motion detectors turn fans on when someone is in the bathroom, and off when they leave. They may not work in some situations, such as for a person soaking quietly in a tub. Heat sensors can increase speed automatically if extra heat is desired.

Variable speed controls match speed to the need for ventilation. Running a fan slower is quieter, and may encourage more frequent use of ventilation.

Fans can be wired to the same switch as a bathroom light. You need a quiet fan if this arrangement is used. To leave the fan running, the light must also be on, wasting energy.

DEMAND-CONTROLLED VENTILATION

Demand-controlled ventilation is possible where there is little or no off-gassing of toxic materials, and where the primary ventilation need is to avoid the build up of carbon dioxide. The amount of outside air is adjusted to achieve an acceptable carbon dioxide level, which reduces (but does not eliminate) the need for ventilation.

HUMIDITY AND MOISTURE CONTROL

Architects and engineers take elaborate and expensive precautions to keep water out of buildings. Rain and snow and their surface runoff cause subsurface water to come in contact with building foundations. People track water into buildings. Condensation, drips, leaks, and spills from piping, plumbing fixtures, cooking, washing, and bathing contribute unwanted water to building interiors.

Drying building materials including concrete, brickwork, tile work, and plaster produce water vapor that can create condensation. Masonry materials all absorb water and transmit it to some extent. Water penetrating into brick, stone, or concrete and freezing causes **spalling**, the chipping of flakes from a surface by expansion of water as it freezes.

Water destroys the insulating value of building materials and can raise humidity to unhealthy levels. Many materials used in building interiors disintegrate when wet, and water causes staining or corrosion of others. Water supports the growth of bacteria, molds, mildew, fungi, plants, and insects. It is a good conductor of heat, increasing thermal conductivity by creating a shorter path for heat flow.

There are many openings through which water can enter a building. These include planned movement joints, joints between exterior cladding materials, and cracks around door and window frames. Unintentional openings including concrete shrinking cracks result from poor workmanship, materials defects, holes for pipes and wires, and cracks and holes due to deterioration.

Water Movement

Gravity causes hydrostatic pressures exerted by water at rest where it accumulates. Air pressure differentials due to wind action can drive water in any direction. Capillary action can pull water through porous materials and narrow cracks.

When water freezes into ice it clogs drainage paths, causing water to pond on the roof or ground. Its expansion creates open paths through the building enclosure.

Excess moisture in building materials can result in peeling paint, rusting metal, and deterioration of structural framing or joists. Damp materials attract dirt, requiring more cleaning and maintenance. Damp spaces foster the growth of many biological pollutants, including bacteria and viruses. They also encourage pests including dust mites and cockroaches, and support mold growth.

The typical family of four produces an average of 4 gallons (15 L) of water vapor per day. The kitchen is one source of excess moisture. Cooking, especially boiling or simmering on a cooktop, produces water vapor. Microwaves and conventional ovens remove moisture from food and vent it into the kitchen. Gas cooking appliances generate water vapor as a combustion product, and can double the amount of moisture released into kitchen air. Running the dishwasher and defrosting the refrigerator also add moisture to the air.

Moisture problems are more likely to occur in smaller, more tightly constructed residences. Hard or nonabsorbent materials, such as glazed tiles, solid surfacing, vitreous china, and engineered stone, dry faster than absorbent ones. Sealers help with absorbent or porous materials such as clay tiles, marble, and grout.

Good air circulation speeds drying. Providing enough towel bars, rings, or hooks for all bathroom users and placing them near heat registers helps.

To control moisture, separate clothes storage from damp areas of the bathroom by a door or partition, and provide good ventilation of the closet area.

Humidity

Water vapor is a colorless, odorless gas that is always present in air. The warmer the air, the more water vapor it can contain. The amount of water vapor in the air is usually less than the maximum possible, and when the maximum is exceeded, water vapor condenses onto cool surfaces or becomes fog or rain.

As indicated in Chapter 1, relative humidity (RH) is the amount of vapor actually in the air at a given time, divided by the maximum amount of vapor that the air could contain at that temperature. Colder air can hold less water vapor. If the temperature drops low enough, it reaches the dew point, which is the point at which the air contains 100 percent RH. The dew point temperature is the temperature at which water condenses out of air. When the RH is raised to 100 percent, as in a gym shower room or a pool area, fog is produced. The vapor condenses only enough to maintain 100 percent RH, and the rest stays in the air as a gas.

People are comfortable within the 20 to 50 percent RH range. In summer, relative humidity can be as high as 60 percent when temperatures rise up to 75°F (24°C), but above that we are uncomfortable, because the water vapor (sweat) does not evaporate off our bodies well enough to help us cool off.

Humidity levels affect interior design materials. Too much moisture causes dimensional changes in wood, most plant and animal fibers, and even in masonry. Steel rusts and wood rots. Surface condensation damages decorative finishes and wood and metal window sashes as well as structural members.

Specific humidity is the amount of water vapor in the air sample measured in weight of water per weight of air. It is indicated in pounds of water vapor per pound of dry air (grams of water per grams of air).

The effect of humidity on thermal comfort occurs only below 20 percent and above 60 to 75 percent. Below the higher limit, the body has no difficulty releasing about a quarter of its heat. A household RH between 40 and 60 percent is generally comfortable but still prevents condensation and mold growth.

For more information on humidity, see Chapter 12, "Principles of Thermal Comfort."

LOW HUMIDITY

Heated winter air can be very dry, causing wood in buildings and furniture to shrink and crack. Wood shrinks primarily in the dimension perpendicular to the grain, leaving unsightly cracks and loose furniture joints.

Low humidity below 20 percent RH causes plants to wither. Our skin becomes uncomfortable and dry, and the mucous membranes in our nose, throat, and lungs become dehydrated and susceptible to infection. Added moisture helps, as do lower air temperatures that reduce evaporation from the skin.

Dry air can generate shocks. Carpeting and resilient flooring is commercially available that incorporate conductive materials that reduce voltage buildup and help to alleviate static electric shocks.

In warm-air heating systems, moisture can be added to the air as it passes through the furnace by using water sprays or absorbent pads or plates supplied with water. Pans of water on radiators are an old-fashioned but effective method of raising humidity in the winter. Boiling water or washing and bathing release steam. Plants release water vapor into the air, and water evaporates from the soil in their pots. Spraying plants with a mist increases the air humidity, and the plants like it, too.

HUMIDIFICATION

Humidification adds moisture to the air without intentionally changing the air temperature. Although this can be accomplished with a humidifier, this is usually not adequate by itself to create thermally comfortable conditions.

Heating with humidifying is often desirable in smaller buildings in cold climates. No single HVAC device produces heating with humidification, so humidifiers are often added.

Electric humidifiers help relieve respiratory symptoms, but may harbor bacteria or mold in their reservoirs if not properly maintained. Task humidifiers are used to relieve symptoms of respiratory illnesses. Prevention of bacterial and mold growth in water reservoirs requires good maintenance. Adding an ultraviolet (UV) lamp may counteract this threat.

Condensation

When hot, humid air comes in contact with a cold surface, condensation forms. For example, when you take a glass of iced tea outside on a hot humid day, little drops of water will appear on the outside of the glass and run down the sides. The water vapor in the air condenses to form visible droplets of water on the cooler surface. In cold climates, water vapor can condense on the cold interior surfaces of windows. Condensation can result in water stains and mold growth.

Even at normally acceptable interior humidity levels, moisture can collect on interior sides of exterior walls behind furniture such as bookcases.

In summer, concrete basement walls, floors, and slabs-on-grade that are cooled by the earth can collect condensation. Rugs on the floor or interior insulation on basement walls inhibit the rise in the concrete slab temperature and make matters worse. Both rugs and insulation may be damaged if the relative humidity is very high or if condensation occurs. Insulating the exterior or below the slab with well-drained gravel can help.

Where a significant difference in interior and exterior temperatures exists, condensation can occur within an exterior wall's insulation. (See Figure 13.23) In winter, air cooled below the dew point fogs and frosts windows, and condensation can cause rust or decay when it collects under window frames. Wintertime condensation collects on cold closet walls, attic roofs, and single-pane windows.

Blowing warm air across perimeter windows artificially warms cold surfaces, and avoids condensation. Ventilating moist air out of the space reduces water vapor in the air. Room arrangements can be designed to avoid pockets of still air and surfaces shielded from the radiant heat of the rest of the room. Interior surfaces should be insulated from the cold outdoors in the winter. Enough air motion should be provided to keep condensation from settling on cold surfaces in the winter. Cold

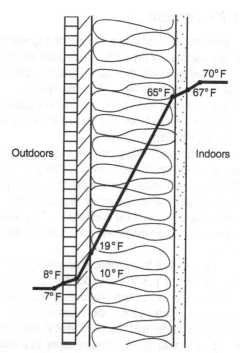

Figure 13.23 Thermal gradient

Source: Redrawn from Norbert Lechner, *Heating, Cooling, Lighting*, Wiley 2009, page 500

water pipes and ductwork should be insulated in the summer. Reducing the amount of water vapor in the air avoids condensation in all seasons.

Insulated curtains that can be moved over windows can contribute to condensation problems, since the interior surface of the window is shielded from the heating source in the room and becomes cold. If warm room air can pass around or through the insulating window treatment, moisture will condense on the window. Thermal window treatments designed to seal out cold air need to be properly gasketed or sealed at the top, bottom, and sides to prevent moist room air from entering the space between the insulation and the glass, where it will condense against the cold window. The insulating material must also be impervious to moisture that might accumulate.

For more information on moisture and the building envelope, see Chapter 2, "Designing for the Environment."

CONCEALED CONDENSATION

When the moisture content of the air rises inside a building, it creates vapor pressure, which drives water vapor to seeking equilibrium by expanding into areas of lower vapor pressure like the exterior walls. When there is moist air on one side of a wall and drier air on the other, water vapor migrates through the wall from the moist side to the drier side. Water vapor will also travel along any air leaks in the wall. Most building materials have relatively low resistance to water vapor.

When the temperature at a given point within a wall drops below the dew point at that location, water vapor condenses and wets the interior construction of the wall. (See Figure 13.24) This condensation causes an additional drop in vapor pressure, which then draws more water vapor into the area. The result can be very wet wall interiors, with insulation materials saturated and sagging with water, or frozen into ice within the wall. The insulation becomes useless, and the heating energy use of the building increases. The wall framing materials may decay or corrode, and hidden problems may affect the building's structure.

The amount of vapor pressure within a building depends on the amount of vapor produced, its inability to escape, and the air temperature. The higher the moisture content in the air, the higher the vapor pressure.

The temperature drops gradually from the warmer surface to cooler one. Temperature drops at different rates through the various layers of construction until the surface of the cool side is just slightly warmer than the cold air temperature.

Indoor air in cold climates may be more humid than outdoor air, and vapors will then flow from the warmer interior to the colder exterior surfaces of the walls, ceilings, and floors. This can leave the building envelope permeated with moisture.

A solid coat of exterior paint that keeps the water vapor from traveling out through the building's wall will trap vapor inside. Vapor pressure can raise blisters on a wall surface that will bubble the paint right off the wall. This is sometimes seen outside kitchens and bathrooms, where vapor pressure is likely to be highest.

By using a vapor barrier as close to the warm side of the building envelope as possible, this creeping water vapor can be prevented from traveling through the wall. The vapor barrier must be between the main insulating layer and the warm side of the wall. (See Figure 13.25)

In a cold climate, the vapor barrier should be just under the plaster or paneling inside the building. In an artificially cooled building in a warm climate, the warm side is the outside. When the interior side of a wall is warmer than the exterior, the warm inside air will release water vapor as it moves to the drier air outside.

In hot humid climates, the problem is to keep the moisture from getting into the interior of the building. A drainage plane inside the exterior surfacing material is safer than using a vapor barrier that may keep the moisture trapped in the wall.

Vinyl wallcoverings or vapor barrier paints on interior surfaces offer some protection, but do not replace the need for a vapor barrier. Where adding a vapor barrier to an older building is not practical, plugging air leaks in walls and applying paint to warm-side surfaces while providing ventilation openings on cool-side surfaces clears moisture from the construction interior. Special vapor retardant interior paints are available for this purpose.

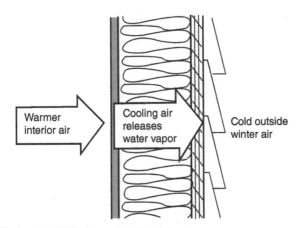

Figure 13.24 Without vapor barrier

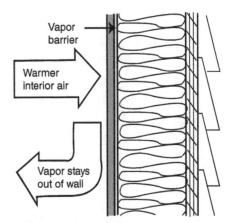

Figure 13.25 With vapor barrier

Dehumidification

Ventilating or dehumidifying the interior space can help prevent concealed condensation. Dehumidification removes moisture without intentionally changing air temperature.

Although lower humidity does not lower air temperature, it does increase comfort. Excessively low humidity can cause discomfort from dry skin.

Sensible cooling reduces air temperature without changing absolute humidity. Some HVAC systems use sensible cooling through part of their operating range. Other systems, such as radiant cooling systems, are specifically designed as sensible cooling devices. Lower humidity feels cooler and drier. Too low may result in skin irritation.

Combining cooling and dehumidification combines sensible and latent cooling processes. Most active cooling systems produce both sensible and latent cooling effects.

Evaporative cooling involves simultaneous sensible cooling and latent heating (cooling plus humidification). Sensible heat is exchanged for latent heat at little net energy cost. Evaporative cooling is very efficient where climate conditions support the process.

Evaporative cooling is covered in more detail in Chapter 14, "Heating and Cooling."

DEHUMIDIFIERS

Refrigerant (mechanical) dehumidifiers are an option for spaces that do not need mechanical cooling but do need to reduce humidity. Dehumidifiers operate on same principle as air conditioning. Dehumidifiers chill air, which lowers the amount of moisture the air can hold and causes water vapor to condense on the cooling coils of the dehumidifier. The condensed water then drops off into the dehumidifier's collection container. Accumulated water must be removed periodically to avoid disease.

Refrigerant dehumidifiers are often free-standing units used in smaller buildings. Refrigerant dehumidifiers do not work well below 65°F (18°C), because frost forms on their cooling coils; they might not be an appropriate choice in a cool basement. They require regular maintenance, and generate both heat and noise. The equipment must be sized to the space it serves.

Dessicant dehumidifiers rely on **desiccants** (porous materials such as silica gel, activated alumina, and synthetic polymers with a high affinity for water vapor) that lower humidity without overcooling the air. In active desiccant systems, desiccants are usually heated with natural gas or solar energy to drive out the moisture that they have removed from the air. Passive systems use the heat from the building's exhaust air to release and vent moisture removed from incoming air. Dessicant dehumidifiers use no refrigerants, and can lower humidity without overcooling the air.

This chapter has echoed back to what was covered in Chapter 2, "Designing for the Environment," and has looked ahead to the contents of our next chapter, Chapter 14, "Heating and Cooling." As you can see, building systems are very much intertwined.

14

Heating and Cooling

In recent years, the approach of architects and engineers to the design of building heating and cooling systems has undergone significant and substantial change.

> During the second decade of the twenty-first century, several trends appear to be bubbling in the world of HVAC design. These are being driven by a desire to produce higher-performance buildings—deep green projects, net-zero energy projects, carbon-neutral projects. One trend. . . is a willingness to let active systems partner with passive systems. No matter how efficient an active HVAC system, an appropriate passive system will use less energy (and renewable energy to boot). One of the more direct paths to net-zero energy is not to reduce energy use for heating and cooling, but rather to eliminate such energy use. Building automation systems make the integration of active and passive systems easier to manage. (Walter T. Grondzik and Alison G. Kwok, *Mechanical and Electrical Equipment for Buildings* [12th ed.], Wiley 2015, pages 556–557)

INTRODUCTION

The design of the HVAC system has major implications for the building's architecture and interior design. The architect must coordinate with all consultants from the beginning of the project to make decisions about the use of HVAC equipment. The design of the mechanical system has become more fully integrated with the architectural and structural design, and is developed concurrently.

In the past, designers tended to close the building, exclude the outdoors, and rely on mechanical equipment. Now they need to clarify whether equipment is used for occasionally modifying environmental conditions, as permanent connectors, or as permanent excluders of the outdoors. Where HVAC equipment is used, spaces requiring quiet, such as bedrooms and conference rooms, should be located as far away from noisy equipment as possible, both horizontally and vertically.

> Instead of approaching design with the expectation that we can depend on complex, energy-consuming mechanical systems to provide acceptable thermal conditions, regardless of how environmentally irresponsible we might choose to be in our design, we must begin with a commitment to maximize the architectural potential to establish comfortable conditions through passive means that do not rely on energy consumption. In this approach, the intention is to use mechanical systems to supplement architectural contributions rather than to correct architectural problems. (David Lee Smith, *Environmental Issues for Architecture*, Wiley 2011, page 189)

Architectural and Engineering Considerations

In most North American climates, the roofs, walls, windows, and interior surfaces of a carefully designed building can maintain comfortable interior temperatures for most of the year. The building's form, along with the climate, produces air motion; mechanical assistance is provided for faster speeds.

The mechanical engineer decides which HVAC system will be used in a large building based on costs, the intended occupancy, floor space for the required equipment, maintenance requirements, and system controls. The architect must communicate and coordinate with the engineer, and work to assure that the architectural and interior design will be integrated with the engineering system.

Together, the architect and engineer evaluate issues that affect the thermal qualities of the building such as insulation and shading that influence the size and fuel consumption of mechanical equipment. Architectural design elements may reduce the operating expenses of the system and decrease the size and initial cost of HVAC equipment, but increase the initial cost of the building's construction. Cost effectiveness analyses help determine the optimal economic balance between passive and active approaches.

INTERIOR DESIGN IMPLICATIONS

The architect and interior designer are both concerned with the locations and dimensions of ductwork and piping. The noise generated by mechanical equipment is an acoustic concern for them both.

Early in the project, the architect must consider the size and location of central HVAC mechanical rooms, which may be either separate spaces or combined together. HVAC requirements can have a substantial impact on the space plan, ceiling heights, and other interior design issues, so getting involved early in the process is a good idea.

The locations and dimensions of piping and ductwork will determine where chases must be located for distribution equipment running vertically between floors. Suspended ceiling grids make access to ductwork easier. With proper early planning, only minor changes in the floor plan will be needed to fit in the final mechanical design.

Space must be allocated along exterior walls for exposed terminal delivery devices, such as registers and diffusers. Their form and position must be coordinated with the interior design, to avoid conflicts between furniture arrangements and the location of grilles or wall-mounted units. Thermostat locations are determined by the engineer and are dependant upon the surrounding heat sources, but affect the visual quality of the interior space.

The HVAC distribution pattern directly affects the interior designer's work. Code-mandated building height limits may limit floor to ceiling heights. Ceiling heights and the transition from higher to lower spaces has a direct impact on interior volumes and relationships. Ductwork lowers ceiling heights, and can affect daylighting design.

Heating and cooling systems continue to evolve, with equipment being replaced by passive building energy design. However, substantial amounts of HVAC equipment are still used in many large buildings. As an interior designer, you should be aware of how heating and cooling equipment works, and of how the equipment will affect your design, energy efficiency, and your client's comfort. Interior design issues such as whether to use an open office plan or private enclosed offices have a significant impact on the mechanical system, and should be shared with the engineer early in the design process.

Commercial office design is going through major changes. Four basic types of office space may be interchanged within a flexible overall plan. These include enclosed offices, open plan offices with dividers at around desk-height, open plan offices with higher partitions for privacy, and open plan offices with some individually designed workstations and partitions of varying heights. The acoustic and privacy challenges of open offices continue to undergo development.

Uniform ceiling heights, lighting placement, and HVAC grille locations increase flexibility in office arrangements and extend the building's useful lifespan. Diversity in ceiling lighting, air handling, and size can make design of connecting corridors, lounges, and other support services more difficult. Diverse design elements require complete and detailed design of a space, but the resulting design may be a more complex and interesting building for designer, builder, and user. Variety aids user orientation and distinguishes spaces from one another.

Some spaces require diverse thermal conditions. In winter, we expect offices to be relatively warmer than circulation spaces that transition from the exterior to the interior. Transitional spaces that are closer to outside temperatures can make key spaces seem more comfortable without extreme heating or cooling, saving energy over the life of the building.

The design of the air circulation and ventilation system interacts with the layout of furniture. Even furniture like filing cabinets and acoustic screens less than 5 feet (1.5 meters) high can impede air circulation. If walls or full-height partitions enclose spaces, each enclosed space should have at least one supply vent and one return or exhaust vent.

HVAC Design Process

Project scope and extent varies with project size and complexity. For a smaller building, HVAC selection and design may be done by the architect alone or with the assistance of a mechanical contractor. Larger, more complex buildings involve consulting engineers, sometimes with other specialists such as fire protection engineers and laboratory consultants.

Once the extent of active HVAC design has been established, the mechanical engineer selects the HVAC system based on performance, efficiency, and initial and life costs of the system. The engineer considers the availability of fuel, power, air, and water, and the means for their delivery and storage. The need for access to outdoor air is taken into account. The system is evaluated for its flexibility to serve different zones with differing demands. The type and layout of the distribution system for heating and cooling is reviewed, with an eye to laying out efficient short direct runs and a minimum number of turns and offsets that minimize friction losses.

HVAC equipment can take up 10 to 15 percent of the building area. The size requirements for a building depend on the heating and cooling loads, and the role of passive design. Some pieces require additional access space for service and maintenance. Codes may require mechanical areas to have noise and vibration control and fire-resistant enclosures. Heavy equipment may need additional structural support.

CENTRALIZED VERSUS LOCALIZED EQUIPMENT

The designer of the mechanical system for a building looks at whether the building's needs are dominated by heating or by cooling concerns. Because climate is such a strong factor in small buildings, and heating and cooling needs may vary room-by-room, localized equipment rather than a centralized system may be the better choice.

Centralized HVAC equipment is often located outside the occupied space to facilitate regular maintenance. Large-scale equipment may promote energy recovery. On the other hand, centralized HVAC systems distribute heating and cooling through distribution trees that take up a lot of space, both horizontally and vertically. (See Figure 14.1) They need to be

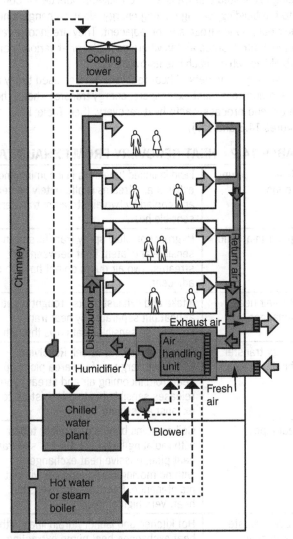

Figure 14.1   Centralized HVAC system

Source: Redrawn from Francis D.K. Ching, *Building Construction Illustrated* (5th ed.), Wiley 2014, page 11.17

coordinated with the lighting, ceiling design, and other interior design elements. The breakdown of a single piece of equipment may affect the entire building. Energy is wasted when the entire system is activated to serve a single zone.

A localized HVAC system is more responsive to the differing but simultaneous needs common to skin-dominated buildings. Distribution trees are shorter, and control systems are simpler. However, equipment produces noise in occupied rooms, where performing maintenance may be disruptive.

History

In the first century CE, the Roman **hypocaust** passed warm air from a central furnace under raised floors and up via flues in walls. The floor was usually topped with poured concrete and finished with tiles or mosaics. Unfortunately, this early central heating disappeared with the fall of the Roman Empire.

In medieval Western Europe, an open fire in the center of a room provided heat, light, and fire for cooking, with smoke exiting through the roof or high windows. The invention of the chimney in the twelfth century made buildings relatively smoke-free but harder to heat, and the resulting fireplace was only about 10 percent energy efficient. Eventually, ceramic stoves, with 30 to 40 percent efficiency, gained popularity in colder parts of Europe.

In a large manor house, everyone usually slept in the main hall with a large central fire on cold nights. Heavy curtains were sometimes used to divide the large heated space into areas for varied thermal conditions and privacy. This evolved into a large bed for the lord and lady, enclosed with curtains hung from the ceiling, and then into a bed with a frame supporting the curtains.

English settlers brought the fireplace to North America. Observing that forests were being cut down near big cities for fuel, Ben Franklin invented a fuel-efficient cast iron stove.

The Industrial Revolution brought steam heat to Europe in the eighteenth century. Steam conveyed in pipes heated public buildings and the homes of wealthy people. The extremely hot surfaces of the steam pipes dried out the air uncomfortably and generated the odor of charred dust.

Homes in eighteenth-century North America had a large coal furnace in the basement that sent heated air through a network of pipes with vents in major rooms. Around 1880, many buildings were converted to steam systems. A coal furnace heated a water tank, and hot air pipes carried both steam and hot water to vents connected to radiators.

By the nineteenth century, central heating was popular in larger buildings, with gravity air and water systems in multistory buildings. Basements held a furnace or boiler next to a wood or coal bin. Heated air or water was moved through the building by natural convection currents.

Beginning in the twentieth century, airflow was controlled by fans in forced air systems, resulting in much longer duct runs that could also move cold air. True active cooling did not appear in buildings until after World War II.

TABLE 14.1 ENERGY CONSERVATION EQUIPMENT

Type	Description
Boiler fuel economizers	Hot gases from boiler stack preheat incoming boiler water.
Runaround coils	Transfer heat between intake and exhaust air ducts.
Economizer cycles	Cool outdoor air aids refrigeration cycle as it cools recirculated indoor air. Filters and tempers outdoor air.
Energy storage	Central water and ice storage tanks in large buildings use daily temperature changes to increase efficiency, reduce consumption.
Geoexchange systems	Geothermal wells tap earth's deep heat sources.
	Ground source heat pumps use near-surface heat.
	Reservoirs and other water bodies used as thermal sinks.

With the increasing use of electric lighting and air conditioning, floor plans became wider and large central internal areas developed. For example, the RCA Building in New York's Rockefeller Center (1931–1932) retained a slab-like floor plan that allowed daylight to reach the working areas arranged around a core containing elevators and other service spaces. Central boilers, chillers, and fan rooms supplied large quantities of forced cooled air through bulky air distribution trees. The advent of glass curtain walls and the slick, two-dimensional modern look in the mid-twentieth century made air distribution trees visually intrusive, and they were pushed to the building's core.

Heating and cooling systems became more complex and specialized in a wide range of building types. Specialization and separation of design responsibilities resulted in a poorly integrated design process, which was not good for owners and put roadblocks in the path to high-performance building outcomes. Today, we are well on the way to a fully integrated design process.

Building Energy Conservation

Architects and engineers are designing buildings with systems intended to wean us from our dependence on nonrenewable fuels. With an eye to protecting natural resources and preserving the environment, building engineers are seeking ways to share heating and cooling tasks between mechanical systems and natural ventilation and daylighting. Energy conservation measures include economizers, runaround coils, energy storage, and geoexchange systems. (See Table 14.1)

HEAT EXCHANGERS

Heat exchangers are widely used today. (See Figure 14.2) They are also referred to as heat recovery ventilators (HRV) or air-to-air heat exchangers. They recover heat from air that is

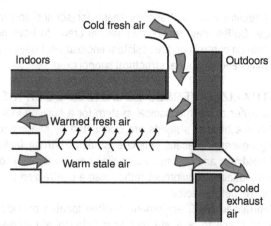

Figure 14.2 Air-to-air heat exchanger

Source: Redrawn from Walter T. Grondzik and Alison G. Kwok, *Mechanical and Electrical Equipment for Buildings* (12th ed.), Wiley 2015, page 152

being exhausted and transfer it to make-up outside air coming into the building, saving heating energy. Heat exchangers need their out and in streams to be adjacent. They are incorporated at the central forced-air fan where possible, or at various points in building, where each has its own fan.

Among the variety of heat recovery devices used today are run-around coils, heat exchangers, energy transfer wheels, heat pipes, and process waste heat recovery. (See Table 14.2, and Figures 14.3 and 14.4)

TABLE 14.2   HEAT RECOVERY FROM EXHAUST AIR

Closed-loop run-around	Fluid in finned-tube coils in incoming and exhaust air streams is alternately heated and cooled by two air streams, transfers sensible heat.
Open run-around	Hygroscopic fluid spray transfers both sensible and latent heat between air streams. Two air ducts do not have to be adjacent.
Air-to-air heat exchanger	Intake and exhaust ducts brought together at heat but separated by heat transfer surface, so sensible heat flows through.
Energy transfer wheel	Regenerative heat wheel: Revolving wheel with large surface area picks up heat from incoming air and stream and releases sensible heat and moisture to exhaust air stream.
Heat pipe	Coil contains bundle of straight tubes with radiating fins. Each tube is separate heat pipe, passive heat exchanger with no moving parts, long life without maintenance. Recovers only sensible heat. Very high efficiency.
Process waste heat recovery	Hot kitchen and laundry drain lines with heat exchanger, heat pump extracting heat at lower temperatures from cold drain lines, other methods.

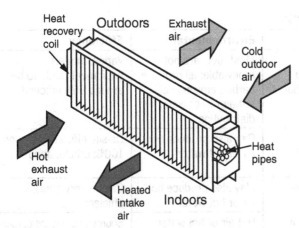

Figure 14.3 Heat pipe heat exchanger

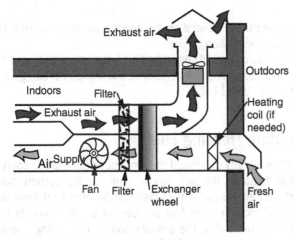

Figure 14.4 Energy transfer wheel

Heating and cooling equipment is given energy efficiency ratings. These include AFUE, COP, EER, IPLV, and SEER ratings. (See Table 14.3)

Codes and Standards

In the United States, it is rare for a code to specifically require that an HVAC system be provided for a building, or that a building be thermally comfortable. However, once a decision has been made to provide an HVAC system, there are many codes, standards, and guidelines that address the design and installation of specific components or equipment, the arrangement and installation of systems, and overall system performance (especially energy efficiency).

The 2013 *ASHRAE Handbook—Fundamentals* (ASHRAE 2013) provides an extensive list of HVAC-applicable codes and standards. Three important and commonly encountered standards and guidelines are:

- ANSI/ASHRAE/IES Standard 90.1-2013—Energy Standard for Buildings Except Low-Rise Residential Buildings
- ASHRAE Standard 62.1-2013—Ventilation for Acceptable Indoor Air Quality

TABLE 14.3 HEATING AND COOLING EQUIPMENT EFFICIENCY RATING SYSTEMS

Rating	Definition
Annual fuel utilization efficiency (AFUE)	Ratio of annual fuel output energy to annual input energy; includes any nonseason pilot input loss. Heating and cooling equipment.
Coefficient of performance (COP)	For cooling systems: ratio of rate of heat removal to rate of energy inputs in consistent units.
	For heat pump system: ratio of rate of heat delivered to rate of energy input in consistent units.
Energy efficiency ratio (EER)	Ratio of net equipment cooling capacity in BTUH to total rate of electric input in watts. When consistent units are used, this ratio is same as COP.
Integrated part load value (IPLV)	Single number figure of merit. Part-load efficiency for air conditioning and heat pump equipment.
Seasonal energy efficiency ratio (SEER)	Cooling in BTUH, divided by total electrical energy input during same period in watt-hours. Total cooling output for an air conditioner during normal annual usage period for cooling.

- ASHRAE Standard 62.2-2013—Ventilation and Acceptable Indoor Air Quality in Low-Rise Residential Buildings

Most building codes now contain minimum requirements for energy conservation that apply to all new construction, including single-family residences.

The 2012 International Energy Conservation Code® (IECC) is a model code that establishes minimum design and construction requirements for energy efficiency. It references standards for both commercial and residential projects.

The International Residential Code (IRC) requires that the interior design temperatures used for heating and cooling load calculations shall be a maximum of 72°F (22°C) for heating and minimum of 75°F (24°C) for cooling. At least one programmable thermostat is mandatory. Other energy-efficiency measures apply.

The many fire protection standards published by the National Fire Protection Association (NFPA) are also important resources. These include standards for heating and air conditioning systems.

HEATING SYSTEMS

There are four means of introducing heat into a building: on-site combustion, electric-resistance, heat transfer from an on-site source, or energy capture. (See Table 14.4) However, there are dozens of specific equipment options.

TABLE 14.4 MEANS OF INTRODUCING HEAT TO BUILDING

Type	Energy Source	Requirements	Environmental	Efficiency
On-site combustion, usually within building	Natural gas, oil, propane, firewood, coal	Needs supply of combustion air, and combustion gas exhaust	Most fuels are not renewable; all produce carbon emissions; hot air or hot water distribution	Varies, modern equipment tends to be around 95% efficient.
Electric-resistance	Offsite electrical utility or on-site PV or wind	Combustion air or exhaust venting not required	Can also produce hot air or hot water	On-site efficiency around 100%; offsite, about 33%.
Heat transfer from on-site source	May use electricity or be self-powered	Heat exchanger, heat pump, etc.	May also produce hot air or hot water	Can be very energy efficient
Energy capture	Solar energy, wind	Energy from nonheat source converted to heat	Hot air or hot water readily produced	Sources tend to be carbon-free, no cost to supply

Central Heating Systems

Central heat sources include electric, fuel, solar radiation, and other sources that transfer thermal energy to a fluid (air or water). As the temperature rises, the fluid changes from a liquid to a gas.

Central heating equipment selection depends largely on capacity and economic factors. System components depend on the types of fuels and space available. When possible, recovering internal heat gains from lighting, people, and equipment reduces the size of the heating plant and its energy use.

Central heating systems are generally classified according to heat-carrying medium (air, steam, water) and the energy source. Combustion systems such as gas, oil, coal, wood, and solid waste require a supply of air for combustion and to cool the mechanical space, plus a flue to remove combustion gases.

They are usually located on the building perimeter or roof for access to ventilation.

The basic types of central heating systems include forced-air, hot water, electric-resistance, radiant, and active solar energy systems. (See Table 14.5)

Building Heating Fuels

We have already discussed the most powerful heating source at our disposal: the sun. Any building heating system must start with an assessment of the available free heat from the sun, and look to other fuel sources as supplements. In the United States today, the primary source of building energy remains fossil fuels. (See Table 14.6)

TABLE 14.5 CENTRAL HEATING SYSTEMS

System Type	Description	Advantages	Disadvantages
Forced-air heating	Air heated in gas, oil, or electric furnace is distributed by fan through ductwork to registers or diffusers. Used to heat houses, small buildings.	Can also ventilate, cool, control humidity, and filter. Quick response.	Bulky ducts, can be noisy. Difficult to install in renovations and to zone in small buildings.
Hot-water heating	Water heated in boiler is circulated by pump through pipes to radiators or convectors. Steam heating is similar.	Compact pipes in walls; radiant floor heating. Very quiet, easy to zone.	Usually only heating, with no ventilation, humidity control, or filtering. May leak.
Electric-resistance heating	Electric conductor resists passage of current and converts electricity into heat. Heating elements are in forced-air heating system furnace or ductwork, in hydronic heating system's boiler, or in space-heating units.	Compact, quick response, easily zoned, quiet, low initial cost.	Expensive to operate (except heat pump), low source energy efficiency.
Radiant heating	Uses heated pipes or tubing with hot water or electric-resistance heating cables embedded in ceilings, floors to raise temperature.	Radiant ceiling panels can heat and cool.	Radiant floors are slow to respond to temperature changes.
Active solar heating system	Absorbs, transfers, and stores energy from solar radiation to heat or cool building. Uses solar collector panels, equipment for heat circulation and distribution, and heat exchanger.	Sustainable heat source.	Requires storage facility.

TABLE 14.6 BUILDING HEATING FUELS

Fuel	Description
Oil	Burned in a central system, then distributed as steam, water, or air. Around 85% efficient. Delivered by truck to building's storage tank.
Natural gas	Around 95% efficient. Does not require storage in building. Distinctive odor added to aid leak detection.
Liquified gas	Propane and butane are petroleum gases that become liquids under moderate pressure. Transported in pressurized cylinders, connected to building's gas piping. Used for small installations in remote locations.
Coal	Rarely used for heating new residential construction; requires complex heating system with high maintenance requirements. Bulky, heavy, dirty.
Wood (cordwood)	Requires storage space. Incomplete combustion can give off dangerous gases. Chimney must be cleaned often to remove creosote (tar residues).
Wood pellets	Dense pellets of sawdust. Efficient, cleaner, saves storage space, produces less pollution than cordwood. Automatically fed into stoves by auger.
Electricity	Requires little distribution space; easy individual control. 100% efficient within space, but considering fuel used to generate electricity, only 30% overall.

Fossil fuels include gas, oil, and coal. The heating system is the largest energy expense in most homes, accounting for two-thirds of the annual energy bill in colder climates. According to the US Department of Energy (DOE) *Buildings Energy Data Book* (March 2012), buildings consume 41 percent of primary energy in the United States.

For every unit generated by electric fossil fuel or nuclear energy, two or three units are discarded into waterways or the atmosphere. Consequently, electric heat is roughly half as efficient in terms of fuel consumption as direct combustion of fuel in heating devices inside a building.

See Chapter 1, "Environmental Conditions and the Site," for more information on fuel sources.

Solar Space Heating

Solar energy conserves fossil fuels and decreases the level of air pollution emissions. It also allows on-site self-sufficiency. The amount of sunlight falling on a building typically carries enough energy to keep it comfortable throughout the year. Most solar heating systems can handle 40 to 70 percent of the heating load for a building.

In a purely passive solar design, the sun is the only energy source, and the building itself is the solar system. This chapter looks at active and hybrid solar heating systems. (See Table 14.7)

For more information on passive solar design, see Chapter 2, "Designing for the Environment."

Residential solar heating system applications are typically used for small-scale, externally loaded buildings. These buildings can be designed to be responsive to climate and site concerns.

Commercial buildings may be large-scale, internally loaded buildings, where climate and site responses are intended primarily to minimize heat gain from the environment during the cooling season. Passive solar heating may not be the best solution in these well-insulated buildings with internal loads and their need for ventilation air. An active or hybrid solar heating system may be the answer.

TABLE 14.7 PASSIVE AND ACTIVE SOLAR HEATING COMPARISON

Passive	Active
The building is the system; no separate collectors	Generally use outdoor collector panels to collect heat
Thermal energy flows naturally by radiation, conduction, and natural convection, without pumps or fans	Heat is transported with fans, pumps, other mechanical equipment
No storage units	Heat transported by water or air from isolated storage unit to building spaces
No mechanical elements; little or no noise	Mechanical distribution system
Lasts as long as building itself; no moving parts	Typically lasts 20 years until some equipment needs replacement
Depends heavily on local site and climate	Better thermal control, easier to retrofit

ACTIVE SOLAR HEATING SYSTEMS

Active solar heating systems offer better control of the environment within the building, and can be added onto most existing buildings. Active solar heating systems absorb solar energy in flat-plate collectors, from which heat is removed by a heat transfer fluid and conveyed to storage. (See Figure 14.5) These systems use pumps, fans, heat pumps, and other mechanical equipment to transmit and distribute thermal energy via air or a liquid. Most systems use electricity continually to operate the system. Many buildings use hybrid systems, with passive solar design features and electrically driven fans or pumps.

There are two general types of collectors used with active solar heating systems: flat-plate collectors (collector panels) and concentrating collectors.

Flat-plate collectors are more common, cost less, and are available in many designs and materials. The sun's rays pass through cover plates and heat up the blackened metal surface of the absorber plate. Fluid circulated through tubes or channels in the plate picks up heat and carries it to a remote storage unit. When the absorber's temperature is greater than the surrounding ambient temperature, the collector loses heat. Glazing over the absorber plate reduces radiation and conduction heat losses. Panels are typically around 4 by 8 feet (1.2 by 2.4 m). The absorber can be painted black or have a high-efficiency selective surface coating.

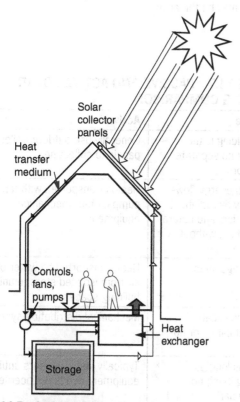

Figure 14.5 Active solar heating system

Source: Redrawn from Francis D.K. Ching and Corky Binggeli, *Interior Design Illustrated* (3rd ed.), Wiley 2012, page 226

Concentrating collectors use only the direct rays of the sun. They can achieve much higher temperatures than flat plate collectors. Concentrating collectors are used for absorption or Rankine cycle cooling, or to produce steam to drive electrical-generating turbines or other equipment. Concentrating collectors generally use optical lenses or reflectors to focus direct-beam solar radiation onto a much smaller point than their receiving aperture, concentrating energy to produce higher temperatures. Most concentrating collectors use sun-tracking mechanisms to receive the maximum amount of direct radiation.

Solar hot air collectors use air rather than water or another fluid primarily for space heating. Air does not boil or freeze, and leaks do no damage. Warm air can be used directly to heat the building. Solar hot air systems require fan power, bulky collectors, and ducts.

Solar collectors can be located anywhere near or on a building that is exposed to direct sunlight, but are usually on the roof. The optimum size of the collector array requires study for each installation.

Backup sources that are provided for long periods of low sunlight or when the system is out of service can also furnish supplemental heat. A heat pump can serve as a backup for both heating and cooling.

The storage facility for active systems holds heat for use at night and on overcast days in an insulated tank filled with water or other liquid, or in a bin of rocks or phase-change salts for air systems.

Heat distribution in an active solar system is similar to that in a conventional heating system, using all-air or air-water delivery. A heat pump or absorptive cooling unit accomplishes cooling.

Fireplaces and Wood Stoves

Wood is popular for heating homes in regions where energy costs are high and local regulations permit burning wood. Fireplaces and wood stoves require maintenance if they are to remain reliable and safe.

Wood-burning fireplaces and stoves burn seasoned wood or manufactured wood logs or pellets. They can add to indoor and outdoor air pollution, emitting carbon monoxide, irritating particles, and sometimes nitrogen dioxide. Wood smoke can cause nose and throat irritation and trigger asthma attacks. To keep chimneys clean and minimize pollution risks, it is advisable to burn small hot fires, not large smoky ones, use seasoned wood, and provide adequate ventilation. A significant amount of dry space is needed for wood storage.

CODES AND STANDARDS

Strict environmental laws may prohibit burning wood on certain days for all but certified clean-burning appliances, which usually means factory-built fireplaces or wood stoves. The Environmental Protection Agency certifies prefabricated fireplaces and stoves for burning efficiency and allowable particulate emissions.

The 2015 International Residential Code (IRC) sets requirements for chimneys and fireplaces, including hearth extensions, lintels, and clearances, among others. There are also requirements for factory-built fireplaces and chimneys, and for masonry heaters.

ANSI/ASHRAE Standard 90.2 requires a fireplace to have a tight-fitting damper, firebox doors, and a source of outside combustion air within the firebox.

Prefabricated fireplaces and woodburning stoves should be certified by the Environmental Protection Agency (EPA) for burning efficiency and allowable particulate emissions.

FIREPLACES

A fireplace is technically a framed opening in a chimney, designed to hold an open fire and sustain the combustion of fuel. Modern fireplaces combine masonry and steel construction; some are almost entirely steel.

Fireplaces should be designed to carry smoke and other combustion by-products safely outside, and to radiate the maximum amount of heat comfortably into the room. The designer must keep the fireplace adequate distances from combustible materials. Multifaced fireplaces are sensitive to drafts in a room, so avoid placing their openings opposite an exterior door.

To burn properly, a fire requires a steady flow of air. Traditional fireplaces draw the air for combustion from inside the house. A standard fireplace is only around 10 percent efficient unless it has a heat exchanger, outdoor combustion air, and doors. With a direct exterior air supply, this may increase to around 20 to 30 percent efficiency.

Metal fireplace inserts for newer fireplaces circulate room air around the firebox. Sometimes a fan is used to increase heat transfer to the circulating room air.

Masonry fireplaces heat the fireplace and chimney masonry, which evens out and prolongs the flow of heat from the fire. Masonry construction can make use of the exterior walls, or employ a wood frame construction with a central masonry shaft collecting the flues.

The parts of a fireplace and its chimney are designated with traditional terms. (See Figure 14.6) The flue creates a draft and carries off the smoke and gases of a fire to the outside. The smoke chamber connects the throat of a fireplace to the flue of a chimney. The smoke shelf at the bottom of a smoke chamber deflects downdrafts from the chimney. The throat is the narrow opening between a firebox and the smoke chamber; it is fitted with a damper that regulates the draft in a fireplace. The firebox is the chamber where combustion takes place. The hearth extends the floor of a fireplace with a noncombustible material such as brick, tile, or stone. A building's chimney may serve more than one fireplace and also vent other heat sources.

Interior designers may be particularly involved in the design of the hearth, which is made of noncombustible material (brick, tile, stone) and extends the floor of the fireplace out into the room to resist flying sparks. The chimneybreast and mantle are

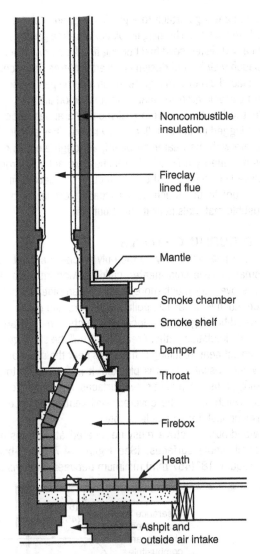

Figure 14.6 Fireplace and chimney section
Source: Redrawn from Francis D.K. Ching and Corky Binggeli, *Interior Design Illustrated* (3rd ed.), Wiley 2012, page 217

often treated as focal points in a room. The chimneybreast is the part of the wall that projects a few inches into the room. A mantle may trim the top of the fireplace.

Wood-burning fireplace types include traditional brick or stone masonry, high-efficiency (HE) or hybrid wood-burning, manufactured prefabricated, Rumford, heat-circulating, zero-clearance, and glass enclosed fireplaces.

Gas logs and gas fireplaces burn natural gas or propane to provide flames that are largely decorative, although some units also provide heat. Gas fireplaces require only a small vent. Direct-vent models can be vented directly through an exterior wall and do not need a chimney.

Ceramic gas logs can be hooked up to a gas starter in a fireplace or a line brought to the firebox. Gas logs should be installed only in a fireplace designed to burn wood and always operated with the damper open, so carbon monoxide combustion gases will vent up the chimney.

Vented gas logs attach to a gas hookup, and can be used in any UL-listed solid fuel burning fireplace. Vent-free gas logs can be used in any UL-listed solid fuel burning fireplace or American Gas Association (AGA) listed design-certified vent-free fireplace. Both can be placed directly on a grate, or on a flame pan covered with a bed of volcanic granules that simulate a wood fire.

Electric fireplaces have a realistic flame appearance. They can be plugged in and installed in any room. They are designed to operate with the heater on or off, although electric radiant heat is not energy efficient. The flames are actually produced by a cleverly designed light bulb. Although the electric fireplace itself is cool to the touch, avoid locating draperies or other combustible materials near its heat outlets.

WOOD-BURNING STOVES

A wood-burning stove may be the only mechanical heat source for a small passive solar heated building. Modern metal wood-burning stoves are much more efficient than older models and produce much less particle pollution indoors and out.

A wood-burning stove's location affects furniture arrangements and circulation paths. Areas that "see" the stove get most of the radiant heat, resulting in hot spots near the stove and cold spots where visual access is blocked. It is imperative to leave circulation paths around hot stove surfaces. Plan space for wood storage, which should be covered, well ventilated, accessible, and large enough for an ample supply.

A wood-burning stove must be located at safe distances from combustible surfaces. (See Figure 14.7) Wood-burning stoves require 18" (457 mm) minimum between an uninsulated

metal chimney and combustible wall or ceiling surfaces. The stove itself must be at least 36" (914 mm) from the nearest wall. This may be reduced to 18" (457 mm) if the wall is protected by a noncombustible heat shield or 1" (25 mm) clear air space.

Wood-burning stoves may be more efficient than heat-circulating fireplaces. Some stoves only radiate heat, while others also heat air passing around the firebox in convection currents. Precision dampers and other controls adjust heat output.

Catalytic combustors reduce air pollution by igniting the wood smoke at a lower temperature, burning up gases and producing more heat and less creosote. Wood-burning stoves with catalytic combustors should burn only natural wood, with small amounts of uncolored paper used to ignite it.

Pellet stoves burn pellets made from densified quality sawdust, a manufacturing byproduct. They are highly efficient, with limited pollution emitted. Pellet fuel is cleaner and takes less space than cordwood to store. An electric auger automatically feeds fuel into the stove to the maintain fire.

A **masonry heater** (masonry heating stove or Russian or Finnish stove) is usually taller than it is wide and deep. (See Figure 14.8) The vertical inner firebox produces an efficient, hot, clean burn. The heat transfers to the exterior masonry surfaces, which radiate a gentle even heat all night after the fire is out. Masonry heating stoves efficiently burn up to 99 percent of combustible wood solids and gases. They radiate energy

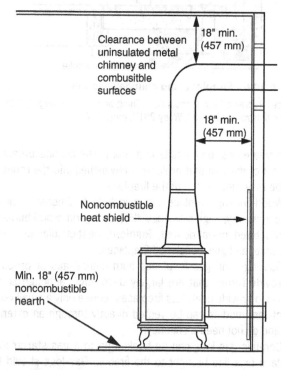

Figure 14.7 Wood-burning stove
Source: Redrawn from Francis D.K. Ching and Corky Binggeli, *Interior Design Illustrated* (3rd ed.), Wiley 2012, page 218

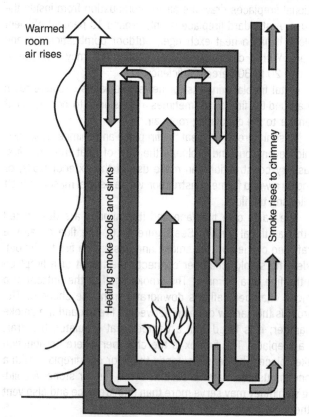

Figure 14.8 Masonry heater
Source: Redrawn from Walter T. Grondzik and Alison G. Kwok, *Mechanical and Electrical Equipment for Buildings* (12th ed.), Wiley 2015, page 454

stored in their thermal mass back into space slowly and evenly over many hours, heating solid objects that then reradiate into the air. They can be used with passive solar heating, and have much lower particle emissions levels than woodstoves.

CHIMNEYS AND FLUES

We usually associate chimneys with fireplaces, but boilers and furnaces have chimneys too. Burning fossil fuels produces carbon monoxide and carbon dioxide. Chimneys carry these gases and other products of combustion up and out of the building.

An unducted chimney sucks in outdoor air through the room, producing a draft. It is better to provide a duct for sufficient outdoor combustion air to be drawn directly from outdoors to the base of the fire without passing through the room.

The flue from a wood stove carries very hot gases. It can be exposed to produce radiant heat, but must be isolated from combustible construction.

Prefabricated chimneys are replacing heavier, bulkier field-built masonry ones. High efficiency boilers and furnaces remove so much heat from the exhaust gases that the flues can be smaller and may be able to be vented through a wall to the exterior, eliminating the chimney.

Mechanical Heating Systems

Mechanical systems required to control the thermal environment typically use purchased energy. The use of mechanical systems affects the building's appearance through rooftop equipment, grilles, ductwork, and interior machinery. Consequently, it is desirable to minimize the amount of active control and to use highly efficient equipment. The capabilities of passive systems should be considered first; they may be used with active systems to mitigate energy and carbon impacts.

Thermal comfort and indoor air quality (IAQ) modification systems involve numerous single-purpose components that are typically only lightly integrated into the overall building fabric. They are normally designed by a consultant other than the architect.

An HVAC system should be able to control the air temperature by heating and/or cooling, as well as the humidity of the air, the air speed and direction, and the indoor air quality.

The term "HVAC" (heating, ventilating, and air conditioning) is used to describe an active control system. The term "air conditioning" is often used as a synonym for HVAC.

Knowing about mechanical equipment helps decision making about interior design. "We cannot wait until the contract document phase of a project to learn that a dropped soffit might be required to accommodate the air ducts or that a vertical chase must be added in the center of a major space." (David Lee Smith, *Environmental Issues for Architecture*, Wiley 2011, page 308)

TABLE 14.8 HEATING EQUIPMENT FOR SMALLER BUILDINGS

Equipment	Description
Gas-fired baseboard heaters	Local heat by convection and radiation. 80% efficiency. Fan directly vented to outside.
Electric-resistance heaters	Local heat. Around 35% effective considering electricity source. Surfaces may be hot; avoid contact with people, combustibles.
Hot-water baseboard and radiator systems	Series loop, 1-pipe, 2-pipe, and 4-pipe layouts.
Local air-to-air heat pumps	Buildings with all-perimeter spaces (motels); user control, equipment noise may provide masking.
Radiant ceiling panels	Electric-resistance wiring; tends to stratify hot air just below ceiling. Used for both heating and cooling.
Hydronic radiant floors	May need panel smaller than floor area. Cross-linked polyethylene (PEX) tubing.
Hydronic perimeter and air	Hot-water heating pipe with overhead air-handling system.
Fan-coil units	Used within space for both heating and cooling.

EQUIPMENT FOR SMALLER BUILDINGS

Having central HVAC system equipment in its own space rather than in each room has the advantage of maintenance without disrupting occupant activity. Central system distribution trees include ducts for air and pipes for water. Ducts are bulky, and are often located above the ceiling and in vertical chases. Pipes are smaller, and can sometimes be integrated with structural columns. Both ducts and pipes can produce unwanted noise.

Skin-load dominated buildings such as motels that have differing but simultaneous needs may require a room-by-room solution to enable quick response to individual room needs. Heating equipment for smaller buildings includes heaters, radiators, heat pumps, radiant and hydronic systems, and fan-coil units. (See Table 14.8)

HOT-WATER AND STEAM HEATING SYSTEMS

Hot-water (**hydronic**) systems heat a building by means of water heated in a boiler and circulated by a pump through pipes to a fin-tube radiator, convector, or unit heater for heating. (See Figure 14.9)

Steam-heating systems use boilers to generate steam that is circulated through piping to radiators. Steam rises from a central boiler to a radiator, where some condenses and drains back to the boiler. Heat is released primarily as the steam condenses into water and flows in sloped pipes back to its point of origin, allowing one pipe to be used for both supply and return.

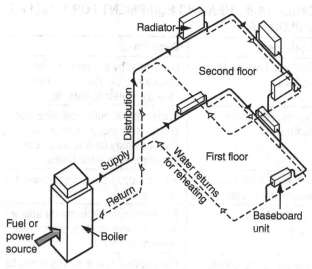

Figure 14.9 Hot-water heating system

Source: Redrawn from Francis D.K. Ching and Corky Binggeli, *Interior Design Illustrated* (3rd ed.), Wiley, 2014, page 227

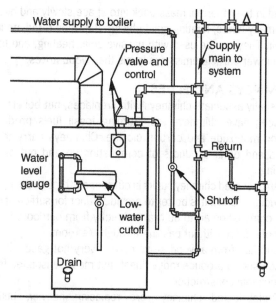

Figure 14.10 Steam boiler

BOILERS

A **boiler** is a closed arrangement of vessels and tubes in which water is heated or steam is generated. A horizontal pipe carries exhaust gas from the boiler, and is connected to a vertical flue section called the stack. Boilers also need ventilation air, with the inlet and outlet on opposite sides of the room.

Boiler systems require a fuel, a heat source, and a pump or fan to move the water. The heat source may be electric-resistance or on-site combustion. A distribution system, heat exchanger, or terminal within the space to be heated, plus a control system complete the equipment. The type of boiler used depends on the size of the heating load, the heating fuels available, the efficiency needed, and whether the boilers are single or made up of multiple modular units.

There are many types of boilers. Two basic types include fire tube boilers in which hot gases are taken through tubs surrounded by water, and water tube boilers in which water is taken through tubes surrounded by fire. Compact boilers have smaller dimensions with high thermal efficiencies and various venting options. All boilers require space for cleaning, general operation, and maintenance.

Small portable steel boilers are assembled of welded steel units and prefabricated on a steel foundation. They are transported as a single unit from the factory. Large boilers are installed in refractory brick settings built on-site.

Boilers are often fueled with oil or gas. Coal boilers require anti-pollution equipment to control fly ash consisting of various sizes of particles as well as flue gas containing sulfur and nitrogen. Flue gases contribute to acid rain. A steam boiler heats water to generate steam, which is distributed through pipes to steam radiators or convectors. (See Figure 14.10)

Boiler efficiencies depend on the type, unit, size, and somewhat on the age of the boiler. Multiple small boilers are sometimes more efficient on average than a single large one.

HEATING DISTRIBUTION

In **steam-heating systems**, steam that is produced in a boiler is circulated under pressure through insulated pipes, and then condensed in cast iron radiators. In the radiator, the latent heat given off when the steam cools and becomes water is released to the air of the room. The condensed water then returns to the boiler through a network of return pipes. The system is reasonably efficient but difficult to control precisely as the steam gives off its heat rapidly.

Electric or hydronic **baseboard heaters** are typically long heating units attached close to baseboards at floor level. They are inexpensive to install and provide localized control, usually using thermostats. Extensive use of baseboard heaters is usually more expensive than a gas or electric central system. A **toe kick heater** is a small electric heater below a cabinet that heats the feet and floor area. (See Figure 14.11) It is provided for when the cabinet is selected and installed.

Hydronic baseboard heaters require a piping system and central hot water service. They can be efficient if used with an efficient boiler, are widely used in Europe, and are usually a little larger than an electric baseboard.

The IRC states that all bathrooms should have an appropriate heat source to maintain a minimum room temperature of 68°F (20°C). The designer should check whether the central heating system can handle the load before adding heat to a previously unheated bathroom or expanding into an unheated area such as

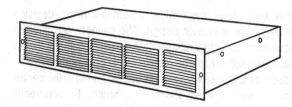

Figure 14.11 Electric toe kick heater

a closet as this may require a supplemental heater. Oversized showers or those with a walk-in entry may require additional heat for comfortable use. The enclosed area of a tub/shower with a door may also require additional heat for comfort.

Heat sources are often located just below windows, even though some of this warmth will be driven through the exterior wall in cold weather. Well-insulated windows and walls stay warmer, and the need for heat at the building's edge is less.

Hot water heating circuits that serve baseboards or radiators come in four principal arrangements, classified according to the number of parallel pipelines connecting the system together. (See Table 14.9) They include series loop, one-pipe, two-pipe, and four-pipe systems. (See Figures 14.12 and 14.13)

TABLE 14.9   HOT WATER CIRCUIT ARRANGEMENTS

Circuit Type	Description
Series loop system	Water flows to and through each baseboard or fin tube in turn. Single zone system. Water cooler at end of circuit. No individual heating element shutoff.
One-pipe system	Fittings divert part of flow into each baseboard. Water cooler at end of circuit. No individual shutoff. Poor thermal control between zones.
Two-pipe reverse-return system	Separate supply and return pipes. Same temperature at each baseboard or radiator. All water and terminals either hot or chilled at any given time. Return flow is reversed and does not mix with supply.
Four-pipe system	Simultaneous heating and cooling in same zone. Hot and cold water do not mix. Two separate water distribution systems with two circulating pumps.

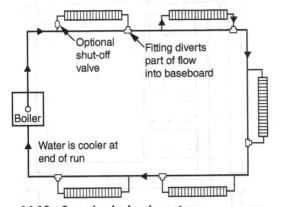

Figure 14.12   One-pipe hydronic system

Source: Redrawn from Walter T. Grondzik and Alison G. Kwok, *Mechanical and Electrical Equipment for Buildings* (12th ed.), Wiley 2015, page 515

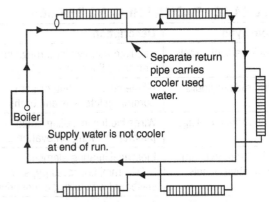

Figure 14.13   Two-pipe hydronic system

Source: Redrawn from Walter T. Grondzik and Alison G. Kwok, *Mechanical and Electrical Equipment for Buildings* (12th ed.), Wiley 2015, page 515

Radiant Heating

As we have seen, the temperature of surrounding surfaces affects thermal comfort. Warm surfaces can maintain comfort even at lower air temperatures. Radiant heating is typically a more comfortable way to warm people than introducing heated air into a space.

Radiant heat can be more energy efficient than hot air systems, by transferring heat directly to objects and occupants without heating large volumes of air first. The warmer surfaces that result mean that more body heat can be lost by convection without the room becoming uncomfortably cold. As a result, the temperature of the air in the space can be kept cooler, and less heat will be lost through the building envelope.

Radiant heating systems have a minimal impact on the appearance of the space, as they do not require any visual expression.

Radiant heating systems use ceilings, floors, and sometimes walls as radiant surfaces. The heat source may be pipes or tubing carrying hot water, or electric-resistance heating cables embedded within the ceiling, floor, or wall construction. (See Table 14.10)

RADIANTLY HEATED FLOORS

Radiant floor heating systems work well in kitchens as they require no wall space, have no vents, and do not interfere with cabinet or furniture placement. They may raise the height of the floor, necessitating door clearance and possibly cabinet adjustments.

Today, one-piece coils of cross-linked polyethylene (PEX) tubing are directly embedded in cast-in-place concrete or other flooring material. (See Figure 14.14) PEX can also be used under wood floors. Mats with integral tubing are also available.

TABLE 14.10   RADIANT HEATING DEVICES

System Type	Description
Hydronic radiant panels	Circulate warm water through tubing. Better for floors than ceilings.
Electric radiant floors	Whole house heating or spot comfort in kitchens and baths.
Electric radiant ceilings	Wires hidden in ceiling could be punctured during renovations.
Manufactured gypsum board heating panels	Electrical heating element in $5/8$" (16 mm) fire-rated gypsum wallboard with wiring connections.
Electric towel warmers	Attach to door hinges or wall, or free standing. Do not locate where can be reached while in water. Hard-wired or plug-in (dangling cords).
Hydronic towel warmers	Water supplied by heating system or hot water tank. Can be installed near a tub or whirlpool.

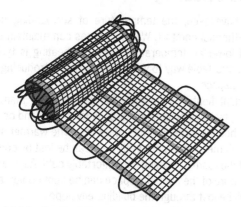

Figure 14.14   Radiant floor mat with tubing

Rugs or carpets over radiant floors interfere with heat exchange. Special under-carpet pads can facilitate heat transfer.

Furniture can reduce radiant heat output, and changes in finish material or layout may have an impact on operation. Tables and desks block infrared (IR) wavelengths from reaching above them. Radiant floor heating may provide only supplemental heat, especially in cold climates. Radiant floors react slowly to small or sudden changes in heating demand within a building.

Active solar collectors can efficiently produce the relatively low water temperatures around 90°F (32°C) required for radiant floor heating.

RADIANTLY HEATED CEILINGS OR WALLS

Radiant ceiling or wall panel heaters consist of electric heating coils behind drywall. Compared to other radiant heat sources, radiant ceiling or wall panels have a lower heat output. Radiant panel systems do not respond quickly to changing temperature demands, and are often supplemented with perimeter convection units. Separate ventilation, humidity control, and cooling system are required for completely conditioned air.

Do not install radiant panels in walls above 48" (1219 mm) to avoid hammering nails into electrical coils when hanging pictures.

Preassembled electric radiant heating panels can be installed in a modular suspended ceiling system or surface mounted to heat a specific areas.

Most of the heat from radiant heating panels flows directly beneath the panel and falls off gradually with greater distance, dropping by about 5°F over the first 6 feet (1.8 m). This may seem like a disadvantage, but some occupants like to find a spot that is relatively cooler or warmer within the room. Proper placement of panels must be coordinated with ceiling fans, sprinkler heads and other obstructions.

Radiant panels avoid some of the problems inherent with forced-air systems, such as heat loss from ducts, air leakage, energy use by furnace blowers, and inability to respond to local zone conditions. Installation costs for energy-efficient radiant panels are considerably less than the cost for a forced-air system, but radiant panels cannot provide cooling as a forced-air system can.

Towel warmers are designed to dry and warm towels, and also serve as a heat source in a bathroom or spa. (See Figure 14.15) They are available in electronic and hydronic models in a variety of styles and finishes.

Figure 14.15   Towel warmer

Electric-Resistance Heat

Small electric-resistance space heaters are low-cost, easy to install, offer individual thermostatic control, and do not waste heat in unoccupied rooms. (See Table 14.11) However, they use expensive high-grade electricity for low-grade space heating, so their use should be limited to spot-heating a small area for a limited time.

Avoid portable heaters in bathrooms as they are dangerous near water and can be a tripping hazard.

Most electric-resistance heating systems consist of baseboard units or small, wall-mounted heaters, both of which contain hot wires. (See Figure 14.16) The heaters are compact, inexpensive, and clean, and do not have to be vented, but lack humidity and air quality controls. Electric-resistance heaters have hot surfaces, and their location must be carefully chosen in relation to furniture, drapery, and traffic patterns. (See Figure 14.17)

TABLE 14.11 TYPES OF SPACE HEATERS

General Type	Use
Toe space unit heater	Installed in low space under kitchen and bathroom cabinets.
Wall unit heaters	Surface-mounted or recessed for use in bathrooms, kitchens.
Fully recessed floor unit heaters	Typically used where glazing comes to the floor, as at a glass sliding door or large window.
Small IR unit heaters	Radiate heat instantly from a small area, beam where needed.
High-temperature heaters	Temperatures greater than 500° F (260° C). Radiant heat for swimming pools, shower rooms, bathrooms. Electric, gas, or oil.
Industrial unit heaters	Suspended from ceiling or roof structure. Industrial buildings, outdoors, loading docks, public waiting areas, garages.
Quartz heaters	Resistance-heating elements sealed in quartz-glass tubes produce IR radiation. Heat within 15 feet (4.6 m); quiet.
Electrical forced air heaters	Blow warm air throughout a room, preferably one closed off.
Ceramic forced air heaters	Ceramic heating element safer than other electric space heaters.
Oil-filled heaters	Electricity heats the oil inside to heat a room.
Portable electric-resistance heaters	Heats small nearby area. Use for building heat may lead to fires from ignition of bedding, drapery, furniture.

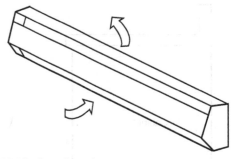

Figure 14.16   Baseboard convector

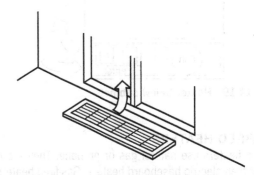

Figure 14.17   Recessed floor unit heater

Wall heaters are typically recessed into the wall with a grate or screen covering the heating elements. (See Figure 14.18) They are typically located in the lower part of a wall, where it is unfortunately easy to back into them and get burned. Recessed or surface-mounted wall unit heaters are used in bathrooms, kitchens, and small rooms. Toe space unit heaters use a fan to blow air into a room from below cabinets. (See Figure 14.19)

Electric-bulb heating units are designed for residential use. They combine a radiant heating element with a fan and a light in a ceiling mounted unit. Some units include a nightlight as well. Bulb heaters provide silent, instant warmth using 250W R-40 infrared heat lamps. They are available vented and unvented, and recessed or surface-mounted.

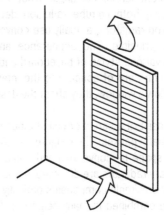

Figure 14.18   Recessed wall unit heater

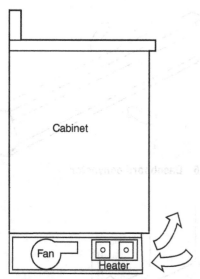

Figure 14.19   Heater below cabinet

GAS-FIRED HEATERS

Gas-fired heaters use natural gas or propane. They are more efficient than electric baseboard heaters. Gas-fired heaters are typically direct-vented with a built-in fan to the outside, and installed on or near an exterior wall.

Vent-free gas heating appliances produce nitrous oxides that cause nose, eye, and throat irritation, along with carbon monoxide. They also generate a great deal of water vapor that can cause condensation, mildew, and rot in wall and ceiling cavities. A nearby window must be kept open a couple of inches for an adequate fresh air supply to prevent oxygen depletion; this results in heat loss. The IRC requires that unvented room heaters be equipped with an oxygen-depletion-sensitive safety shutoff system. Unvented gas heaters are prohibited in homes in many states and cities throughout the United States and Canada.

Natural Convection Heating Units

Radiators and convectors are used to supply heat in residential and small light-commercial buildings. What we usually call radiators, including both fin-tube radiation devices and old fashioned cast iron radiators, actually use convection as their primary heating principle. The appearance and the space occupied by the various styles of baseboard and cabinet convection heating units are of concern to the interior designer. When located below a window, they affect the design of window treatments.

Radiators consist of a series or coil of pipes through which hot water or steam passes. (See Figure 14.20) The heated pipes warm the space by convection and somewhat by radiation. Contemporary radiators are designed in a variety of colors and styles based on simple components typically 2¾″ (70 mm) wide that can be combined in many heights and widths. (See Figure 14.21)

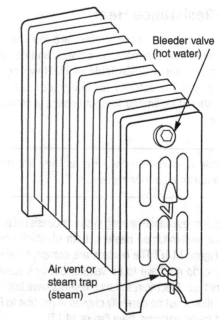

Figure 14.20   Cast iron radiator

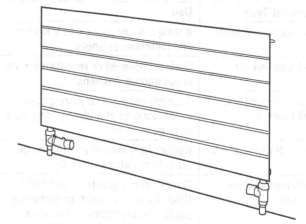

Figure 14.21   Wall panel radiator

Radiators are easier to control in hot water than in steam systems. A very even release of heat to the air is achieved by regulating the temperature and rate of circulation of the water. Hydronic systems are silent when properly installed and adjusted, and produce comfortable heat.

For pumped water distribution, cast iron radiators have generally been replaced by linear transfer devices that are smaller and take up less interior space. Fin-tube radiators (also called finned-tube convectors) are basically a pipe on which fins are placed. Fin-tube radiators are usually used along outside walls and below windows, where they raise the temperatures of the glass and wall surfaces. There are also electric-resistance fin-tube units with an electrical element instead of the copper tubing. Although covered convectors are easier to clean and have a more finished appearance, some buildings still have exposed fin-tube convectors. (See Figure 14.22)

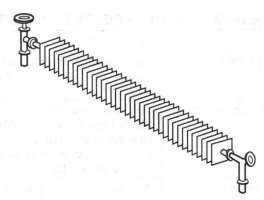

Figure 14.22 Exposed baseboard fin-tube convector

Baseboard radiation units are a smaller form of fin-tube radiation often used in residential applications, where they are installed at the base of walls. Their height depends on whether there are one, two, or three tiers of elements. Baseboard fin-tube enclosures usually run the length of the wall, but the element inside may be shorter. They tend to be less conspicuous than cabinet-style units.

Convectors are useful in stairwells or along large glazed openings. They consist of elements enclosed in a cabinet at least 2 feet high by 3 feet wide (0.6 m by 0.9 m). Cabinets may be free-standing, wall-hung, or recessed. A variety of enclosure styles are available for finished spaces, plus simple utilitarian covers for protection only. The most effective placement is along an outside wall. Air must flow freely around the units in order to be heated.

Unit heaters are used in large open areas such as warehouses, storage spaces, and showrooms. They are made up of factory-assembled components including a heating mechanism in a casing. Unit heaters supplement convection with a fan that blows forced air across the unit's heating element and into the room. The casing has an air inlet and vanes for directing the air out. Units are usually suspended from the roof structure or floor mounted and located at the building's perimeter. Smaller cabinet models are available for use in corridors, lobbies, vestibules, and similar auxiliary spaces. Heat sources include steam, hot water, electricity, or direct combustion of gas or oil.

Cabinet heaters are used in entry vestibules. Natural convection releases heat form the coil at low rates when the entry is not in use. When the outside door is opened, the fan can quickly increase airflow. They need an electrical connection for the fan and a thermostat for fan control.

Warm-Air Heating

The central warm-air system began as a large stove in the center of basement, with grilles in the floors above. Adding ductwork improved uneven temperatures and airflows. Adding a fan to drive the air reduced the size of the ducts, and adding filters at the furnace cleaned the air and resulted in a better room air mix.

HISTORY OF WARM-AIR HEATING

Around 1900, warm-air heating systems began to take the place of fireplaces. In the original warm-air systems, an iron furnace in the basement that was hand-fired with coal was attached to a short duct that delivered warm air to a large grille in the middle of the parlor floor above, with little heat going to other rooms.

Over time, oil or gas furnaces that fired automatically replaced coal furnaces, and operational and safety controls were added. Air was ducted to and from each room, which evened out temperatures and airflow. Fans were added to move the air, making it possible to reduce the size of ducts. Adjustable registers permitted control within each room. Filters at the furnace cleaned air as it was circulated. Eventually, by adding both fans and cooling coils to the furnace, both hot and cold air could be circulated.

During the 1960s, fewer homes were being built with basements, and sub-slab perimeter systems took the place of basement furnaces. The heat source was located in the center of the building's interior, where heat that escaped would help heat the house. Air was delivered from below, up and across windows, and back to a central high return grille in each room. The air frequently failed to come back down to the lower levels of the room, leaving occupants with cold feet. In addition, water penetrating below the house could get into the heating system, causing major problems with condensation and mold.

Eventually electric-resistance heating systems became popular as they eliminated combustion, chimneys, and fuel storage. Horizontal electric furnaces were located in shallow attics or above furred ceilings. Air was delivered down from the ceiling across windows, and taken back through door grilles and open plenum spaces between the suspended ceiling and structural floor above. Today, heat pumps have mostly replaced less-efficient electric-resistance heating.

FORCED-AIR HEATING

Forced-air heating works by heating air in a gas, oil, or electric furnace and distributing it by a fan through ductwork to registers or diffusers in inhabited spaces. (See Figure 14.23) Forced-air heating is the most versatile widely used system for heating houses and small buildings, and can also provide cooling. Fresh air is typically supplied by natural ventilation.

The system can include filtering, humidifying, and dehumidifying devices. **Supply registers** are often located in the floor, typically below glass areas. A separate system of exhaust ducts draws cool air back through return air grilles to be reheated and recirculated. **Return grilles** can be located to minimize ductwork for return air circulation. Sometimes there is no separate ductwork for the return air, and return grilles are then placed in the suspended ceiling to collect return air.

Heating equipment for forced air includes furnaces and unit heaters. (See Table 14.12) Advantages of forced-air heating include control of air temperature and air volume for comfort and redistribution; this is especially welcome in tall spaces. Forced-air heating incorporates filtration, humidification, ventilation,

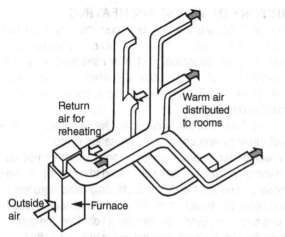

Figure 14.23 Forced-air heating system

Source: Redrawn from Francis D.K. Ching and Corky Binggeli, *Interior Design Illustrated* (3rd ed.), Wiley 2012, page 225

TABLE 14.12 FORCED-AIR HEATING EQUIPMENT

Equipment	Description
Hot-air furnace	Heat exchanger keeps combustion air from mixing with room air. Blower and filter standard, humidifier and cooling coil optional. Some gas pulse-type furnaces are 95% efficient.
Wall furnace	For heating single space, no ducts required. When gas powered, can draw air and vent directly through wall.
Unit heaters	Utility spaces with high ceilings. Gas, electric, or hot water. Ceiling/wall hung.

and cooling. Disadvantages involve bulky ductwork and the need for constant maintenance to avoid circulating dust. Forced air can also be noisy, especially at high velocity.

FORCED-AIR DISTRIBUTION

Perimeter heating is the term for a layout distributing warm air to registers placed in or near the floor along exterior walls. A perimeter loop system consists of a loop of ductwork, usually embedded in a concrete ground slab, for distribution of warm air to each floor register. A perimeter radial system uses a leader from a centrally located furnace to carry warm air directly to each floor register. Extended plenum systems run supply ducts between joists. (See Table 14.13)

Outlets for forced-air ducts for residential central heating systems need to be located so that they do not interfere with kitchen cabinet and appliance design.

TABLE 14.13 FORCED-AIR DISTRIBUTION SYSTEMS

System	Description
Loop perimeter system	Slab on grade in cold climates. Provides most comfort, but high initial cost.
Radial perimeter system	Slab on grade with hot air. Less cost but less comfort. Suitable for crawl space construction. Horizontal furnace can be used in crawl space or attic.
Extended-plenum system	Buildings with basements. Supply ducts run between joists, saving space and headroom. Health and energy efficiency concerns.

FURNACES

The term **furnace** is generally used for residential-sized equipment. Larger buildings use the term **air handling unit (AHU)**.

Systems using air for primary distribution have a furnace as a heat-generating source, rather than the boiler used for water or steam. (See Table 14.14) Warm-air furnaces are usually located near the center of the building. Today's furnaces are up to 95 percent efficient, compared to around 65 percent for older furnaces. Cooling coils are commonly added to warm-air furnaces.

In residential design, the furnace burner is controlled by a thermostat, usually in or near the living room. The thermostat should be in a location where the temperature is unlikely to change rapidly, protected from drafts, direct sun, and the warmth of nearby warm air registers. The blower continues after the burner stops, until the temperature in the furnace drops below a set point. A high limit switch shuts off the burner if the temperature is too high. Programmable thermostats that are easy to use are readily available.

TABLE 14.14 FURNACE TYPES

Furnace Type	Description
Forced-air gas furnace	Gas fed to burner tubes, lighted by electric spark or pilot light flame. Air warmed in heat exchanger is circulated by the furnace blower. Requires room for maintenance.
Oil-fired forced air furnace	Efficient and durable. Oil pumped into combustion chamber is atomized and ignited by spark to heat a heat exchanger that warms air, which is circulated by a blower.
Electric forced-air furnace	Very high efficiency/local efficiency, but only 33% considering off-site generation. Clean, simple, few problems.

DUCTWORK

Ductwork transports air from the furnace or air handling unit (AHU) to the conditioned spaces at a specified velocity and then back. Ducts are usually constructed of galvanized sheet metal or fiberglass. Fiberglass duct liner is used with metal ducts to reduce heat loss or gain, prevent condensation, and control air noises.

Early coordination can allow ducts to be integrated within joist spaces and roof trusses, and between bulky recessed lighting fixtures.

Ducts are either round or rectangular. Round ducts may be preferable where exposed, but require larger clearances. Flexible ducting is used to connect supply air registers to the main duct-work to allow adjustments in the location of ceiling fixtures, but is not permitted in exposed ceilings. Duct dimensions on construction drawings are usually inside dimensions; add 2" (51 mm) to each dimension for the duct wall thickness and insulation.

Ductwork can be concealed or exposed. Concealed ductwork permits better isolation from the noise and vibration of equipment and from the flow of air. Surfaces are less complicated to clean. As the ducts are less visible, construction can be less meticulous, and construction costs are lower. It can cost more to install visually acceptable exposed ductwork than to construct a ceiling to hide standard ducts. Concealed ductwork provides better architectural control over the appearance of the ceiling and wall surfaces. Access panels and doors or suspended ceilings may be needed to provide access for maintenance.

DAMPERS

Dampers balance and adjust the distribution system. Supply registers should have dampers. Narrow damper openings, especially in long runs, can potentially produce damper blade vibration or whistling air that becomes a noise problem.

Large commercial structures have fire-rated partitions, floors, and ceilings that confine fires for specified periods of time. Air ducts through a fire barrier are required by codes to have **fire dampers** made of fire-resistive materials.

For more information on fire dampers, see Chapter 18, "Fire Safety Design."

REGISTERS, DIFFUSERS, AND GRILLES

Air for heating, cooling, and ventilation is supplied through registers and diffusers. The selection and placement of supply and return openings requires architectural and engineering coordination, and has a distinct effect on the interior design of the space. Registers, diffusers, and grilles are selected for airflow capacity and velocity, pressure drop, noise factors, and appearance. (See Table 14.15 and Figures 14.24, 14.25, and 14.26)

TABLE 14.15   REGISTERS, DIFFUSERS, AND GRILLES

Type	Description
Grilles	Rectangular openings with fixed vertical or horizontal vanes or louvers through which air passes. Supply grilles have adjustable vanes that control the direction of air entering room and do not have dampers.
Registers	Wall grilles with a damper directly behind the louvered face to regulate the amount and direction of airflow.
Diffusers	Slats set at angles to deflect warm or conditioned air. Mix air supply from ceiling. Round, rectangular, or linear. Dampers adjust volume supplied.
Perforated metal faceplates	Placed over standard ceiling diffusers to create a uniform perforated ceiling. Can also be large perforated ceiling panels.
Local supply-air outlets	Used to avoid creating stagnant or drafty areas. Make sure not blocked by beams or other objects. Place lower when heating is the major problem.
Air supply units	Designed to distribute air perpendicular to the surface. Formerly round, now usually rectangular, often square.
Return grilles	Louvered, eggcrate, or perforated. Connected to a duct, lead to plenum above ceiling or transfer air directly. May be called either grilles or registers.
Slotted diffusers	Long, continuous linear slots under glass doors, windows. Tend to collect dirt, often obstructed. Also available for ceilings.
Return air inlets	For heating systems, usually located near floor and across room from supply outlets. For cooling, located in ceilings or high on walls.
Exhaust air inlets	Usually located in ceilings or high on walls, and are almost always ducted. Supply outlets can also be used as return grilles.

A register is a wall grille with a damper, but a ceiling diffuser with a damper is still called a diffuser.

Supply air cannot enter a space unless a comparable amount of return air is returned or exhausted. This can be a problem in a residence where there is only one central common return and doors to individual rooms remain closed. Undercut doors transmit sound, so it is preferable to have a return air grille for each closed space. A space may require a number of supply air outlets, but often only a single return air inlet.

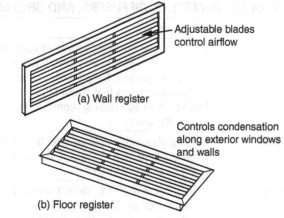

(a) Wall register — Adjustable blades control airflow

(b) Floor register — Controls condensation along exterior windows and walls

Figure 14.24 Registers

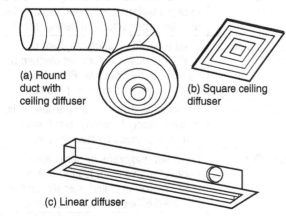

(a) Round duct with ceiling diffuser

(b) Square ceiling diffuser

(c) Linear diffuser

Figure 14.25 Diffusers

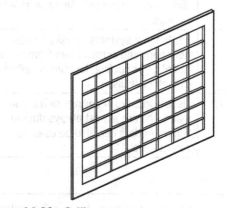

Figure 14.26 Grille

COOLING SYSTEMS

According to the 2008 DOE Energy Data Book, HVAC use amounts to 39 percent of total energy use for residential buildings, and 32 percent for commercial buildings. The 2009 Residential Energy Consumption Survey (RECS) indicates that 87 percent of US households are now equipped with air conditioners (AC), and most of those have central AC equipment. Almost 90 percent of new homes are built with central air conditioning. Residential air conditioning adds about 100 million tons of carbon dioxide to the atmosphere from the electric power generation stations that use fossil fuels.

An overheated environment can be improved through shading, increasing airflow, evaporative cooling, and thermal storage. Architects, engineers, and interior designers can make decisions about whether to open or close the building, or to do both at different times of day. We can employ strategies to keep direct sun out of the building, and allow daylight in the winter without overheating in the summer. The building structure can be used by day to absorb heat, and then flushed with cooler air at night. Providing shade from the sun is essential for passively cooled buildings, and for passively heated buildings that might become overheated in hot weather.

The decision can be made to cool using outdoor air rather than mechanical air conditioning. Use of mechanical refrigeration can be concentrated during the coldest (night) hours.

Precooling reduces the size of the cooling system needed for a building with relatively short but intense peak cooling load periods. Cooling equipment precools the space, then shuts off when occupants arrive and the thermostat is reset upward.

Engineers use some common terms when discussing air conditioning.

- Cooling load: the rate at which heat needs to be removed from air.
- Capacity: the ability of equipment to remove heat.
- Heat gain: total load on a cooling system; almost the same thing as its cooling load, although from the engineer's viewpoint there is a technical difference.

History of Cooling

The earliest known home air-cooling systems were used around 3000 BCE, when Egyptian women put water in shallow clay trays on a bed of straw at sundown. The rapid evaporation from the water's surface and the damp sides of the tray combined with the night temperature drop to produce a thin film of ice on top, even though the air temperature was not below freezing. The low humidity aided evaporation and the resulting cooling brought the temperature down enough to make ice.

Evaporative cooling was also used in ancient India. At night, wet grass mats were hung over openings on the westward side of the house. Water sprayed by hand or trickling from a perforated trough above the windows kept the mats wet through the night. When a gentle warm wind struck the cooler wet grass, evaporation cooled temperatures inside as much as 30 degrees.

By the end of the nineteenth century, large restaurants and public places were embedding air pipes in a mixture of ice and salt and circulating the cooled air with fans. The Madison Square Garden Theater in New York City used four tons of ice per night. However, none of these systems addressed how to remove humidity from warm air.

The term "air conditioning" is credited to physicist Stuart W. Cramer, who presented a paper on humidity control in textile mills before the American Cotton Manufacturers Association in 1907. Willis Carrier, an upstate New York farm boy who won an engineering scholarship to Cornell University, produced the first commercial air conditioner in 1914.

By 1919, Chicago had its first air-conditioned movie house. The same year, the Abraham & Strauss department store in New York was air-conditioned. In 1925, a 133-ton air conditioning unit was installed in New York's Rivoli Theater. By the summer of 1930, over 300 theaters were air conditioned, drawing in hordes of people for the cool air as well as the movie. By the end of the 1930s, stores and office buildings claimed that air conditioning increased workers' productivity enough to offset the cost. Workers were even coming in early and staying late to stay cool.

Passive Cooling

In many climates, the right combination of properly implemented natural methods can provide cooling equivalent to mechanical air conditioning. At the very least, natural cooling allows installation of smaller cooling equipment that will run fewer hours and consume less energy. In most of the eastern United States, passive cooling can replace or reduce the need for air conditioning for much of the summer.

Passive cooling strategies have long been used in hot, dry climates, where traditional buildings employ few and small windows, light surface colors, and massive adobe, brick, or stone construction. Wind scoops and towers were used in Egypt around 1300 BCE and are still used in the Middle East today. Earth sheltering has traditionally provided thermal mass to aid cooling from Cappadocia, Turkey, to Mesa Verde, Colorado.

Hot, humid climates seek natural ventilation. Much of Japan has hot, humid summers. Traditional Japanese house post and beam construction uses movable lightweight paper wall panels with large overhanging roofs that protect the panels and create outdoor space.

Temperate climates may have very hot summers and very cold winters, a difficult combination to design for. The first building with a central mechanical system for both space heating and cooling was the Milam Building in San Antonio, Texas, designed in 1928 by architect George Willis. The design retained the requirement to provide direct access to light and air in all occupied spaces. The first building designed to be truly responsive to a central air conditioning system was the PSFS Building in Philadelphia, Pennsylvania, designed by George Howe.

Steps toward passive cooling begin with minimizing heat gain through shading, orientation, color, vegetation, insulation, daylight, and control of internal heat sources. A **heat sink** is likely needed to remove heat from a building. Passive cooling can involve a shift in the comfort zone to include higher indoor temperatures, along with modifying humidity, mean radiant temperature (MRT), or air speed. The mechanical equipment required is relatively small, using modest amounts of energy. Hybrid systems use some fans and pumps.

Kitchen equipment emits heat; well-insulated and efficient appliances minimize heat. Exhaust ventilation helps minimize residential cooling loads.

Radiant cooling in a hot and dry climate makes use of traditional buildings with deep courtyards and narrow alleys that expose massive walls to only a few hours of direct sunlight, with the whole wall radiating heat to the cold night sky. Clouds limit the amount of radiation, but on a clear night, the effective sky temperature can be 20° to 30°F (11° to 17°C) lower than the ambient air temperature. Roof ponds are probably the most effective method to achieve this.

SOLAR COOLING

Solar energy can be stored as hot water as it comes out of collectors, or as chilled water. Solar air conditioning systems operate by absorption, by the Rankine cycle in which solar steam turns a turbine to power an air conditioner, or by **desiccant cooling** that uses dehumidification to cool. The equipment for solar cooling is expensive, and is used only where cooling loads cannot be avoided by good building design. New technology, mass production, and rising energy costs may change that.

HIGH THERMAL MASS COOLING

High thermal mass cooling works well in places with warm, dry summers. The thermal mass of the building absorbs heat and stays cool during the hot daytime, and is flushed by cool air at night. Such buildings use thermal mass on the floors, walls or roofs. Fans are often used with high thermal mass systems.

In a high thermal mass design, the building needs a heat sink, a place from which heat is ejected at night. Earth contact can provide that heat sink, keeping walls, floors, and earth-covered roofs cool.

Mechanical Cooling

Mechanical cooling systems were originally developed as separate equipment, to be used in conjunction with mechanical heating equipment. (See Table 14.16) Today, cooling equipment is often integrated into HVAC systems, which we will discuss later.

In order to understand the discussions that take place among the client, the architect, the engineers, and the contractors, the interior designer should have a basic understanding of air conditioning.

TABLE 14.16 MECHANICAL COOLING SYSTEM TYPES

System	Description
Packaged systems	Everything is packaged except site-installed ducts. Low installation, maintenance, and operating costs. Rooftop units are most common.
Split systems	Used in most homes and many small- to medium-sized buildings. Compressor and condenser coils in outdoor unit, AHU with evaporator indoors.
Ductless split systems	Used to retrofit existing and historic buildings. Only two small copper refrigerant lines. Compact indoor units are unobtrusive, very quiet.
Chilled water system	Reciprocating (small), centrifugal (large), or rotary water chillers.

COOLING EQUIPMENT

There are two basic refrigeration processes, but there are dozens of equipment options. We briefly look at both processes and equipment.

An air conditioner normally cools by removing sensible heat from the air. It requires a difference in temperature between the warmer and cooler areas for heat transfer to take place. If the surface temperature falls below the dew point temperature of the air, condensation occurs, which requires dehumidification to remove latent heat and almost always necessitates drainage.

Cooling equipment uses either compressive refrigeration or absorption refrigeration. A heat pump can provide both heating and cooling. Evaporative cooling is another method that can be used where the air is dry.

COMPRESSIVE REFRIGERATION

In simple terms, the **compressive refrigeration** cycle pumps heat out of the chilled water systems into the condenser water system. The reverse of this cycle is what happens in a heat pump.

Compressive refrigeration produces cooling by the vaporization and expansion of a liquid refrigerant. (See Figure 14.27) The **refrigerant**, a liquid able to vaporize at a low temperature, absorbs heat from a cooling medium and changes state from a liquid to a vapor or gas. The compressor reduces the volume of the vapor or gas and increases its pressure. The condenser further reduces the vapor or gas to a liquid form, releasing heat to air or water. Finally, an expansion valve lowers the pressure and evaporation temperature of the refrigerant as it flows back to the evaporator to repeat the process.

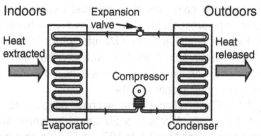

Figure 14.27 Compressive refrigeration cooling cycle

Source: Redrawn from Francis D.K. Ching, *Building Construction Illustrated* (5th ed.), Wiley 2014, page 11.16

ABSORPTION REFRIGERATION

Absorption refrigeration is more complicated. An absorber and a generator are used to transfer heat and produce cooling with absorption refrigeration, instead of a compressor. (See Figure 14.28) Heat is extracted from a space by a heat exchanger, producing chilled water for cooling. Water vapor from this process moves to the absorber. The generator heats the saline solution, producing more water vapor. The absorber draws water vapor from the evaporator, which cools the remaining water. The steam that is removed by the generator goes to the condenser, where waste heat is extracted. Water from the condenser is then routed to the evaporator, and the process begins again.

In the absorption cycle, water cools rapidly as it is evaporated into the evaporator vessel. The water vapor from the evaporator vessel is attracted to a concentrated salt solution that absorbs water and dilutes the salt solution. The diluted salt solution is drawn off from the vessel continually, sprayed into a generator that boils excess water off, and returned to repeat the absorption cycle. The steam that boils off from the generator goes to a condenser with cool water or air where it condenses, and then travels to the evaporator vessel. The cooled water left in the evaporator is tapped through a heat exchanger as a source of chilled water for cooling.

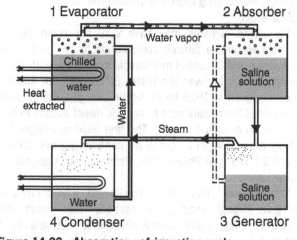

Figure 14.28 Absorption refrigeration cycle

Source: Redrawn from Francis D.K. Ching, *Building Construction Illustrated* (5th ed.), Wiley 2014, page 11.16

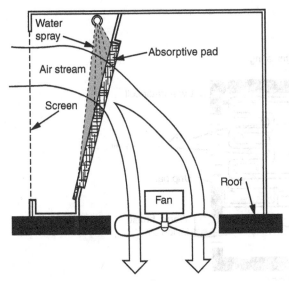

Figure 14.29 Evaporative cooler

TABLE 14.17 PACKAGED TERMINAL AIR CONDITIONERS (PTACS)

Equipment	Description
Unitary (unit) air conditioner (window air conditioner)	Individual apartments, motel rooms. If used only as needed, can save energy. Noisy. Minimize installation costs, portable, limited capacity, no heating. Exterior grille may admit ventilation air.
Through-the-wall unit	Each unit is essentially a compressive refrigeration machine. Condenser coil, compressor, and one noisy fan on outside of interior partition within equipment. Evaporator coil and fan to blow air over it on inside. Permanently mounted.
Heat pumps	Use electricity efficiently to heat and cool, reversing functions seasonally.

EVAPORATIVE COOLING

Moisture evaporated into dry air lowers its temperature by using up some sensible heat to evaporate water. This is balanced by an increase in latent heat energy so that the total energy content of the air mix remains unchanged.

An evaporative cooling system precools outside air before passing it through a space. (See Figure 14.29) The air first passes through a wet pad from which it evaporates water, increasing the air's water content (latent heat) and reducing its dry-bulb temperature (sensible heat). Evaporative cooling will work in most parts of the United States, as long as high humidity and rapid outside airflow rates can be tolerated. Evaporative cooling does not need a drain, but requires a continuous water supply.

When the outdoor air is at 105°F (41°C) and the relative humidity is a low 10 percent, evaporative cooling can produce indoor air at 78°F (26°C) and 50 percent relative humidity with only the power necessary to operate a fan. However, the fans that drive evaporative coolers are noisy, and the aroma of the wetted cooler may be unpleasant.

COOLING EQUIPMENT TYPES

Packaged terminal air conditioners (PTACs), also called incremental units, are self-contained units that are factory assembled and located in the space served. They usually have no ductwork. PTACs include window air conditioners, through-wall room units, and heat pumps. (See Table 14.17) PTACs are used in apartment buildings, hotels, motels, and some office buildings. They minimize HVAC space requirements and disruption for renovations.

Unit air conditioners are not as efficient as larger central units. (See Figure 14.30) Unit air conditioners are noisy, and due to high air velocity, can cause drafts. Sometimes the noise is welcome, as it can mask street noise. In moderate climates, air can be circulated either through cold-side or hot-side coils, using the unit as a heat pump to cool in hot weather and heat in cool weather. This does not work economically in very cold weather, when there is not enough heat outdoors.

EFFICIENCY RATINGS FOR COOLING EQUIPMENT

Energy efficiency rating systems for heating and cooling equipment were introduced earlier in this chapter. We are elaborating on ratings for cooling equipment here.

Room air conditioners measure energy efficiency with the energy efficiency ratio (EER), which is the ratio of the cooling capacity in BTUs per hour to the power input in watts. The higher the EER rating, the more efficient the air conditioner. US DOE standards effective June 2014 require room air conditioners to have an EER ranging from 9.0 to 11 or greater, depending on the type and capacity. The efficiency of air conditioning equipment is listed on an EnergyGuide label on the unit.

Central air conditioners circulate cool air through ductwork. They are rated according to their seasonal energy efficiency ratio (SEER), which indicates the relative amount of energy needed to provide a specific cooling output. The minimum SEER allowed (2012) is 13.

In addition to the EER and SEER, several other ratings apply, depending upon the size and type of equipment. The annual fuel utilization efficiency (AFUE) rating is a ratio of the annual fuel output energy to annual input energy. The coefficient of performance (COP) assesses the rate of heat removal for cooling equipment, and the efficiency of heat pump systems for heating. The integrated part load value (IPLV) expresses the efficiency of air conditioning and heat pump equipment.

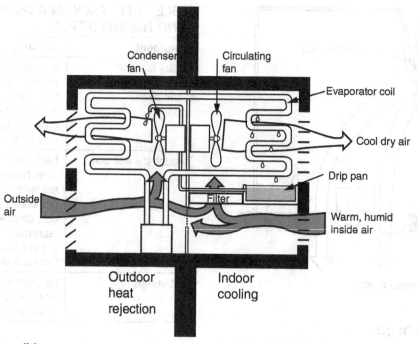

Condenser fan

Circulating fan

Evaporator coil

Cool dry air

Drip pan

Filter

Warm, humid inside air

Outside air

Outdoor heat rejection

Indoor cooling

Figure 14.30 Unit air conditioner

Source: Redrawn from Walter T. Grondzik and Alison G. Kwok, *Mechanical and Electrical Equipment for Buildings* (12th ed.), Wiley 2015, page 511

HEAT PUMPS

Both compression cycle and absorption cycle air conditioners have hot and cool sides. For cooling, air or water is circulated past the cold side. When the hot side is used as heat source, the equipment is called a **heat pump**. The heat pump uses a relatively small amount of energy to pump a large amount of heat from the colder side (the ground or outside air) to the warmer air inside the building.

A heat pump is an efficient way to heat with electricity. Earth-coupled heat pumps are more efficient than air-to-air ones.

Heat pumps use the compressive refrigeration cycle to absorb and transfer excess heat outdoors in hot weather. (See Figure 14.31) They also draw heat from outdoor air for heating by reversing the cooling cycle and switching the heat exchange functions of the condenser and evaporator. (See Figure 14.32) Heat pumps work best where the heating and cooling loads are almost the same.

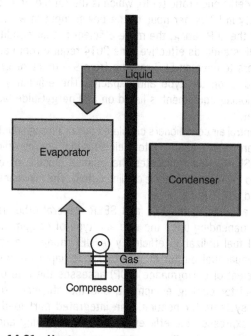

Liquid

Evaporator

Condenser

Gas

Compressor

Figure 14.31 Heat pump summer cooling

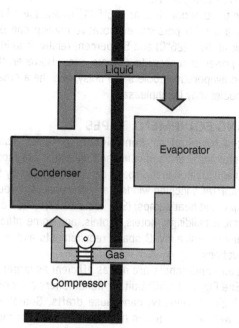

Liquid

Condenser

Evaporator

Gas

Compressor

Figure 14.32 Heat pump winter heating

Heat pumps are available in sizes from window units, to units to heat or cool a house, to units for larger buildings. A heat pump is often part of a total energy system that concentrates waste heat from an electric generating system to heat same building served by electric generator. Types of heat pumps include air-to-air, air-to-water, water source (hydronic), water-to-air, water-to-water, and ground source heat pumps.

Other types of cooling equipment include fan-coil units, chilled beams, radiant panels, cooling coils, central residential air conditioners, and chillers (See Table 14.18)

Cooling systems are often classified by the fluids used to transfer heat from habitable spaces to the refrigeration machine. Four major categories include direct refrigerant, all-air, all-water, and combination air-water. (See Table 14.19)

Most large multistory buildings use central systems on the roof or in the basement, with systems divided in many buildings. Very high buildings may have intermediate mechanical floors to minimize space lost to vertical air ducts.

TABLE 14.18   OTHER COOLING EQUIPMENT

Equipment	Description
Fan-coil unit (FCU)	Cabinet with heating/cooling coil, fan, air filter. Room air circulated through unit by fan. Vertical, horizontal, and stackable configurations.
Chilled beams	Manufactured device at ceiling (not a structural beam). Radiant cooling plus convective heat transfer; heating also possible. Passive, or active with induced airflow.
	Passive chilled beams: all-water terminals that exchange heat via radiation and convection.
	Active chilled beams: air-water terminals with convective transfer predominating.
Prefabricated metal radiant panels	Can cool and heat. Chilled water is circulated through tubing at back of the panel, absorbing excess room heat and carrying it away.
Cooling coils (direct expansion [DX] coils)	Provides cooling via room air passing across coil containing circulation refrigerant. Can add to warm air furnaces. Refrigeration cycle indoors, noisy compressor/condenser unit outdoors.
Central residential air conditioners	Cool the entire house with a large compressor unit outside.
Chiller (any refrigeration machine used to chill water)	In a chilled water system, entire cycle takes place in chiller. Prefabricated assembly contains compressors, condenser, and evaporator.

TABLE 14.19   COOLING SYSTEM CATEGORIES

Cooling System	Description
Direct refrigerant (Direct expansion or DX) systems	Simplest, most basic refrigeration machine plus two fans. Indoor air blown directly over evaporator coil. Small to medium-sized spaces that require their own separate mechanical units.
All-air systems	Air blown across cold evaporator coil, then delivered by ducts to rooms to be cooled. Ventilates, filters, dehumidifies air. Bulky ductwork, large fans.
All-water systems	Water chilled by evaporator coil, then delivered to fan-coil units in each space. Pumps less energy than fans. Ventilation by windows.
Combination air-water systems	Bulk of cooling by water and fan-coil units. Small air system completes cooling and ventilates, dehumidifies, and filters air. Air ducts can be small.

A well-designed air conditioning system must eliminate both the heat and humidity unintentionally leaking into the building or generated within it, and that introduced with air for ventilation. Engineers try to design air conditioning systems that are large enough to assure adequate comfort, but not so large that they cycle on and off too frequently, which would wear out the equipment faster. With some equipment, excessive cycling on and off results in decreased efficiency and more energy use.

HEATING, VENTILATING, AND AIR CONDITIONING (HVAC) SYSTEMS

A heating, ventilating, and air conditioning (HVAC) system integrates mechanical equipment designed to provide thermal comfort and air quality throughout a building. The difficulty of doing this is apparent when we consider that a building may be hot from the sun on one side, colder on the other, and warm in its interior, all at the same time on a winter day.

In the 1960s when energy costs were low, architects, engineers and building owners did not worry about how easily heat was transmitted through the building envelope. Dramatic architectural effects like all-glass buildings took precedence over energy conservation. Omitting roof and wall insulation minimized initial building costs. The HVAC system designer made the building comfortable by using as much mechanical equipment as necessary.

With increased fuel costs, energy has become one of the largest expenses in the building's operating budget. Architects and engineers are moving away from equipment-intensive mechanical systems and toward passive and hybrid system design. The earlier, energy-intensive systems are being modified or replaced.

Because of all these changes, we confine our investigation of HVAC systems to basic information.

There are four means that HVAC systems use to move heat to a conditioned space. The most common in the recent past has been ducted warm air with supply and return paths, ventilation, and air cleaning and mixing. The others are piped hot water, piped steam, and electricity. Cooling is delivered using chilled water or cool air.

All-air systems include single-duct constant-air-volume (CAV), single-zone, multizone, single-duct variable-air-volume (VAV), dual duct, and terminal reheat systems. (See Table 14.20) All-water systems are typically either two-pipe or four-pipe systems. (See Table 14.21)

See 2012 ASHRAE Handbook—HVAC Systems and Equipment for detailed information on water and steam systems.

TABLE 14.20   ALL-AIR HVAC SYSTEMS

System Type	Description
Single-duct, constant-air-volume (CAV) system	Constant temperature, air delivered at low velocity air to spaces
Single-zone system	Single thermostat regulates temperature for whole building
Multizone system	Separate ducts from central air handling unit serve each zone
Single-duct, variable-air-volume (VAV) system	Terminal outlets control airflow for each zone or space
Dual duct system	Separate ducts deliver warm and cold air to mixing boxes that blend it before distributing to each zone or space
Terminal reheat system	Supplies air to terminals that heat or cool temperature for each zone or space

TABLE 14.21   ALL-WATER HVAC SYSTEMS

System Type	Description
Two-pipe	One pipe supplies hot or chilled water to each fan-coil unit (FCU), another returns it to boiler or chilled water plant. FCU draws mixture of room air and outside air over heating or cooling coils, blows it back into space.
Four-pipe	Two separate piping circuits, one for hot water, one for chilled water, provide simultaneous heating and cooling to various zones.

An air-water system supplies primary conditioned air from a central plant to each zone or space. There it mixes with room air and is heated or cooled in induction units. The primary air draws room air through a filter and passes it over coils that are heated or chilled by water from a boiler or chilled-water plant. Local thermostats control the air temperature.

Packaged systems are self-contained, weatherproof systems powered by electricity or a combination of electricity and gas. They are mounted on roofs (where multiple units may serve long buildings) or on concrete pads along the exterior wall of a building. Connected to vertical shafts, packaged systems can serve buildings up to four or five stories in height. Split packaged systems combine an outdoor compressor and condenser, and indoor heating and cooling coils and fan.

All but local HVAC systems are field-assembled from large numbers of components from a variety of suppliers and manufacturers. Most are unique and resemble but do not replicate other systems. As a result, things can go wrong, so a commissioning process highly recommended.

HVAC Zones

The numbers and types of HVAC zones required for thermal comfort influences the selection of centralized versus local HVAC equipment systems. A zone placed away from the building envelope may not have the access to outdoor air required for a localized system. Space must be available for equipment within a local zone.

As seen in Chapter 12, large multipurpose buildings conventionally use a system of 16 zones. Each function (for example, apartments, offices, and stores) has five zones: north, east, south, and west sides, and a central core. Underground parking adds a sixteenth zone. Each of these zones may encompass more than one floor. Adding in scheduling considerations may increase the number of zones. If apartments need individual controls and have varied usage patterns, each apartment may become one zone. Some small commercial buildings have no interior zone.

HVAC System Components

Although interior designers do not design the HVAC system, they will have to deal with the space that components take up, their noise, their terminal outlets in occupied rooms, and the access space needed for repair and maintenance.

HVAC systems provide heating, cooling, and ventilation. (See Table 14.22) HVAC systems consist of three main parts: the equipment that generates the heating or cooling, the medium by which the heat or cooling is transported, and the devices by which it is delivered. For example, a building might use an oil-fired boiler to generate hot water. The water is the medium that carries the heat throughout the building, and the pipes and radiators are the delivery devices.

TABLE 14.22 BASIC HVAC SYSTEM TASKS

System	Intake and Exhaust	Movers, Converters, Processors	Distribution	Results
Heating	Intake fuel, combustion air	Boilers, furnaces, pumps	Pipes, ducts, electrical conduits	Warm air or surfaces. Air motion often controlled. May need humidity control.
	Exhaust heat and CO_2	Fans, filters, heat pumps	Diffusers, grilles, radiators, dampers, thermostats, valves	
Cooling	Intake air, water, fuel. Exhaust air, water vapor, heat, CO_2	Evaporative coolers, heat pumps, chillers, cooling towers, coils, pumps, fans, filters	Pipes, ducts, diffusers, grilles, radiators, thermostats, valves, dampers	Cool air or surfaces. Air motion usually controlled. Humidity control usually provided.
Ventilation	Air	Fans, filters	Ducts, diffusers, grilles, switches, dampers	Fresh air. Air motion control. Air quality control.

The comprehensive term air-handling unit (AHU) covers a number of separate pieces of equipment, which may be combined into an HVAC system. Types of air-handling units include unitary air-handling units, computer room units, and central air-handling units.

Large buildings often use combination systems. Water or steam is piped to an AHU, passed over a heat exchanger coil, and heat or cooling is transferred to air and ducted to conditioned spaces.

HVAC Distribution

Distribution of air in an HVAC system is typically accomplished through ductwork. Underfloor distribution is another option.

AIR SUPPLY DUCTWORK
The amount of resistance to airflow varies with the length of the duct run and the velocity of airflow. Higher velocity results in greater resistance.

Horizontal air distribution above corridors is very common, where the reduced headroom is not usually a problem. Circulation spaces are usually away from windows, so daylight is not affected. Corridors provide logical connections from one space to another, and are good paths for distribution trees. The change from lower service spaces to higher ceilings in office spaces enhances the openness of the higher spaces.

AIR SUPPLY PRESSURIZATION
In some spaces, like shopping malls, the corridors of apartment houses, and stair towers, more air is introduced into the space than is mechanically removed. These spaces are kept under positive pressure, so that air tends not to flow into them. This pressurization helps prevent unheated or uncooled outdoor air or smoke from a fire from entering these spaces. Higher air pressures also reduce discomfort from drafts and uneven temperatures from infiltration of air through the building envelope.

UNDERFLOOR AIR DISTRIBUTION (UFAD)
Underfloor air distribution (UFAD) with displacement ventilation provides fresh air cooled to just below the design room temperature, at a low velocity. Using the area under a raised floor as a plenum, high volumes of low-velocity air are distributed without ductwork. (See Figure 14.33)

Fresh air enters the underfloor distribution space, and rises upward as it picks up heat from occupants, office equipment, and lighting. Eventually, it stratifies at the ceiling. The provision of cooler, fresher air near the floor and warmer, staler air at the ceiling provides better IAQ and thermal comfort than ceiling supply and return systems. Individual control of floor-mounted registers increases comfort.

The raised floor is typically supported on a 2-foot (610 mm) square module. This is also often the location of power and data cabling outlets. The units allow multiple orientations and ease of reorganization.

Codes may restrict the height of the raised floor, and wiring is often required to have a special wear-resistant coating. A ceiling at least 9 feet (2.7 meters) high is necessary for proper air stratification; this height is also the minimum to accommodate daylighting design. Without a suspended ceiling, lighting, fire safety, and other building systems are exposed; some of these may be moved to under the floor above.

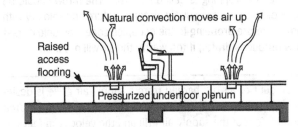

Figure 14.33 Underfloor air distribution

Source: Redrawn from Francis D.K. Ching and Corky Binggeli, *Interior Design Illustrated* (3rd ed.), Wiley 2012, page 230

TABLE 14.23 HVAC DELIVERY FUNCTIONS AND DEVICES

Function	Devices
Heating only	Natural convection devices, unit heaters, usually radiation devices
Heating, cooling, dehumidification, and air filtering	Fan coil units
Heating, cooling, dehumidification, and air filtering; conditioned ventilation	Unit ventilator, packaged terminal air conditioners, heat pumps, air handling systems that terminate at air delivery devices

ASHRAE's 2013 UFAD Guide: Design, Construction, and Operation of Underfloor Air Distribution (UFAD) Systems provides more information on this technology.

See Chapter 16, "Electrical Distribution," for information on underfloor systems for wiring distribution.

Terminal Delivery Devices

HVAC terminal devices deliver heating or cooling to a conditioned space. (See Table 14.23) Selection criteria include the heating, cooling, dehumidification, filtering, and ventilation functions required; their location in the room; the energy source; and the type of any distribution system to which the terminal device is connected.

AIR DELIVERY DEVICES

As air moves from a supply outlet, room air moves toward it. This can produce a narrow band of discoloration around the outlet on smooth ceilings; on textured ceilings, smudging occurs over a larger area. Beveled mounting frames and other special designs minimize smudge problems.

Supply outlet placement may affect the installation of lighting fixtures, cabinetry, or other components.

The horizontal distance that the air stream travels from an outlet before dropping is called its throw. The throw should be long enough to mix the supply air temperature and velocity with room air before dropping to the occupied zone. Too cold or fast a flow produces drafts; if too warm, the air will not drop.

The amount of air discharged from a supply air grille is designated in cubic feet per minute (cfm), based on the open area in square feet of the supply air grill and the velocity in feet per minute of air passing through it. The SI unit is cubic meters per minute (m³/min).

Workstation delivery systems are also known as personal comfort delivery systems. Systems mix outdoor air and recirculating indoor air together as the primary air brought by duct from a main duct or floor plenum to each workstation mixing box. Each worker can adjust the supply air temperature, mixture of primary and locally recirculating air, air velocity and direction, and radiant supplementary heat located below desk level. The system can dim task lighting and adjust the masking sound level. An occupant sensor shuts down the system when a workstation is unoccupied, maintaining minimum airflow.

WATER DELIVERY DEVICES

Delivery in water-to-air or all-water HVAC system involves introducing heating or cooling indirectly from water to air within the room. Most hot water systems use convectors rather than the cast iron radiators of the past. Today, convectors include fin tubes or fin coils working by natural convection. Unobtrusive baseboard, cabinet units under windows, or underfloor devices are used.

FAN COIL UNITS

A fan coil unit (FCU) is a factory-assembled unit with a heating and/or cooling coil, fan, and filter. (See Figure 14.34) A FCU is similar to an AHU but is a terminal device, serving a single room or a small group of rooms. Units are available as 1-, 2-. 3- or 4-pipe systems that offer varying levels of control of changeover from heating to cooling. An FCU with a cooling coil also dehumidifies.

Wall, ceiling, and vertical stacking models are available. Some designs are concealed in custom enclosures, semi-recessed into the wall, or installed as floor consoles with various cabinets. Recessed units are often found along corridors.

Ceiling models are available in cabinets for exposed locations, or without a cabinet for concealed mounting. They should not be mounted above solid ceilings, as their condensate drains are prone to clogging and the drain pans can overflow, requiring maintenance. One ceiling unit can be ducted to supply several adjacent small spaces.

Vertical stacking units are used in multiple floor apartment buildings, condominiums, office buildings and hotels. They eliminate the need for separate piping risers and runouts.

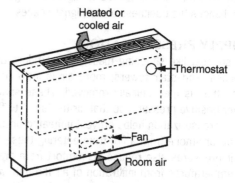

Figure 14.34 Fan-coil unit

FCUs provide a good solution for buildings with a large number of small, individually controlled, and variously occupied rooms, such as hotels, motels, apartment buildings, nursing homes, and medical centers. There is no mixing of air from the conditioned spaces. Each space requires operable windows or infiltration for ventilation air, or alternatively, a ducted ventilation system for the whole building.

FCUs must be maintained and serviced within occupied areas. Fans on older units tend to become noisy. A condensation drain line is required for each unit that provides cooling, and bacteria tend to grow in the drain pans. Small and inefficient filters require frequent changing to maintain proper airflow.

A fan-coil unit adjusted to supply a significant amount ventilation air, for example in a classroom, is typically called a unit ventilator and has a somewhat larger outside intake grill than a fan-coil unit. A unit ventilator offers individual temperature control in each room of a building by circulating chilled or heated water from central unit to each room.

Control Systems

Most HVAC systems are actuated and regulated by automatic controls. Most controls in larger buildings are computerized, and are usually set to maintain a range of conditions, rather than having a specific set point.

Control functions include maintaining the desired thermal comfort conditions, increasing energy efficiency by promoting optimum operation, avoiding human error, and promoting occupant satisfaction.

Safety devices are designed to limit or override mechanical or electrical equipment. Power sources are electrical, including analog and direct digital control, or self-contained including passive controls.

Building automation systems (BAS) are programmed to integrate the many aspects of building control into one decision-making unit. They are used in most large buildings today.

See Chapter 20, "Communications, Security, and Control Equipment," for more information on building automation systems.

THERMOSTATS
In a residence, a thermostat is usually located in or near the living room at a thermally stable location protected from cold drafts, direct sunlight, and nearby warm-air registers. Programmable thermostats learn your schedule, program themselves, and can be controlled from your phone. (See Figure 14.35) They note occupancy and automatically turn to an energy-efficient temperature when you are gone.

Thermostats also control the flow of water to radiators and convectors. The fans that circulate warm air use thermostats. The thermostat triggers a low-limit switch, which turns fans and pumps on for the heat distribution system when a preset low temperature is reached. An upper-limit

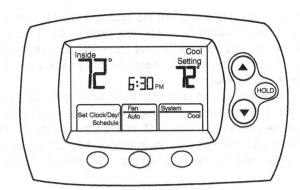

Figure 14.35 Programmable thermostat

switch shuts off the furnace when the specified temperature is reached. A safety switch prevents fuel from flowing to the heating plant if the pilot light or fuel-ignition system is not working.

The HVAC controls for small buildings are usually thermostats. The mechanical engineer determines the location of a thermostat based on the location of surrounding heat sources. To work properly, the thermostat must be mounted on an inside wall away from doors and windows, so that it will not be affected by the outside temperature or by drafts. Do not place lamps, appliances, TV sets, or heaters under the thermostat, as their heat will affect furnace operation. The interior designer should make it a point to know where the thermostat is to be located as it could end up in the middle of a featured wall, right where the designer had planned to hang a piece of art.

WIRELESS CONTROL SYSTEMS
Wireless control systems are becoming increasingly common. They eliminate the need for power and data wiring at hundreds of locations in the building, reducing the cost of the control system and making retrofit installations feasible.

Systems that provide remote wireless control and switching have design implications for architects and building engineers who must make provisions to provide for facilities to accommodate them.

Wireless residential building management systems regulate temperature and activate equipment in advance of the occupant's arrival, and provide real-time building condition status reports. They can control door and window locks, security cameras, lighting, and appliances.

Neural networks involve automation systems capable of learning from use, to predict usage patterns and anticipate needs while conserving energy. They can adjust operations in advance without needing specific commands from occupants. In retail and commercial spaces with highly predictable patterns, neural networks quickly learn to anticipate needs and conserve energy.

BUILDING COMMISSIONING
Building commissioning is a formal start-up and testing process that identifies and corrects operational deficiencies, saves

energy, and helps ensure that the building owner receives the intended performance from all building systems. It is a process for verifying and documenting that the performance of a building and its various systems meets the design intent and the owner's operational needs.

As engineers and architects move away from designing buildings that are equipment driven, interior spaces will reflect new energy-efficient priorities. Older buildings will gradually be refitted with new systems. Understanding how existing buildings work and how they can be made more resource-efficient is important to interior designers working as part of a building design team.

Our reliance on electricity has serious implications for environmental quality and resource conservation. Existing twentieth-century buildings were rarely designed for daylighting, and rely strongly on electrical lighting. This in turn adds heat to the building's interior and increases energy use for air conditioning. In Part VI, "Electrical and Lighting Systems," we examine alternatives to heavily wired, energy-intensive, and artificially illuminated buildings.

PART

VI

ELECTRICAL AND LIGHTING SYSTEMS

Our reliance on electricity has serious implications for environmental quality and resource conservation. Much of the energy produced from coal, petroleum, or nuclear sources ends up as electricity distributed for building use.

Twentieth-century buildings were rarely designed for daylighting, and relied strongly on electrical lighting. According to the U.S. Energy Information Administration (2012), lighting consumes about 17 percent of the energy used in residential and commercial buildings. This in turn adds heat to the building's interior and increases energy use for air conditioning.

> Historically, usable energy was most often produced by burning a fossil fuel such as coal or oil...Only since the end of the nineteenth century, however, has [the heat produced been] in turn used to create... electricity. Nuclear reactors, geothermal resources, and concentrating solar collectors may also be used to produce heat for electricity generation...It is well to remember that, in terms of consumption of fuel resources, electricity is an expensive form of energy because the efficiency of overall heat-to-electricity conversion, on a commercial scale, rarely exceeds 40%. (Walter T. Grondzik and Alison G. Kwok, *Mechanical and Electrical Equipment for Buildings* (12th ed.), Wiley 2015, page 1219)

Existing buildings contain miles of wiring, much of which is no longer in use. Wireless technologies and underfloor distribution are becoming the dominant ways to power equipment and send data. Interior designers specifying appliances and lighting fixtures need to have an understanding of the basics of electricity. The risk of fires caused by electrical failures is a safety issue. In Part VI of this book we will be looking at these issues and others.

Chapter 15, "Electrical System Basics," introduces electrical principles and looks at electrical power sources and energy conservation, as well as circuit design and safety issues.

Chapter 16, "Electrical Distribution," addresses electrical wiring in commercial buildings and residences, including electrical service, distribution, devices, and loads.

Chapter 17, "Lighting Systems," looks at light and vision and at daylighting, then probes the basics of electrical lighting design, light sources, and fixtures.

The effect of electrical systems and particularly of lighting on interior design is immense, and has undergone many changes in recent years. This part of *Building System for Interior Designers, Third Edition* seeks to explain the developing perspectives of architects and engineers and enlighten interior designers as to recent developments.

15

Electrical System Basics

Electricity is the most prevalent form of energy in a modern building. Electricity supplies electrical outlets and lighting fixtures. Ventilation, heating, and cooling equipment depend upon electrical energy. Electrical devices help to provide acoustic privacy and hearing. Elevators and material transporters, along with signal and communication equipment, rely on electricity for energy.

Interior designers are involved in making sure that electrical power is safe and available where the user needs it. Lighting design is critical to the function and ambience of an interior space. Understanding the principles involved in these tasks is important in an interior designer's work.

> The electrical system of a building supplies power for lighting, heating, and the operation of electrical equipment and appliances. This system must be installed according to the building and electrical codes in order to operate safely, reliably, and effectively. (Francis D.K. Ching, *Building Construction Illustrated* (5th ed.), Wiley 2014, page 11.30)

Before we get into the details of electrical distribution and devices, this chapter addresses what electricity is and how it works. Energy conservation and safety are also included in this chapter.

INTRODUCTION

Architects and electrical engineers design the electrical system of a building. They often call on equipment consultants and lighting designers for support. Their efforts have major impacts on the building's interior design.

Working with the building's architect and engineers, the interior designer is responsible for seeing that power is available where needed for the client's equipment, and for making sure that the lighting and appliances are safe, appropriate, and energy efficient.

History

In 1752, Benjamin Franklin famously concluded that lightning must be made of electricity, and invented the lightning rod to protect buildings. In 1792, Alessandro Volta invented the battery, producing electric current.

By the end of the nineteenth century, research by Andre Ampere (1776–1836), Georg Ohm (1780–1854), Heinrich Hertz (1857–1894), and others resulted in the conclusion that electricity is a stream (current) of electrons flowing from negative to positive charges.

Thomas Edison developed a functioning electric lamp in 1879. In 1882, Edison opened the first centralized electrical utility, the Pearl Street Station in New York, which provided **direct current (DC)** to nearby homes and shops. (See Figure 15.1) Although Edison's central generating station provided electrical energy for New York, widespread distribution remained a problem.

Inventor Nicola Tesla tried to convince Edison of the superiority of **alternating current (AC)** but failed to do so, and subsequently, sold his patents to George Westinghouse. This resulted in two competing electrical industries in the United States, Westinghouse (AC) and Edison (DC).

Up through the early twentieth century, nearly all large buildings and groups of buildings supplied their own on-site or local

Figure 15.1 Pearl Street generating station, 1882
Source: T.K. Derry and Trevor I. Williams, *A Short History of Technology*, Dover Publications, Inc., 1993, page 617

DC power. On-site generators supplied electricity for elevators, ventilators, call bells, fire alarms, and lighting. The low voltages produced by these local DC power plants lost too much voltage when distributed over long distances. Although AC power ultimately prevailed, DC systems persisted in some urban areas throughout the twentieth century.

AC power allowed electricity to be transmitted at high voltages over long distances. High-voltage transmission reduced power losses through the wires. The high voltage was transformed down to useable voltages at the point of use. By the 1920s, the construction of larger central power stations led to lower costs.

In 1978, the **Public Utilities Regulatory Policies Act (PURPA)** decreed that utilities must buy on-site generated electricity that small private power producers wanted to sell. Local energy generation systems could now be connected to the electric utility's grid. Techniques for generating and storing energy on-site are advancing, and we are seeing a significant increase in locally produced electrical energy.

Currently, large centralized electrical generating plants are usually powered by water or steam turbines. The steam is most often generated by coal, but also by oil, gas, or nuclear fuel. In electrical generation plants, large quantities of heat energy go up the chimney and into waterways rather than as energy to transmission lines. Further losses occur in the transmission lines to the user, so that the electrical energy we receive is only about one third of the initial energy available from the fuel.

Electrical System Design Process

Electrical engineers start the process of designing an electrical system by estimating the total building electrical power load. They then plan the spaces required for electrical equipment. The engineer and the utility determine the point at which the electrical service enters the building. The engineer looks at how all areas of the building will be used and the type and rating of the client's equipment. The electrical engineer gets the electrical ratings of all the equipment from the HVAC, plumbing, elevator, interior design, and kitchen consultants.

The electrical engineer is responsible for determining the location and estimated size of all required electrical equipment spaces. The architect must then reserve spaces for electrical equipment.

The electrical engineer, the architect, the interior designer, and the lighting designer design the lighting for the building. Often the lighting plan is separate from the layouts for receptacles, data, and signal and control systems. Underfloor, under-carpet, over-ceiling wiring, and overhead raceways are usually shown on their own plans. The engineer then prepares a lighting fixture layout.

For more information on lighting design see Chapter 17, "Lighting Systems."

All electrical apparatus is located on an electrical plan, including receptacles, switches, and motors. Data processing and signal apparatus is also located, as is communications equipment and fire and smoke detectors. Control wiring and building management system panels are also indicated.

Next, the engineer designs the circuiting for all lighting, electrical devices, and power equipment. The engineer prepares **riser diagrams** that show how wiring is run vertically, and designs the panels, switchboards, and service equipment.

Interior designers are also responsible for showing electrical system information on their drawings, usually on a power plan. (See Figure 15.2) The electrical engineer may use the interior design drawings to help design the electrical system. The interior design drawings often indicate all electrical outlets, switches, and lighting fixtures and their type. Large equipment and appliances should be indicated along with their electrical requirements. Communication system equipment, like public phones, phone outlets and related equipment, and data outlets are shown.

The interior designer should be familiar with the location and size of the electrical panels, and of building systems that affect the type of wiring used, such as plenum mechanical systems. He or she must know the locations of existing or planned receptacles outlets, switches, dedicated outlets, and **ground-fault circuit interrupters (GFCI)**. The interior design must coordinate lighting fixtures, appliances, equipment, and emergency electrical systems. The interior designer may need to coordinate the location of equipment rooms with other requirements, and should be aware of the presence of an uninterrupted power supply or standby power supply.

Codes and Standards

Electricity has the capacity to cause shocks and fires. Codes and standards are critical to using electricity safely.

CODES

The electrical system must be installed according to building and electrical codes in order to operate safely, reliably, and effectively. The National Electrical Code (NEC) defines fundamental

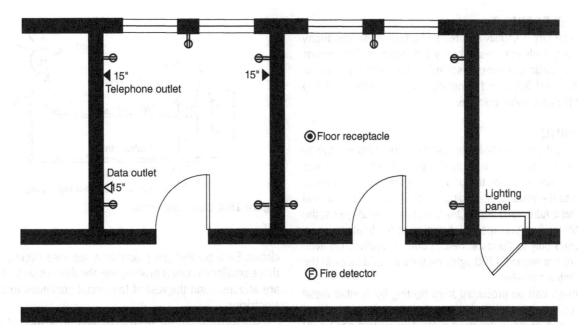

Figure 15.2 Sample interior design power plan

safety methods that must be followed in selection, construction, and installation of electrical equipment and systems. It is used by all inspectors, electrical designers, engineers, contractors, and operating personnel. The NEC has been incorporated into OSHA, and effectively has the force of law. The Canadian Electric Code (CEC) is very similar to the NEC.

Many large cities (including New York, Boston, and Washington DC) have their own electrical codes. They incorporate many NEC provisions with their own special requirements.

STANDARDS

Standards for electrical and communications systems are set by the American National Standards Institute (ANSI) and the National Electrical Manufacturers Association (NEMA). In addition, the standards of the utility supplying electric service apply.

UL establishes standards and tests and inspects electrical equipment. The organization publishes extensive lists of inspected and approved electrical equipment. UL listings are universally accepted, and many local codes only accept electrical materials bearing the UL label of approval.

Electrical material or equipment without a UL label should not be permitted on any project.

PRINCIPLES OF ELECTRICITY

Electricity is a form of energy that occurs naturally only in uncontrolled forms like lightning and other static electrical discharges, or in natural galvanic reactions such as those that cause corrosion. Until the end of the nineteenth century, electricity was viewed as somewhat beyond our comprehension. The existence of electrons was not discovered until 1897. (See Figure 15.3)

Today, we accept that **electrical current** involves a flow of electrons along a conductor. It is actually the flow of energy, with limited physical movement of the electrons themselves. In many materials, such as wood, plastic, glass, and ceramics, electrons stick tightly with their atoms. These materials are **insulators**, and conduct electricity poorly if at all.

Electrical current transmits electrical energy from one point to another. Electricity needs a **conductor** in order to move. The electrons in most metals can detach from their atoms; these are called free electrons. The loose electrons allow electricity to flow easily through these materials, making them good electrical conductors.

Types of Electricity

There are two types of electricity, static electricity and current electricity. Lightning is a type of static electricity that can damage buildings. Current electricity is productively used in buildings.

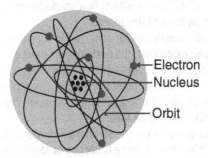

Figure 15.3 Electrons in atom

STATIC ELECTRICITY

Static electricity is usually created by friction. Static electricity creates very high voltages but very low currents. This means that, when static electricity discharges, the resulting spark is extremely short lived, and generally harmless unless the tiny spark ignites a combustible gas.

LIGHTNING

A lighting strike is a massive electric discharge between the atmosphere and an object on earth. Current induced by a lightning strike does not only take the path of least resistance, it takes all paths to the ground. According to the National Oceanic and Atmospheric Administration (NOAA), over the past 20 years, the United States has averaged 51 lightning strike fatalities annually, second only to floods in deaths due to weather. On average, there are around 4300 lightning fires in residences in the United States annually.

Buildings can be protected from lighting by pointed metal rods that connect directly with the earth through heavy electrical conductors. A properly designed protection system can be 99 percent effective. A good system will include surge arrestors at the main electrical panel and wherever telephone and other communication wiring enters the building. Protection of a building against lighting strikes should be done completely and properly, with UL-labeled equipment and a UL-approved installer.

Electrical Current

As indicated above, electrical current involves a flow of electrons along a conductor. **Voltage** (electromotive force or electrical pressure) is electrical potential. The electricity used in buildings consists of voltage and current maintained over time.

The function of an **electric circuit** is to deliver power (measured in **watts**). The power output of an electric circuit is a function of the electromotive force (in **volts**) and current (in **amperes**).

CIRCUITS

An electrical circuit is any closed path followed by an electrical current. Electrons flow along a closed path (a wire, for example) from a point with a negative charge to one with a positive charge. An electrical circuit is a complete conduction path that carries electrical current from a source of electricity to and through some electrical device (load) and back to the source. (See Figure 15.4) Current will not flow unless there is a complete (closed) circuit back to the source.

Electrical circuits can be arranged in a couple of different ways. In a **series circuit**, the parts of the circuit are connected one after another, and the resistances and voltages add up. The current is the same in all points of a series circuit. Failure of any one load, such as a burned-out lamp, will open the circuit and shut off power to all of its loads.

Parallel circuits are the standard arrangement in all building wiring. When two or more branches or loads in a circuit are connected between the same two points, the result is a parallel

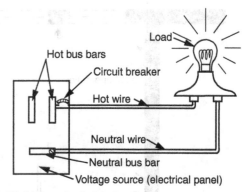

Figure 15.4 Electrical circuit

circuit. Each parallel group acts as a separate circuit. If one of these smaller circuits is broken, only the devices on that section are affected, and the rest of the circuit continues to circulate electricity.

Sometimes, due to worn insulation on a wire or another problem, an accidental connection is made between points on a circuit. This connection shortens the circuit and lets the electricity take a shortcut back to the source. This is called a **short circuit**. (See Figure 15.5) The electricity does not encounter the resistance that would be in the normal wiring, and the current rises instantly to a very high level. If the flow of electricity isn't stopped by a fuse or circuit breaker, the heat generated by the excessive current can start a fire.

Power in an electrical circuit is expressed in watts or kilowatts and time in hours, so units of energy are watt-hours (Wh) or kilowatt-hours (kWh).

AMPERAGE

Electricity flows at a constant speed, and moves virtually instantaneously. The process of electricity flowing along a circuit is called electrical current or **amperage**. It is measured in

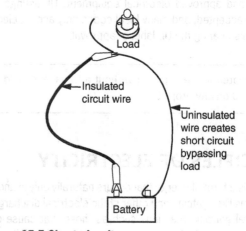

Figure 15.5 Short circuit

amperes (abbreviated **amps** or simply "A"), named after French mathematician and physicist André Ampère. The amount of current (number of amperes) determines the size of the wire needed for a particular use.

VOLTAGE

Electron movement and its energy—in other words, electricity—occurs when there is a higher positive electrical charge on a conductor than at another point on that same conductor. For example, in an ordinary battery, chemical action causes positive (+) charges to collect on the positive terminal and negatively charged (-) electrons to collect at the negative terminal. Even when the battery is not connected to any load, the electrified particles at the positive and negative terminals tend to flow; this tendency or force is called potential difference or voltage. When a conductor runs between the positive and negative terminals, the voltage between the terminals causes current to flow in the conductor. The more voltage in a system, the more current flows, the more electrons move along the conductor each second, and the more amps are measured in the circuit.

A unit of voltage is called a **volt (V)**, after Italian physicist Alessandro Volta (1745–1827). A volt is defined as a unit of electrical potential. It is a measure of electron movement and its associated energy caused by creating a higher electric charge at one point on a conductor than exists at another point on the same conductor.

To get an idea of how much power is in a volt, try building up a static charge by scuffing your feet on a wool carpet. You can generate about 400 volts by doing this, enough to make a visible spark jump between your finger and a metal object—or another person. A static shock has an extremely tiny current flow (amperage), so only a limited number of electrons are available to make the jump. Its effect is startling rather than harmful, despite the high voltage. However, the current flow available from the utility grid is almost unlimited, making our 120-volt household systems powerful and dangerous. Without insulation, the electrical current could easily melt all the wiring in your home.

RESISTANCE, CONDUCTORS, AND INSULATORS

Simply described, **electrical resistance** is a result of impurities in the conductor and disturbance of the structure of the conductor's electrons by heat. Electric current always flows through the path of least resistance. Electron flow through a material that has resistance generates heat. Higher voltage produces higher current for a given resistance.

Electrical resistance is measured in units called ohms. An **ohm** is equal to the resistance of a conductor in which a potential difference of one volt produces a current of one ampere. The ohm was named after German physicist George Simon Ohm (1787–1854).

Good conductors are materials in which there are a lot of electrons that are free to move around, so that there is not a lot of resistance to the electrons moving. Materials with low resistances are very useful as they conduct electricity more efficiently and lose less energy to heat. Metals generally have the least resistance to electrical current, and are good conductors. The best conductors are silver, gold, and platinum, with copper and aluminum only slightly inferior.

Insulators are materials that offer so much resistance that they virtually prevent the flow of any electricity at all, so they are used to contain electricity in its path. Glass, mica, and rubber are very good insulators, as are distilled water, porcelain, and certain synthetic materials. Rubber and plastic are used for wire coverings, and porcelain is used for electrical sockets.

WATTS

The watt is named after James Watt (1736–1819), the Scottish engineer who invented the modern steam engine. A **watt (W)** is defined as one ampere flowing under the electromotive force of one volt. It is used to measure how many electrons are passing a point, and how much force is available to move them.

In physics, **energy** is technically defined as the work that a physical system is capable of doing in changing from its actual state to a specified reference state. We define power as the ability to do work, or the rate at which energy is used in doing work. A watt is a unit of electric power that represents the rate at which energy is being used at any given moment, and 1000 watts equal one kilowatt (kW).

Power is energy used over time. Electrical power is expressed in watts or kilowatts, and time is expressed in hours, so units of energy are watt-hours or kilowatt-hours (kWh). One kWh equals one watt of power in use for 1000 hours. Utility meters measure electrical power use in kWh. The amount of energy used is directly proportional to the power of a system (the number of watts) and the length of time it is in operation (hours).

One Kilowatt (kW) = 1000 watts (W). 1 megawatt = 1,000,000 W. 1 gigawatt = 1,000,000,000 W.

DIRECT CURRENT AND ALTERNATING CURRENT

As indicated above, there are two types of electrical current. Direct current (DC) has a constant flow rate from a constant voltage source, like a battery in which one terminal (pole) is always positive and the other always negative. (See Figure 15.6) The flow is always in the same direction (**polarity**). Any current in which each wire is always of the same polarity, with one wire always positive and one always negative, is a direct current. Direct current is produced in batteries and photovoltaic equipment.

The principal DC application in buildings today is charging storage batteries for emergency power. Most electronic devices such as computers run on low-voltage DC, but have their own power supply that converts 120V AC line voltage to both low voltage and DC. The wide use of plug-in power supplies is creating interest in providing low-voltage DC outlets in buildings. This would eliminate the need for the power supplies and increase ease of use of power sources such as wind, solar, fuel cells, and batteries, all of which supply DC.

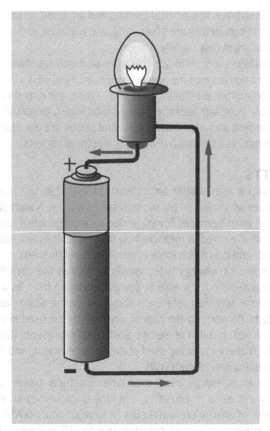

Figure 15.6 Direct current

With alternating current (AC), the voltage difference between the two points reverses in a regular manner. This means that the electrical current changes direction back and forth at a fixed electrical **frequency**. The change from positive to negative to positive again is called one **cycle**, and the speed with which the cycle occurs is the frequency of the current. Commercial power from utility companies in the United States and Canada is AC, typically supplied at 60 cycles per second, or 60 hertz (Hz). Many other countries supply commercial power at 50 Hz. The hertz was named after German atomic physicist Gustav Hertz (1887–1975).

Production of alternating current is more complex than direct current, involving an electrical **generator** with a metal loop that rotates, changing the magnitude and direction of the induced voltage (and current). One complete rotation of the loop produces one complete cycle in voltage and in the current.

Equipment made for one frequency is not compatible with any other frequency. Motors will not perform as desired at the wrong frequency, and may overheat, burn out, or have a shortened life.

In AC circuits, resistance is measured in ohms and is called impedance.

The advantage of AC over DC is the ease and efficiency with which the level of voltage can be changed by transformers. Generators put out currents at many thousands of volts. **Trans-**

Figure 15.7 Electrical transmission substation

formers at the generating plant further increase the voltage before the electricity is passed to the main transmission lines, to keep amperage at a minimum. When the amperage is kept low, large amounts of energy can be transmitted through small wires with minimum transmission energy losses.

The electricity passes through substations on its way to local transmission lines. (See Figure 15.7) Once the electrical energy has reached the local area, it is reduced in voltage at another transformer for distribution to buildings. The local lines have higher transmission losses per mile than the main lines, but are much shorter.

The voltage that reaches the building is still too high for consumer use, so each building or group of buildings has a small transformer to reduce the voltage still further before it enters the building. Electrical service for small buildings is provided at 230 or 240 volts. You may have seen black transformers on utility poles that reduce the voltage for small buildings. The voltage is again reduced to around 120 volts for household use. Some older homes have only 120-volt service. Near large cities, the supply may be 120/208 volts.

Large buildings and building complexes often buy electricity at the local line voltage and reduce it themselves with indoor transformers before use. The transformer steps down 4160V service to 480V for distribution within the building. A second transformer in an electrical closet steps 480V down to 120V for receptacle outlets.

The electricity within a home may not be exactly 120 and 240 volts. Typically, a city dweller might have 126 volts at an outlet, while a suburbanite may receive only 118 volts. Outlets at the far end of a branch circuit have lower voltages than those near the service entrance panel, but the wiring in a home should not vary by more than four volts. The minimum safe supply required to avoid damage to electrical equipment is 108 volts.

ELECTRICAL POWER SOURCES

According to the US Energy Information Administration (2013), coal is used to generate 39 percent and natural gas 27 percent of the electricity in the United States. (See Figure 15.8)

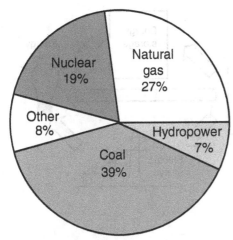

Figure 15.8 Percent of 2013 US electrical energy generation by source
Source for data: US Energy Information Administration, 2013

Regardless of the fuel source, most of the electric energy generated by steam turbines is lost as the steam condenses. There are additional distribution losses even with high voltage distribution.

IP units of power include horsepower, Btu per hour, watt, and kilowatt. SI units of power include joule per second, calorie per second, watt, and kilowatt.

Photovoltaic (PV) Power

Photovoltaic (PV) technology converts sunlight directly into electricity. It works any time the sun is shining, but more electricity will be produced the more intense the light and the more direct the angle of the light. Unlike solar energy systems for heating water or interior spaces, PV technology does not use the sun's heat to make electricity. Instead, it produces electricity directly from electrons freed by the interaction of sunlight with certain semiconductor materials in a **PV array**.

Light exhibits both characteristics of a wave and of a stream of energetic particles called **photons**. When a photon strikes a photoelectric metal surface, it dislodges a single electron from its normal orbit. Silicon exposed to an intense stream of photons, as in sunlight, dislodges a large number of electrons from their orbits. The electrons proceed to wander about the crystal lattice structure of the PV material's silicon crystals.

The photovoltaic electricity generated is direct current (DC), which is either stored in a battery system or converted to alternating current (AC) for use in commercial and residential buildings. For large electric utility or industrial applications, hundreds of solar arrays are interconnected to form a large utility-scale PV system.

Systems are not limited to sunny tropical areas. A solar electric system in Boston, Massachusetts, will produce over 90 percent of the energy generated by the same system in Miami, Florida. PV can use diffuse light quite well, and functions at around 80 percent on partly-cloudy days, 50 percent on hazy humid days, and 30 percent even on extremely overcast days.

PHOTOVOLTAIC CODES AND STANDARDS

The National Electrical Code's *NFPA 70, Article 690—Solar Photovoltaic (PV) Systems* sets standards for photovoltaic systems. If the system is connected to the electrical grid, the local utility will have additional interconnection requirements. In most locations, building and or electrical permits are required from city or county building departments to install a photovoltaic system. After the PV system is installed, it must be inspected and approved by the local permitting agency (usually the building or electrical inspector) and often by the electric utility as well.

PHOTOVOLTAIC HISTORY

The observation that many metals emit electrons when light shines on them is called the photoelectric effect. It was discovered in 1839 by Alexandre Edmond Becquerel (1820–1891). Johann Elster (1854–1920), and Hans Geitel (1855–1923) invented the first practical photoelectric cells.

In 1954, Bell Labs developed the first crystallizing silicon **photovoltaic cell**. It was put to use in 1958 to satisfy the US space program's need for an extremely light and reliable source of electricity for satellites.

In 2013, manufactured PV cells made of amorphous silicon ranged from 6 to 9 percent efficient. Newer methods of generating silicon crystals continue to increase their efficiency.

PV SYSTEMS

PV cells collect the sun's energy, which is then converted from DC to AC current. From there, energy is sent to storage (battery), put immediately to work, or transferred to an electric utility and sent out over the electrical grid. (See Figure 15.9) PV installations can be integrated with solar hot water collectors, without batteries or grid backup required.

Site-generated PV produces direct current. The DC power is then converted to AC power and tied to the central electrical energy grid. During periods of low supply, the energy grid provides backup energy. When extra production is available on-site, the electrical meter runs backward, effectively selling the extra energy as a credit to the grid. Systems can be paired with a generator or other back-up systems for emergency power.

Some utilities are augmenting central power plants with large centralized PV farms. Other utilities are setting up smaller PV fields near electrical users.

STAND-ALONE AND GRID-CONNECTED SYSTEMS

There are two basic types of PV systems for buildings: stand-alone and grid-connected. **Stand-alone PV systems** are not connected to the utility's electrical grid. They are used for

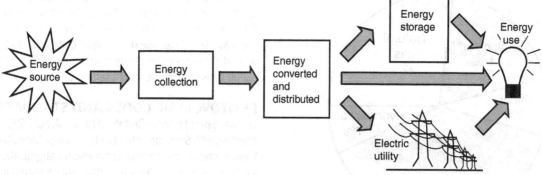

Figure 15.9 Photovoltaic system

remote or unattended loads, including isolated small residences. Most stand-alone systems use storage batteries to store excess energy from peak hours for use during cloudy days and at night.

Hybrid stand-alone systems add a fuel-powered generator for larger loads, especially when the peak load is periodic. Hybrid systems can be used with wind power, or employ the generator as backup for use a few times per year.

To connect a PV system to a utility grid, one or more PV modules are connected to an inverter that converts the modules' DC electricity to AC. **Grid-connected PV systems** require an inverter to change DC current from the PV array to AC at the correct voltage of the grid and feed it to the utility through the electric meter. Some systems include batteries to provide back-up power in case the utility suffers a power outage.

When the PV system generates more electricity than is needed at the site, excess energy can be fed directly onto electric lines for use by other electric customers on the utility's grid. Through a **net metering** agreement with the electric utility, PV system owners are credited for the excess power they produce.

PV CELLS AND MODULES

Photovoltaic (solar) cells are made from a very pure form of silicon, an abundant element in the earth's crust that is not very difficult to mine. (See Figure 15.10) PV cells provide direct electrical current. When enough heat or light strikes a cell connected to a circuit, the difference in voltage causes current to flow. No voltage difference is produced in the dark, so the cell only provides energy when the sun shines.

Individual PV cells are wired together to produce a **PV module**, the smallest PV component sold commercially. PV modules range in power output from about 10 watts to 300 watts. There are many module sizes, but they are rarely greater than 3 feet wide by 5 feet long (1 by 1.5 meters). Some modules can be designed directly into the roof and act as both a roofing material and an electricity generator.

Some modules are combined to form solar panels. (See Figure 15.11) Individual panels are usually mounted onto a roof. Panels can be combined into PV arrays.

PHOTOVOLTAIC ARRAYS

PV system arrays are complete connected sets of modules mounted and ready to deliver electricity. Building mounted arrays are stationary and usually consist of flat plates mounted at an angle. Tracking arrays follow the motion of the sun, providing more contact with the solar cells.

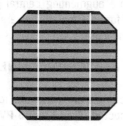

Figure 15.10 Photovoltaic cell

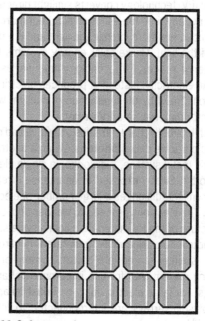

Figure 15.11 Solar panel

BUILDING INTEGRATED PV ELEMENTS

Building-integrated photovoltaic (BIPV) elements use thin-film solar cells made from amorphous silicon or nonsilicon materials such as cadmium telluride. Because of their flexibility, thin-film solar cells can be incorporated into the roof, walls, or windows of a building as a source of electrical power, often replacing conventional building materials.

BIPV elements are available in many sizes, finishes, and colors. Although most silicon cells are blue, many thin-film PV modules are dark brown, and gold, violet, and green cells are being developed. They can be round, semicircular, octagonal, square, or rectangular. Custom modules and panels can be produced for large projects.

A type of semitransparent PV modules can be used like tinted glazing. Opaque cells can be mounted on clear glazing, with the spacing of cells determining the ratio of clear to opaque. These modules are especially appropriate for clerestories or skylights, where view is not a factor.

NET METERING

Net metering is a policy of some public utilities that promotes investment in renewable energy-generating technologies by allowing customers to offset their consumption over a billing period when they generate electricity in excess of their demand.

As mentioned previously, the 1978 Public Utility Regulatory Policy Act (PURPA) requires that electric utilities buy electrical power from small suppliers at price equal to the costs they avoid by not having to produce that power. Most states have adopted net-metering laws that require the utility to buy power during peak PV generation periods at the same rate at which they sell power. The energy that the customer generates and uses is credited at the rate the utility would otherwise charge that customer. When the PV user buys from the utility, they pay at the conventional utility rate.

Net-metering benefits both the customer and the utility. Connection to the electrical grid removes the need for expensive battery installation to provide energy during periods of low PV production.

Other Electrical Energy Sources

Other sources of electrical energy include fuel cells, biopower, hydroelectric, and wind power. (See Figure 15.12 and Table 15.1)

ELECTRICITY AND SAFETY

The main electrical danger to a building results from fires caused by faulty equipment or wiring. Injury or death by electrical shock is also a danger.

Electrical Shocks

People can be killed or injured by electrical shock due to electrical current flowing through their body. The risk of electri-

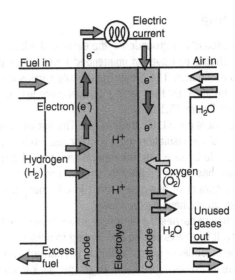

Figure 15.12 Fuel cell

cal shock is negligible either if voltage is low enough or if resistance is high enough. A person's resistance to electrical shock can vary greatly, depending on the layers of internal body tissue and skin. A wet floor conducts electricity much more readily than a dry floor in contact with rubber-soled shoes.

Touching a "hot" electrical device with one hand while touching a good ground (such as a water faucet) with the other can transmit a shock through your body. A situation that would be only unpleasant at 120 volts may be deadly at 240 volts. The magnitude of the injury also depends on the amount of time the current flows. Ground fault circuit interrupters (GFCIs) greatly reduce exposure time by opening a damaged circuit in milliseconds.

TABLE 15.1 OTHER ELECTRICAL ENERGY SOURCES

Source	Description
Fuel cells	Use electricity to extract hydrogen from water to make DC electrical energy. Can provide a compact, safe, efficient source of energy for electricity, heat and water purification. About 40% efficient for power production.
Biopower	Process of using plant and other organic matter (biomass) from agricultural or wood waste to generate electricity.
Hydroelectric	Energy produced from moving water. Large dams may cause environmental damage. Small hydroelectric generators can be used with running water.
Wind power	Large wind farms supply large amounts of power. Small residential windmill turbines are becoming more common.

Grounding

Electrical circuits are grounded in the event that a faulty circuit allows electricity to travel an unintended route rather than returning to its source. The grounding literally carries electricity into the ground rather than letting it travel other less desirable routes. (See Figure 15.13)

An electrical circuit has three wires. The hot wire, which is covered by black insulation (or any color but white, green or gray) runs side by side with the neutral and ground wires. The neutral wire has white or gray insulation. The ground wire is either bare copper or has green insulation. Homes built before 1960 often do not have a ground wire.

The hot wire carries the electrical power generated by your local utility. It is always poised and waiting to deliver its charge from inside an outlet or behind a switch, but current will not flow and release its power until it has a way to get back to its source to close the loop of the circuit. The neutral wire closes the loop. When you throw a switch to turn on an electric device, you are essentially connecting the hot and neutral wires together and creating a circuit for electricity to flow.

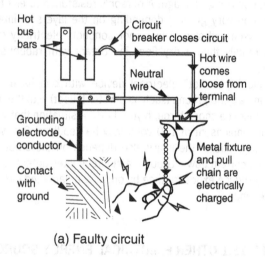

(a) Faulty circuit

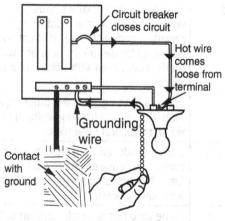

(b) Properly grounded circuit

Figure 15.13 Grounding

The hot wire immediately senses this path and releases its energy. If nothing impeded the current flow, most of that energy would go unused. An electrical device standing in the path between the hot and the neutral wire uses up virtually all the energy available in the hot wire, leaving little for the neutral wire to carry back to the source. This is why you are more likely to get a shock from touching the hot wire than the neutral one, even when current is flowing.

When you get a shock from a hot wire, your body can act like a neutral wire and complete the circuit to the ground you are standing on. This is because the earth itself is also an excellent path that leads back to the power source and closes the loop.

In fact, the electrical system uses the earth as an alternate path for safety purposes. The neutral wire is connected to the ground at the main service panel. From the main service panel, a wire goes to a copper-coated steel rod driven deeply into the earth next to the building. All building wiring is grounded, so that unless you are in contact with damp ground either by touching it directly or through wires, metal pipes, or damp concrete in contact with the soil, you will not get a shock.

Your body is not as good an electrical path as a wire; your skin thickness, muscle, and other body parts make you a poor path for electrical current. Even so, your body is very vulnerable to electrical shocks. This is because shocks kill by stopping your heart. A steady beating heart relies on tiny electrochemical nerve pulses that carry a current in the range of .001 amps. Even a charge as small as .006 amps can shatter the hearts micro-circuitry and disrupt its beating rhythm. Often the nerves can not stabilize quickly enough to restore the circuitry and save your life.

Fortunately, it takes a fairly high voltage to push a significant amount of current through us. Generally, you will not get a shock from circuits under 24 volts. Even within this range, however, a shock can disrupt the heartbeat of a person with a pacemaker.

The NEC has introduced three features that make electrical systems safer: the equipment ground, the ground fault circuit interrupter, and polarized plugs.

EQUIPMENT GROUND

An **equipment grounding conductor** is a bare or green-insulated third wire that does not ordinarily carry current. It is energized only momentarily when there is a fault between an ungrounded conductor and metal electrical equipment that could cause a shock. Grounded equipment has a plug with a third prong that goes into the half-round hole in an outlet.

Faulty equipment grounds are most likely to occur where vibration and other types of movement wear out a wire's insulation or break the wire itself. Old refrigerators and washing machines, which vibrate a lot, are typical culprits. So are lamps whose insulated cords harden as they age. When such leaks occur, a hot wire can be exposed or an entire metal appliance can be electrically charged. Such a fault could connect the metal case of the appliance with the electrical power circuit. If you touch the now electrified metal case and a ground, like a water pipe, you will get a nasty 120-volt shock. If your hands are wet when you make contact, the resulting shock could be fatal. Consequently, appliance manufacturers recommend that appliance cases be grounded to a cold water pipe and supplied with three-wire plugs. Two of the three wires connect to the appliance, and the third to the metal case.

The ground wire runs alongside the hot and neutral wires and is attached to the metal parts of electrical boxes, outlets, and electrical tools and appliances that could carry an electrical charge should a leak occur. The ground wire siphons off those leaks by providing a good path back to the main service panel, exactly like the neutral wire. Any leak that's picked up by the ground wire will probably blow a fuse or trip a circuit breaker and shut the circuit down, signaling that you have a serious problem somewhere in the system.

To accept the three-prong plugs that accommodate the ground wire and to provide a safe ground path, the NEC requires that all receptacles be of the grounding type, and that all wiring systems provide a ground path separate and distinct from the neutral conductor. Electrical codes require that each 120-volt circuit have a system of grounding. This prevents shocks from contacts where electricity and conductive materials come together, including parts of the electrical system like metal switches, junction and outlet boxes, and metal faceplates.

Where wiring travels through the building inside armored cable, metal conduit, or flexible metal conduit, the conductive metal enclosure forms the grounding system. When a metal enclosure is not used, a separate grounding wire must run with the circuit wires. Nonmetallic or flexible metallic wiring (Romex or BX) are required to have a separate grounding conductor. Nonmetallic cable already has a bare grounding wire within it. Insulated grounding conductors must have a green covering.

Cable and conduit are covered more fully in Chapter 16, "Electrical Distribution."

Replacing old two-slot outlets with the now-standard three-slot, grounded types will allow a three-prong cord to be plugged in anywhere may seem like a good solution. However, since old two-wire systems do not have an equipment ground, the ground prong on the plug does not really go to ground. Installing a proper ground wire for these outlets is time consuming and costly, but if it's not done, you have created the illusion of a grounded outlet that is not really safe.

Another way around the problem is the three-prong/two-prong adapter, more popularly known as a cheater plug. The NEC accepts this device provided you insert the screw that attaches the cover plate through the equipment ground tab. This screw connects to the metal yoke, which in turn connects to the metal electrical box. However, unless the metal electrical box has been grounded to earth, this again only creates a false impression of a safe, grounded system. This false sense of security puts you one step closer to receiving a dangerous or even fatal shock.

Electrical Fire Safety

The National Electrical Code (NEC) of the National Fire Protection Association (NFPA) defines fundamental safety measures that must be followed in the selection, construction, and installation of all electrical equipment. All inspectors, electrical designers, engineers, contractors, and operating personnel use the NEC. The NEC is incorporated into OSHA, the Occupational Safety and Health Act, and has the force of law.

UL establishes standards and tests and inspects electrical equipment. UL publishes lists of inspected and approved electrical equipment. Many local codes state that only electrical materials bearing the UL label of approval are acceptable.

INSPECTIONS

An electrical permit is usually required when doing electrical work. It ensures that the work is reviewed with the local building inspector in respect to national and local code requirements.

Inspections are made to determine whether design, material, and installation techniques meet code requirements. Inspections are required after raceways have been installed (roughing-in) and before closing-in of walls, and after the entire job is complete. Commissioning of electrical systems to Owner's Project Requirements is recommended.

Circuit Protection

Because the amperage available from the utility grid is almost unlimited, the electrical current in a 120-volt household system could easily melt all the wiring in your home. Special devices that limit current are located in the main service panel. If you open up the door of your electrical panel, you will find either **fuses** or **circuit breakers** (and sometimes both), each rated to withstand a certain amount of current, usually 15 amps. If the current exceeds the listed amount, the fuse will burn out (blow) or the breaker will trip, shutting off the current and protecting the wiring system from an overload. When this happens, it is a signal that you are trying to draw too much power through the wires.

Electrical service equipment such as the main service panel is described in Chapter 16, "Electrical Distribution."

FUSES AND CIRCUIT BREAKERS

Overloaded and short-circuited currents can result in overheating and fires. Circuit protective devices include fuses and circuit breakers that protect insulation, wiring, switches, and other equipment from these dangers by providing an automatic way to open the circuit and break the flow of electricity. Overcurrent protective devices are designed to disconnect a circuit automatically whenever the circuit reaches a predetermined value that would cause a dangerous temperature in a conductor due to a short circuit, excess current draw from too much load, or

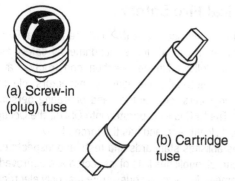

(a) Screw-in (plug) fuse

(b) Cartridge fuse

Figure 15.14 Fuses

a sudden surge in the power supply. Fuses and circuit breakers protect against this possibility by cutting off power to any circuit that is drawing excessive power.

The key element in a **fuse** is a strip of metal with a low melting point. When too much current flows the strip melts, or blows, thereby interrupting power in the circuit. When the fusible strip of metal is installed in an insulated fiber tube, it is called a cartridge fuse. When encased in a porcelain cup, it is a plug fuse. (See Figure 15.14)

It is dangerous to replace a fuse with a higher-rated fuse or a solid conducting metal piece.

A **circuit breaker** is an electromechanical device that performs the same protective function as a fuse. (See Figure 15.15) It acts as a switch to protect and disconnect a circuit. A strip made of two different metals in the circuit breaker becomes a link in the circuit. Heat from an excessive current bends the metal strip, as the two metals expand at different rates. This trips a release that breaks the circuit. They frequently have solid-state electronic tripping control units that provide adjustable overload, short-circuit, and ground fault protection.

Circuit breakers can be reset after each use, and can be used to manually switch the circuit off for maintenance work. Circuit breakers are easily installed as needed for various circuits in the building.

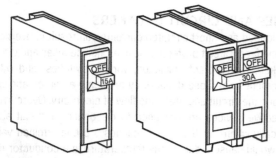

Figure 15.15 Circuit breakers

Circuit breakers are built to withstand momentary power surges, but standard fuses are not. A time-delay or slow-blow fuse can help cope with brief surge demands when a circuit frequently blows a fuse, for example when an appliance such as a refrigerator or room air conditioner is turned on. Both plug and cartridge fuses are available in slow-blow designs that safely allow temporary overloads. Whether a fuse or a circuit breaker is the better choice depends on the application and other technical considerations.

GROUND FAULT CIRCUIT INTERRUPTERS

A ground-fault circuit interrupter (GFCI) protects a person from a potentially dangerous shock. A GFCI is required in areas where any leaking electricity would cause extreme hazard, such as where it is possible to make contact with an electric ground in a bathroom, kitchen, or laundry, or where a standing surface is connected to the ground, as in a garage, basement, or outdoors. A GFCI has a quicker and more sensitive response than a fuse or circuit beaker on a circuit. GFCI devices can be part of a circuit breaker or installed as a separate outlet. A GFCI can be located within the electrical device itself, the receptacle to which it is connected, a receptacle from which the connected receptacle derives its current, or as part of the circuit breaker.

If you leave a hair dryer with a frayed cord in a little spilled water that is in contact with a sink's metal faucet, you risk accidentally touching an exposed hot wire in the cord while at the same time turning off the faucet with your other hand. Even though the dryer is turned off, an electric current can immediately flow from the cord, through your body, through the plumbing system, and eventually to ground. It will not cause the circuit breakers or fuses in the main service panel to break the circuit, and the current will continue to flow through your body. A GFCI instantaneously senses misdirected electrical current and reacts within one-fortieth of a second to shut off the circuit before a lethal dose of electricity escapes.

Another function of GFCIs is to detect small **ground faults** (current leaks) and to disconnect the power to the circuit or appliance. The current required to trip a circuit breaker is high, so small leaks of current can continue unnoticed until the danger of shock or fire is imminent.

GFCIs permit the easy identification of ground faults. If the GFCI senses any leakage of current from the circuit, it will disconnect the circuit instantly and completely. The GFCI does this by precisely comparing the current flowing in the hot and neutral legs of the circuit. If the amount of current is different, it means that some current is leaking out of the circuit.

GFCIS do wear out from use, and need to be tested regularly and replaced as necessary. It is recommended that you test your GFCIs every week and replace them immediately if they are not working properly. To make sure that GFCIs are working, manufacturers added the "test" and "reset" buttons that you see on them. (See Figure 15.16) Pushing the test button creates a small electrical fault, which the GFCI should sense and immediately react to by shutting off the circuit. The reset button restores the circuit. Repeated action by the GFCI to protect a leaking circuit will eventually wear out the GFCI.

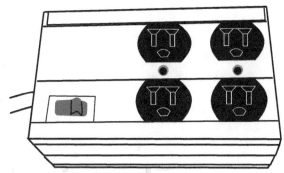

Figure 15.17 Surge protector

Figure 15.16 Ground fault circuit interrupter

The NEC requires GFCIs in specific locations, including outdoors and in kitchens, where GFCIs are recommended for all appliance circuits. The NEC requires a GFCI in all bathroom receptacles. Most codes also require a minimum of one GFCI installed within 36" (914 mm) of the outside edge of lavatory, located on a wall or partition adjacent to a lavatory basin, on a countertop (but not face up), or installed on the side or face of a basin cabinet not more than 12" (305 mm) below the countertop. Verify requirements and placement with local codes.

ELECTRONIC EQUIPMENT PROTECTION
The sudden power increases called surges that momentarily disrupt a building's steady power flow can destroy many of our electrical appliances and other devices. Electrical power can jump from its normal 120V up as high as 500V. Fortunately, most such surges are small and do not cause much damage, ex-

cept for the massive surges caused by a direct lightning strike, and the best protection from that is to unplug your equipment.

Microprocessors found in computers and an increasing number of devices and home appliances are sensitive to power surges. Each microprocessor's built-in power supply converts 120 volts of electricity to about 5 volts. Small changes in power, even a split-second surge, can scramble the microprocessor's electrical signals. A surge that slips past the power supply can destroy delicate chips and burn out circuits.

All computer installations, even the smallest home office, need to be protected from line transients with a **surge suppressor**. (See Figure 15.17) Multitap plug-in strips with built-in surge suppressors are typically inadequate unless they meet specifications for surge current, clamping voltage, and surge-energy suitable for the particular installation. Major data processing installations require additional types of treatment including voltage regulators, electrical noise isolation, filtering, and suppression, and surge suppressors. Surge stations are large surge protectors that offer better voltage protection and power conditioning. An ordinary **uninterruptable power supply (UPS)** will provide a high level of protection, but you should still use a surge protector. A continuous UPS will give you a few minutes to save your work and shut down your computer.

We have now explored the basics of electrical systems. In Chapter 16, we look at how electricity is distributed through a building.

16

Electrical Distribution

The electrical distribution system supplies power for lighting, heating, and operation of electrical equipment and appliances to a building. The design of the electrical distribution system has implications for both the architect and interior designer. The location of electrical equipment within the space affects both its function and appearance.

> Even though the electrical system of a building is compact, flexible, mostly hidden, and small in actual size, it is still a concern for the architect. Besides allocating space for the electrical equipment, the architect must be aware of situations where the electrical system has aesthetic as well as functional impact.
>
> Although wires and electrical boxes are small, they can be unsightly in open spaces and in situations where the structure is exposed unless the architect considers them in his or her design. It is especially important to address the difficult problem of providing power and communication outlets in open-plan spaces. As always, difficulties are easier to address when anticipated, and problems are best avoided by thoughtful design. (Norbert Lechner, *Plumbing, Electricity, Acoustics: Sustainable Design Methods for Architecture* (3rd ed.), Wiley 2012, page 28)

INTRODUCTION

There are two separate electrical systems in most buildings. The **electrical power system** distributes electrical energy through the building. The **electrical signal or communication system** transmits information via telephone, cable TV wires, or other separate data lines.

The electrical signal or communication system is addressed in Chapter 20, "Communications, Security, and Control Equipment."

Components of building electrical systems include service entrance equipment (transformers, service disconnect, fuses, circuit breakers, and meters), interior distribution equipment (conductors and raceways), and loads (lighting, motors, and miscellaneous outlet devices).

Electrical System Design Procedure

Electrical engineers start by making an electrical load estimate. With the local utility, they decide on the location and type of service entry equipment.

The electrical engineer, architect, interior designer, and other consultants work with the client to determine proposed electrical power usage and information about client-furnished equipment. Early on, the engineers determine locations and sizes of electrical equipment spaces so that the architect can reserve appropriate spaces.

Lighting design and daylighting are parts of this design process, and are addressed in Chapter 17, "Lighting Systems."

ELECTRICAL DRAWINGS

The electrical system designer locates all electrical apparatus and equipment on drawings. Data-processing and communications apparatus are also indicated on plans. The process involves making decisions for control wiring and equipment. The engineers design circuits for all lighting devices and power equipment, compute panel loads, and prepare riser diagrams. They also check their work and coordinate the electrical work with other trades and the architectural plans.

Because the location of outlets and switches are dependent on the layout of furniture and intended use of the room, they are often shown on the interior design drawings. Power requirements and locations for special built-in equipment are also usually indicated there. The information the interior designer supplies is integrated into the electrical engineer's drawings. (See Figure 16.1)

The circuits for signal outlets for fire alarm, telephone and intercom, data and communication, radio and TV, and other equipment are not usually indicated on the floor plan, but on a separate power plan. Lighting fixture outlets are usually included with wiring devices, unless this leads to a cluttered drawing, in which case they are represented on their own drawing. Underfloor, under-carpet, and over-ceiling wiring and raceway systems are usually shown on a separate plan. Motors, heaters, and other fixed, permanently installed equipment are shown and identified on power plans. Equipment with a cord and plug is not usually represented, but receptacles for plug-in equipment are shown and identified.

ELECTRICAL SERVICE EQUIPMENT

Lines carrying electricity from the power company run from a transformer through the meter and into the **main service panel**. In smaller buildings, service is usually provided at 230V or 240V. Most homes have three-wire service, with two hot conductors each supplying 115V or 120V, and one neutral conductor. The actual voltage supplied can vary between 115V and 125V within a given day.

In order to provide the amount of energy required at the location desired, and to do this safely, electrical conductors are isolated from the structure of the building except at the specific points, such as wall receptacles, where you want contact. This is accomplished by insulating the conductors and putting them in protective **raceways**.

The National Electrical Code (NEC) sets minimum standards for electrical design for construction. Minimum electric service for a single-family residence is set at 100-amp, 120/240V, single phase, 3-wire service.

The quality of the installation is the responsibility of the contractor. The designer should be wary of equipment substitutions by the contractor, whose bid was submitted on the basis of the plans and specifications. The contractor should be required to supply the equipment that is specified.

Electrical Power Distribution Systems

Once the building is connected to the electrical power source, you need to locate the places where you want the electricity to be available and provide a way to turn it on and off safely. During the design of a building, the electrical engineer or electrical contractor will design the circuiting and wiring. The interior designer should be familiar with the basic principles of power supply and distribution in order to be able to coordinate interior design issues with the rest of the design team. The interior designer will also want to have a say in the appearance of cover plates and other visible electrical devices.

OVERHEAD AND UNDERGROUND SERVICE

The electrical power service from the utility line may come into the building either overhead or underground. The length of the service run and type of terrain as well as installation costs affect the decision of which to use. Service voltage requirements and the size and nature of the electrical load also influence the choice. Other considerations include the importance of appearance, local practices and ordinances, maintenance and reliability criteria, weather conditions, and whether some type of inter-building distribution is required.

In a large building, electrical power from the utility line is stepped down to building use levels by transformers. Controls and protection devices are installed at the main building **switchboard**. From there, power goes directly to large equipment and also to distribution panels and individual lighting and appliance panels. It is at this point that branch circuit wiring carries electrical power to its end uses. (See Figure 16.2)

SERVICE ENTRANCE

Service equipment connects a building to the utility's electrical service. The main service panel is usually located where the power line enters the building. In a residence, the service panel

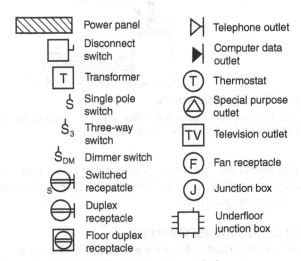

Power panel	Telephone outlet
Disconnect switch	Computer data outlet
Transformer	Thermostat
Single pole switch	Special purpose outlet
Three-way switch	Television outlet
Dimmer switch	Fan receptacle
Switched receptacle	Junction box
Duplex receptacle	Underfloor junction box
Floor duplex receptacle	

Figure 16.1 Typical electrical plan symbols

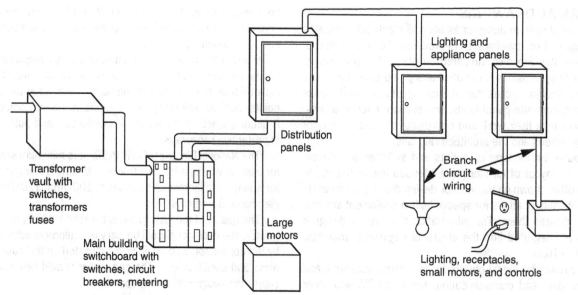

Figure 16.2 Electrical power distribution system

is usually located in the basement or in a utility room. In larger buildings, it is usually mounted on a main switchboard near the entrance of the service conductors. It is the origin of the network of wires carrying electrical current through a building.

Three wires run from the transformer to the building in small buildings. One is neutral, with no electrical potential existing between this wire and the ground. The neutral wire is connected to one or more long copper-covered steel grounding rods driven into the soil near point where the wire enters building. The other two wires are hot. They have a potential of 230V between them, but only 115V between either one of them and the neutral wire.

The hot wires go through an **electric meter** before entering the building, which measures consumption in kilowatt-hours (kWh). The three wires then enter the main service panel inside building.

The **main disconnect (service) switch** on the main service panel disconnects normal service to the building. It must be in a readily accessible spot near where service enters building, and access to it must not be blocked. (See Figure 16.3)

The neutral wire is connected to the steel box housing the panel and to the copper or aluminum bar to which all building circuits are grounded. Each of the hot wires (black and red) is connected to a copper or aluminum bar fitted with connectors for attaching circuit breakers. Breakers can be easily installed as needed to connect various circuits in building. Each breaker serves as an on/off switch for circuit maintenance and protection.

ELECTRICAL EQUIPMENT RATINGS

The **rated voltage** of an electrical equipment device is the maximum voltage that can safely be applied to a unit continuously. It frequently, but not always, corresponds to the voltage applied in normal use. For example, an ordinary wall receptacle is normally rated at 250V maximum, but only 120V is applied

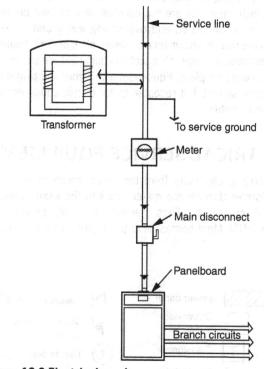

Figure 16.3 Electrical service

Source: Redrawn from Francis D.K. Ching and Corky Binggeli, *Interior Design Illustrated* (3rd ed.), Wiley, 2012, page 238

to it in normal use. The rating is determined by the type and quantity of insulation used and the physical spacing between electrically energized parts.

BUILDING TRANSFORMERS

As we have described earlier, a transformer is a device that transforms alternating current (AC) of one voltage to AC of another voltage. A transformer is used when the building voltage is different from the service voltage. Step-down transformers lower voltage, and step-up transformers do the opposite.

A unit substation (transformer load center) is a single package combining an assembly of transformer-related equipment that combines a step-down transformer with complete switchboard and meter(s). It transforms and distributes electrical energy to a lower voltage that can be utilized in facility.

METERING

A watt-hour electric meter measures and records the quantity of electric power consumed over time. (See Figure 16.4) It is supplied by the utility and placed ahead of the main disconnect switch so that it cannot be disconnected. Even with remote readers, the meter must be available for inspection and service.

In single-occupancy buildings or where the landlord pays for electrical service, there is one meter. For multitenant buildings, banks of meters are installed so that each unit is metered separately. A single meter is not allowed in new multiple dwelling constructions by federal law, as tenants tend to waste energy when they do not have to pay for it directly.

Submetering shows how energy is used for different types of service or different tenants or customers. Submetering is desirable in apartment houses and rental office buildings, where individual accountability of energy usage encourages conservation efforts.

Smart meters can reduce peak electrical demands. They allow the utility to signal individual buildings to turn off nonessential appliances such as hot water heaters during peak electrical demand.

SWITCHBOARDS AND SWITCHGEAR

A switchboard is a large, free-standing assembly of switches and fuses and/or circuit breakers that distributes the electricity from the utility service connection to the rest of the building. It distributes bulk power into smaller packages and provides overcurrent protection for that process.

Some switchboards are referred to as switchgear. There is no clear distinction between the terms switchboard and switchgear. Generally a switchboard has lower voltage, with large circuit breakers and high-voltage (above 600V) referred to as switchgear.

Electricity is measured, controlled, and distributed in an electrical room by switchgear, then sent by feeder cables to lighting and power panels placed throughout the building. Panels may be located in electrical closets, public spaces such as corridors, or utility rooms.

PANELBOARDS

A **panelboard** is an electrical panel, smaller scale than switchboard but with the same function. It accepts a relatively large block of power and distributes it in smaller blocks. A panelboard comprises main buses to which are connected circuit breakers or fuses that feed smaller branch circuits. Small electrical panels, especially in residential work, are often referred to as **load centers**.

Panelboard components are mounted inside an open metal cabinet called a backbox, which has prefabricated knockouts at its top, bottom, and sides for conduits carrying circuit conductors. The backbox is closed with a protective front panel with an access door. (See Figure 16.5) Each fused switch or circuit

Figure 16.4 Watt/hour meter

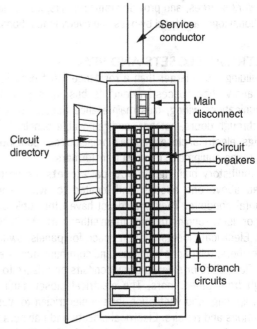

Figure 16.5 Electrical panelboard

Source: Redrawn from Francis D.K. Ching, *Building Construction Illustrated* (5th ed.), Wiley 2014, page 11.33

breaker is connected to busbars of the panelboard and feeds outgoing branch circuit.

Panelboards must be surface-mounted on solid walls or structural columns, or recessed flush with a finished wall surface. Lighting panels are often flush-mounted in finished areas such as corridors. Lighting panels in two-story-high commercial spaces may be vertically stacked and recessed into a corridor wall. Buildings six and more stories high use an electrical closet to accommodate panel and riser conduits.

In residences, the service equipment and the building's panelboard are combined in one unit. The panelboard is often located in the garage, a utility room, or the basement, as close to major electrical loads as feasible. Sometimes, an additional subpanel is added near the kitchen and laundry. In apartments, an electrical panel may be located in the kitchen or in an adjacent corridor, where it can be used as the code-required means to disconnect most fixed appliances. In small commercial buildings, it may be recessed into a corridor wall.

The NEC specifies the minimum working space required in front of electrical equipment.

Panelboards are often labeled to identify the circuit or equipment served. Labels also may indicate area served, floor number, and abbreviation for a zone.

Intelligent panelboards are compact, centralized programmable microprocessors that provide electrical load control and switching functions directly within the panelboard, eliminating external devices and the associated wiring. The intelligent panelboard can also accept signal data from individual remote or network sources, and provide status reports, alarm signals, operational logs, and local by-pass and override functions.

ELECTRICAL CLOSETS AND SPACES

Any building much larger than a residence must provide horizontal and vertical spaces for conduits, bus ducts, panels, and communications wiring, plus maintenance access for electricians through doors, hatches, and removable panels. In large buildings, electrical closets contain branch circuit panelboards, submetering equipment, or small transformers.

In multistory buildings, **electrical closets** are vertically stacked above one another and located to avoid blocking horizontal conduits. They should not have other utilities like piping or ducts running through them either horizontally or vertically. Electrical closets contain space for panels, switches, transformers, telephone cabinets, and communications equipment. Floor slots or sleeves allow conduits and risers to pass through from other floors. The electrical closet must have space, lighting, and ventilation for the electrician to work on installations and repairs. Electrical closets and cabinets must be fire-rated, as they are common places for fires to start, and they should not be located next to stairwells or other main means of egress.

The electrical engineer is responsible for locating electrical closets, and their location will have implications for the interior designer's space plan. Before asking an engineer to move an electrical closet, consider its vertical and horizontal location requirements, with locations on outside walls or adjoining shafts, columns, and stairs being poor choices. Sufficient wall space for equipment is required, plus coordination with underfloor and over-the-ceiling raceway connections.

An electrical closet with a panelboard needs at least 4 feet (1.2 m) clearance in front of the panel.

ENERGY CONSERVATION CONSIDERATIONS

The selection of materials and equipment for a building's electrical distribution system involves considerations that affect energy conservation. After setting the building's energy budget, the engineers can choose to set an energy reduction goal of 10 to 20 percent.

Energy conservation methods for electrical distribution systems include providing electric load control (demand control) equipment and individual user metering in multitenant residential buildings. Using the highest available service voltages can lower line losses and result in smaller panelboards at the branch circuit level. Providing metering points throughout the system aids accurate analysis of energy use.

Other options for energy conservation include systems that conform to an ideal energy use curve automatically. **Energy management control systems (EMCS)** have microprocessor-based controllers that can be preprogrammed to automatically reschedule or disconnect electrical loads for demand management.

Engineers need a preliminary load estimate for planning transformer rooms, conduit chases, and electrical closets. A load-control analysis can then be performed, which affects the maximum demand. This is done with or followed by a building energy consumption analysis of electrical load types. (See Table 16.1)

TABLE 16.1 ELECTRICAL LOAD TYPES

Load Type	Description
Lighting	Often the greatest load
Miscellaneous equipment	Data processing, computers, and peripherals; convenience outlets, plug-in heaters, water fountains, other electrical power users
HVAC and plumbing equipment	Electrical motors and switches
Transportation equipment	Elevators, escalators, material handling equipment, dumbwaiters, trash and linen systems
Kitchen equipment	In restaurants, hospitals, offices, educational buildings
Special loads	Laboratory equipment, shop loads, display areas and windows, flood lighting, canopy heaters, industrial processes

INTERIOR DISTRIBUTION

Electricity must have a complete circuit from its source, through a device, and back to the source. Interrupting the circuit, as with a switch, stops the flow around the circuit's loop. When an appliance is turned on, AC electricity flows both ways in the loop, changing direction 60 times a second (60 cycles, or 60 hertz).

Circuits are introduced in Chapter 15, "Electrical System Basics."

In a simple single-family dwelling system, branch circuits spread out directly from the service entrance. Buildings with larger capacity and larger number of loads distribute **feeders** from the service entrance to subpanels located nearer to the loads.

Feeders typically lead to local distribution points from which smaller capacity circuits branch out. They generally extend as near as possible to the center of the area served to minimize the length of branch circuits. **Home runs** (the distance from the first outlet on a circuit to the panelboard) are kept short to minimize voltage drops, preferably to less than 100 feet (30.5 m).

Branch Circuits

Branch circuits carry the electrical power throughout the building to the places where it will be used. After passing through the main service disconnect, each hot conductor wire connects to one of the metal **bus bars** that conduct electricity within a switchboard. They accept the amount of current permitted by the main fuses or circuit breaker, and allow the circuit to be divided into smaller units for branch circuits. Each branch circuit attaches to one or both hot bus bars via fuses or circuit breakers.

Each 120V branch circuit has one hot and one neutral conductor. The hot conductor originates at the branch circuit overcurrent protective device (fuse or circuit breaker). A 240V circuit uses both hot conductors, and originates at a branch overcurrent protective device connected to both hot bus bars.

All the neutral conductors are in direct electrical contact with the earth through a grounding conductor at the neutral bus bar of the service entrance panel. An overcurrent protective device never interrupts the neutral conductors, so that the ground is maintained at all times. The effect of this arrangement is that each branch circuit takes off from an overcurrent protective device and returns to the neutral bus bar.

In order to decide how many branch circuits to specify and where they should run, the electrical system designer takes into account the variety of different loads. (See Fgiure 16.6) Once the electrical power requirements of various areas of the building are determined, the electrical engineer lays out wiring circuits to distribute power to points of use. Each circuit is sized according to the amount of load it must carry, with about 20 percent of its capacity reserved for flexibility, expansion, and safety. To avoid excessive drops in voltage, branch circuits should be limited to less than 100 feet (30.5 meters) in length.

Electrical engineers may need specific information about lighting fixtures, appliances, and other equipment from the interior designer to lay out branch circuits. The client may have preferences about which equipment shares a given branch circuit. Manufacturers specify load requirements for lighting fixtures and electrically powered appliances and equipment, and the interior designer may be responsible for getting these

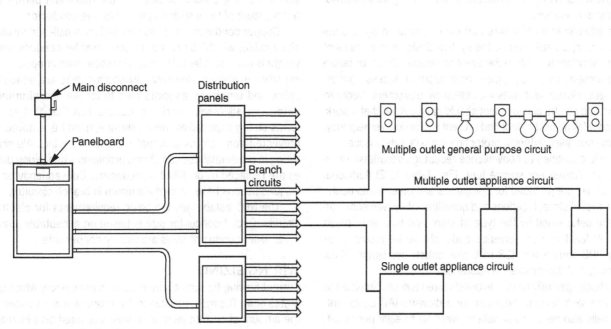

Figure 16.6 Branch circuits
Source: Redrawn from Francis D.K. Ching and Corky Binggeli, *Interior Design Illustrated* (3rd ed.), Wiley, 2012, p. 239

specifications to the engineer. The design load for a general-purpose circuit depends on the number of receptacles served by the circuit, and on how the receptacles are used.

Branch circuits with multiple general-purpose-type 20A outlets or multiple appliance-type outlets have a maximum 50A capacity. Single-type outlets for specific pieces of equipment may be 200A to 300A, depending upon the equipment needs. A branch circuit may supply a single large load such as a motor or heating element, or serve a group of smaller devices. Limiting circuit loads on 15A and 20A circuits allows for building load expansion.

Lighting, convenience receptacles, and appliances should each be grouped on separate circuits. Audio equipment may need to be on the same ground as the room it is serving, to avoid interference problems. Similarly, dimmers may need to be shielded to protect sensitive electronic equipment.

Circuits are arranged so that each space has parts of different circuits in it. If receptacles within a space are located on more than one circuit, the loss of one circuit will not eliminate all power to the space.

Branch circuit distribution panels may be shallow enough to install flush in stud partitions. In large installations, electrical closets contain branch circuit panelboards.

CIRCUIT DESIGN GUIDELINES

The layout of the branch circuits, feeders, and panels is designed for flexibility in accommodating all probable patterns, arrangements, and locations of electrical loads. Laboratories, research facilities, and small educational buildings require much more flexibility than residential, office, and fixed-purpose industrial installations. With the rapid changes in wireless equipment, it can be difficult to anticipate future uses and requirements. Overly specific designs waste money and resources, both during initial installation and in operation.

In addition to making sure that the electrical design is compliant with applicable codes, the system designer must prevent electrical safety hazards in the event of misuse, abuse, or failure of equipment. Large equipment may obstruct access spaces, passages, closets and walls with electrical equipment. Doors to rooms with electrical equipment should open out, so that a worker cannot fall against a door and prevent rescue in an emergency. In some buildings, lightning protection is also a safety issue.

Basic quantities of convenience receptacles (outlets) for offices are allotted per square foot. (See Table 16.2) Additional receptacles are provided for computers and other equipment.

In a retail store, locations and quantities of convenience outlets are determined by the type of store and their anticipated uses. At least one convenience outlet should be provided for every 300 square feet (28 m²), plus outlets for lamps, show windows, and demonstration appliances.

Schools generally need 20A outlets wired two per circuit at the front and back of each classroom for audiovisual (AV) equipment. Side walls also need similar outlets, wired six to eight per circuit. Appropriate receptacles should be provided for special equipment in school laboratories and shops, and for cooking equipment.

TABLE 16.2 OFFICE CONVENIENCE OUTLET REQUIREMENTS

Office Type (sf = square feet, m² = square meters)	Receptacles per square foot. Wall length in linear feet (lf)
Offices less than 400 sf (37 m²)	One per 40 sf (3.7 m²) or one per 10 lf (3 m) of wall, whichever is greater
Larger offices	One per 100 to 125 sf (9.3 to 11.6 m²) up to 6 to a 20A branch circuit
Each computer terminal	One 20A duplex receptacle at each desk
Office corridors	One 20A, 120V outlet each 50 lf (15.3 m) for cleaning and waxing machines

Public areas and corridors in schools require heavy-duty devices and key-operated switches, plastic rather than glass lighting fixtures, and vandal-proof equipment wherever possible. All electrical panels must be locked, and should be in locked closets.

Electrical Wiring and Distribution

Conductors extend from the circuit breaker boxes to individual switches, lights, and outlets. Conductors are rated in amps for their capacity to carry current. Conductors are surrounded by insulation that provides electrical isolation and physical protection. A jacket over the insulation gives added physical protection. **Conductor ampacity** (capacity in amperes) increases with increasing conductor size and the maximum permissible temperature of the insulation protecting the conductor.

Copper conductors are usually used for small- and medium-sized cable, which can be run through smaller conduits where weight is not a problem. Aluminum is lighter than copper, reducing labor expenses. However, aluminum cables are difficult to splice and terminate, as joints tend to loosen and **aluminum oxide**, an adhesive, poorly conductive film that that rapidly forms on any exposed aluminum surface, must be removed and prevented from reforming to make long-lasting joints. Aluminum wiring in residential use can create problems when wiring devices are replaced by unskilled homeowners. Consequently, some US jurisdictions have banned aluminum in branch circuitry.

The NEC establishes the basic requirements for electrical circuits. Circuit conductor size is based on acceptable ampacity. Smaller conductor sizes are usually copper wire.

WIRING SIZING

Current flowing through a wire produces resistance, which generates heat. The more resistance, the more heat is produced. As the amount of current increases, wire size must also increase. The circuit breaker or fuse in each circuit limits current to the rated ampacity of the wiring for that circuit.

American wire gauge (AWG) is the US wire and cable industry standard for round cross-section conductors. A larger AWG number indicates a smaller size. A single conductor No. 8 AWG and smaller is termed a wire. Outside the United States, conductor sizes are simply given by their diameter in millimeters.

A cable is a single insulating conductor No. 6 AWG or larger, or several conductors of any size twisted together to give more carrying capacity and more flexibility. Smaller (usually 16 AWG) cable is also used for appliance power cords.

As indicated previously, a conductor's ampacity is its current-carrying capacity, determined by the maximum safe operating temperature of the insulation used on conductor. It is standard practice to install larger ampacity conductors (which are also more efficient) in an electrical system to accommodate future loads as this is more difficult to do once walls, ceilings, and floors are finished.

Interior Wiring Systems

Interior wiring systems use **cables** and raceways to distribute and protect wiring. Cables are almost always used for residential wiring due to easy installation and lower cost. Conductors and enclosures are also combined in manufactured assemblies.

Insulated conductors for general wiring are called **building wire**. They usually consist of a copper conductor covered with insulation and sometimes a protective jacket.

As an interior designer, you may find yourself involved in discussions about the type of electrical cables that can be run in a project. Be aware that the type of cable that is permitted has a very significant effect on the cost of the electrical work.

EXPOSED INSULATED CABLES

Exposed insulated cable types include NM (**Romex®**) and AC (**BX armored cable**), along with some other types with their own electrical insulation and mechanical protection. (See Table

TABLE 16.3 EXPOSED INSULATED CABLE TYPES

Cable Type	Description
Flexible armored cable (type AC); smaller sizes known by trade name BX	Assembly of insulated wires bound together and enclosed in protective spiral-wound interlocking steel tape armor. Residences, rewiring existing buildings; dry locations. Generally restricted to dry locations.
Armored or nonmetallic sheathed cable NM and NMC (Romex®)	Plastic outer jacket for NM. NEC limits use to residential one- and two-family dwellings not over three floors in height; typically wood frame buildings.
Metal-clad (MC) cable	Exposed or in cable trays. With moisture-proof jacket, in wet and outdoor locations. Similar to BX, with additional green ground wire.

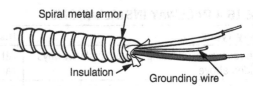

Figure 16.7 Armored BX cable
Source: Redrawn from Francis D.K. Ching, *Building Construction Illustrated* (5th ed.), Wiley 2014, page 11.34

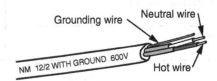

Figure 16.8 Nonmetallic sheathed Romex cable
Source: Redrawn from Francis D.K. Ching, *Building Construction Illustrated* (5th ed.), Wiley 2014, page 11.34

16.3) Romex cables are easier to handle but more vulnerable to physical damage than BX cables. (See Figures 16.7 and 16.8) BX cable is often used to connect fluorescent lighting fixtures in suspended ceiling grids to allow for flexibility in relocation. Its use may be restricted in some jurisdictions even where the NEC allows it.

INSULATED CABLES IN RACEWAYS

A raceway is any channel for supporting and protecting conductors. (See Table 16.4) In commercial, industrial, and institutional buildings, running conductors in a raceway makes frequent replacements easier and less expensive.

All types of facilities use insulated conductors in closed raceways. The raceway is generally installed first, with wiring pulled in or laid in later. The layout of raceways should be visually coordinated with the physical elements of the space.

An open raceway (cable tray) provides continuous open support for approved cables. Insulated cables in open raceways are used in industrial applications, and rely on the cable and tray for safety.

Individual floor raceways are sometimes installed to get power for computers, telephone lines, and other equipment at locations in open plan offices away from the structure. They may involve labor-intensive processes such as channeling the concrete floor, installing conduit in the opening (or chase), connecting the wiring to the nearest wall outlet, and patching the chase. Wireless communications are having a major impact on the use of floor raceways.

Metal or plastic underfloor raceways can be added onto the building structure and covered with concrete fill. The raceway is generally installed first, with wiring pulled in or laid in later. They are expensive, and today their use is limited to facilities where other alternatives are not available.

TABLE 16.4 RACEWAY INSTALLATIONS

Raceway Types	Installation
Surface raceways, including conduit and wireways suspended above	Attached to structure, used for electrical additions where concealed raceway is too costly or difficult to install.
Conduit in floor slab or underfloor duct buried in structure	Parallel metal plastic raceways on concrete slab, covered with concrete fill. Used with under-carpet wiring.
Cellular metal floors as integrated structural/ electrical system	Fully accessible floor partially or completely electrified. Separate floor cells and header ducts for electric power, data-transmission wiring, and phone/ signal systems.
Cellular concrete floor raceways as integrated structural/electrical system	Channel wiring through cells, voids for versatile layout of partitions, furniture in open plans, exhibit halls, merchandising areas. Floor finish options limited to exposed metal cover plates or carpet tiles.
Ceiling raceways	Wiring for lighting, power, telephones, outlets for floor above. Used with suspended tile ceiling, power poles.

Electrical fittings that poke through the floor are wired from underneath with power, telephone, signal, and data cables. The NEC requires that electrical penetrations in fire-rated floors, walls, ceilings, and partitions maintain their fire ratings, so newer poke-through fittings have been developed to preserve the fire rating.

COMBINED CONDUCTORS AND ENCLOSURES

Combined conductors and enclosures include all types of factory prepared and constructed integral assemblies of conductor and enclosure. They include all types of busway, busduct, and cablebus, as well as flat cable, and manufactured wiring systems. (See Table 16.5)

The terms **busway** and **busduct** are often used interchangeably for assemblies of copper or aluminum bars in a rigid metallic housing. (See Figure 16.9) They are preferred when it is necessary to carry large amount of current that can be tapped at frequent intervals along its length. Light duty busduct or busway is used for feeder or branch circuits. Connections can be changed easily. The NEC does not allow busways and busducts

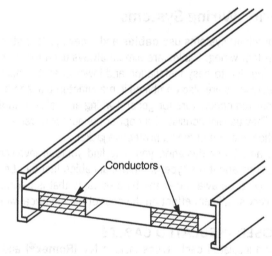

Figure 16.9 Busduct

to be concealed. They are usually located in equipment rooms or industrial applications.

PREFABRICATED ASSEMBLIES

Prefabricated assemblies encompass light-duty plug-in busway along with flat-cable assemblies and lighting track. Prefabricated assemblies act as branch circuit plug-in electrical feeders.

Flat cable assemblies allow lights, small motors, unit heaters, and other equipment to be served without hard wiring. (See Table 16.6) The assembly consists of two to four conductors field installed in a rigidly mounted square structural channel. Power-tap devices are installed as required, and connect to the device directly or to an outlet box with a receptacle.

Flat cable layouts are separate and distinct from the wire and conduit system, and are usually shown on a separate electric plan. Factory-assembled flat cable is approved for floor installation only under carpet squares, with accessories to connect to 120V power outlets. (See Figure 16.10)

TABLE 16.5 BUSWAY, BUSDUCT, CABLEBUS AND BUSBAR

Type	Application Example
Busduct	Copper or aluminum bars assembled in a rigid metallic housing. Feeder busduct has no plug-ins. Used to carry large amounts of current.
Light duty busway	Copper or aluminum bars in rigid metallic housing. Frequent taps into conductor along length. Plug-in light duty busway used for direct connection of light machinery and industrial lighting.
Cablebus	Uses insulated cables instead of busbars, rigidly mounted in open space-frame. Open mounting may have higher rating, but difficult to make taps.
Busbar	Used when many wires must be connected to each other.

TABLE 16.6 FLAT CABLE ASSEMBLIES

Type	Description
Under-carpet wiring: Flat conductor cable (NEC type FCC)	Small factory-assembled flat cable with 3 or more flat copper conductors enclosed in insulation and grounded with a metal shield. Also requires bottom shield, usually heavy PVC or metal.
Over-the-ceiling flat cable assemblies	For lighting power, electrical power, and telephone; can provide outlets for floor above. Typically used with lift-out suspended ceiling panels.
Lighting track	Flat cable assembly with conductors for 1 to 4 circuits permanently installed in the track. Tap-off devices carry power to attached lighting fixtures anywhere along the track.

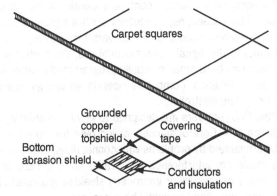

Figure 16.10 Flat cable

Lighting track may be used only to feed lighting fixtures. Taps to feed convenience receptacles are not permitted.

CONDUIT

Individual wires are run in protective conduits. **Conduit** is required for fire-resistant construction; this may affect the interior designer's decision to remove a suspended ceiling and expose the wiring and other equipment above, as the cost of rewiring into conduit can be substantial. Wires are installed in the conduits after the conduit system has been inspected and approved.

Today, electrical conduit may be required for all electric wiring in nonresidential occupancies. It may also be required in a single-family residence within the occupied zone where wiring is exposed (rather than behind finished construction such as gypsum wallboard). There are various types of conduit, including RMC, IMC, EMT, and Greenfield among others. (See Table 16.7 and Figure 6.11) Conduit may also be buried in a floor slab, exposed on a wall surface, or hung from a ceiling.

TABLE 16.7 CONDUIT TYPES

Type	Description
Rigid metal conduit (RMC)	Metal conduit, usually steel. Protects enclosed wiring from damage, provides ground, supports conductors.
Intermediate metal conduit (IMC)	Steel tubing heavier than EMT but lighter then RMC; may be threaded.
Electrical metallic tubing (EMT)	Thin-wall conduit commonly used instead of galvanized rigid conduit in commercial and industrial buildings. Usually steel, may be aluminum.
Flexible metal conduit: Greenfield and BX	Interlocked, spiral-wound steel tape. NEC limits use, but used for equipment connections and around obstacles. Liquid tight available.
	Greenfield is empty conduit, wires pulled through later.
	BX includes insulated wires.
Rigid nonmetallic conduit	PVC or other material. Moisture resistance, lightweight, easy to install. Used in low hazard installations, often underground, sometimes flexible.

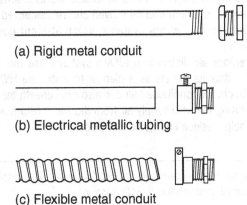

(a) Rigid metal conduit

(b) Electrical metallic tubing

(c) Flexible metal conduit

Figure 16.11 Conduit

Source: Redrawn from Francis D.K. Ching, *Building Construction Illustrated* (5th ed.), Wiley 2014, page 11.34

Conduit is sometimes exposed in renovation and addition projects to avoid the cost of cutting and patching finished surfaces.

Full Access Floors and Wiring

Underfloor delivery of power, data, and telecommunications services is becoming a much more cost-effective option for many offices, both because of the greater ease with which wires and cables can be distributed and the flexibility that full access floors permit. The system allows rapid and complete access to an underfloor plenum. Full access floors provide space for both air supply and cabling. Electrical conduits, junction boxes, and cabling are run below the full access flooring panels for computer, security and communications systems.

Usually, lightweight die cast aluminum panels are supported on a network of adjustable steel or aluminum pedestals. The panels are typically 24" (610 mm) but vary in size, typically between 18" and 36" (457 and 914 mm), and the floor depth is normally between 12" and 24" (305 and 610 mm. Without air requirements, where the floor is primarily used for cabling, the pedestals can be as short as 4" (102 mm).

The panels are made of steel, aluminum, or a wood core encased in steel or aluminum, or of lightweight reinforced concrete. They are finished with carpet tile, resilient tile, or high-pressure laminates. Fire-rated and electrostatic-discharge-control coverings are also available. Seismic pedestals are available to meet building code requirements.

The construction is usually completely fire-resistant. The ceiling height must be adequate to accommodate the raised floor. Raised floors may require steps or ramps for level changes.

Access flooring systems are used in spaces with heavy cabling requirements, especially where frequent recabling and reconnection are required. They provide accessibility and flexibility in the placement of desks, workstations, and equipment. Equipment can be moved and reconnected fairly easily with modular wiring systems, which also cut down on labor costs.

Underfloor air distribution (UFAD) systems use the space under the flooring panels as a plenum to distribute HVAC air supply. Ducts for conditioned air can also run beneath the floor. By separating the cool supply air from the warm return air, the system helps reduce energy consumption.

See Chapter 14, "Heating and Cooling," for more information on underfloor air distribution (UFAD) systems.

Low-Voltage Wiring

Low-voltage circuits carry alternating current below 50V, supplied by a step-down transformer from the normal line voltage. These circuits are used in residential systems to control doorbells, intercoms, heating and cooling systems, and remote lighting fixtures. Most telephone and communication wiring is low voltage, with the current provided by the communications company.

Low-voltage wiring is usually 12V to 14V, and is used for thermostat circuits and for switching of complex lighting circuits or remote-control panels. Low voltage wiring cannot give dangerous shocks or cause fires, and can be run through the building without cable or conduit.

See Chapter 20, "Communications, Security, and Control Equipment," for more information on communications wiring.

Low-voltage switching is used when a central control point is desired from which all switching may take place. The low-voltage switches control relays that do the actual switching at the service outlets.

Power Line Carrier Systems

Adding sophisticated building management controls to an existing building would be very expensive if all new wiring had to be installed. **Power line carrier (PLC) systems** use existing or new electrical power wiring as conductors to carry control signals for energy management controls in existing large, complex facilities. Low-voltage, high-frequency control signals are injected into the power wiring. Only receivers tuned to a particular code react to the signals. In residential use, the control signal generator can be a small manually programmed controller. In commercial facilities, computers operate an energy management or lighting controller.

Most PLC receivers are designed to fit into an ordinary wiring device box. Control signals carried over the power wiring can be attenuated by problems with connections, grounds, and faulty insulation. Interference from radio noise generated by faulty power equipment or improperly shielded grounded electronic equipment can frequently be overcome.

Electrical Emergency Systems

Most buildings except for single-family homes and a few other small building types are required by code to have emergency energy sources to operate lighting for means of egress, exit signs, automatic door locks, and other equipment in an emergency. Emergency systems supply power to equipment that is essential to human life safety on the interruption of the normal supply.

Emergency power can be provided by batteries or by on-site generators. Larger installations of batteries are used in an uninterruptible power system (UPS) to provide reliability for computer facilities, microprocessor-based demand controllers, and wherever a momentary power interruption could be a disaster. The UPS filters out any aberrations in the power supply and keeps batteries charged continuously.

There are three classes of systems: emergency, legally required standby, and optional required standby systems. (See Table 16.8) Possible emergency power supply arrangements depend on the requirements of local codes; this also applies to standby systems.

TABLE 16.8 EMERGENCY SYSTEMS

System	Code Reference	Description
Emergency systems	NEC Article 700	Essential to human safety, such as lighting for emergency egress, power for elevators, fire alarm systems, and fire pumps.
Legally required standby systems	NEC Article 701	Provide power as for emergency systems, but does not involve immediate danger to human life. Firefighting systems, health hazard controls, long-term rescue operations.
Optional required standby systems	NEC Article 702	Intended to minimize cost and meet project requirements established by owner/operator.

EMERGENCY SYSTEM CODES

The determination as to whether an emergency system is required is made by the applicable jurisdictional authorities, usually as a response to the requirements of NFPA 101. Equipment and installation must comply with NEC requirements. Relevant National Fire Protection Association (NFPA) codes include:

- NFPA 70 National Electric Code
- NFPA 99 Standard for Health Care Facilities
- NFPA 101 Life Safety Code
- NFPA 110 Standard for Emergency and Standby Power Systems
- NFPA 111 Standard on Stored Electrical Energy Emergency and Standby Power Systems

EMERGENCY LIGHTING

Emergency lighting systems illuminate areas of assembly to permit safe exiting and prevent panic. Codes require that artificial lighting must be present in all exit discharges any time a building is in use, with exceptions for residential occupancies.

See Chapter 17, "Lighting Systems," for more information on emergency lighting.

WIRELESS SYSTEMS

Wireless systems eliminate the need to connect lights, fans, and other equipment directly to the building's hard wiring. Wireless systems are widely used for telephone and computer data transmission, and also for building control switching.

Wireless communication transfers information between two or more points using electromagnetic energy, most commonly radio waves. Each switch and sensor has a small transmitter that sends a radio-frequency signal to the device it controls and to an area controller, all of which have radio receivers (detectors). Sensors can be powered by replaceable batteries, light falling on photocells, or thermal energy. Wireless switches can also be controlled by throwing or pushing a switch. Existing switches can be replaced by wireless switches operated by remote sensors.

See also Chapter 20, "Communications, Security, and Control Equipment," for more information on wireless systems.

ELECTRICAL DESIGN FOR RESIDENCES

Residential electrical requirements are set by *NFPA 70A— Electrical Code for One & Two Family Dwellings*, which establishes the distances for electrical outlets and mandates the use of ground fault circuit interrupters (GFCIs) in certain locations. Ranges and ovens, open-top gas broiler units, clothes dryers, and water heaters have their own specific code requirements or standards.

Codes as well as manufacturers' instructions may not permit electrical outlets to be installed directly above baseboard heating units. Verify requirements with local authorities.

The service equipment and building panelboard for a private residence is usually single unit. The main disconnect is usually the main switch or breaker of the panel. In a residence, the electrical panel is normally located in the garage, utility room, or basement, and placed as close to major electrical loads as practical to minimize the voltage drop. A smaller subpanel frequently feeds from the main panel to kitchen and laundry loads.

In apartments, electrical panels are usually located in the kitchen or a corridor immediately adjoining it, so that the panel circuit breaker can act as the NEC-required disconnecting means for most fixed appliances.

Residential Code Requirements

The NEC requires that no point on a residential wall be greater than 6 feet (1.8 m) from an electrical receptacle, so outlets are placed about 12 feet (3.7 m) on center. Where a doorway or other obstruction occurs, the receptacle is located within 6 feet (1.8 m) of either side of it. This arrangement results in a minimum of four receptacles, one on each wall.

Residential Branch Circuits

For residential work, it is important for the designer to determine the size of the electric service and the number of circuits that the service will support. Residential receptacles can be shown on an electrical power plan. (See Figure 16.12) Many existing homes have service between 60A and 100A, where 200A service may be more appropriate. The electrician can evaluate the service, and it should be stamped on the fuse box or circuit breaker panel.

The number of branch circuits required for a residence, including an allowance for expansion, can be estimated by allotting one 15A circuit per 400 to 480 square feet (37 to 45 square meters), or one 20 A circuit per 530 to 640 square feet (49 to 60 square meters), plus an allowance for expansion, with more branch circuits provided as needed. A rule of thumb for receptacle circuits is 12 receptacles for a 15-amp circuit and 16 receptacles for a 30-amp circuit.

Interior design residential power plans are frequently left uncircuited with outlets shown for information.

RESIDENTIAL CIRCUIT GUIDELINES

Lay out convenience receptacles so no point on wall is more than 6 feet (1.8 m) from an outlet. Use 20A, grounding-type receptacles only. Do not combine receptacles and switches into a single outlet except where convenience of use dictates high mounting of receptacles.

Circuit lighting and receptacles so each room, including basements and garages, has parts of at least two circuits. Avoid placing all lighting in a building on a single circuit. In rooms without overhead lighting, provide switch control for one-half of a strategically placed receptacle for a lamp. Provide switch control for closet lights, rather than cheaper but bothersome pull chains.

Individual circuits should be planned for electric resistance heaters or electric floor heaters. Wiring also needs to be planned for ceiling heaters and ventilation systems.

Countertop outlet spacing is intended to allow small appliances to be plugged in with short cords, and to discourage the use of extension cords. The NEC requires a minimum of two 20A miscellaneous appliance branch circuits to feed all receptacle outlets in kitchen, pantry, breakfast, and/or dining

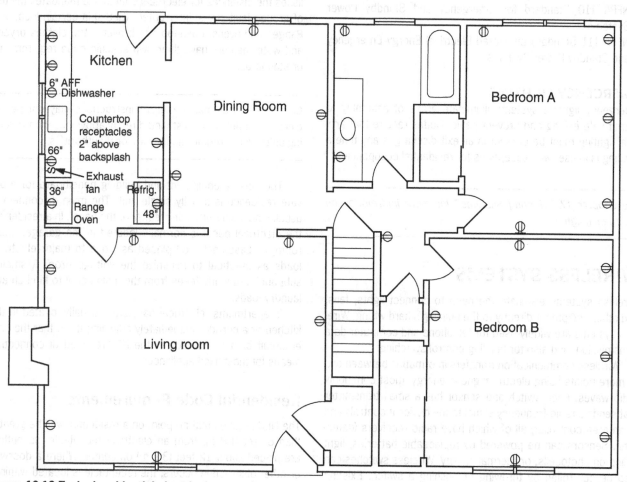

Figure 16.12 Typical residential electrical power plan

Source: Redrawn from Walter T. Grondzik and Alison g. Kwok, *Mechanical and Electrical Equipment for Buildings* (12th ed.), Wiley 2015, page 1364

room and similar areas, and only these outlets; all are potential appliance outlets and must be fed and circuited as such. Permanently installed appliances such as a garbage disposer, dishwasher, or fan hood may not be connected to these circuits. All kitchen outlets intended to serve countertop areas must be fed from at least two of these appliance circuits, so that not all countertop workspace will be de-energized by failure of a single circuit.

All countertop convenience outlets must be of the GFCI type. GFCI outlets were formerly only required when located within 6 feet (0.5 m) of the sink.

The NEC requires that two additional 20-amp circuits that do not serve any other areas be dedicated for kitchen countertops, and spaced so that no point along the wall line is more than two feet (610 mm) measured horizontally from a receptacle. The length of the counter occupied by a sink or range is not included. Receptacles are required to be located above countertops by a maximum of 20" (508 mm). Any kitchen island or peninsular countertop greater than 24" wide by 12" deep (610 by 305 mm) is required to have at least one electrical receptacle. (See Table 16.9 and Figure 16.13)

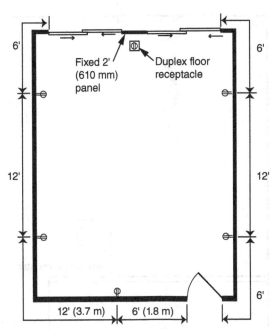

Figure 16.13 IRC 2015 residential electrical receptacle locations

Source: Redrawn from 2015 International Residential Code, Figure E3901.2 General Use Receptacle Distribution

TABLE 16.9 RESIDENTIAL BRANCH CIRCUIT DESIGN GUIDELINES

Spaces	Guidelines
Kitchen (see text also)	Provide readily accessible means to disconnect electric ranges, cooktops, and ovens within sight, often with small panel recessed in kitchen wall.
	120V circuits for garbage disposer, dishwasher, microwave oven, refrigerator, exhaust fans.
	240V circuits for range, cooktop, on-demand water heaters, wall ovens.
Workshop spaces (garage, utility room, basement)	20A receptacles on 20A circuits with no more than four such receptacles per circuit. Must be GFCI type, with some NEC exceptions.
Bedroom	Additional circuit for window air conditioner similar to appliance circuits for one outlet in each bedroom without central air conditioning.
	Two duplex outlets either side of likely bed location for electric blankets, clocks, radios, lamps, etc.
Home office (study and workroom or large bedroom)	Equip to double as home office. Minimum six duplex 15A or 20A receptacles on at least two different circuits with adequate surge protection. 2 of these receptacles with separate insulated and isolated ground wire for isolated ground (IG) receptacles. Two phone jacks.
Laundry	NEC requires minimum one 20A appliance circuit to supply only laundry outlets. Where electric clothes dryer anticipated, distinct individual branch circuit to serve this load via heavy-duty receptacle.
Bathroom	At least one 20A wall-mounted GFCI receptacle adjacent to each bathroom lavatory, fed from 20A circuit for only bathroom receptacles. No receptacles within tub or shower space. Bathroom lighting, exhaust fan, heaters, or other outlets should not be connected to bathroom receptacle circuit. Steam shower, sauna, and tub heaters may need 240V circuits.
Outside	Provide at least two GFCI-protected and waterproof receptacles on outside front and rear of house. Switch control from inside is convenient.

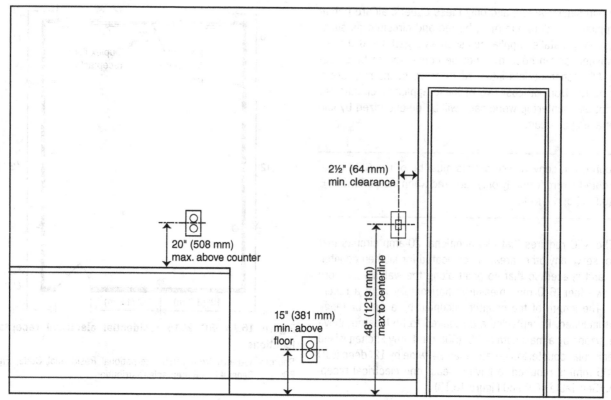

Figure 16.14 Recommended switch and outlet heights

Source: Redrawn from Francis D.K. Ching and Corky Binggeli, *Interior Design Illustrated* (3rd ed.), Wiley 2012, page 240

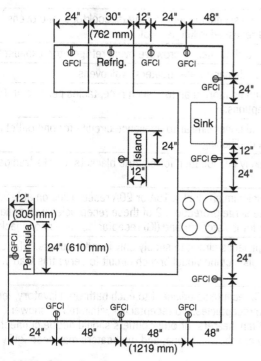

Figure 16.15 IRC countertop receptacles

Source: Redrawn from 2015 International Residential Code, Figure E3901.4 Countertop Receptacles

Although the 2010 ADA does not set required heights for electrical switches and outlets, some other codes do. Projecting objects such as wall sconces do have ADA clearance requirements. (See Figure 16.14)

Locations for countertop receptacles are covered by the IRC. (See Figure 16.15)

WIRING DEVICES

Whether you are turning on a wall switch, plugging an appliance into a receptacle, or using a dimmer for the lights, you are using a **wiring device** installed in an **outlet box**. The **attachment plugs (caps)** and **wall plates** are also considered to be wiring devices, as are a variety of other electrical devices.

Outlet boxes are also used where light fixtures are connected to the electrical system. Low voltage lighting control devices are considered wiring devices. **Junction boxes** are enclosures for housing and protecting electrical wires or cables that are joined together in connecting or branching electrical circuits.

Electrical equipment is given ratings for voltage and current. As described earlier, voltage ratings indicate the maximum voltage that can be safely applied to the unit continuously. An ordinary electric wall receptacle is rated at 250V, but is only supplied with 120V in normal use. The type and quality of insulation used and the physical spacing between electrically energized parts determines the voltage rating. Wiring devices are usually rated at 300 amps or less and frequently at 20 amps. They can be mounted in a small wall box.

Manufacturers classify wiring devices by grades to indicate their quality and expected use. (See Table 16.10) However, their grades are not standard across manufacturers, so if you specify by grade without a manufacturer named you have little control over what you may actually get.

At each electrical receptacle, lighting fixture, or switch, a metal or plastic box is securely fastened to the building structure to support the wiring device and protect its connections. Each cable or conduit is clamped tightly to the box where the wires enter, preventing wires from being pulled from their connections in case a cable or conduit is disturbed. The bare neutral wire is connected to the box and to the frame of the wiring device to prevent shocks should the device become faulty. After the wiring device is screwed securely to the box, a cover plate is attached.

Cover plates may be made of metal, plastic, or glass, and are available in a variety of colors and finishes. Coordination with the electrician helps interior designers control their appearance.

Outlet and Device Boxes

Outlet and device boxes are made of galvanized stamped sheet metal. Nonmetallic boxes may be allowed in some wiring installations with NM and NMC cable and nonmetallic conduit. Square and octagonal 4" (102 mm) outlet boxes are used for fixtures, junctions, and electrical devices. The most common sizes are 4" (102 mm) square and 4" (102 mm) octagonal boxes used for fixtures, junctions, and devices, and 4" by 2⅛" (102 by 54 mm) boxes used for single devices where no splicing is required. Depths range from 1½" to 3" (38 to 76 mm).

Outlet boxes for electrical receptacles are wall- or floor-mounted. For lighting fixtures, they are wall or ceiling mounted. Floor boxes of cast metal can be set directly into the floor slab.

Junction boxes are designed to fit within the dimensions of stud walls. (See Figure 16.16) They are also used with exposed

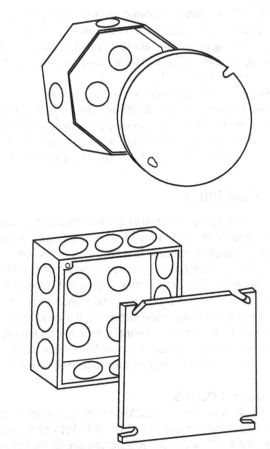

Figure 16.16 Junction boxes

TABLE 16.10 WIRING DEVICE GRADES

Types	Grades	Description
NEMA and UL grades	Hospital grade; highest quality and price	Green dot on device face. Highest quality, built to withstand severe abuse while maintaining reliable operation, must meet UL requirements for their grade.
	Federal specification grade	Roughly equivalent to industrial (premium) and commercial specification grades; less stringent than hospital grade.
	UL general purpose grade. Lowest quality and price	Corresponds roughly to residential grade, and is the least demanding quality.
Manufacturer grades	Hospital grade	Must meet industry standards. Approximately same quality among manufacturers.
	Premium or industrial specification grade	Roughly correspond to federal specification grade. Industrial and high-grade commercial construction.
	Commercial specification grade	Usually used in most educational and good residential buildings and commercial work.
	Standard or residential grade	Usually used in low-cost construction of all types, but not necessarily in all residential work.

steel conduit. Switch boxes are typically mounted on a stud within the wall to control a lighting outlet box. Installation of an electrical box usually just penetrates the wall surface, and does not require firestopping. The opening in the wall cannot allow a gap greater than ⅛" (3 mm) between the box and the gypsum wallboard.

The NEC specifies the minimum number of electrical boxes allowed for some building types, especially dwelling units. The NEC and building codes typically specify that no more than 100 square inches (645 cm²) of electrical boxes can be installed per 100 square feet (9.3 m²) of wall surface.

When an existing electrical box is not being used, it must either have a cover plate or be totally removed, including the box and all its wiring, with the wall opening properly patched. In fire-rated walls, when boxes are used on opposite sides of the same walls, they must be separated by 24" (610 mm) horizontally.

Electrical Plugs

Electrical power plugs are used to connect electrically operated equipment to the AC power supply in a building. Plugs are usually movable connectors attached to a device's power cable. They are intended for attachment to a fixed socket connected to an electrical circuit. They differ in their voltage and current rating, shape, size, and type of connectors, and there are many types (including obsolete ones) in use throughout the world today.

Technically, the term **"wall plug"** is the name for the cap on the wire that carries electricity to an appliance—the part on the end of a line cord that is plugged into a wall. (See Figure 16.17)

POLARIZED PLUGS

Polarized plugs have in the past had an enlarged neutral prong to make sure that the connection is hot to hot and neutral to neutral. The small slit and projecting prong are for the hot wire, and the large slit and prong for the neutral wire. A grounded or polarized plug should be used when a plug does not include a ground prong and also is not polarized; this also reduces chances of a crossed-connection in the event that an installation is faulty.

Since grounded receptacles have been required since 1962, they are generally the only type available for purchase

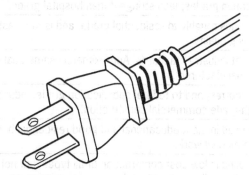

Figure 16.17 Wall plug

in hardware stores, and often installed without a ground connection to replace an ungrounded receptacle. Consequently, it is necessary to verify that an apparently grounded outlet in an older structure is actually grounded. The NEC requires grounding all circuits for new work.

GROUNDED PLUGS

A grounded plug is required by the NEC for use on all branch circuits. It assures that the connection is properly polarized, and provides a means to open an electrical circuit if any current is leaking.

Adapters to connect a grounded plug into an ungrounded receptacle can override safety controls. Attaching the ground wire or metal tab on a plug to the screw on a receptacle cover plate does not help if there is no ground connection for the receptacle. There is an exception to this with an older receptacle with a metal electrical box connected to a grounded metal conduit system.

An appliance ground (3-prong plug) directly connects the metal housing of an appliance to the ground. Since the neutral is also connected to the ground, a fault in the appliance causes current to flow from hot wire to ground and from ground to neutral wire. The resulting fault current trips the circuit breaker to prevent fire or electric shock. This safety system will fail if the ground path broken accidentally, or intentionally by breaking off grounding prong.

An inexpensive testing tool will reveal several possible electrical problems, including an open ground path.

Electrical Receptacles

The NEC definition of a **receptacle** is "a contact device installed at the outlet for the connection of a single attachment plug." This is usually a common wall outlet, or may be a larger and more complex device.

CONVENIENCE RECEPTACLES

The electrical outlet that we plug an electrical cord into is technically known as a **convenience receptacle** outlet, a receptacle outlet, or a convenience outlet. "Receptacle" is a single term; a normal wall convenience receptacle that accepts two attachment plugs is called a duplex convenience receptacle or duplex convenience outlet, commonly shortened to **duplex receptacle** or duplex outlet. (See Figure 16.18)

The different types of receptacles are identified by the number of poles (prongs) and wires, and whether they have a separate grounding wire or not. Grounded receptacles are used on standard 15A or 20A branch circuits. The equipment grounding wire is separate from and not to be confused with the neutral wire. Receptacles are typically 20A, 125V, but are available from 10A to 400A, and from 125V to 600V. Locking, explosion proof, tamper proof, and decorative design receptacles are

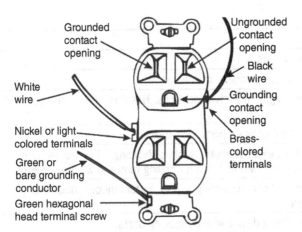

Figure 16.18 Duplex convenience receptacle wiring
Source: US Department of Labor

available. Some units, such as range receptacles, are designed for specific uses. Split-wired receptacles have one outlet that is always energized, and a second controlled by a wall switch. These are sometimes used for a lamp, with the switch near the entrance to the room.

Receptacles are normally mounted 12" or 18" (305 or 457 mm) above the finished floor. Receptacles above kitchen counters are typically mounted 48" (1219 mm) above the finished floor. In shops, laboratories, and other spaces with tables against walls, they are often mounted at 42" (1067 mm) above the floor.

Universal design seeks to accommodate all users. To conform to the universal design reach range, receptacles should be less than 44" (1125 mm) from the floor. This places them only 8" (203 mm) above a standard 36" (914 mm) counter height. Accessible receptacles should not be mounted more than 12" (305 mm) below a counter top. Receptacles below a countertop should not be located where the countertop extends more than 6" (152 mm) beyond its support base. Receptacles installed for a specific appliance such as a clothes washer should be within 6 feet (1.8 m) of the intended location of the appliance.

NONRESIDENTIAL RECEPTACLES

The NEC establishes requirements for convenience receptacles in commercial spaces. The code seeks to ensure that there are enough outlets to prevent a spaghetti-like tangle of extension cords, while respecting the total energy use in the space.

Code requirements change. The NEC is revised every three years, so check the most recent edition.

General design considerations include flexibility for certain types of buildings, including laboratories, research facilities, and small educational buildings. Convenience receptacles should be designed for building and electrical service expansion, while

avoiding over-design that wastes money and resources. Safety code requirements should be observed. Obstruction of access spaces, passages, closets, and surfaces with electrical equipment should be avoided.

SURGE SUPPRESSION AND EQUIPMENT GROUNDING

Modern electronic equipment is very sensitive to the random, spurious electrical voltages called electrical noise. Two special receptacles help eliminate electrical noise. Receptacles with built-in surge suppression protect equipment from over-voltage spikes. Receptacles with an insulated equipment-grounding terminal separating the device ground terminal from the system (raceway) ground eliminate much of the unwanted electrical noise. This latter type is connected to the system ground at the service entrance only, and is identified by an orange triangle on the faceplate.

See Chapter 15, "Electrical System Basics," for more surge suppression and grounding information.

Switches

You can turn a light on and off either by plugging a lighting fixture into an electrical receptacle and turning the switch on the lighting fixture on and off, or by hard wiring the lighting fixture directly into the building's power supply system and using a wall switch to control the power.

Switches up to 30A that can be outlet-box mounted are considered wiring devices. Switches are installed in hot wires only. They disconnect an electrical device from the current, leaving the device with no voltage going through it, and eliminating the possibility of shock when the switch is open.

There are many types of switches used in buildings. (See Figure 16.19 and Table 16.11.) A common wall light switch is a type of general duty safety switch called a **contactor**.

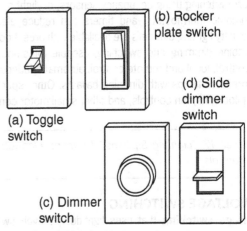

Figure 16.19 Wall switch types

TABLE 16.11 SWITCH TYPES

Type	Description
General duty safety switches	Normal use in lighting and power circuits.
Heavy-duty (HD) switches	Frequent interrupting, high fault currents, ease of maintenance.
Three-way switches	Used with another switch to control lights from two locations. Has three terminals, so on/off positions can change.
Four-way switches	Used with two 3-way switches to control lights from 3 locations.
Remote-control (RC) switches	Used for switching blocks of lighting, exterior lights, whole or buildings. Electromagnets operate fixtures from a distance without wiring. Used for ceiling fans and window treatments.
Solid-state switches	Electronic device with conducting and nonconducting states. Instantaneous, noiseless change by control signal in voltage.
Time-controlled electronic switch	Solid-state switch with electronic timing device with no moving parts.
Programmable time switch with memory circuit	Fits wall outlet box for lighting, energy management, automated building control, and clock and program systems.
Programmable controller	Microprocessor responds with particular control plan to given input signals for specific type of function.
Automatic transfer switch	Emergency and standby power arrangements. Automatically transfers to emergency service and back to normal service.

Contactors close an electrical circuit physically by moving two electrical conductors into contact with each other, allowing the power to flow to the lighting fixture. The contactor physically separates the two conductors to open the circuit, which stops the electrical flow and shuts off the light. Contactors are good for remote control by manual or remote pushbutton, or for automatic devices such as timers, float switches, thermostats, and pressure switches. They are found in lighting, heating, and air conditioning equipment, and in motors.

Contactors can be operated by hand, electric coil, spring or motor. A typical wall switch is an example of a small, mechanically operated contactor. A relay is a small electrically operated contactor. The operating handles for contactors may be toggle, key, push, touch, rocker, rotary, or tap-plate types.

Switches can have multiple control points. For four or more control locations, low voltage, PLC, or wireless controls are probably a better option.

Smart switching using occupancy sensors, daylight sensors, programmable thermostats, and timers can reduce electrical use. Specialty lighting dimmers are available with preset controls that combine dimming and switching, useable for smooth fan speed control, local and remote control, automatic fadeout, and with fluorescent lamps with dimming ballasts. Other specialties include pilot lights, fan controls, and other small motor controls.

See Chapter 17, "Lighting Systems," for more information on lighting controls.

LOW-VOLTAGE SWITCHING

Low-voltage switching that uses light-duty, 24-volt switches is also called remote control switching and low-voltage control.

Low-voltage switches control relays that do the actual switching. This offers many advantages over full-voltage switching, including flexibility of control location, and makes possible individual load control override by local control devices such as occupancy sensors and photocells, as well as group load override by central control devices including timers, daylight controllers, and energy management systems, both of which promote energy conservation. Low-voltage, low-current wiring does not require conduit and is less expensive than conventional building wiring. Individual load status can be monitored at a centralized control panel.

Dimmers, switches, receptacles, fan controls, cable TV and telephone jacks are all available in a variety of colors. Be aware that cover plates specified in a bold color may have to be replaced if the wall color or other décor changes.

ELECTRICAL LOADS

Common types of building electrical loads include lighting, motors for HVAC and plumbing systems, convenience receptacles, heating elements, and electronic devices. Special process equipment such as lab equipment and x-ray machines also create electrical loads.

Lighting loads are covered in Chapter 17, "Lighting Systems."

Convenience devices are rated according to total anticipated amperage requirements of all connected loads operating concurrently. Household electric ranges and clothes dryers operate at 240V, and larger window air conditioners normally require 240V. These devices must be either directly wired into a circuit or plugged into a special receptacle designed for that purpose.

Many electrical devices consume power in standby mode even when turned off or in sleeping mode. For example, these so-called **vampire loads** occur with battery chargers, televisions, and computers. In homes, vampire loads comprise about 10 percent of energy consumed. Almost one-half of all employees leave their computers on when they go home, which can present a security risk as well as waste energy.

Residential Appliances

Electrical equipment comes with several ratings that indicate wiring requirements. As described earlier, the voltage listing is a rating that gives the maximum voltage that can safely be applied to the unit continuously. An ordinary wall receptacle has a 250V rating, although it normally uses only 120V. It is therefore safe to use the receptacle even if the voltage should go above 120 volts.

Another wiring rating for appliances, the current-carrying ability, is measured in amps. The maximum operating temperature at which components of a piece of equipment (for example, a toaster) can operate continuously determines the number of amps. Equipment with better electrical insulation inside can carry more current safely.

Newer models of appliances that previously used 120V may now require 240V. These may include clothes washers that heat water to higher temperatures, or small ovens with large power requirements.

ENERGYGUIDE LABELS
Yellow EnergyGuide labels are found on dishwashers, refrigerators, freezers, and clothes washers. (See Figure 16.20) These labels give the yearly estimated energy costs for an appliance, based on US national average energy costs, along with estimated yearly usage in kilowatt hours. The label contains a graph showing the range of energy costs to operate similar appliances, and how the appliance with the EnergyGuide label compares to the range.

The black and white Canadian EnerGuide label is similar to the EnergyGuide label. An EnerGuide label is required on all new electrical appliances manufactured in or imported into Canada. The label indicates the amount of electricity used by the appliance. The use of the EnerGuide label requires third-party verification that the appliance meets Canada's minimum energy performance levels.

For more information on residential appliances, see Chapter 11, "Fixtures and Appliances."

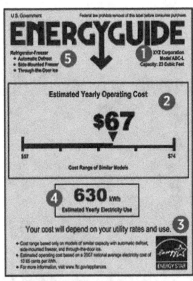

Figure 16.20 EnergyGuide label
Source: US Department of Energy

1. Maker, model, and size
2. Cost range helps to compare energy use of different models.
3. Energy Star logo
4. Estimated annual energy use based on typical use
5. Key features

When helping a client decide what type of equipment to purchase, it is important to determine the most efficient solution to the problem at hand. For example, a window fan may be a more energy efficient choice than an air conditioner. Equipment that uses a timer, thermostat, or sensor can save energy without decreasing comfort. Photocells are available to control day and night operation. A timer on an air conditioner can turn it on or off depending on the occupant's presence, outside temperature changes, or on a preprogrammed time schedule.

KITCHEN APPLIANCES
Kitchen appliances are primarily, but not exclusively, found in the home. (See Table 16.12) Interior designers who work on commercial office projects find themselves designing office kitchen and coffee areas. (See Figures 16.21 and 16.22) Office kitchens may combine the functions of a catering kitchen, a coffee break area, and a staff lunchroom. If not easy to maintain and share, they can become a source of employee dissatisfaction. Even beauty salons and similar retail-oriented spaces

TABLE 16.12 RESIDENTIAL KITCHEN EQUIPMENT

Appliance	Volts	Amps	Outlets on Circuit
Range	115/230	60	1
Oven (built-in)	115/230	30	1
Cooktop	115/230	30	1
Dishwasher	115	20	1
Waste disposer	115	20	1
Microwave oven	115	20	1 or more
Refrigerator	115	20	Separate circuit recommended
Freezer	115	20	

Figure 16.21 Office kitchen
Source: Courtesy of Herb Fremin

Figure 16.23 Medieval kitchen
Source: *Medieval Life and People CD-Rom & Book*, Dover Publications 2007, Image 007

may require kitchen and laundry appliances. Designs for kitchen spaces include outlets at several levels and plumbing for sinks and dishwashers, along with bracing for wall-hung units.

Kitchen design has come a long way since the open fires of medieval kitchens, where everything from butchering to meal preparation took place. (See Figure 16.23)

REFRIGERATORS AND FREEZERS

Refrigerators have been greatly improved in terms of energy efficiency. The average refrigerator made today uses about 700 kWh per year, one-third of the energy of a 1973 model, and is larger and has better controls as well. Newer refrigerators have more insulation, tighter door seals, a larger coil surface area, and improved compressors and motors.

The larger the refrigerator, the more energy it will use, but one large refrigerator will use less energy than two smaller ones with the same capacity. The most efficient refrigerators are in the 16 to 20 cubic foot range. Side-by-side models are

Figure 16.22 Office kitchen electrical rough-in
Source: Courtesy of Herb Fremin

less energy-efficient than styles with the freezer on top. Built-ins sometimes use more energy than free-standing models. Automatic icemakers and through-door dispensers also use more energy.

When selecting the refrigerator style, consider the depth required when the door is open 90 degrees. The side-to-side models take up less space in front of the refrigerator. Refrigerators typically use a 115V, 60 Hz, AC, and 15 ampere grounded outlet. Some manufacturers offer counter-depth refrigerators that align with cabinets smoothly, providing maximum storage capacity without taking up a lot of space in the kitchen.

Chest-style freezers that load from the top are 10 to 25 percent more efficient than upright, front-loading freezers, thanks to better insulation and the fact that cold air does not spill out as readily when they are opened. However, chest freezers are more difficult to organize. Manual defrosting is more common than automatic defrosting for freezers, but automatic defrosting may dehydrate food and cause freezer burn. Icemakers that produce up to 50 pounds of ice cubes every 24 hours are available in built-in or free-standing models. They require a water supply, and some models include a factory-installed drain pump.

Refrigerators and freezers should be located away from heat sources such as dishwashers and ovens, and out of direct sun. Allow a 1" (25 mm) space on each side for good air circulation.

STOVES AND OVENS

Cooktops are available as standard ranges or as separate units, and may be either electric or gas. New gas ranges are required to have electric ignition rather than wasteful pilot lights. Electric cooktops typically require 240/208 V, 60 Hz, 20 or 40 amp grounded electrical service.

Critical dimensions for installation of a cooktop include the size and type of the cooktop itself, the horizontal dimensions between cabinets above and below the cooktop, vertical clearances above the cooktop, and the size of the counter opening required. (See Figure 16.24)

Slide-in ranges provide a more solid, streamlined impression by removing frames and crevices. Free-standing and slide-in electric ranges require 120/240 V, 60 Hz service on a separate 50-ampere grounded electric circuit.

Electric cooktops are available with various types of heating elements. (See Table 16.13)

Gas cooktops should have electric pilot lights, and often have sealed gas burners. They typically require single-phase AC, 120 V, 60 Hz, 15 ampere grounded electric service. Gas cooktops are available in combination with gas grills. Some manufacturers offer dual fuel models that have sealed gas burners with an electric convection oven. When a customer purchases a gas appliance, delivery and installation must include the services of a plumber to connect the gas line. (See Figure 16.25)

Gas cook stoves consume oxygen from room air and release carbon dioxide and water vapor. A fan must be installed

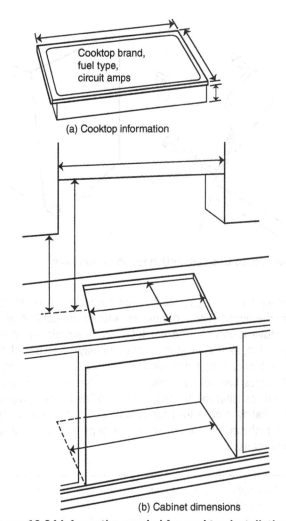

Figure 16.24 Information needed for cooktop installation

(a) Cooktop information

Cooktop brand, fuel type, circuit amps

(b) Cabinet dimensions

TABLE 16.13 ELECTRIC COOKTOP HEATING ELEMENTS

Element type	Description
Exposed coils	Heat up fast, difficult to clean.
Solid disk elements	Attractive and easy to clean. Heat up slowly, increasing their energy consumption.
Radiant elements under ceramic glass	Very cleanable. Heat up faster than solid disks; more energy efficient than coils or solid disks
Halogen elements	Heat food by means of the contact of the pan on the ceramic glass surface. Energy savings, but higher price.
Induction elements	Transfer electromagnetic energy directly to pan. Very energy efficient but work only with ferrous metal cookware and not with aluminum pans. When the pan is removed from the heating element, almost no heat lingers on the cooktop. Tend to be expensive

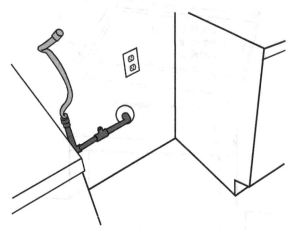

Figure 16.25 Gas connection for stove and oven

(a) Single oven   (b) Double oven   (c) Microwave above oven

Figure 16.26 Oven cabinets

to make sure that gas combustion products are eliminated from the house. The fan must exhaust to the outside when the stove is in use, and not just recirculate air. Avoid too large a fan, as it will waste energy and may cause backdrafting problems. Large downdraft ventilation fans in cooktops and ranges can suck so much air that the house is depressurized, causing the heating system to fail to vent properly and creating a backdraft for combustion gases. Large fans need makeup air ducts, which are supplied by some fan manufacturers.

For information on gas exhaust equipment, see Chapter 11, "Fixtures and Appliances."

Conventional ovens with self-cleaning features have more than the usual amount of insulation, and are more energy efficient unless the self-cleaning feature is used more than once a month. A window in the door saves wasting energy when the door is opened to look at the progress of the food being cooked. Built-in electric ovens typically require a separate 208/240V, 60 Hz, 30 ampere grounded circuit. A time delay fuse or circuit breaker is recommended. Double ovens located one above the other get a lot of oven into a small amount of floor space. Gas ovens use less energy than electric ovens; in addition, gas may be less expensive. Cabinets for wall-mounted ovens and microwaves must be designed to accommodate the equipment specified. (See Figure 16.26)

Convection ovens are more energy-efficient than conventional ovens. Heated air continuously circulates around the food for more even heat distribution. Food can be cooked faster and at lower temperatures, saving about one-third of energy costs.

There are an increasing number of accessory cooking appliances available today. Warming drawers keep contents in the 90° to 225°F (3° to 107°C) range. They offer removable serving pans, and the drawer itself may be removable. The electrical

service required is 120V, 60 Hz. Rotisseries for residential use require electrical and gas connections, as well as an exhaust hood. They are heavy, weighing about 130 kg (290 pounds), and must be supported adequately.

MICROWAVE OVENS

Microwave ovens use very high frequency radio waves to penetrate the food surface and heat up water molecules in the food's interior. Using a microwave can reduce energy use and cooking times by around two-thirds over a conventional oven, especially for small portions and for reheating leftovers. They also contribute less waste heat to the kitchen. Temperature probes, controls that shut off when the food is cooked, and variable power settings also save energy.

Microwave ovens are available in countertop, over-range, over-counter, and built-in models. Most specify a 120V, single phase, 60 Hz, AC, 15 ampere 3-wire grounded circuit. Built-in microwaves may offer a drop-down door and a convection hood, and may require 120V or 240V, 20-ampere service. Wall mounted microwaves should not be hung directly from the cabinets above; they usually require the support of a mounting bracket.

Microwave combination ovens come with a convection oven below, and may require 120/240V, 60 Hz, 40 ampere grounded service. Microwaves, toaster ovens, and slow-cook crock-pots that combine an insulated ceramic pot with an electric heating element, can save energy when used to prepare smaller meals.

For more information on microwaves, dishwashers, and laundry equipment, see Chapter 11, "Fixtures and Appliances."

DISHWASHERS

Heating water makes up 80 percent of the energy use of an automatic dishwasher, so using less water saves energy. Electrical connections are typically a 120V, 60 Hz, 15 amp or 20 amp fused electrical supply on a separate circuit. A time-delay fuse or circuit breaker is recommended.

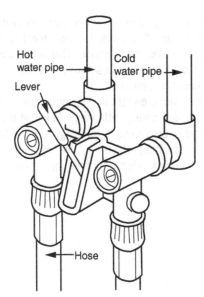

Figure 16.27 Single shutoff lever for washer

LAUNDRY EQUIPMENT

Clothes washers require both hot and cold water connections, as well as a drain that meets code requirements. A lever that shuts off the water supply when the machine is not in use can help prevent flooding. (See Figure 16.27)

A dedicated circuit is needed for a clothes washer and/or clothes dryer. (See Table 16.14) An electric dryer requires a

TABLE 16.14 RESIDENTIAL LAUNDRY EQUIPMENT

Appliance	Volts	Amps	Outlets on Circuit
Washing machine	115	20	1
Dryer	115/230	20	1
Hand iron	115	20	1

TABLE 16.15 RESIDENTIAL LIVING-AREA EQUIPMENT

Appliance	Volts	Amps	Outlets on Circuit
Workshops	115	20	1 or more, separate circuit for heavy-duty appliances
Portable heater	115	20	1
Television	115	20	1 or more
Audio center	115	20	1 or more
Personal computer and peripherals	115, isolated ground may be required	20	1 or more, surge protection recommended

TABLE 16.16 RESIDENTIAL UTILITIES

Appliance	Volts	Amps	Outlets on Circuit
Fixed lighting	115	20	1 or more
Window air conditioner ¾ hp	115	20 or 30	1
Central air conditioner	115/230	40	1
Sump pump	115	20	1 or more
Heating plant (forced-air furnace)	115	20	1
Attic fan	115	20	1 or more

240V circuit. Some European clothes washers require a 240V circuit for heating water. Other living-area and utility-related equipment also have circuiting requirements. (See Tables 16.15 and 16.16)

Appliance Control and Energy Conservation

Networked appliances are available today, some of which communicate with each other through the home electrical system, while most use wireless or internet connections. These appliances are designed to perform intelligent tasks such as collecting recipes from the Internet, tracking food inventory, and downloading new wash cycles for different kinds of clothes.

Kitchen appliances use nearly one-third of the electricity used in a home. (See Figure 16.28) Interior designers can play an important role in selecting energy-efficient residential appliances. (See Figure 16.29)

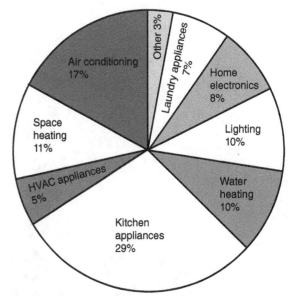

Figure 16.28 Residential electricity use

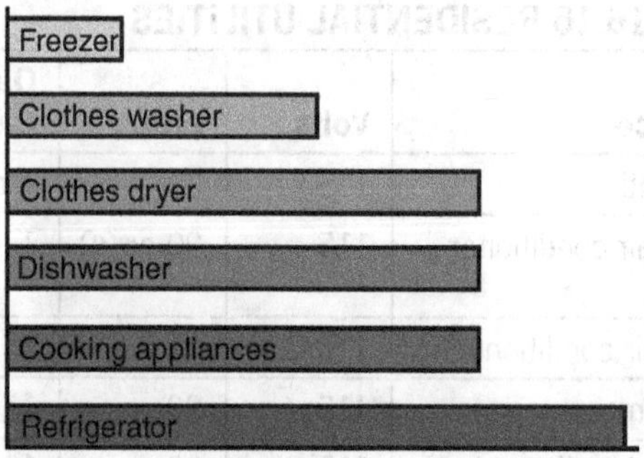

Figure 16.29 Comparative appliance energy use

Inside many refrigerators is an energy-saver switch that can be shut off when humidity is low. The switch activates a heater that reduces moisture condensation. This works well in humid environments, but in dry climates it will use more energy than it saves, and should be turned off.

As we have seen, electrical distribution has implications for both safety and energy efficiency. Much of the electrical equipment and energy used in buildings is used for lighting. In Chapter 17, "Lighting Systems," we see how daylighting can be combined with electrical lighting for round-the-clock use.

17

Lighting Systems

Lighting has been the major user of electrical energy in most buildings. In commercial buildings, motors for HVAC systems, plumbing pumps, elevators, and most industrial processes are the second heaviest user of electrical energy.

Many interior design programs offer a full-semester course in lighting design. The physics of light is another substantial topic. We do not attempt this level of completeness, but we do cover the major points of vision, color perception, and other topics as they affect interior daylighting and electrical lighting.

The chief drawback of daylighting is its inconstancy, especially its total unavailability between sundown and sunup. Artificial electric light is instantly and constantly available, is easily manipulated by a designer, and can be controlled by the occupants of the building. This suggests strongly that daylighting and artificial lighting make good partners, with artificial lighting being used mainly for nighttime illumination and as a daytime supplement when daylighting alone is insufficient. (Edward Allen, *How Buildings Work* (3rd ed.) Oxford University Press 2005, page 134)

INTRODUCTION

The US Energy Information Administration estimates that in 2012, about 12 percent of total US electricity consumption was used for residential and commercial lighting. Good lighting design can save up to half of the electrical power used for lighting.

Lighting is a major contributor to the building heat load. One watt of lighting adds one watt (3.4 BTUH) of heat gain to the space. It takes about 0.28W of additional energy to cool one watt produced by lighting in the summer; however, the added heat may be welcome in the winter. Reducing the lighting power energy levels to below 2 watts per square foot (0.09 m^2) in most areas results in less impact on the HVAC system from light generated heat.

History of Lighting

The earliest gas lighting used plant and animal (including whale) oils, beeswax, or similar fuels. In the late eighteenth century, gas lighting made streets safer, and made it possible for factories in Great Britain to run 24 hours a day.

Natural gas began to replace coal-gas from the nineteenth century through the early twentieth century. (See Figure 17.1) The electric incandescent lamp was patented by Thomas Edison in the United States in 1880.

As is indicated in Chapter 15, "Electrical Systems Basics," "lamp" is the technical name for what we commonly call a light bulb.

For many years, an unwarranted division existed in the field of lighting design, dividing it into two disciplines: architectural lighting and utilitarian design. The former found expression in design that took little cognizance of visual task needs and displayed an inordinate penchant

Figure 17.1 Gas lamp
Source: *Victorian Goods and Merchandise CD-Rom & Book*, Dover Publications, Inc., 2006, Image 652

for incandescent wallwashers, architectural lighting elements, and form-giving shadows. The latter saw all spaces in terms of illuminance levels and cavity ratios, and performed its design function with footcandles (lux) and dollars as the ruling considerations. That both of these trends have generally been eliminated is due largely to the efforts of thoughtful architects, engineers, and lighting designers, assisted in part by the energy consciousness that followed the 1973 oil embargo. That last event spurred research into satisfying vision needs within a framework of minimal energy use. That research, and its resulting energy codes and continuing development of higher-efficiency sources, are today motivated by environmental concerns. (Walter T. Grondzik and Alison G. Kwok, *Mechanical and Electrical Equipment for Buildings* (12th ed.), Wiley 2015, page 583)

Until 1973, daylighting was considered part of architectural design, not part of lighting design. Since an artificial lighting system had to be installed anyway, the practice was to ignore daylight, even to the extent of shutting it out completely. However, since the energy crisis, the extensive use of electrical energy

in nonresidential buildings for lighting has driven designers to integrate the cheapest, most abundant, and in many ways most desirable, form of lighting, daylighting.

Lighting Design Team

The best lighting designs blend seamlessly into an overall interior design. Differences inherent in the objectives of the interior designer and the electrical engineer may lead to difficulties in achieving this goal. These differences have their roots in the training and functions associated with each profession. The interior designer may focus on aesthetics and strive for an interior space that supports the client's image and work process. The electrical engineer's perspective may focus on technical issues and accentuate standardization and energy efficiency. Twenty-first century lighting design is facilitating an improved team approach.

Lighting Calculation Methods

Lighting should be designed for what people perceive, not for what meters measure.
Norbert Lechner, *Heating, Cooling, Lighting: Sustainable Design Methods for Architecture* (3rd ed.), Wiley 2009, page 377

With these wise words in mind, we take a brief look at how engineers measure lighting. Engineers calculate spaces requiring overall uniform illumination by the lumen method, averaging illuminance. Spaces requiring local or local plus general lighting are calculated by the point by point or another method.

The lumen method uses zonal cavity calculations of the reflectance coefficients of room surfaces to model the how light reflected from the walls and the ceiling contributes to useful illumination at the working level of the room. Today, computer programs do the calculations, and three-dimensional computer models help the design team and the client visualize the result.

The point-to-point method indicates illumination in lumens per unit area at a particular location. It can be used to calculate illumination levels from several light sources, with the results added together for the overall illumination level.

Daylighting diagrams aid in daylighting design. They consider skylight orientation and light control and northern orientation for consistent lighting. Daylighting diagrams can aid in the integration of daylight with electric light sources and with other environmental issues.

An alternative approach called **brightness design** involves using a computer perspective of a space with the level of luminances desired; the lighting is then designed to meet these lighting levels. The computer model provides the designer with a picture of the space's brightness patterns and encourages experimentation with window configurations and treatments, light sources and shadows. Brightness design relies heavily on the experience of the designer and is more intuitive—and less quantifiable—than analytical lighting design.

The interior designer and the engineer may have differing perspectives on their relationship with their client. The interior designer typically works with the client's executive management team to blend business objectives, work processes, and corporate image. The engineer frequently works with the facility manager, who may also be one of the interior designer's contacts. Facility managers are looking for a lighting scheme that is flexible, efficient, and low maintenance. A third client group is the users, including employees, whose needs focus on comfort and productivity.

Unless the interior designer and the electrical engineer understand the needs of these three distinct client groups, they may not be able to work together effectively. When the interior designer and electrical engineer work well together, they help the client recognize and prioritize each client group's objectives, and achieve a design that integrates each discipline's strengths and meets the client's overall needs.

Professional lighting designers can help bridge the gap. With expertise in the technical aspects of lighting and strong resources in the aesthetic and functional aspects of lighting design, the lighting designer is able to see both perspectives. Added to this is extensive knowledge of the fixtures available on the market and the ability to speak the electrician's language.

Thoughtful architects, engineers, and lighting designers are leading research into ways to satisfy vision needs with minimal energy use. The Illuminating Energy Society (IES) is a research, standards, and publishing organization that develops stable scientific bases for lighting, while remaining aware of its artistic aspects. The combination of science and art make lighting design a truly architectural discipline.

Lighting Design Process

The designer of the lighting for a project begins by considering the integration of daylight with electric light, and the energy use relationships between electric lighting, daylighting, and heating and cooling. The interior designer should be involved in the coordination of the interior layout of the space with the lighting design. Decisions are made regarding the sources, characteristics, and equipment requirements for electric lighting. The visual requirements for specific tasks and occupants are assessed.

Three basic types of lighting are employed to achieve these effects. **Focal and task lighting** focuses our attention and helps us identify what is important. (See Figure 17.2) **Ambient lighting** uses the glow of diffuse and indirect lighting to illuminate space. (See Figure 17.3) The **sparkle** of decorative light sources and reflections draws attention to visually exciting elements. (See Figure 17.4)

"Luminaire" is another name for a lighting fixture. A luminaire houses and provides electricity for the light source and directs its output.

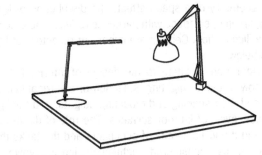

Figure 17.2 Task lighting

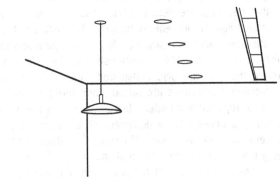

Figure 17.3 Ambient lighting

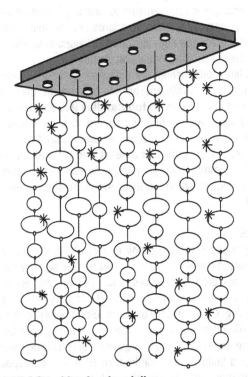

Figure 17.4 Sparkle of a chandelier

The geometry of a space affects the distribution of light. Breaking up an open space with partitions reduces the efficiency of the illumination. Ceilings are important reflectors of light to tasks below.

Considerations relating to the design of interior lighting include how to use daylighting, selection of electrical lighting fixtures and their spacing and mounting heights, and the light reflectance levels of interior surfaces. The role of decorative lighting and the visual needs of occupants and the tasks they perform are also considered. Lighting design is involved in determining the location, relationships, and psychological effects of brightness and shadow patterns. The color of both light sources and the surfaces they illuminate are also important aspects.

The process of lighting design includes several phases similar to other design processes. The first phase involves establishing project lighting and energy budgets. Task analysis then investigates the nature of tasks, including their repetitiveness, variability, and duration. The characteristics of the individuals performing the task are also considered. During the design stage, detailed suggestions are raised, considered, modified, accepted, or rejected in an interactive process with the architect, lighting designer, interior designer, and engineers to produce a detailed, workable design. The evaluation stage involves analyzing the design solutions for cost and energy. The results of the process go to the architectural group for final overall project evaluation.

Lighting Codes and Standards

As we indicated earlier, in recent years, there have been significant changes in requirements for building lighting energy efficiency. Energy codes and standards set minimum efficiency requirements for new and renovated buildings. In addition, LEED v4 give credits for interior lighting strategies.

Building codes generally establish minimum levels of illumination. It is usually appropriate to keep the general area illumination at less than 30 **footcandles (fc**, equal to 300 lux) with additional task lighting as required. In some jurisdictions, all habitable and occupiable spaces are required to have daylight or electrical lighting of at least 6 fc (60 lux) on average at 30" (762 mm) above floor level. Average illumination provided by electric lighting in toilet rooms must be at least 3 fc (30 lux).

CODES

Building codes regulate lighting energy efficiency and specify the permitted maximum number of **watts per square foot (W/m²)** of floor area. Energy restrictions commonly apply to all buildings over three stories, and to all building types except low-rise housing. Codes usually allow trade-offs between energy efficient building envelope components and energy use by HVAC or lighting. Interior lighting energy use can usually be calculated by either a building area method, or on a space-by-space basis. Code requirements typically apply to new construction and additions, and do not require alteration or removal of existing systems, although some efforts at relamping existing fixtures may be required.

The National Electrical Code (NEC) has strict requirements for access to electrical components. All electrical boxes, including those for lighting, must allow access for repairs and wiring changes at any time. All lighting fixtures must be placed so that both the lamp and the fixture can be replaced when needed. This is especially important when they are used with architectural elements such as ceiling coves.

The process of meeting the code requirements involves extensive calculations and reporting, provided by the electrical engineer or lighting designer using special software; contractors sometimes provide the documentation. Since the lighting energy allowance meshes with building envelope and HVAC requirements, the entire architectural and engineering team is involved in meeting code requirements. With good design, lighting levels even lower than limits set by codes can be achieved. Energy efficiency code requirements mandate automatic shut-off provisions for interior lighting.

Some types of facilities are not included in the total energy use calculations, for example spaces specifically for the visually impaired, enclosed retail display windows, display lighting in galleries or museums, lighting integral to advertising or directional signage, and lighting for theatrical purposes.

The 2015 International Residential Codes (IRC) requires that every habitable room and bathroom, hallway, stairway, attached garage, and outdoor entrance must have a minimum of one lighting outlet controlled by a wall switch. A single wall switch in rooms other than a kitchen and bathroom can control one or more receptacles for plugging in lamps. One lighting outlet is required in each utility room, attic, basement, or under-floor space that is used for storage or that contains equipment that may require servicing.

Chapter 40 of the 2015 IRC covers electrical devices and luminaires and their installation. These include requirements for protecting energized parts from contact, and restrictions on luminaires in wet or damp locations and clothes closets. Track lighting requirements are also covered.

Fire-tested and labeled lighting fixtures should be used on interior projects. Only certain types of fixtures are allowed in fire-rated assemblies. When a lighting fixture is placed in a wall or ceiling of combustible material, the mechanical part of the fixture must be fully enclosed. Usually, noncombustible materials must be sandwiched between the fixture and the finished surface.

The method of attachment of a fixture to the outlet box is related to its weight and the type of fixture. The outlet box can usually support fixtures up to 50 pounds (22.7 kg). Heavier fixtures require additional support.

Jurisdictional authorities are principally involved in energy budgets and lighting levels. There are a variety of codes and standards related to lighting. (See Table 17.1) Authorities may include the US Department of Energy (DOE) and General Services Administration (GSA) if federal funds involved.

TABLE 17.1 CODES AND STANDARDS RELATED TO LIGHTING

Type	Title
Codes	International Energy Conservation Code (commercial and residential buildings)
	NFPA 70 National Electrical Code®
	NFPA 900—Building Energy Code based on ASHRAE Standards 90.1 and 90.2
Standards	ANSI/ASHRAE/IES Standard 90.1—Energy Standard for Buildings Except Low-Rise Residential Buildings
	ANSI/ASHRAE Standard 90.2—Energy Efficient Design of Low-Rise Residential Buildings
	ANSI/ASHRAE / USGBC / IES Standard 189.1—Standard for the Design of High-Performance Green Buildings

STANDARDS

Lighting standards are set by a variety of authorities, depending upon the type of building, whether it is government owned or built, and where it is located. Agencies that publish standards and code requirements for lighting include:

- US Department of Energy (DOE)
- General Services Administration (GSA)
- National Fire Protection Association (NFPA), including the National Electrical Code (NEC)
- American Society of Heating, Refrigeration and Air Conditioning Engineers (ASHRAE)
- Illuminating Engineering Society (IES)
- National Institute of Science and Technology (NIST)

Energy budgets and lighting levels set by these standards affect the type of lighting source, the fixture selection, the lighting system, furniture placement, and maintenance schedules.

ACCESSIBILITY AND SAFETY REQUIREMENTS

Lighting fixtures are marked for wall mounting, under-cabinet mounting, ceiling mounting, or covered ceiling mounting. Wall-mounted lighting fixtures (**sconces**) mounted between 27" and 80" (686 and 2032 mm) above the finished floor must meet the ADA maximum 4" (102 mm) projection limit where applicable. (See Figure 17.5)

Each lighting fixture is manufactured and tested for a specific location. Those approved for damp locations are labeled "Suitable for Damp Locations." Lighting fixtures listed for a wet location can also be used in damp locations.

LIGHT AND VISION

Light is electromagnetic radiation that the unaided human eye can perceive. Light has the characteristics of a wave, but is released in quantum units. It radiates equally in all directions

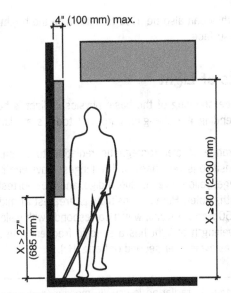

Figure 17.5 ADA limits of protruding objects

Source: Redrawn from 2010 ADA Standards for Accessible Design, Figure 307.2

and spreads over a larger area as it emanates from its source. As it spreads, it diminishes in intensity. Light causes electrons to move, a phenomenon that is used by a **light meter** to indicate the amount of incident electromagnetic radiation. (See Figure 17.6) These characteristics of light apply to both daylighting and electric lighting.

A light meter converts radiant energy to an electrical current and indicates the intensity of illumination. Light meters are typically used to measure the level of illumination incident on a

Figure 17.6 Light meter

surface; they can also be used to determine the brightness of a diffuse surface.

Physics of Light

Some understanding of the basic physics of light is helpful in comprehending the changing nature of today's available light sources.

The range of electromagnetic radiation to which human eye is sensitive is very narrow. The longest wavelength is red, then orange, yellow, green, blue, indigo, and the shortest visible wavelength, violet. Human sensitivity is greatest at mid-range, around 550 **nanometers**, which corresponds with yellow light. Each wavelength of light has a specific frequency, usually referred to as cycles per second or hertz (Hz).

Electromagnetic radiation is usually measured in nanometers or angstroms. There are 10 million nanometers per meter.

Visible light is emitted by the vibration of electrons. The temperature difference determines the rate of vibration; as the temperature rises, the rate of both the amount and energy level of molecular activity and radiation emitted increases.

Radiation within the visible range is referred to as incandescent radiation, and is emitted in bands of varying frequencies. A light source producing over the entire visible spectrum in approximately equal quantities produces white light.

Our eyes adjust to light. Light sources with large differences in **chromatic** (relating or produced by color) content may all appear white after a short accommodation period. Only when sources differing widely in chromaticity are view sided by side is variety in whiteness noticed on colored objects and neutral surfaces.

Although waves bend, light does not. We usually think of light as rays that travel in a straight line until they encounter some object where they are absorbed, transmitted, or reflected. Generally, all three occur at any given surface, but one or two usually predominate. The proportion of light absorbed, transmitted, or reflected depends on the type of material and the angle of incidence of the light rays as measured from the perpendicular to the surface. Grazing angles higher than 55 degrees tend to reflect light. Even clear glass or plastic behaves as a mirror when the angle is high enough.

REFLECTANCE

The angle of incidence equals the angle of reflection. **Reflectance** is defined as the ratio of reflected light to incident light. The amount of absorption and reflection depends on the type of material and angle of light incidence. **Specular** reflection is reflection on a smooth surface such as polished glass or stone, where the angle of incidence equals the angle of reflection. Most materials reflect light in both specular and diffuse reflections.

The **reflectance factor (RF)** indicates how much of the light falling on a surface is reflected. It is equal to reflected light divided by incident light. Reflected light is always less than incident light, so the RF is always less than 1. A little light is always reflected, so the RF is always greater than 0. The RF of a white surface is around 0.85, and of a black surface around 0.05.

Wall reflectance is rarely equal to that of the surface finish of objects in a space. The average of reflectances should be weighted by the area of all reflecting surfaces, including doors, window treatments, whiteboards, and cabinets.

Floor reflectance is not very important for desktop height tasks, but gives the impression of more and better-distributed light. It contributes to illumination of free-standing tasks such as inspecting the sides or undersides of objects, where otherwise additional lighting would be needed. Floor reflectance also contributes to general visual comfort as it reduces brightness contrasts.

Visual comfort depends on the relationship (luminance ratio) between tasks or other elements and their surroundings. (See Table 17.2)

To control reflected glare from desktops, use light-colored, diffusely reflecting surfaces with 25 to 40 percent reflectance. Dark polished wood or a glossy finish is most likely to cause glare.

TRANSMITTANCE

Luminous transmittance is a measure of a material's capability to transmit incident light. The **transmission factor** (transmittance or coefficient of transmission) is the ratio of the total transmitted light to the total incident light, and is used only for materials that transmit various component colors equally. Clear glass has a transmission factor between 80 and 90 percent; frosted glass, 70 to 85 percent; and solid opal glass, 15 to 40 percent, with rest of the light absorbed or reflected. Translucent materials result in diffuse transmission.

Visible transmittance (VT) quantifies the amount of visible light that passes through glazing. It ranges from 0.9 for very clear glass to less than 0.1 for highly reflective or tinted glass.

TABLE 17.2 MAXIMUM LUMINANCE RATIOS FOR COMFORT

Educational/Commercial Surfaces	Average Ratios
Between task and adjacent surroundings	1 to 1/3
Between task and more remote darker surfaces	1 to 1/10
Between task and more remote lighter surfaces	1 to 10
Between luminaires or fenestration and adjacent surfaces	20 to 1
Anywhere within normal field of view	40 to 1

TABLE 17.3 LIGHTING UNITS

IP unit	SI unit	Quality Measured	Comments
Lumen (lm)	Lumen (lm)	Supply of light	Unit of measurement of the flow of light energy, or luminous flux. Measurement applies to source only, not as eye sees light.
Candlepower (cp) or candela (cd)	Candela (cd)	Luminous intensity	Candlepower describes intensity of a light beam in any direction. One candela has approximate horizontal luminous intensity of a wax candle.
Footcandle (fc)	Lux (lx)	Illuminance	Unit of lighting intensity. One footcandle is equal to 10.76391 lux.
Candela per square foot	Candela per m^2	Luminance	Objective measurement of a light meter, the amount of light reflected off an object's surface that reaches the eye.
Lumens per square foot	Usually in lumens per m^2	Luminous exitance	Total lumens flux density leaving (exiting) a surface, irrespective of directivity or viewer position.

MEASURING INTENSITY

The human eye's visual sensitivity is measured based on a logarithmic progression; a ten-times increase in intensity is perceived as a doubling in lighting level. Doubling the actual energy level of lighting produces a barely perceptible change.

Lighting is measured in IP units including lumens, candlepower, and footcandles, among others. Some of these are also SI units, while others have SI equivalents. (See Table 17.3)

Perception of brightness is a function of an object's actual luminance, adaption of the eye, and brightness of adjacent objects. A person judges the brightness of an object relative to the brightness of its surroundings, rather than the absolute luminance as measured by a light meter. It is more important to design for people than for light meters. Consequently, it is generally more important to consider perceived brightness than objective luminance.

Brightness constancy is the ability of the brain to ignore differences in luminance under certain conditions. For example, the ceiling of a room with windows at one end appears to have a constant brightness, when it actually varies. Lights generally appear brighter at night than by day, but register the same luminance on light meter. Interior designers can use **brightness ratios** for designing interior lighting and surface reflectances. (See Table 17.4)

TABLE 17.4 MAXIMUM RECOMMENDED BRIGHTNESS RATIOS

Ratio	Description	Example
3:1	Task to immediate surroundings	Paper to worksurface
5:1	Task to general surroundings	Paper to wall nearby
10:1	Task to remote surroundings	Paper to distant wall
20:1	Light source to large nearby area	Lighting fixture to adjacent ceiling

Vision

Vision is our ability to gain information through light entering our eyes. The eyes convert light into electrical signals that our brain processes. What we perceive is the work of the eye and brain, plus our associations, memory, and intelligence. The way in which an object is illuminated is critical to how we perceive it visually. What we see is the electromagnetic radiation rejected by an object and reflected to our eyes. We identify the color of an object by the color it reflects.

What an individual perceives visually depends on his or her own subjective experience. However, people do share common experiences, especially within a culture, so similarities exist.

Designers can manipulate a visual message to make it consistent with or somewhat independent from the actual physical reality. Four factors that affect vision can be combined to create the desired effect. These include illumination level, contrast, size, and time of exposure.

An adequate level of illumination for most visual tasks is between 10 and 20 fc (100 and 200 lux). The benefit of illumination above 30 fc (300 lux) decreases, and visual effectiveness may decline above 120 fc (1200 lux).

The contrast between a visual object and its background provides important information. Strong contrast (for example, 100:1) is useful for emphasis.

The size of a visual object also matters. When the object size is reduced, it is necessary to increase illumination level, contrast, or time of exposure.

Because of the orientation of our eyes in our body, vertical walls are the dominant surfaces in most spaces. In larger spaces, the ceiling and floor become more visually dominant.

HUMAN EYE AND BRAIN

The eye focuses light on the light-sensitive rod and cone cells of the retina. (See Figure 17.7) **Cone cells** give the eye its ability to see detail sharply, and also to detect color. The clarity of our center of focus is due to the cone cells packed into a

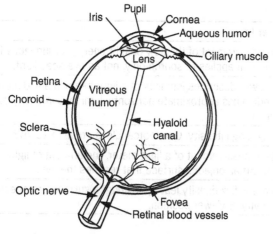

Figure 17.7 Human eye
Source: Redrawn from Schematic_diagram_of_the_human_eye_en.svg.
png public domain.png

tiny area near the fovea. **Rod cells** are found farther from the fovea. Rods can operate at lower light levels than cones, but are less sensitive to bright light; they help us see at night. Because rods lack the color sensitivity of cones, the eye perceives color poorly in dim light.

It takes a large increase in light for the eyes to notice a small increase in brightness. The eye adapts very well to gradual changes in brightness associated with daylighting, but full adaptation from dark to bright can take over 40 minutes.

The human eye is most sensitive to yellow-green light, and less sensitive to colors such as red or blue.

VISUAL ACUITY

Factors affecting **visual acuity** (sharpness of vision) include the task, the lighting condition, and the observer. (See Table 17.5) Basic visual tasks often require perception of low contrast, fine detail, and gradations of brightness.

TABLE 17.5 VISUAL ACUITY FACTORS

Type	Primary Factors	Secondary Factors
Task	Size, luminance (brightness), contrast (including color contrast), exposure time needed or given	Type of object (required mental activity, familiarity), degree of accuracy required, task (moving or stationary), peripheral patterns
Lighting condition	Illumination level (illuminance), disability glare, discomfort glare	Luminance ratios, brightness patterns, chromaticity
Observer	Condition of eyes (health, age), adaptation level, fatigue level	Subjective impressions, psychological reactions

SUBJECTIVE BRIGHTNESS

The human eye detects luminance over a range greater than 100 million to one. Vision at lower illumination levels improves after an adaptation time of 2 minutes for cones and up to 40 minutes for rods; the process is much faster going from dark to light.

An increase in task brightness due to increased illumination at first results in significant improvement in visual performance, but additional increases have smaller and smaller benefits. When detail discrimination is not needed, improved illumination will not necessarily improve performance. Consequently, high brightness is an expensive way to improve visual performance. It is possible to increase performance by reducing background brightness, which increases the relative brightness of the task by increasing the eye's sensitivity to light. This method is used in museum lighting, where keeping lighting levels low protects objects on display.

CONTRAST AND ADAPTATION

Contrast is important for detailed visual tasks such as reading print on paper. Visual efficiency falls off rapidly when the surrounding brightness is greater than that of the work. Optimum work efficiency is achieved when background brightness is between 10 and 100 percent that of the work itself. Areas surrounding task locations should average one-third the level of the task lighting, but not less than 20 fc (215 lux). Circulation and seating areas adjacent to task spaces but without visual tasks can have an average illumination level as low as 10 fc (108 lux).

Having the same levels of background and surface luminance is best, although a ratio of 3:1 is acceptable in most circumstances. High background luminance makes an object look darker, providing better outline detail discrimination. However, it also makes surface examination more difficult.

With lower background illumination levels, supplementary lighting provides visual interest and balance in a space. Supplementary lighting can be used for selective highlighting of artwork and accent wall segments. It can define the boundary of a space and help to create a visually coherent environment. Slight increases in the illumination of surroundings can brighten an otherwise gloomy space and enhance the appearance of architectural materials.

As the brightness of surrounding surfaces increases, the eye becomes more sensitive to brightness differences. Reducing brightness differences as the general level of brightness is increased aids visual comfort. Low adaptation levels are useful in theaters, lecture halls, storage, and restaurants, where bright lights tend to be annoying.

AGING AND THE EYE

Visual performance of healthy eyes decreases with age. People in their twenties tend to have visual performance four times better than people in their fifties, and eight times better than people in their sixties. Lens rigidity causes an increase in the minimum focusing distance with age.

In the normally aging eye, clouding of the cornea causes decreased sensitivity, resulting in opacity that dims and blurs the viewed image. Light entering the lens is scattered by internal reflections from opaque particles. The eye develops a heightened sensitivity to glare, intolerance of the blue through ultraviolet (UV) end of the spectrum, and an overall requirement for higher illuminace levels.

Lighting design for the aging eye requires very careful selection and placement of luminaires, increasing use of indirect lighting, and special attention to the spectrum of light sources used. More light is needed, and sources of glare and peripheral light should be eliminated.

Color and Light

What the eye perceives as color is related to the wavelength of the electromagnetic radiation. As mentioned previously, red is the longest wavelength in the visible spectrum and violet the shortest. White light is a mix of various wavelengths of visible light. Clear skylight, especially from the northern sky (in the Northern Hemisphere), is rich in the blue end of spectrum, making it excellent for rendering of cool colors.

When red, green, and blue colored lights are mixed, they combine together in an additive process to create white light. Combining complementary colors (red plus blue-green, blue plus yellow, green plus magenta) produces white or gray light. Unlike colored light, colored pigments are mixed by a subtractive process.

COLOR TEMPERATURE

Color temperature is based on the temperature to which a light-absorbing body (blackbody) must be heated to radiate a light similar in color to the color of the source in question. The temperature is measured in degrees Kelvin.

The Kelvin temperature scale is introduced in Chapter 12, "Principles of Thermal Comfort."

Color temperature is only used for light sources that produce light by heating, such as incandescent lamps. Other sources are assigned a **correlated color temperature (CCT)**, for which there is no relation between their operating temperature and the color produced.

Color temperature is an expression of the dominant color, not the spectral distribution of various wavelengths. Two sources with the same color temperature can have noticeably different appearances.

Color temperature is mostly used to describe the warmness or coolness of a light source. (See Figure 17.8) Low-color temperature (warm) light sources tend to render red colors well. High-color temperature (cool) light sources tend to render blue colors well.

People prefer white light sources between 3000°K (warm white) and 4100°K (cool white). Warmer colors are preferred

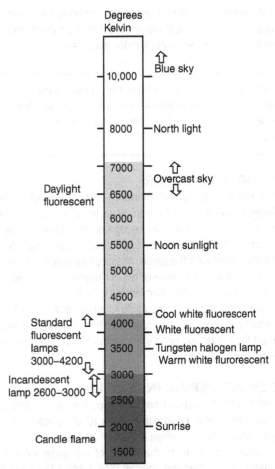

Figure 17.8 Approximate color temperatures of common light sources

Source: Redrawn from Walter T. Grondzik and Alison G. Kwok, *Mechanical and Electrical Equipment for Buildings* (12th ed.), Wiley 2015, page 628

when illumination levels are low and to complement skin tones. Cooler colors are preferred at high light levels and in hot climates. Use 5000°K (cold white) light when very accurate color judgments must be made.

OBJECT COLOR

The color of a surface is due to the selective reflectance of the light falling on it. For example, red reflects most of the red light and small amounts of other colors. This produces problems when the source of illumination has a poor mix of colors. Where object color is very important, light source selection is critical. A full spectrum light source or one rich in the desired color should be used.

Light transmission through colored glass or plastic is similar to reflection. The dominant light color (for example, red) is primarily transmitted, with most other colors absorbed. Consequently, white light passing through red appears the dominant color (red).

Black absorbs all light equally and reflects very little. White reflects all light equally and absorbs very little. An object that does not exhibit selective absorption is considered colorless.

The perceived color of an object depends heavily on the illuminating source. The illuminant must contain the color of the object in order for the object's color to be seen.

The color of a light source affects furnishings, paints, and pre-finished construction materials such as carpets and floor tiles. Interior designers need to choose colors under the proper illumination type.

REACTIONS TO COLOR

As described earlier, color constancy is the ability of the brain to eliminate some of the differences in color due to differences in illumination, so that we can recognize objects under the varied colors of lighting sources. Light colors appear lighter than dark colors even with the same measured luminance. Color can define spaces within an area of equal illumination. All colors appear less saturated when illumination is high.

Color constancy is not possible when more than one type of light source is used simultaneously, so it is best not to mix very different light sources, or to use clear window glazing adjacent to tinted window glazing.

COLOR RENDERING INDEX (CRI)

The effect of light sources on color appearance is called color rendition. The **color rendering index (CRI)** indicates the degree to which the perceived colors of objects illuminated by a test source conform to the colors of the same objects as illuminated by a reference source. (See Table 17.6) The CRI is a measure of how closely the illuminant approaches daylight of the same color temperature. The CRI values of two lighting sources cannot be compared unless their color temperatures are equal or quite close.

The color rendering index has some limitations, and should be used with care. It can only compare light sources of the same color temperatures. Even then, a specific color may not appear natural. Actual tests, examining the objects with the type of light source by which they will be illuminated, work best for color selection or matching.

The CRI is unreliable when applied to light emitting diodes (LEDs), which today can reproduce color better than incandescent lamps. New measurements and classifications are currently

TABLE 17.6 CRI RATINGS

CRI Rating	Description
CRI 100	Daylight; perfect match with standard
CRI greater than 90	Where color is important
CRI 90	Quite good
CRI above 80	Wherever possible
CRI 70	May be acceptable for some purposes
CRI below 60	Unacceptable in most situations

being considered to address this issue. The National Institute of Standards and Technology's Color Quality Scale (CCQS) is more rigorous than CRI; other new measurements are also achieving acceptance.

Quantity of Light

Lighting design involves establishment of lighting power budgets. IES energy standards encourage the use of daylighting and combined task/ambient lighting design where high levels of task lighting are required. (See Table 17.7) IES illumination standards include recognition of fatigue and task familiarity as factors in determining illumination levels. Differences in surface reflectance have a great impact. A single illumination scheme is often inadequate where there are widely different visual tasks.

Glare

Glare is a result of excessive contrast, or of light coming from the wrong direction. It can occur due to daylighting or as a result of lighting fixture selection and location. Glare results in discomfort and eye fatigue as the eye repeatedly readjusts from one lighting condition to another.

Direct (discomfort) glare is glare caused by light sources in the field of vision. (See Figure 17.9) Reflected glare (veiling reflection) is reflection of a light source in a viewed surface. The severity of glare is affected by the adaptation level of the eyes, the apprehended size of the glare source, luminance ratios, room size and surface finishes, and the size and position of lighting fixtures and windows.

Direct sunlight or reflected sunlight from bright, shiny surfaces can be disturbing or even disabling, and should not be permitted to enter the field of view of the building's occupants. The contrast between the bright outside environment viewed through a window and the darkness of the interior space can produce glare. Windows or skylights within the normal field of vision of the building's occupants can appear distractingly bright next to other objects.

TABLE 17.7 IES RECOMMENDED ILLUMINANCE TARGETS IN FOOTCANDLES

Area or Activity	Under 25	25 to 65	Over 65
Passageways	2	4	8
Conversation	2.5	5	10
Grooming	15	30	60
Reading/Study	25	50	100
Kitchen Counter	37.5	75	150
Hobbies	50	100	200

Source: *Light in Design—An Application Guide*, IES Consumer Publication CP-2-10

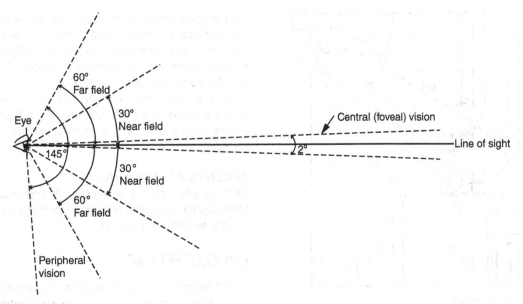

Figure 17.9 Fields of vision
Source: Redrawn from Walter T. Grondzik and Alison G. Kwok, *Mechanical and Electrical Equipment for Buildings* (12th ed.), Wiley 2015, page 599

There are two basic methods to avoid glare and reduce brightness contrasts: sensitive interior design and daylighting controls.

Daylighting controls and shading are covered in Chapter 6, "Windows and Doors." Issues involving glare are introduced there as well.

Interior design strategies to control glare begin with assigning tasks where adequate natural light is available. Surface brightness should change gradually from outside to inside. Use light colors and high reflectances for window frames, walls, ceilings, and floors. Incoming direct and reflected sky light hits the floor first, while reflected sunlight hits the ceiling first. Ideally, furniture should be oriented so that daylight comes from the left side or the rear of the line of sight. Varying the direction and intensity of light sources and creating soft, overlapping shadows helps to diffuse light.

Use supplemental electric lighting in dark areas, rather than oversupplying daylight elsewhere for an adequate level. Using light colors and high reflectances far from openings increases light in dim areas.

DIRECT GLARE

Direct glare occurs when the light source is within the field of vision. Direct glare depends on the brightness, size, and position of each light source in the vision field. It is a response to the attraction of the eye to the highest luminance (brightness) in a scene. The repeated alternation of the eye from the overall scene to the brightest point causes visual fatigue.

Direct glare is generally controlled by keeping the brightness of luminaires, ceiling areas, and walls below certain values. (See Table 17.8) Unavoidable glare sources should be located as far from workstations or the field of vision as possible. (See Figure 17.10) Light sources should be spread out while maintaining necessary illumination levels.

REFLECTED GLARE

Reflected glare (veiling reflection) happens when a light source is reflected in a surface within view. Reflected glare becomes a problem when the viewer is looking downward at a worksurface. Veiling reflections involve both the source and the task, with the principal light sources near observer being the main contributors. Veiling reflections are worse when the incident angle equals the viewing angle.

The terms "reflected glare" and "veiling reflections" can be used interchangeably. People often use the former term when referring to specular surfaces, and the latter when referring to glare on dull or matte finish surfaces.

TABLE 17.8 MINIMIZING DIRECT GLARE

Surface	Reflectance
Ceiling	80 to 90 percent
Walls	40 to 60 percent
Floors	20 to 40 percent
Furnishings and equipment	25 to 45 percent

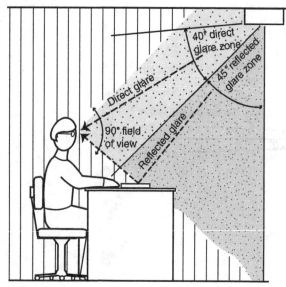

Figure 17.10 Glare zones

Source: Redrawn from Walter T. Grondzik and Alison G. Kwok, *Mechanical and Electrical Equipment for Buildings* (12th ed.), Wiley 2015, page 612

Figure 17.11 Reflected glare optimal position

To reduce veiling reflections, minimize reflected glare through physical arrangement of sources, task, and observer. The optimal position for the light source is usually to the left and slightly forward of the task. (See Figure 17.11) Minimizing objectionable brightness eases eye adaptation. Selecting light sources designed for minimum glare reflectance also helps as well as controlling the quality or nature of the task to accommodate reflections.

Lighting Effects

Patterns of light and shadow resulting from illumination sometimes create unexpected lighting effects. Computer modeling for daylighting can help reveal these.

DIFFUSION
Diffusion deals with the degree to which light is shadowless. It is a function of the number of directions from which light

impinges on a particular point and their relative intensities. The level of diffusion can be judged by the depth and sharpness of shadows. Well-diffused illumination from multiple sources and reflections from high room surfaces produce soft multiple shadows that do not obscure visual tasks.

Adding some directional lighting increases interest by producing shadows and brightness variations. Highly directional grazing lighting is required to examine textures or surface imperfections.

ACCENTS AND SPARKLE
Visual attention can be drawn by high brightness. Sparkle is generated by small, high-brightness sources creating points of interest and visual excitement.

DAYLIGHTING

Daylighting improves energy efficiency by minimizing electricity use for lighting and reducing the associated heating and cooling loads. Prioritizing the use of daylight is an important element of sustainable design, acting to limit global warming by reducing carbon emissions.

Research shows that daylight is an important factor in human behavior, health, and productivity. Windows add views and connect us with outdoor conditions. However, steps must be taken to control glare.

Daylighting can save around half of all lighting energy used by buildings, and up to 70 percent for offices and schools, providing significant reductions in total energy consumption.

European building codes require access to both views and daylighting, as do LEED credits.

History of Daylighting

Early artificial light sources were historically both poor quality and expensive. Until around 1950, when inexpensive fluorescent lighting became available, all buildings relied on daylight.

Medieval European structural innovations such as Gothic vaults made very large windows possible. Until the twentieth century, buildings with E- and H-shaped floor plans were rarely more than 60 feet (18 m) deep so that no point was more than 30 feet (9 m) from windows that provided ventilation and daylight. Today's lower ceilings allow daylight to reach about 15 feet (4.5 m).

In the twentieth century, electric lighting seemed easier to design and use, allowing the architect to ignore the impact of window locations for daylighting. The 1970s energy crisis led to a re-examination of the potential of daylighting.

Characteristics of Daylight

Sunlight is a highly efficient source of illumination, and a comparatively cool color source. Daylight varies with the season, the time of day, the latitude, and weather conditions. More sunlight is available in summer than in winter, and the day's sun peaks at noon. An overcast day is very different from a day with a clear sky, and conditions can change several times during a day. Most US climates have enough suitable days to design for both overcast sky and clear sky with sunlight. Of course, sunlight is unavailable until dawn and after dusk.

On a bright day, sunlight provides illumination levels fifty times as high as the requirements for artificial illumination. Direct sun may be desirable for solar heating in winter, but the glare from direct sun must be managed. Indirect sunlight produces illumination levels between 10 and 20 percent as bright as direct sun, but still higher than needed indoors.

HUMAN FACTORS AND DAYLIGHTING

Daylight helps us relate the indoors to the outdoors. Colors appear brighter and more natural in daylight. The variations in light over the course of a day and in varying weather conditions stimulate visual interest. People need full spectrum light, which is a main characteristic of daylight. Without daylight, people tend to lose track of time, are unaware of weather conditions, and may feel disoriented. Views allow us to look into the distance to avoid eye fatigue. The most preferred view includes sky, horizon, and ground.

Studies of classrooms, windows, daylight, and performance by the Heschong Mahone Group (*Windows and Classrooms: A Study of Student Performance and the Indoor Environment*, CEC Pier 2003) have shown that students with more daylighting in their classrooms progressed faster on math and reading tests than those with less daylighting. Sources of glare negatively impact student learning, so control of sun penetration with windows and blinds is important. Research on retail stores found that daylight correlated with increased monthly sales (Heschong Mahone Group, *Skylighting and Retail Sales*, PG&E 1999).

Daylighting Design

"Daylighting design is part of the fundamental building design from the first line drawn." (Norbert Lechner, *Heating, Cooling, Lighting: Sustainable Design Methods for Architecture* (3rd ed.), Wiley 2009, page 394)

Daylighting has a greater impact on building architecture than electric lighting, affecting fenestation, building orientation, and shape. Historically, building orientations and configurations worked with window openings and interior finishes to ensure sufficient daylight in interior spaces.

There are three components of daylight in a building: the sky component, the externally reflected component, and the internally reflected component. The work of interior designers is critical to the internally reflected daylight in a building.

The quality and quantity of daylight depends on sky conditions and determines how it can be used. When the sun does not penetrate a room directly, the amount of natural light available depends on how much of the sky can be viewed through windows and skylights, and the relative brightness of those parts of the sky. The sky at the horizon is around one-third as bright as directly overhead, so more light is available nearer to the ceiling. The ground may reflect significant amounts into windows directly or by reflection on interior surfaces.

The shape and surface finishes of a space have an impact on daylighting. Tall, shallow spaces with high surface reflectances are brighter than low, deep rooms with dark, cold surfaces and windows only at the narrow end. It takes fewer bounces off the walls for light to get deep into a room when the windows are high on the wall. High windows distribute light more evenly to all walls and allow light to penetrate into the interiors of large, low buildings. The ceiling and back wall of the space are more effective than the side walls or floor for reflecting and distributing daylight. Tall objects such as office cubicle partitions or tall bookcases can obstruct light both direct and reflected light.

The level of daylight illumination diminishes as it goes deeper into the interior space. In order to reduce glare, you need to design a gradual transition from the brightest to darkest parts of the space. The amount of light about 5 feet (1.5 meters) from the window should not be more than ten times as bright as the darkest part of the room. In a room with windows on only one wall, the average illumination of the darker half of the room should be at least a third of the average illumination level of the other half with windows. It is best to allow daylight from two directions for balance, preferably at the end of the room farthest from the main daylight source. (See Figure 17.12)

Light shelves are used to reflect sunlight into building interiors. (See Figure 17.13) They are part of the exterior architecture of a building, and must be planned as such. They work best in spaces with high ceiling reflectances.

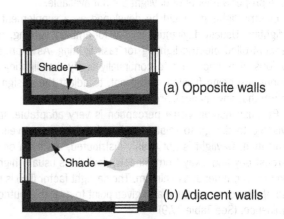

Figure 17.12 Bilateral daylighting

Source: Redrawn from Walter T. Grondzik and Alison G. Kwok, *Mechanical and Electrical Equipment for Buildings* (12th ed.), Wiley 2015, page 251

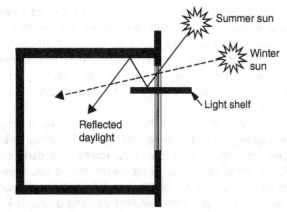

Figure 17.13 Light shelf

DAYLIGHTING DESIGN PROCESS

Planning for daylighting involves a complex systems integration process. During the conceptual phase, the architect designs the building's form, orientation, layout, and major apertures. In the design development phase, the architect, interior designer, and engineers specify materials and interior finishes and zoning for integration with electric lighting. They also coordinate control systems with occupancy schedules and commissioning procedures. The process continues during occupancy with fine-tuning and maintenance of systems. Post-occupancy evaluation verifies levels of satisfaction, visual comfort, and lighting system performance.

Many daylighting simulation programs provide realistic visual daylighting output, with varying degrees of accuracy.

Daylighting must be integrated with the view, natural air movement, acoustics, heat gain and loss, and electric lighting. Daylighting will not save energy unless lights are turned off, so lighting zones must be circuited separately, with lights turned off or dimmed when the natural light is adequate and left on where proper amounts of daylight are not available.

Electric lighting should be developed as a supplement to daylighting. Usually daylighting is used for ambient lighting, with user-controlled electric lighting for task lighting. As a project develops, it is important to continually verify that lighting intentions are being fulfilled, and if not, to adjust the design or confirm that any changes are appropriate.

Because human visual perception is very adaptable, it is advisable to design to relative rather than absolute levels of illumination. Daylight is not evenly distributed, but even on an overcast day and away from window, daylight is usually higher than the recommended minimum. The daylight factor (DF) is the ratio of interior illuminance at a given point to available outdoor illuminance. (See Table 17.9)

Lighting designers use daylight factor rather than footcandles when referring to daylight illumination.

TABLE 17.9 DAYLIGHT FACTORS

Types of Spaces	Recommended Minimum DF
Offices, classrooms	2
Corridors, bedrooms	0.5
Lobbies, reception areas	1
Art studios, galleries	4 to 6
Industrial, laboratories	3 to 5

An open space plan helps bring light into the building's interior. Glass partitions help this, but may not provide adequate visual privacy. Translucent materials, venetian blinds, or glass located only above eye level can remedy this.

Light colors bring light further into the interior and diffuse to reduce dark shadows, glare, and excessive brightness. In descending order importance, high reflectivity is important on the ceiling, back wall, side walls, floor, and small pieces of furniture, in order to bring daylight into the space.

Daylighting and Fenestration

In a daylighting design, heat and light are controlled through the form of the building. For example, in a Middle Eastern mosque located in a sunny climate, limited sunlight enters the building through small windows high in a decorated ceiling, and then is diffused as it bounces off interior surfaces. The large windows in a Western European cathedral, on the other hand, flood the interior with light colored and filtered through stained glass. Direct sun can cause a large quantity of heat gain through unshaded windows; east and west windows create glare at low sun altitudes in the morning and evening, and add unwanted summer heat gain. To take advantage of sunlight without an excess of heat or glare, the building should be oriented so that windows are on the north and south sides. American architect Frank Lloyd Wright shaded the west and south sides from the most intense sun with deep overhangs.

True daylighting is actually passive solar design, involving the conscious design of building forms for optimum illumination and thermal performance. It is most challenging in workspaces with varied and demanding tasks, and least challenging in public spaces where comfort standards are less stringent and controlled lighting is less important.

For information on passive solar design, see Chapter 2, "Designing for the Environment." For more information on windows and skylights see Chapter 6, "Windows and Doors."

Interior spaces need high ceilings and highly reflective room surfaces for the best light distribution. The sun is constantly changing in direction and intensity. Light bouncing off the ground outside ends up on the ceiling. Ground reflected light is both bright and diffuse, but adjacent buildings or trees often shade the ground.

Daylighting is generally broken into two categories: sidelighting through windows in walls, and toplighting through skylights in roofs and clerestory windows high up on walls.

SIDELIGHTING

Sidelighting admits light and allows views of the exterior during the day, and of the interior at night. Sidelighting varies greatly depending on its orientation and the season. Sidelighting is often best for desk tasks as veiling reflections can be avoided with proper worker orientation.

Lighting from two walls evenly distributes light and reduces glare. Light from high on a wall penetrates more deeply into a space for more uniform distribution. Placing windows close to adjacent walls reduces contrast at the window edge and reflects light further into the space; it may also reveal desirable sunlight patterns and colors. Splaying the walls of an aperture causes light to wash across a longer or rounder surface area around a window, reducing contrast, increasing visual comfort, and offering less potential for glare.

Once inside the building, reflected light can reach indoor points directly or by reflecting off other interior surfaces. These successive reflections can bring daylight more deeply into the space.

As a rule of thumb, you can use daylighting for task lighting up to a depth of two times the height of the window. The ideal height from the top of the window to the floor should be about one-half the depth of the room if you want to get maximum daylight penetration and distribution without glare. (See Figure 17.14)

Highly reflective surfaces absorb less light at each reflection, and pass more to the room's interior. Surface brightness should change gradually from the outside to the inside. Light colored exterior surfaces and window frames gather more reflected light in through windows. Frames that are splayed at an angle help reduce uncomfortable contrasts between the bright outdoor views and darker interiors.

Select interior surface colors and reflectances according to the primary source of incoming light. Direct and reflected sky light hits the floor first, while reflected sunlight hits the ceiling first. Light colors and higher reflectances on surfaces far from openings help increase the light in dim areas. By placing windows adjacent to light colored interior walls, reflected light goes through a series of transitional intensities rather than creating an extremely bright opening surrounded by unlit walls.

TOPLIGHTING

Lighting from above offers the best distribution of diffuse skylight, with deeper penetration and better uniformity of daylight. Lighting through the roof using skylights produces fairly uniform illumination over large interior areas. Horizontal openings receive much more light than vertical ones, but with greater intensity in summer than winter; they are also more difficult to shade. Vertical glazing on the roof using clerestories, monitor windows, and sawtooth arrangements works better.

Toplighting provides more light per square foot with less glare than sidelighting. However, where the direct sun enters a skylight, it may strike surfaces and produce glare and fading. Unshaded skylights gain heat from summer sun and lead to winter night heat loss.

Clerestories provide balanced daylight throughout the changing seasons better than skylights. Clerestories use standard weather-tight window constructions. In the Northern Hemisphere, south-facing clerestories provide the most heat gain in winter.

SKYLIGHTS

Skylights allow daylight to enter an interior space from above. Skylights are typically metal-framed units preassembled with glass or plastic glazing and flashing. They come in stock sizes and shapes or can be custom fabricated. (See Figure 17.15)

Splaying the sides of a skylight diffuses daylight into a space. Higher ceilings with skylights yield more surface area for light to diffuse; this works best with a skylight well above the field of view. Position a reflector below a skylight, clerestory, or roof monitor window to redirect sunlight and diffuse light onto another surface.

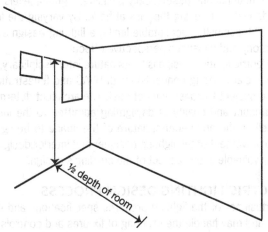

Figure 17.14 Window proportions for daylighting

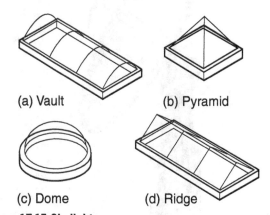

(a) Vault   (b) Pyramid

(c) Dome   (d) Ridge

Figure 17.15 Skylights

Source: Redrawn from Francis D.K. Ching, *Building Construction Illustrated* (5th ed.), Wiley 2014, page 8.36

A well-designed and installed horizontal skylight with a domed surface can bring daylight into an interior space and provide a view of sky. Skylights can be mounted flat or angled with the slope of a roof. It is important to install them correctly to avoid leaks. Controlling brightness and glare may require louvers, shades, or reflector panels.

Angled skylights on a north-facing (Northern Hemisphere) or shaded roof avoid the heat and glare associated with direct sun. They bounce sunbeams off angled interior ceilings to further diffuse brightness. Angled skylights may offer a view of sky and trees from the interior.

Horizontal skylights get less of the low-angle winter sun than angled skylights. They admit the most heat in summer and need shades to avoid adding to the cooling load. Horizontal skylights do not collect much solar heat in winter, may be covered by snow, and may leak. They work best in overcast conditions.

Other toplighting options include tubular skylights (light pipes), roof monitors, roof windows, and sloped glazing walls.

See Chapter 6, "Windows and Doors," for information on tubular skylights and roof monitor windows and other fenestration options.

HELIOSTATS AND TRACKING DEVICES

A heliostat is a dish-shaped mirror that focuses sunlight onto a stationary second mirror. (See Figure 17.16) A heliostat dynamically readjusts the primary mirror to track the sun and maximize the capture and use of sunlight at all times of the day. Collected light is distributed, often with a light pipe. A heliostat requires maintenance to prevent dirt and dust accumulation from affecting its performance.

Figure 17.16 Heliostat

See Chapter 6, "Windows and Doors," for exterior and interior shading strategies and window treatments.

ELECTRIC LIGHTING

Electric lighting design has significant thermal and visual aspects. Design also considers the integration of lighting with architecture, acoustics, structural concerns, HVAC systems, electrical systems, wayfinding, and budget.

It is not the purpose of this book to try to cover all the facets of electric lighting design to the degree that a full semester lighting course would. Instead, we look at how current lighting design practices affect the relationships among architects, engineers, lighting designers, and interior designers. We also cover the selection of lighting sources and controls, and consider fixture requirements and lighting system maintenance.

History of Electric Lighting

British inventors had been experimenting with electric lights for more than 50 years before Edison invented the light bulb. By the end of 1882, over two hundred Manhattan individual and business customers were using more than three thousand electric lamps, each with an average bulb life of only 15 hours. Thanks to decreasing electric rates and word of mouth, by 1900, ten thousand people had electric lights. By 1910, the number was over three million.

In 1934, Dr. Arthur Compton of General Electric developed the first practical fluorescent lamp in the United States. By 1954, energy-saving fluorescent tubes had edged out incandescent lamps for commercial use.

Today, inefficient, heat-producing incandescent lamps are being replaced by solid-state light emitting diodes (LEDs) and long fluorescent tubes are yielding to compact fluorescents. Energy-efficient LEDs are now available in virtually all types of fixtures.

Electric Lighting Design

The goal of electric lighting design is to create an interior that is both functional and aesthetically pleasing. Lighting levels must be adequate for seeing the task at hand. By varying the levels of brightness within acceptable limits, a lighting design avoids monotony and enhances perspective effects.

Electric lighting design is by its nature interdisciplinary, with an especially strong connection with HVAC and fenestration. It is also subject to constraints of cost. The architect determines the amount and quality of daylighting admitted to the interior as well as the architectural nature of the space to be lighted. The involvement of the lighting designer and interior designer is highly valuable in this aspect of the building's design.

ELECTRIC LIGHTING DESIGN PROCESS

The designer of the lighting prepares specifications and drawings, and may handle the ordering of fixtures and controls. It is often necessary to address several stages of the design process simultaneously. Fixtures often have long delivery times, so advance planning is necessary. When fixtures arrive, their many parts must be inspected and accounted for. Although fixtures may be one of the last things to be installed, they often need to be one of the first things to be planned and ordered.

Electric lighting design begins with task analysis to assess difficulty, time factors, type of user, cost of errors, and any special requirements. Preliminary design evaluates the need for general, local, or supplementary lighting, choices of sources and systems, architectural lighting elements, daylighting, and ambiance. Detailed design involves decisions on fixtures and ceiling systems, degrees of nonuniform lighting, and use of fixed or movable lighting. It also involves making detailed calculations and considering maintenance requirements. Evaluation looks at the design's energy use, construction and operating costs, and life cycle costs.

The interior designer or lighting designer typically prepares a lighting plan and schedule that indicates fixture locations and selections. (See Figure 17.17) They then must coordinate their selections with the HVAC engineers, who will monitor power loads. The resulting detailed design may still involve relocating a space or changing lighting or HVAC system details.

After installation, light levels must be fine-tuned. The individuals responsible for day-to-day use must be trained in using controls, and those involved in maintenance must understand lamp types and replacement requirements. The designer of the lighting can expect to make additional visits after installation until everything is working properly.

LIGHTING DESIGN CONSIDERATIONS

High levels of illumination today are typically only used for areas where tasks are actually performed, with ambient lighting for general levels of illumination. Illumination for general lighting should be at least one-third as high as task levels. Accent lighting levels that provide focus on a specific object should not be greater than five times the ambient level.

Expectations of occupants may lead to a room with low general illumination to be visually perceived as inadequately lit, even when the task is adequately lit.

Exposed electric lamps can delineate space, but their brightness can be a problem. Indirect lighting conceals bright sources from view, while bouncing light off a room surface (often the ceiling) and reflecting it onto the desired plane.

Increasing the illumination level can result in an apparent increase in the physical dimensions of a space. Accenting a wall can change the apparent room proportions.

Task lighting is often provided by small, portable lighting fixtures that can be moved and turned on and off as needed. Worker-adjustable task lights increase a user's feeling of control and comfort. Designing task lighting as an integral part of the lighting system requires complex coordination of illumination requirements, task location and orientation, and lighting equipment. Sometimes it is more energy-efficient to look at improving the way a difficult visual task is done than to provide higher levels of lighting.

When a space is used for two or more totally distinct activities with widely varying lighting requirements, it is necessary to design a separate lighting system for each function, using as much equipment in common as possible.

The height and angle of the work plane affects the design of task lighting. The work plane is usually horizontal, but some merchandising displays and library stacks are vertical, and some work surfaces are at other angles.

The size of luminaires should correlate with the room size and ceiling height. Fluorescent fixtures greater than 2 by 4 feet (610 by 1219 mm) should not be used in ceilings lower than 10 feet (3.1 m) without incorporating a surface pattern to minimize their apparent size.

A regular spacing pattern produces uniformity of illumination. Circulation areas and waiting areas in transportation terminals and other large spaces can employ perimeter lighting with supplemental lighting to serve as directional markers for wayfinding.

Wall lighting patterns are always in the direct line of sight. Avoid creating scallops, spots, irregular gradients, and unnecessary points of sparkle that can become unwanted dominant visual elements.

Concentrated pools of light in spaces with low ambient light isolate illuminated areas. This can help define territories in restaurants, work, and school areas within larger spaces.

ELECTRIC LIGHT SOURCES

Probably the most important decision when designing electric lighting is the light source. The choice of lamp affects color, heat generation, energy use, maintenance, and cost.

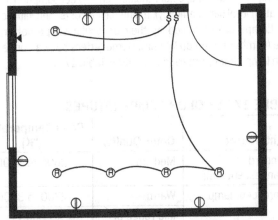

Figure 17.17 Typical interior design electrical and lighting plan

Source: Redrawn from Francis D.K. Ching and Corky Binggeli, *Interior Design Illustrated* (3rd ed.), Wiley 2012, p. 241

Characteristics of Sources

Fixture efficiency is directly affected by temperature. **Efficacy** is the ratio of lumens per watt of electricity used. (See Table 17.10) It measures the relationship between the amount of light and the amount of heat produced by both daylight and electric light sources. Daylight introduces less heat per lumen than electric sources.

The color appearance of any object is strongly influenced by the illuminating lamp's spectrum and color temperature. Visible light and the colors it comprises make up only a small amount of the electromagnetic spectrum. (See Figure 17.18)

Chromaticity, the relative proportions of each of the three primary colors (red, green, and blue) required to produce a given illuminant color, is the basis of the **CIE color system**, the internationally accepted standard for designating illuminant color.

A lamp's color temperature indicates its own color appearance, for example yellow, white, or blue-white. The color temperature is also usually a guide to the colors in which it has the most energy. Color temperature determines whether the light source is regarded as warm, mid-range, or cool; the higher the color temperature, the cooler the source. (See Table 17.11)

Today, LEDs are designed to produce almost any color, including several types of white. Some can even change colors.

Selecting Light Sources

Until recently, incandescent lamps were the most common light sources in residential interiors. Fluorescent lamps were dominant in commercial and institutional spaces. (See Table 17.12) Today, **light emitting diodes (LEDs)** are taking over much of the market, along with **compact fluorescent lamps (CFLs)**.

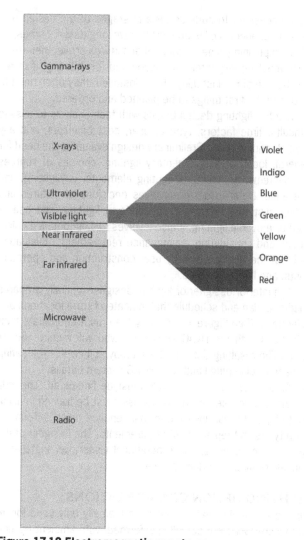

Figure 17.18 Electromagnetic spectrum

TABLE 17.10 EFFICACY OF ELECTRICAL LIGHT SOURCES

Lamp Type (*with ballast)	Efficacy (lm/W)
Incandescent (15 to 500W)	8 to 22
Tungsten-halogen (50 to 1500W)	18 to 22
Fluorescent (15 to 215 W)*	35 to 80
Compact fluorescent*	55 to 75
Mercury vapor (40 to 1000W)*	32 to 63
Metal halide (70 to 1500W)*	80 to 125
High pressure sodium (35 to 100W)*	55 to 115
Induction with power supply	48 to 70
Sulfur with power supply	90 to 100
LED screw base	55 to 102
LED Par 20	29
LED par 30	60

Considerations involved in selecting light sources include the lighting effect desired, color rendition, energy consumption and illumination level (lamp efficacy), maintenance costs, and initial costs.

Lamps are identified by their particular shape (letter) and size (number based on ⅛" unit). Some letters have a connection with lamp shape; others do not. (See Table 17.13)

TABLE 17.11 COLOR TEMPERATURES

Light Source	Color Quality	Color Temperature (°K)
Standard incandescent lamp	Medium	2600 to 3000
Fluorescent lamp	Warm	2700 to 3000
	Mid-range or mixed palettes	3500
	Cool palettes	4100
	Cool white	4250

TABLE 17.12 LAMP TYPES AND CHARACTERISTICS

Lamp Type	Description	Wattage Range	Hours of Life	Color Rendition
Incandescent	Standard 40 to 100W	10 to 1500	750 to 3500	Excellent
	Tungsten halogen	100 to 1500	2000 to 12,000	Excellent
Fluorescent	Standard	15 to 100	9000 to 20,000+	Very good
	High-output	60 to 215	9000 to 20,000+	Very good
High-intensity discharge (HID)	Standard mercury vapor	40 to 1000	12,000 to 24,000+	Poor
	Metal halide	175 to 1500	7500 to 20,000	Very good
	High-pressure sodium	35 to 1000	12,000 to 24,000+	Fair to good
	Low-pressure sodium	18 to 180	12,000 to 18,000	Poor

TABLE 17.13 LAMP SHAPE IDENTIFIERS

Letter	Lamp Shape
A	Discontinued for incandescents in the United States; similar available in LEDs and CFLs
BR	Bulged reflectors; replacing less-efficient R lamps; flood or spot
G	Globe
MR	Mini-reflector; flood or spot
PAR	Parabolic reflectors with heavy glass bulb for inside or outside use; flood or spot
PS	Pear shape
R	Reflectors with thin glass bulb; flood or spot
T	Tube

INCANDESCENT LAMPS

The light from an open fire, a candle, or an oil lamp is incandescent, as is the glowing filament of a traditional light bulb. Up to 90 percent of the electrical energy used by an incandescent lamp is lost to heat, and only the remaining 10 percent is emitted as light. The rejected heat increases the building's cooling load. Incandescent lamps generally have short lives, with about 750 hours being standard. Common incandescent lamp shapes are included here as they continue to be used for various light sources. (See Figure 17.19)

As part of the Energy Independence and Security Act of 2007, incandescent lamps that produce 310 to 2600 lumens have been phased out in the United States as of 2014, with lamps outside that range exempt. By 2030, all lamps less than 70 percent efficient will be phased out.

Several classes of specialty lights are exempt from the Act, including appliance lamps, rough service bulbs, three-way, colored lamps, stage lighting, plant lights, globes, candelabra lights under 60 watts, outdoor post lights less than 100 watts, nightlights, and shatter resistant bulbs. By 2020, additional restrictions become effective requiring all general-purpose bulbs

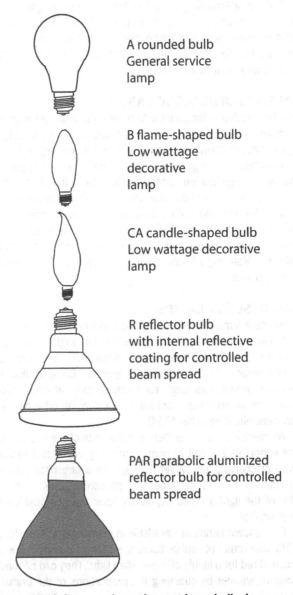

A rounded bulb
General service
lamp

B flame-shaped bulb
Low wattage
decorative
lamp

CA candle-shaped bulb
Low wattage decorative
lamp

R reflector bulb
with internal reflective
coating for controlled
beam spread

PAR parabolic aluminized
reflector bulb for controlled
beam spread

Figure 17.19 Common incandescent lamp bulb shapes

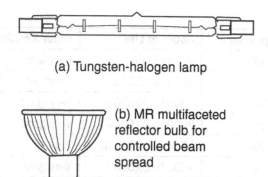

(a) Tungsten-halogen lamp

(b) MR multifaceted reflector bulb for controlled beam spread

Figure 17.20 Tungsten-halogen lamp shapes

to produce at least 45 lumens per watt, which is similar to current compact fluorescent lamps.

PAR tungsten-halogen and reflector lamps began to be phased out starting in 2012. (See Figure 17.20) PAR or reflector bulbs 40 watts or higher are no longer manufactured for use in the United States. Certain reflector flood lamps are affected by the Act, while others are exempted.

GASEOUS DISCHARGE LAMPS

Gaseous discharge lamps include fluorescent and high-intensity discharge lamps. They function by producing an ionized gas in a glass tube or container. They offer long life and high efficacy.

Gaseous discharge lamps require ballasts to trigger the lamp with a high ignition voltage and to control the electric current for proper operation. The ballast is essentially an electrical transformer, and usually contains coiled wires that adjust voltage and limit current. Electronic ballasts have replaced the older magnetic ballasts that may still be in use in some existing buildings. Matching of ballast to lamp is critical to successful lamp operation.

FLUORESCENT LAMPS

Fluorescent lamps come in two basic types: linear and compact. Linear fluorescent lamps are tubular. High output and standard T8 and T5 (26 and 16 mm) lamps being most popular, with the older T12 (38 mm) lamp still available. Compact fluorescents are available as single-tube, double-tube, and triple-tube lamps that fit dedicated sockets. Other specialized types are also available. (See Table 17.14)

An electrical discharge between the ends of the fluorescent tube vaporizes a small amount of mercury vapor and excites it into discharging ultraviolet (UV) light to a phosphor coating the inner surface of the tube. The phosphor glows, with the color of the light it emits depending upon the composition of the phosphor.

Fluorescent lamps are available in many colors. (See Table 17.15) Trichromic phosphor fluorescent lamps combine green, blue, and red for a highly efficient white light. They can be made cooler or warmer by changing the proportions of the primary colors.

TABLE 17.14 FLUORESCENT LAMPS

Type	Description
Standard 1½" (38 mm) diameter, 48" (1219 mm) long	Newer narrower lamps are more efficient.
T12, T8, and T4: 1½", 1", and ⁵⁄₈" (38, 25, 16 mm) in diameter	Lengths 15" and 18" (381 and 457 mm) and one-foot increments from 2' to 8' (0.6 to 2.4 meters).
U-shaped and circular	To fit square fixtures. Similar to standard.
Compact fluorescent lamps (CFLs)	Lamp with integral ballast. Many forms.
High output (HO): ⁵⁄₈" (16 mm) diameter	Twice the output of T8. Narrow diameter used in low-profile indirect luminaires.
Low-energy fluorescent lamps	As replacements for standard lamps. Higher cost, need special ballasts, shorter life, dimming problems.
Ecologically friendly lamps	Low mercury content; can be recycled.

Fluorescent lamp output requires only about 25 percent of the energy of a comparable incandescent lamp. Fluorescent lamps last up to 24,000 hours, decreasing the need for replacement. They operate best between 41° and 77°F (5° and 25°C). Compact fluorescent lamps fit into fixtures originally designed for incandescent lamps. (See Figure 17.21)

Today's fluorescent lamps are either rapid-start and instant-start; older preheat lamps with a separate starter are now considered obsolete. The starting sequence and continued operation of a fluorescent lamp depends on its ballast. Fluorescent lamps do not turn on immediately, and there is a delay until the full output is reached.

Dimming a fluorescent lamp system lowers energy consumption and allows flexibility in lighting levels. Dimming ballasts

TABLE 17.15 FLUORESCENT LAMP COLORS

Color	Description	CRI	Color temperature
Soft white	Comfortable, pleasant for kitchen, bath	78 to 86	3000K
Neutral	Balanced general purpose lighting for offices	70 to 85	3500K
Cool White	Efficient task lighting; garage, basement	62 to 85	4100K
Natural light	Simulates natural outdoor light; any room	70 to 98	5000K
Daylight deluxe	Cool; garage, workshop, laundry	78 to 85	6500

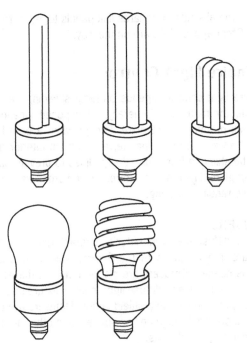

Figure 17.21 Compact fluorescent lamps

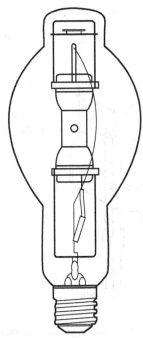

Figure 17.22 Metal halide lamp

Source: Redrawn from Walter T. Grondzik and Alison G. Kwok, *Mechanical and Electrical Equipment for Buildings* (12th ed.), Wiley 2015, page 662

are available for some fluorescent lamps, including T8 and T5 lamps. Analog and digital dimmers and wireless infrared transmitters are used for dimming fluorescent lamps.

All fluorescent lamps contain small amounts of mercury to generate UV energy that energizes the phosphor and produces light. Mercury is a toxic material has been regulated as a hazardous material since 1980. Fluorescent lamps are difficult to dispose of due to their mercury content. Fluorescent and HID lamps fail the Environmental Protection Agency (EPA) Toxicity Characteristic Leaching Procedure (TCLP).

In 2000, the EPA established the Universal Waste Rule, which requires that building owners and management dispose of lamps containing mercury in an environmentally sound manner. Conforming to the rule is easier if fixtures are relamped in a group rather than singly. Facilities may need to provide a storage area for storing spent lamps prior to disposal. Residential users can recycle fluorescents at the stores where they were purchased.

HIGH INTENSITY DISCHARGE LAMPS

High intensity discharge (HID) lamps generate light by discharging electricity through a high-pressure vapor. They include mercury vapor, metal halide, and high pressure sodium lamps. HID lamps used in commercial and institutional buildings today are primarily metal halide, with color-corrected high-pressure sodium used in some gymnasiums and large public areas.

Metal halide lamps typically have high efficacy, rapid warm up, rapid restrike time, and poor color rendering. (See Figure 17.22) Metal halide lamps are used in stores, offices, industrial plants, and outdoors. They are available in many types and sizes, and require a ballast. They contain mercury.

Metal halide lamps require two to three minutes for their initial startup, and eight to ten minutes or longer for restrike. They are high-efficacy, have a long life, and offer small size for optical control.

SOLID STATE LIGHTING AND LEDS

Solid state lighting (SSL) is developing very rapidly. Solid-state lighting sources are extremely resistant to physical abuse and very long-lasting. The most popular SSL lighting source available today is the light-emitting diode (LED).

Light emitting diodes (LEDs) are widely used as lighting sources. Electric current is passed through a solid that has been made into a semiconductor, emitting particular wavelengths of radiation. This direct release of radiation from the flow of current is a very efficient way of producing light without generating significant amounts of heat.

LEDs have been used since the 1960s for many applications. Their architectural illumination applications have included signage, retail displays, emergency signage, and pathway accent lighting. Their development for lighting has increased enormously in recent years.

The LED lamp is typically an assembly of several individual LEDs, each rather small with low lumen output. The lamp can take a linear form or be a tight-packed point source. LED lamps are tiny light sources; a single fixture may use hundreds of individual lamps.

LEDs are easy to install and long-lasting (estimates of 50,000 hours). They radiate very little heat and are highly

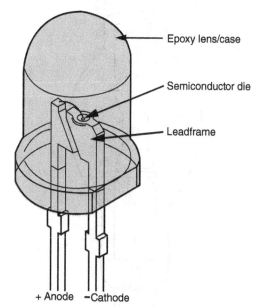

Epoxy lens/case

Semiconductor die

Leadframe

+ Anode   −Cathode

Figure 17.23 LED

Source: Redrawn from LED,_5mm,_green_wikipedia public domain.png

energy-efficient. LEDs operate on DC voltage, which is transformed into AC voltage within the fixture. LEDs can be designed to focus light, and are widely used for task lights. They do not contain mercury as do fluorescent lamps. LEDs do produce some heat, and need access to air to dissipate it.

High-powered white-light LEDs are used for illumination. They are insensitive to temperature, are vibration and shock resistant. The tiny ⅛" (3 mm) lamps can be combined into larger groups to mix colors and increase illumination. (See Figure 17.23)

Today's LEDs can produce any color of light, with CRI values greater than 90. Current LED technologies use phosphor-coated blue LEDs or red, green, and blue (RGB) LEDs to produce wavelengths across much of the visual spectrum. This improves the accuracy of the human eye's perception in any setting, and is much closer to natural daylight than the CRI values of typical fluorescent lamps.

LED luminaires are now available that fill demands for downlight, pendant, panel, ceiling, and other fixture types. Fixtures can be small, facilitating more effective control of light distribution. Other applications for LEDs include wall-grazing channels for stairwells, floors, and walls, and LED strips for closet rods and handrails. Inexpensive LED strip lighting can be used in ceiling coves. Combined LED and sensor technologies produce interactive illuminated surfaces and projection systems.

LEDs are available as A-shaped lamps to replace incandescent lamps.

Transparent organic light emitting diodes (OLEDs) are already commercially available in textile wallcoverings; electrically

powered metal strips or squares link panels together to form a circuit. Other applications are on the way.

Luminaire Light Control

Luminaire light control depends on lamp shielding, reflectors, and reflector materials. Generally, all exposed incandescent lamps will produce direct glare. Except when a bare lamp is selected as a source of sparkle, all lamps in interior fixtures should be shielded from normal sight lines to prevent direct or disabling glare. Fluorescent fixtures placed crosswise to line of sight also require shielding.

DIFFUSERS

Luminaire diffusers vary in material, characteristics, and uses. Some are integral parts of the fixture, while others can be added as needed. Diffusers are placed between the lamp and the illuminated space to diffuse light, control fixture brightness, redirect light, and hide and shield lamps. Diffusers are used in corridors, stairwells, high-ceiling spaces, and other areas without demanding visual tasks.

Translucent diffusers do a good job of hiding the lamp inside, although direct glare and veiling reflections can be problems. Transparent to translucent diffusers are made of glass, acrylic, polycarbonate, or polystyrene.

Prismatic batwing diffusers are lenses, usually of molded or extruded acrylic used for linear or radial distribution. They consist of a series of prisms that produce light pattern angles to the left and right of the observer with little direct downward illumination. They have good efficiency, low direct and reflected glare, and good diffusion. However, the lens creates a dust trap that requires frequent cleaning.

LOUVERS, BAFFLES, AND EGGCRATES

Louvers and baffles are usually rectangular metal or plastic. Their primary use is to shield the light source and diffuse its output. Louvers have average overall efficiency, but efficiency may be lower for parabolic louver designs. White louvers can produce glare. Specular aluminum or dark colored louvers produce little direct glare, but are less efficient and may cause serious veiling reflections.

Parabolic louvers are parabolic wedges with a specular finish. They distribute light almost straight down, and are used to prevent glare on vertical monitors. Parabolic louvers do not illuminate vertical surfaces well or prevent veiling reflections on horizontal surfaces.

LENSES AND REFLECTORS

Many fluorescent fixtures that use large lamps rely on a lens to control light distribution. Prismatic lenses are efficient, with good diffusion and wide permissible spacing. The produce little direct glare, but veiling reflections may be a problem.

The surfaces of clear plastic sheets can be formed into lenses and prisms for good optical control. The refracted light has more downward distribution, which reduces direct glare.

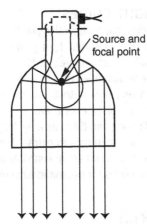

Source and focal point

Figure 17.24 Parabolic reflector

Source: Redrawn from Walter T. Grondzik and Alison G. Kwok, *Mechanical and Electrical Equipment for Buildings* (12th ed.), Wiley 2015, page 701

Fresnel lenses for round fixtures can concentrate or disperse light. They have a smaller housing without a reflector, yet still maintain beam control. A Fresnel lens has poor lamp hiding power, but high efficiency and visual comfort. Fresnel lenses are made from flat sheets with beveled grooves.

Reflectors for light sources finished in white gloss paint diffuse light well. Specular anodized aluminum sheets are highly reflective when new and clean, but become more diffuse with aging, high temperatures, and dirt.

Parabolic reflectors shield the lamp and redirect light rays with the desired output spread for increased illumination levels. (See Figure 17.24) Deep parabolic reflectors have high efficiency, low reflective glare, and low to very low brightness. Placement of the long axis is best parallel to sight lines.

Light Source Controls

Technology for controlling light sources continues to improve. Daylight harvesting and multiple-level switching both respond to changes in available daylight. Occupancy controls respond to use patterns. Many US energy codes require light-level reduction controls in enclosed spaces of certain occupancies.

DAYLIGHT HARVESTING

Daylight harvesting uses photosensors to detect daylighting levels and automatically adjusts the output level of electric lighting to create the desired level of illumination for a space. The lighting control system can automatically turn off all or a portion of the electric lighting or dim the lighting, and immediately reactivate the lighting if daylighting falls below a preset level.

Daylight harvesting controls can be integrated with occupancy sensors as well as having manual override controls. Some control systems can also adjust the color balance of the light by varying the intensity of individual LED lamps of different colors installed in the overhead fixtures.

MULTIPLE-LEVEL SWITCHING

Bi-level switching is a lighting control system that provides two levels of lighting power in a space. The switching system may control alternate ballasts or lamps in a luminaire, alternate luminaires, or alternate lighting circuits independently. The system uses photosensors that detect daylight levels, occupancy sensors that detect users, time-based control panels, or manual switches.

Multilevel switching is a form of bi-level switching in which multiple lamps in a single light fixture can be switched on and off independently of each other. This creates one or two steps between full output and zero illumination while maintaining the required uniform distribution of light. Multilevel switching provides greater flexibility and lessens the abrupt changes in light level of bi-level switching.

CONTINUOUS DIMMING

Continuous dimming maintains the desired level of illumination by modulating the output from electric lamps and fixtures in proportion to the amount of available daylight detected by light-level sensors. Continuous dimming systems minimize the abrupt changes in light level created by bi-level and multilevel switching systems.

OCCUPANCY CONTROLS

Occupancy controls are automatic lighting control systems that use motion or occupancy sensors to turn lights on when human activity is detected and off when a space is vacated. They can replace wall-mounted light switches or be mounted remotely. Retaining the normal switching for use as override switches allows the lighting to be kept off even when the space is occupied.

Remote Source Lighting

Remote sources can transmit light to locations where it is needed. Fiber optics, light guides, and prismatic optical film are all used to transfer light from remote sources. Fiber optic and light guides use very small, effectively invisible light sources.

Remote source lighting is used for display lighting for light- and heat-sensitive objects, and where objects are sensitive to UV. It is also used where relamping is major problem, such as spaces with high ceilings, clean rooms, security access areas, and spaces that cannot be interrupted for maintenance. Using remote-source lighting makes sense where lamp heat is a problem, as in retail windows and refrigerated displays. It is also used where electrical wiring is undesirable, such as for patient-controlled hospital bed lighting and in children's devices, and in other applications.

FIBER OPTICS

Fiber optics can be used to replace recessed ceiling downlights, track and display case museum lighting, and lighting for pools or spas, and for supermarkets and other commercial buildings. Fiber optic lighting uses a single remote source to supply a large number of relatively small point-source lights, making star-sprinkled ceilings feasible. Fiber optic lighting can be used to show direction and create patterns or signs for wayfinding.

Optical fiber conducts light by total internal reflection. The fiber has an inner core of transparent silica, glass, or plastic, and an outer coating with a lower refractive index. Light rays are reflected at the interface of the two materials, and can proceed almost unimpeded down the core. Any loss of energy is very low.

Ultraclear glass fibers are used for communications, with plastic optical fiber used for lighting. The illuminator is usually a metal box containing the light source and accessories, color filters, and usually a fan or blower to cool the source lamp. Illuminators for fiber optic systems utilize a variety of lamp types. No heat is produced where light exits the fiber, and there is no UV transmission through the fiber. The fiber bundle that carries the light can be buried in almost any substrate.

There are many configurations for fiber optic lighting, including axial-mode light bars and discrete sources that emit light from their ends, and lateral-mode devices that emit light along their length. (See Figure 17.25)

Axial-mode linear devices comprise a bundle of fibers in a long narrow enclosure called a **light bar**. Individual fibers or small groups of fibers are brought out of the enclosure as a light-emitting point. Light bars are used in retail display, accent, decorative, and directional applications. Large bundles of axial-mode fibers can be used to make point-source lighting fixtures that do not generate heat in the space and are low maintenance and higher efficacy than the lamps in their illuminators.

Lateral-mode fibers are used for linear lighting tasks such as illuminating stair nosings, path lighting, and decorative trim lighting. Color filters in the illuminator can produce dramatic outlining effects.

Nomenclature for remote-source tubular lighting varies. These devices are referred to as hollow light guides, light pipes, hollow light pipes, prismatic light guides or pipes, hollow prismatic light guides, and remote-source hollow light guides.

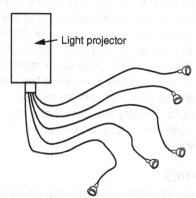
Light projector

Figure 17.25 Fiber optic lighting system
Source: Redrawn from Francis D.K. Ching and Corky Binggeli, *Interior Design Illustrated* (3rd ed.), Wiley 2012, page 267

PRISMATIC LIGHT GUIDES

Prismatic light guides made with a thin plastic prismatic film use total internal reflection. Prismatic light guides, pipes, and lighting fixtures have all the advantages of fiber optic lighting, and can handle very large quantities of light without color distortion. When coupled to a very high output source, such a fixture can illuminate large interior or exterior areas.

Using optical lighting film (OLF) as a lighting fixture requires special techniques and materials to extract light from the pipe. High-bay industrial and commercial installations are typical interior lighting applications for prismatic film light-pipe fixtures.

LUMINAIRES

As noted above, "luminaire" is another name for a lighting fixture. A luminaire's purpose is to house and provide electricity for the light source and to direct its output. The many variations available in luminaire design can make choosing the correct fixture for a given use a difficult task. In addition to the light they contribute to a space, luminaires can serve as focal points, adding sparkle, style, and visual texture. No single luminaire is ideal for a majority of applications, and each application has its own requirements for light control.

By grouping tasks with similar lighting requirements together and placing the most intensive visual tasks at the best daylight locations, you can use fewer fixtures and consequently less lighting energy. Design with effective, high-quality, efficient, low-maintenance, and thermally controlled fixtures.

Glare is an important issue in luminaire selection. A high-efficiency installation can result in high fixture luminance that might produce glare. Wall lighting with luminarire output at a high angle may produce direct glare. Low-angle lighting minimizes direct glare but retains the possibility of veiling reflections. A high shielding angle gives good visual comfort but reduces efficiency.

Characteristics of Lighting Fixtures

Selection of a lighting fixture involves consideration of the characteristics of electric lighting sources, including color composition, the physical size of the generating device, the lamp's length of life, electrical requirements, and operational efficiency. **Photometry** is the science of the measurement of light, in terms of its perceived brightness to the human eye. The luminaire's manufacturer should be able to furnish complete photometric test data from a reliable independent testing lab.

The configuration of a fixture changes the way light is distributed. Spacing fixtures more widely apart reduces the number needed. High-efficiency fixtures deliver more light directly to the work plane, with less bounced off other surfaces.

ARCHITECTURAL LIGHTING

Lighting equipment that is an integral part of a building is referred to as **architectural lighting**. (See Table 17.16) **Coves,**

TABLE 17.16 TYPES OF ARCHITECTURAL LIGHTING

Type	Description
Cove lighting	Indirect lighting of ceiling from continuous wall-mounted fixtures. Cove placed high enough to shield light source. Source must be far enough from ceiling to prevent hot spots. Use high reflectance inside of cove, upper walls, and ceiling.
Cornice lighting	Ornamental molding around wall of room just below ceiling illuminates ceiling only.
Coffer lighting	Recessed panels in ceiling. Large coffers can be illuminated by cove lighting around bottom edges. Small coffers use recessed luminaires in center.
Valance on wall bracket	Shielding board illuminates wall above and below. Must completely shield light sources from common viewing angles. Minimum 12" (305 mm) below ceiling to prevent excessive brightness.
Valance at ceiling	All illumination downward only, ceiling may look dark. Source may be visible unless shielded with cross louvers.

cornices, **coffers**, and **valences** are usually designed as part of the interior architecture, and become important parts of the form of the space. (See Figures 17.26, 17.27, and 17.28) Coffers and valances may also be added to the interior surfaces. Cove, valance, and cornice lighting provide a soft, indirect glow and are often used to highlight ceiling details or wall textures.

Cove lighting can be installed around a raised ceiling opening to create a soft glow in the central space. (See Figure 17.29) Louvers below the light source diffuse light. (See Figure 17.30)

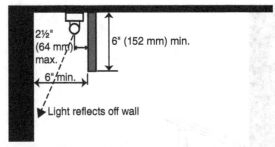

Figure 17.27 Cornice lighting

Source: Redrawn from Francis D.K. Ching and Corky Binggeli, *Interior Design Illustrated* (3rd ed.), Wiley 2012, page 272

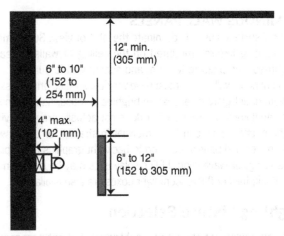

Figure 17.28 Valance lighting

Source: Redrawn from Francis D.K. Ching and Corky Binggeli, *Interior Design Illustrated* (3rd ed.), Wiley 2012, page 272

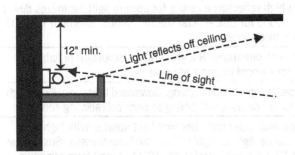

Figure 17.26 Cove lighting

Source: Redrawn from Francis D.K. Ching and Corky Binggeli, *Interior Design Illustrated* (3rd ed.), Wiley 2012, page 272

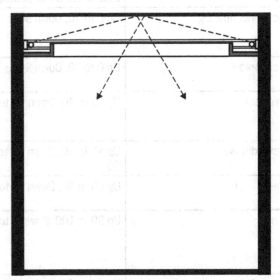

Figure 17.29 Cove lighting around recessed coffer

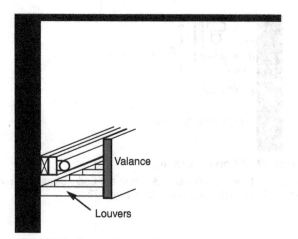

Figure 17.30 Ceiling valance with louvers

LUMINOUS WALL PANELS

Vertical surfaces usually dominate the interior view. Supplementary lighting fixtures mounted on the ceiling or walls increase brightness, emphasize texture, and accent wall features.

Luminous wall panels must have very low surface brightness to prevent direct glare or excessive brightness ratios. As a luminous wall panel implies a window but does not actually provide a view to the outside, they can be somewhat frustrating for the viewer. Light boxes and panels are used to backlight graphics for display, advertising, or wayfinding. LED display units may be less than an inch thick; thicker fluorescent light boxes are also available.

Lighting Fixture Selection

Lighting equipment should be unobtrusive, but not necessarily invisible. Fixtures can be chosen to compliment the architecture and to emphasize architectural features and patterns. Decorative fixtures can enhance the interior décor.

A lighting fixture consists of one or more electrical lamps with all parts and wiring necessary for their support, positioning, and protection as well as to connect them to a power supply and distribute the light.

A **candlepower distribution graph** illustrates the vertical distribution of light from a luminaire. Some manufacturers supply candlepower distribution graphs for each of their lighting fixtures.

Uniformity of illumination assures that points distant from the fixture centerline obtain the same illumination as those below fixture. Uniformity permits wider spacing of fixtures.

High-efficiency designs direct fixture output to the work surface. As we have noted previously, direct lighting fixtures may create glare problems.

Diffuseness allows light to reach work surfaces from multiple directions, although illuminance is reduced by multiple reflections off surfaces. Indirect and semi-direct lighting fixtures produce diffuse light. Ceiling illumination from indirect and up-lighting fixtures results in good diffuseness without hot spots.

The luminance of each lighting source creates a point of visual attention. Quantities of large or very bright luminaires will draw attention from other surfaces, especially when arranged in striking patterns. This property can be used to emphasize an area or object to avoid monotony.

Luminaire size should correlate with the room size and ceiling height. Fluorescent fixtures greater than 2 feet by 4 feet (610 by 1219 mm) are not generally used in ceilings less than 10 feet (3 m) high unless their size is disguised by a smaller luminaire surface pattern.

LIGHTING SYSTEM DISTRIBUTION

A **lighting system** is a particular fixture type applied in a particular way. (See Figure 17.31) Lighting systems vary according to how they control or distribute light. (See Table 17.17) They differ primarily in the proportion of light directed upward or downward.

TABLE 17.17 LIGHTING SYSTEM DISTRIBUTION TYPES

Type	Distribution Percentage	Description
Direct-concentrating	Up 0 to 10, Down 90 to 100	Light directed downward. Ceiling illumination reflected from floor and room surfaces. Light distribution pattern concentrated.
Direct-spread	Up 0 to 10, Down 90 to 100	Light directed downward. Ceiling illumination reflected from floor and room surfaces. Light distribution pattern spread.
Semidirect	Up 10 to 40, Down 60 to 90	Using high reflectance ceiling for upward light minimizes direct glare. Good for offices, classrooms, shops. Minimum 12" (305 mm) fixture stems.
General diffuse	Up 40 to 60, Down 40 to 60	Light in all directions. Roughly equal distribution of light produces bright ceiling and upper wall.
Semi-indirect	Up 60 to 90, Down 10 to 40	Diffuse, low glare room lighting. Downward light typically through translucent diffuser, with ceiling as principal radiating source.
Indirect	Up 90 to 100, Down 0 to 10	Ceiling and upper walls become light source. With high reflectance, light is highly diffuse and shadowless. Suspend at least 12" (305 mm), preferably 18" (457 mm) from minimum 9'-6" (2.9 m) ceiling.
Direct-indirect	Up 40 to 60, Down 40 to 60	Good diffuseness. Vertical-plane illumination, little horizontal. Bright ceiling, upper wall. Minimum 12" (305 mm) stems.

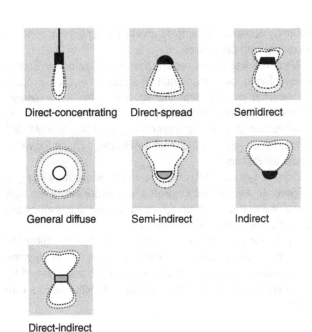

Figure 17.31 Lighting distribution types

Source: Redrawn from Francis D.K. Ching and Coky Binggeli, *Interior Design Illustrated* (3rd ed.), Wiley 2012, page 279

In a direct lighting system with a dark-painted ceiling, pendant fixtures can be used to hide piping and ductwork or lower the apparent ceiling height.

MOUNTING

Recessed lighting fixtures are an unobtrusive way to illuminate circulation paths or to increase light levels in a specific area. (See Figure 17.32) Multiple **downlights** provide ambient light for large spaces or focus light on a floor or work surface. A recessed downlight shields a point light source. The addition of black baffles obscures the lamp and limits its spread, but absorbs much of the light, and so are less efficient. Some recessed lights will appear as black holes in a light-colored ceiling when turned off. Semi-recessed fixtures are partially recessed into the ceiling or wall construction, with part of the fixture projecting beyond the surface.

Wallwashers are designed to illuminate a matte vertical surface uniformly. Individual wallwashers should be placed one-third of the wall height away from the wall and an equal distance from each other to avoid scalloped light spots.

Surface-mounted lighting fixtures include cove, cornice, and valance fixtures as well as wall sconces. Ceiling fixtures are usually positioned to spread light over a broad area above people and furnishings. Wall-mounted fixtures can be used to provide task lighting. The light reflected from wall-mounted fixtures onto a wall or ceiling adds to the general illumination of the space. Their locations must be coordinated with windows and furnishings.

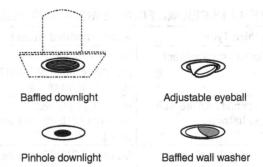

Figure 17.32 Recessed fixtures

Source: Redrawn from Francis D.K. Ching and Corky Binggeli, *Interior Design Illustrated* (3rd ed.), Wiley 2012, page 270

Indirect lighting fixtures (uplights) may be suspended from the ceiling, mounted on top of tall furniture, or attached to walls, columns, or floor stands. **Pendant fixtures** can distribute light up, down, or at an adjustable angle. Pendant-mounted fixtures hang below the ceiling on a stem, chain, or cord.

Track-mounted fixtures comprise adjustable spotlights or floodlights mounted on a recessed, surface-mounted, or pendant-mounted track that conducts the electrical current. (See Figure 17.33) The fixtures can be moved along the track and adjusted to distribute light in multiple directions. Low-voltage fixtures have a transformer either on the track or individually on each fixture. They are attached to ceiling supports for runs of greater than 20 feet (6 meters).

Building energy codes may require that each head on the track be counted as a separate fixture.

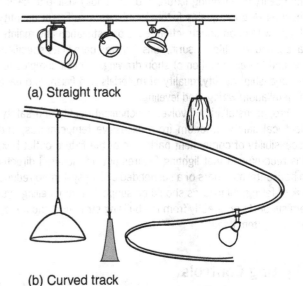

(a) Straight track

(b) Curved track

Figure 17.33 Track-mounted fixtures

Source: Redrawn from Francis D.K. Ching and Corky Binggeli, *Interior Design Illustrated* (3rd ed.), Wiley 2012, page 273

TABLE 17.18 CEILING FIXTURE MOUNTING HEIGHTS

Luminaire Type	Recommended Height
Indirect and semi-indirect luminaires	Minimum 18" (457 mm), preferably 24" to 36" (610 to 914 mm) from ceiling
Single–lamp luminaire with inverted batwing	Have very wide distribution. Minimum 12" (305 mm) from ceiling
Direct-indirect and semi-direct fluorescent fixtures	Minimum 12" (305 mm) for two-lamp units
	Minimum 18" (457 mm) for 3- and 4-lamp units

Decorative lights serve as accents within the space. Their contribution to illumination may be less important than their appearance.

Portable plug-in lighting fixtures are commonly referred to as lamps, and their light sources as light bulbs. Desk and task lamps provide focused light where needed, allowing lower ambient light levels. Portable lamps are usually operated at the fixture itself, giving users control over their environment.

Ceiling illumination should produce good ceiling coverage without hot spots and with good diffuseness. (See Table 17.18) Using a fixture hanger that is too short results in concentrated hot spots on the ceiling and uneven illumination. For fixtures with an upward component, a lower mounting height controls ceiling brightness and provides good light utilization.

CONSTRUCTION AND INSTALLATION

The quality of a lighting fixture is often readily visible to casual observer. A good quality fixture combines a sufficient quantity of light with good construction, ease of installation and maintenance, and long life. Assuring this requires care in its specification and the examination of shop drawings. Check samples for workmanship, rigidity, quality of materials and finish, and ease of installation, wiring, and leveling.

Proper installation involves mechanical rigidity and safety, electrical safety, freedom from excessive temperatures, and accessibility of component parts and of the fixture outlet box. It is recommend that lighting fixtures not be mounted directly to horizontal members of a suspended ceiling system to reduce risk of falling. All fixtures should be supported from ceiling system supports or directly from the building structure, and not by ceiling system itself.

Lighting Controls

A good lighting control design allows a variety of lighting levels and lighting patterns while conserving energy and money. In order to meet building code energy budget requirements, it is frequently necessary to employ energy-saving lighting controls that dim or shut off fixtures.

Many building codes require that a lighting power budget be established by a mandated procedure for each project. The designer must design the lighting within these energy constraints. The nationally accepted standard that defines the establishment of a lighting power budget is ANSI/ASHRAE/IES Standard 90.1—*Energy Standard for Buildings Except Low-Rise Residential Buildings.*

Control systems decisions are made when the lighting is designed, to assure that controls are appropriate to the light source and that the system arrangement and accessories are coordinated with the control scheme.

Light zones are defined to accommodate the scheduling and functions of various spaces. Ambient, task, and accent lighting are considered in laying out the zones. Each zone should be separately circuited, and each task light should have its own switch.

In a complex multi-use space, it is important to talk to the people who will use the space daily and are aware of common problems. Often controls are located in places that are hard to reach during an event, or prevent the result of an adjustment in lighting level from being seen while making it.

Lighting controls may be manual, automatic, or combined, usually with a manual override of the automatic controls. They may be stand-alone controls, or part of larger energy management system and/or building automation system. The designer of the lighting control system selects the number of lighting elements to be switched together and establishes the number of control levels.

Manual lighting controls generally give employees a sense of control, leading to a feeling of satisfaction and increased productivity. With remote-control dimming systems, occupants can adjust the lighting fixtures closest to their workstations without disturbing other employees; this can help them reduce glare.

Wireless lighting control systems for conference and meeting rooms give a presenter control of lights, motorized window shades, and projection screens at the touch of a button. Systems designed for use in classrooms and lecture halls where presenters may not be as familiar with complex audiovisual equipment should be easy to use.

OCCUPANCY SENSORS

Occupancy sensors are used to turn off or dim office lights after a minimum of 10 minutes. Types of occupancy sensors include passive infrared, ultrasonic, and hybrid sensors. (See Table 17.19 and Figures 17.34 and 17.35) They can also turn off fan-coil air units, air conditioners, and fans. Relighting may be instantaneous, delayed, or manually operated by the occupant.

TABLE 17.19 OCCUPANCY SENSORS

Type	Description
Passive infrared (PIR) sensors	React to heat source moving through beam pattern. May not detect small or very slow movements, so lights may shut off when a person is sitting quietly. Heat source must not be blocked by furniture, etc. May be dead spots under beams if units are not designed or located properly.
Ultrasonic sensors	Emit energy above range of human hearing. Detect small movements, may need to reduce sensitivity to avoid false sensing, but this also decreases coverage.
Hybrid (dual-technology) sensors	React to turn lights on, and reaction in either technology keeps lights on. Placement of sensors very important. Ten-minute minimum or manual shut off. Best in individual rooms and workspaces. Wall or ceiling mounted or in wall-outlet box with wall switch.

See Chapter 20, "Communications, Security, and Control Equipment," for use of occupancy sensors as intrusion detectors.

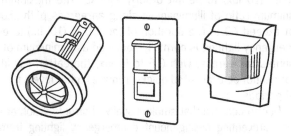

(a) Ceiling-flush-mounted (b) Wall switch and PIR (c) Surface-mounted

Figure 17.34 Passive infrared occupancy sensors

Source: Redrawn from Walter T. Grondzik and Alison G. Kwok, *Mechanical and Electrical Equipment for Buildings* (12th ed.), Wiley 2015, page 724

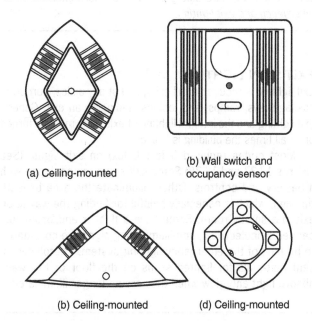

(a) Ceiling-mounted (b) Wall switch and occupancy sensor (b) Ceiling-mounted (d) Ceiling-mounted

Figure 17.35 Ultrasonic occupancy sensors

Source: Redrawn from Walter T. Grondzik and Alison G. Kwok, *Mechanical and Electrical Equipment for Buildings* (12th ed.), Wiley 2015, page 725

DAYLIGHT COMPENSATION

Daylight compensation reduces artificial lighting in parts of a building when daylight is available to meet illumination needs. Photocells trigger automatic dimming as required. Daylight compensation dimming can reduce energy use in perimeter areas by up to 60 percent.

Rapid changes caused by the constant switching on or off of lamp levels can by very annoying to the space's occupants, and is damaging to the lamps. Full-range automatic dimming for daylight compensation can avoid this problem.

Tuning and Maintenance

Designing and specifying lighting is complex, and it is rare that the system functions perfectly in the field as designed. The system is tuned in the field to adjust to these changes and achieve the designer's goals.

Lighting system tuning can reduce energy use by 20 to 30 percent. Tuning often results in the reduction of lighting levels in nontask areas as spill light is frequently adequate for circulation and other functions. Lighting system tuning is also required when the function of an entire space is changed, or furniture movement or changes in tasks alter a single area. It can help with glare reduction and result in improved task visibility. It is a good idea to include the lighting system tuning process as part of the lighting designer's complete scope of services.

A reduction in illuminance can be accomplished by field adjustments of fixtures and by replacing lamps and ballasts. Wall switches can be replaced with time-out units, programmable units, or dimmer units.

Maintenance is often the last consideration when selecting a lighting fixture, yet it is the one most likely to result in long-term negative feedback from building owners and facility managers. A lighting system maintenance problem may not lie in its design but in poor installation and maintenance practices. The lighting system may be operated and maintained by someone with little training or experience with the latest equipment. Fixtures should be simply and quickly relampable, resist dirt collection, and be simple to clean. Short lamp life necessitates that it be easy to replace lamps; difficult replacement indicates that long life lamp should be specified. Replacement parts should be readily available.

Emergency Lighting

Emergency lighting provides power for critical lighting systems in the event of a general power failure, the failure of the building electrical system, an interruption of current flow to the lighting unit, or the accidental operation of a switch control or circuit disconnect. It is customary for battery-powered units to be hard-wired into the building's electrical system, so that the battery can be recharged by the building power.

The general goals of emergency lighting are to avoid distress or panic and to provide lighting for egress from the building. The level of lighting required is related to the level of normal illumination and to the degree of hazard.

Emergency lighting should be essentially uniform, avoiding instantaneous drops from bright to low levels as it takes around five minutes to accommodate fully. Arrange bright, spot-type heads very carefully to avoid disabling glare and distorting shadows. The emergency lighting illuminance in specific areas should be related to the area's normal illuminance and the degree of hazard in that area. (See Table 17.20) The selection of emergency lighting fixtures has been expanded to include LED fixtures. (See Figure 17.36)

For facilities that require an entire emergency power system (standby generators, central battery systems), selected portions of the normal lighting system are usually designated as emergency lighting and served by emergency power. Otherwise, self-contained battery-powered lighting packages are strategically located to operate during power outages.

EMERGENCY LIGHTING CODES AND STANDARDS

NFPA 101—Life Safety Code specifies locations that require emergency lighting and the level and duration of the lighting.

NFPA 70—National Electrical Code (NEC) mandates system arrangements for emergency light and power circuits, including egress and exit lighting. The NEC also discusses power sources and system design. Additional local code requirements may be applicable.

NFPA 99—Standard for Health Care Facilities dictates special emergency light and power arrangements for these occupancies. US Occupational Safety and Health Administration (OSHA) requirements are primarily safety oriented and cover exit and egress lighting.

Industry standards are defined in IES and IEEE publications, including IEEE Standard 446-1995—IEEE Recommended Practice for Emergency and Standby Power Systems for Industrial and Commercial Applications.

Most codes and authorities require a minimum average illuminance at floor level throughout a means of egress of 1.0 fc (10 lux) to permit orderly egress. The maximum to minimum ratio of illuminances along an egress path should not exceed 40:1. The duration of the 1.0 fc (10 lux) level of emergency lighting is normally specified as a minimum of 90 minutes for egress, with 0.6 fc (6 lux) thereafter. Facilities that cannot be evacuated quickly require higher levels for varied periods.

Due to concern that smoke usually obstructs vision at eye level, preventing ceiling-mounted emergency lighting from illuminating the floor in smoky areas or make matters worse, codes may require baseboard-level egress lighting, plus avoidance of widely separated bright sources.

See Chapter 18, "Fire Safety Design," for more information on emergency and exit lighting.

EXIT SIGN LIGHTING

Exit lighting is required at all exits, and at any aisles, corridors, passageways, ramps, and lobbies leading to an exit. General exit lighting and the integral lighting of exit signs must be turned on at all times the building is in use.

Most codes required 5 fc (50 lux) on exit signs. (See Figures 17.37 and 17.38) Some exit signs are equipped with a battery and controls. Others illuminate the area beneath the sign, which is especially helpful for finding the way to an exit in a smoky room. Some have a flasher and/or audible beeper. Nonelectrical, self-illuminating signs are considered to be part of the emergency lighting system. Photoluminescent materials for lighted strips on the floor or low walls absorb light and glow when they are no longer illuminated.

In the United States, an exit sign is usually the word "EXIT" in red, but in most of the world it is green and often depicts a running man.

TABLE 17.20 EMERGENCY LIGHTING

Area	Footcandles (lux)
Exit area	5 (50)
Stair	3.5 to 5 (35 to 50)
Hazard areas	2 to 5 (20 to 50)
Other spaces	1 (10)

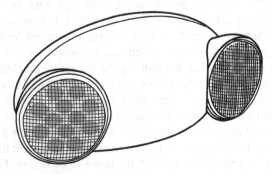

Figure 17.36 LED emergency lighting fixture

Figure 17.37 Typical US exit sign

Figure 17.38 International Organization for Standardization (ISO) exit sign symbol

LIGHTING DESIGN APPLICATIONS

With the advent of solid-state light sources and greater energy efficiency, the application of lighting design to specific types of occupancies is undergoing major changes. The information that follows provides some general guidelines.

Dark walls tend to make a space look smaller, while the high reflectance of white walls fosters a sense of spaciousness.

Bright lighting fixtures detract from walls and diminish the perceived size of the space. Feelings of relaxation and comfort are encouraged by downlighting and color highlights. The inability to identify the source of indirect lighting and very low brightness fixtures may cause discomfort. All lighting fixtures and components should have a UL or Canadian Standards Association (CSA) label or equivalent as required by code.

Residential Applications

It is important to begin the design with daylighting. Each area of a home should have multiple light levels. (See Table 17.21) A multilayered design approach involving task, ambient, and accent lighting results in an environment that allows for a variety of settings or moods. The best approach is to first provide for the task lighting requirements, and then determine what ambient lighting, if any, is required to supplement this layer. Finally, provide appropriate accenting and highlighting to enhance artwork and architectural elements.

Provide for low-level lighting in all rooms with switching and dimmers. Use local task lighting for demanding visual tasks. Low-voltage or wireless remote control produces energy savings.

Commercial Applications

Each type of commercial, institutional, or industrial space has its own lighting requirements. These spaces all include toilet rooms, some of which are open to the general public.

OFFICE LIGHTING

Designing lighting for the office means designing for change. Design for flexibility in both the overall layout and the degree of control that the individual employee has over his or her workspace. Daylighting should be included, both as an energy-efficient light source and for the benefits of keeping in touch with nature. In

TABLE 17.21 RESIDENTIAL LIGHTING TYPES

Type	Kitchens	Bathrooms	Other Spaces
Ambient and General	Traffic paths and to see into cabinets, also large pantries, laundry areas.	General lighting probably adequate in powder rooms. Dimmers desirable.	Family and living rooms.
Task	Cooking surface, at sink, over counters, under cabinets, and over table.	Grooming at various locations. Additional light at mirrors.	Downlights, sconces, table lamps for visual tasks. Reading in bed. Desks in study and home office areas.
Accent	Above wall cabinets for display. Open shelves, glass door cabinets, artwork, architectural features.	Lighting at vanity toe kick for nightlight. Keep lamps away from water sources.	Artwork, sculpture, architectural features.
Decorative	Chandelier in small kitchen, also decorative pendants	In small space, will usually serve multiple purposes.	Adds sparkle, enhances ambience where desired.

TABLE 17.22 RETAIL LIGHTING

Store Type	Lighting Recommendations
Small stores	Use fixed-location spotlights rather than track and flood lighting. Avoid random fixture layouts and shiny or dark surfaces. Lower levels of ambient lighting adequate to examine merchandise. Put focus on important elements.
Medium-sized stores	Supplement ambient lighting with limited accent lighting to set products apart, create highlights, enhance texture and displays.
High-end retail stores	Establish image and enhance product with lighting. Decrease ambient light levels to increase highlights and focus.
High-activity retail stores	Light all objects uniformly and economically, with good visibility for reading labels.

larger offices or open-plan spaces, use more than one type of light fixture, each with specific distribution characteristics.

RETAIL LIGHTING

Good retail lighting can enhance a store's image, lead customers inside, focus their attention on products, and ultimately increase sales. Different types of retail spaces have different lighting requirements. (See Table 17.22) Retail lighting must have good color, contrast, and balance. This can all be done with energy-effective lighting that is energy code compliant.

PUBLIC TOILET ROOM LIGHTING

Good lighting is essential for keeping public toilet rooms clean and pleasant. Well-lit toilet rooms encourage the maintenance staff to keep the space bright and clean, and assure the users that it is safe and hygienic. The toilets, urinals, and lavatories benefit from a task-lighting approach; other areas may be much less brightly lit. All fixtures have to withstand the potential abuse of a public (but often isolated) space.

We have covered a lot of ground in this chapter, but there is much more to learn about lighting. New technologies and products are constantly becoming available. Lighting continuing education courses are consistently popular with interior designers.

In Part VII, we look at the building systems that deal with fire safety, conveyance (elevators, escalators, and materials handling), and communications. Chapter 18 explores how building fires can be prevented, and how the death and destruction they cause can be limited.

VII

FIRE SAFETY, CONVEYANCE, SECURITY, AND COMMUNICATIONS

Part VII concludes the *Building Systems for Interior Designers, Third Edition* with coverage of a variety of topics that affect multiple building systems. Fire safety is a key element of the design of both residential and commercial projects. Conveyance equipment including elevators and escalators are essential means of moving people in tall buildings and providing access for people using mobility aids. Security and communications systems are undergoing rapid development in this age of digital devices and wireless communications.

According to the 2013 report on home fires by the National Fire Protection Association (NFPA), on average, seven people died each day in home fires in the United States (US). Cooking equipment remains the leading cause of home structure fires and injuries, but smoking materials remain the leading cause of deaths. The two leading items in home fire deaths remain upholstered furniture and mattresses and bedding, although the number of deaths from these sources of first ignition continues to decline.

High-rise buildings require great numbers of persons to travel vertically down stairs in order to evacuate. NFPA reports that "in the evacuation of the World Trade Center high-rise office tower following the first terrorist bombing in 1993, tens of thousands of building occupants successfully and safely traversed some five million person-flights of stairs." (www.nfpa.org, accessed August 11, 2014). Since that time, provisions have been made for the use of elevators as part of building evacuation plans.

Communications, security, and control equipment add to building function and safety. Wireless communications are changing the way this equipment works and what is required as part of the building systems.

Chapter 18, "Fire Safety Design," covers how building interiors are designed to prevent fires and help people escape. This is perhaps the most valuable information for interior designers to know about building systems.

Chapter 19, "Conveyance Systems," covers building systems that move people and materials, including elevators, escalators, and materials handling equipment.

Chapter 20, "Communications, Security, and Control Equipment," brings our coverage of building systems to a close with a review of these rapidly evolving systems.

18

Fire Safety Design

The extreme danger that fire represents to a building and its occupants requires a holistic approach to the architectural, mechanical, plumbing, and signal systems. It is important to remember that the technology for fire protection is still being developed, as is information about the effectiveness of each method.

Vaughn Bradshaw, *The Building Environment: Active and Passive Control Systems* (3rd ed.), Wiley 2006, page 383

Chapter 18 addresses basic firefighting principles and fire safety codes. Means of egress protect the building's occupants and directly affect interior design. Compartmentation and construction assemblies are designed to protect the building itself. Interior materials must be selected and tested for fire safety. Smoke management aids firefighters and occupants. Fire detection, alarm, and suppression systems have highly visible elements that must be coordinated with the interior design of the building.

INTRODUCTION

According to the National Fire Protection Association (NFPA), there were 480,500 structure fires in the United States in 2012, resulting in 2470 civilian deaths and 14,700 injuries, and $9.8 billion in property damage. Between 2009 and 2011, civilian fire fatalities in residential buildings accounted for 82 percent of all fire fatalities (FEMA Topical Fire Report Series, *Civilian Fire Fatalities in Residential Buildings 2009–2011*, April 2013).

It is important to consider the overall concept of fire safety rather than just looking at individual systems. The whole building needs to be assessed on basis of its specific conditions. Fire safety strategies start with prevention and passive fire protection. The many topics of fire protection design include design of means of egress, early detection and alarm systems, compartmentation, smoke control and its relationship to material choice and building design, fire suppression systems, and emergency power.

Fire protection requires the coordination of the building's architecture and interiors, mechanical, electrical, and plumbing systems, and signal system. (See Table 18.1) Building systems can integrate fire detection and alarm systems with the HVAC system for energy management. Signaling can be combined with security and intercom functions.

There may be conflicts between fire resistance and features including daylighting, passive cooling, and the HVAC system. Potential conflicts also exist between safe evacuation of people and suppression of fire.

History

Before the Great Fire of London in 1666 left tens of thousands of people homeless, the city had no organized fire protection system. Afterwards, insurance companies created fire brigades, which fought only fires in buildings their companies insured. New buildings were built of stone rather than wood.

In 1736, Benjamin Franklin was instrumental in establishing a volunteer fire department in Philadelphia, the first in the American colonies. The first paid government fire departments did not appear until the mid-nineteenth century.

The Great Chicago Fire in 1871 destroyed 3.3 square miles (9 k^2) and much of the wood construction in the central business district. The disaster resulted in stricter fire regulations and produced some of the best firefighters in the United States.

TABLE 18.1 BUILDING COMPONENTS AND FIRE SAFETY

Component	Ordinary Condition	Fire Condition
High ceilings and low partitions	Aid daylighting and natural ventilation	Without sprinklers, allow fire and smoke to spread through open floor plans.
Interior finishes	Aid comfort, aesthetics, and maintenance	May burn readily and/or give off toxic gases.
HVAC systems	Heat, ventilate, and cool	May be pathways for smoke and fire. Intakes can purge smoke with outside air.
Windows	Provide daylight, view, and fresh air	Give firefighting access, escape routes, dilute smoke with fresh air.
Elevators	Provide accessible access to upper building floors	Apertures can allow vertical spread of fire and smoke. May aid firefighters and egress.
Escalators	Connect retail shopping levels, or hotel lobbies to ballrooms	Apertures can allow vertical spread of fire and smoke. May allow egress.

Design for Fire Safety

To react safely to a fire emergency, a building occupant needs early warning, the means to extinguish a small fire, and at least two ways out of the building. Once a fire has started in a building, there may be only a few minutes to get out safely. A fire can spread at the rate of 15 feet (4.6 meters) per second. Smoke spreads fast and can overcome people in moments, obscuring vision and causing difficulty breathing. People in the building may panic and need to get to a door immediately. The design of the building is critical to getting them out safely.

Fire safety design for buildings considers both common problems and rare problems that could result in many deaths and great property damage. This necessitates design for redundancy of systems.

In older buildings, the goal of fire safety design decisions was to keep the fire from spreading to other buildings. With increasing fire-resistant construction and control by building codes, fires are now usually confined within a single building. Today, the most common fire safety objectives (in order of importance) are protection of life, protection of the building and its contents, and continuity of operation.

Whole-building conflagrations are becoming relatively rare in North America due to the widespread use of automatic fire detection and fire-extinguishing systems. The emphasis is shifting to minimizing water and smoke damage.

Fire suppression systems contain fires to one or two floors or to a single room. A single sprinkler head can extinguish a small fire in about four minutes. Only one to five sprinkler heads are activated in the great majority of residential fires, limiting damage from water as well as fire.

Fire builds up faster in small, enclosed rooms that retain heat. Most thermally massive materials do not burn easily, and thermal mass also benefits passive heating, cooling, and acoustic isolation of airborne sound. (See Figure 18.1 High ceilings allow large quantities of smoke to collect before reaching the occupant's level and allow smoke and flames to be seen from a greater indoor distance.

Figure 18.1 Thermal mass around wood-burning stove

In some buildings, continuing operations during a fire is a priority. Special fire alarm and suppression systems are available for especially critical operations areas such as control rooms. Floors should be waterproofed for speedy removal of water dumped on the fire by the sprinkler system. Waterproofing should continue 4" to 6" (102 to 152 mm) up walls, columns, pipes, and other vertical elements.

BASIC PRINCIPLES

Fire prevention begins with an awareness of fire risks. An understanding of the process of combustion aids in the use of fire prevention methods.

Fire Risk

We tend to be most aware of fires in large, high-profile buildings. However, about 85 percent of all US fire deaths in 2009 occurred in homes (M.J. Karter, NFPA Fire Analysis and Research Division, *Fire Loss in the United States during 2010*, 2011). The groups at greatest risk include children ages 4 and under, adults ages 65 and older, African Americans, Native Americans, and poor people. (Centers for Disease Control and Prevention, *Fire Deaths and Injuries Fact Sheet*, 2011)

When people are caught in a fire, their lungs and respiratory passages may be burned by hot air and their skin severely damaged by thermal radiation. Some deaths occur when panic causes people to push, crowd, and trample others. Other times, panicking people make irrational decisions like running back into a burning building to save belongings.

Three times as many fire deaths result from asphyxiation due to dense smoke as from burns. Victims of fires are often suffocated by air depleted of oxygen and full of poisonous gases. Oxygen content below 17 percent is sufficient to sustain fire, but reduces the ability to breathe and oxygenate blood.

Buildings concentrate fuel that can sustain a fire. Wood building structures, wood paneling, and plastic insulating materials all will burn. Buildings often contain oil, natural gas, gasoline, paints, rubber, chemicals, or other highly flammable materials.

Buildings offer many possible sources of ignition for a fire. (See Figure 18.2) Defective furnaces, sparks from a fireplace, leaky chimneys, and unattended stoves can all start fires. Loose electrical connections and overloaded electrical wiring are common sources of fires. Many home fires are started by poorly maintained furnaces or stoves, cracked or rusted furnace parts, or chimneys with creosote buildup. The largest identified cause of fires in the home is cooking. These are often small fires that are put out without injury to occupants.

According to the US Fire Administration, misuse of wood stoves, portable space heaters, and kerosene heaters are common risks in rural areas. Wood stoves cause over 4000 residential fires every year. Although only two percent of heating fires in homes involve portable heaters, they are involved in 45 percent of all fatal home fires. Of these fires, 52 percent are due to portable heaters too close to combustible materials.

A building is like a stove in that it contains the fire and encourages its growth. The building concentrates heat and flammable combustion gases. A fully developed room fire has temperatures over 1100°F (593°C). Vertical passages through the

Figure 18.3 Firefighting ladder

building that are open to the fire create strong convective drafts that fan the flames. Fire can spread a rate of 15 feet (4.6 m) per second. As the fire spreads up through the building, it finds new sources of fuel.

The loss of firefighters' lives is a tragic result of fires. The design of the building may restrict their ability to escape and serve as a barrier to firefighters. Firefighting ladders can only reach up seven stories, so in tall buildings, firefighters must use the stairs. (See Figure 18.3) In addition, very broad, low buildings can put the fire beyond the reach of fire hoses. Firefighters are exposed to excessive heat, poison gases, and explosions. They are in danger of falls from great heights, toppling walls, and collapsing roofs and floors.

Combustion

Oxidation is a process in which molecules of fuel are combined with molecules of oxygen. The result is a mixture of gases and the release of energy. Oxidation is how our bodies turn food into energy. Rust is the oxidation of iron. In a fire, molecules of fuel are rapidly combined with molecules of oxygen, releasing energy as heat and light. Gases are also released, and smoke is produced when incompletely burned particles are visibly suspended in the air.

FIRE TRIANGLE

For a fire to exist, you need three things: fuel, oxygen, and high temperatures. Fires begin when supplies of fuel and oxygen are

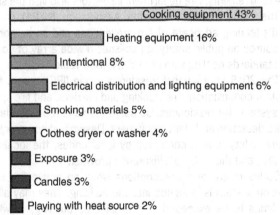

Cooking equipment 43%

Heating equipment 16%

Intentional 8%

Electrical distribution and lighting equipment 6%

Smoking materials 5%

Clothes dryer or washer 4%

Exposure 3%

Candles 3%

Playing with heat source 2%

Figure 18.2 Leading causes of fires in homes

Source: Redrawn from NFPA US Home Structure Fires Fact Sheet, 2007–2011, www.nfpa.org

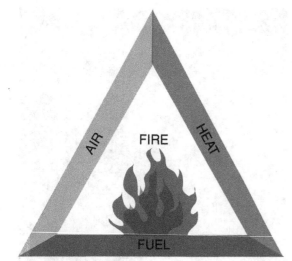

Figure 18.4 Fire triangle

brought together at a sufficiently high temperature for combustion. The fire consumes fuel and oxygen as it burns, and gives off gases, particles, and large quantities of heat.

The **fire triangle** is a graphic representation of the three things necessary for a fire. (See Figure 18.4) Limiting one element of the fire triangle (fuel, oxygen in air, or high temperature) prevents the fire from starting or puts it out.

Fire suppression systems that cover the fuel or that displace the oxygen with another gas limit the supply of oxygen. High temperatures can be controlled by cold water from sprinkler systems. However, the primary way that we strive to prevent and control fire in a building is by controlling the fuel: the building's structure and contents.

Building codes and zoning ordinances regulate the combustibility of materials in different areas of a city, and also the conditions for storage of flammable and explosive substances in or near buildings. Firefighters and fire underwriters (insurers) inspect buildings periodically, looking for accumulated combustible materials. Heating devices, chimneys, electrical systems, electrical devices, and hazardous industrial processes are controlled especially tightly. Smoking is now prohibited by law in many places.

PRODUCTS OF COMBUSTION

The thermal products of a fire are flame and heat. These are responsible for burns, shock, dehydration, heat exhaustion, and the blocking of the respiratory tract by fluid, and result in about one-quarter of the deaths from building fires.

Most fire deaths result from the nonthermal products of combustion including smoke, which can usually be seen or smelled, and the wide range of gases, liquids, and solids it contains. Gases without visible smoke are often difficult to detect; some are toxic and all are dangerous because they displace oxygen. Building fires commonly release gases, including carbon monoxide and carbon dioxide. Smoke is made of droplets of flammable tars and small particles of carbon suspended in gases. It irritates the eyes and nasal passages and sometimes can blind or choke a person.

Most plastics used in furniture, carpeting, draperies, wall coverings, plumbing systems, electrical wiring, and other products are petrochemicals that often burn faster and hotter than other materials. Smoke gases and particulates from plastics may contain hundreds of different chemicals, many of which exist only a few minutes before mixing with other chemicals and turning into something else. Combustion gases include carbon dioxide, carbon monoxide, hydrogen cyanide, hydrogen sulfide, and sulfur dioxide. Hydrogen chloride is produced by burning polyvinyl chlorides found in electronic equipment and cable jackets, as well as in some wall coverings, flooring, and other interior materials.

A fire with decomposing plastics can cause lung and pulmonary damage, and may cause disorientation and loss of the sense of smell. Respiratory failure can follow. Toxic chemicals at levels below lethal can be deadly in combination. Repeated exposure is especially dangerous. Some chemicals remain a danger after a fire is out.

FIRE SAFETY CODES

Modern building codes evolved first as responses to devastating fires, and have gradually come to include many other health and safety issues. The goal of fire code requirements is to protect the building from fire and to contain the fire long enough for people to evacuate the building safely. Building codes set height and area limits in relation to the occupancy or use of the building and the type of construction. They establish structural standards for the construction of walls, floors, and roofs. Codes detail requirements for fire protection systems and means of emergency egress.

Fire safety codes govern how spaces are planned and how materials are used. They dictate the location and number of fire alarms and exit signs, and set sprinkler system requirements that affect the layout of ceiling designs and lighting.

Fire department officials review the plans before a building permit is granted in many communities. Inspections during construction by the building inspector verify that the construction meets code requirements. Fire department inspectors also visit the site.

The National Fire Protection Association (NFPA) is the world's leading advocate of fire preventions and an authoritative source on public safety. It publishes a wide array of codes and standards dealing with fire safety. (See Table 18.2)

The 2015 International Residential Code (IRC) covers fire-resistant construction, fire blocking and barriers, and fire sprinkler systems for residences. UL publishes numerous standards for fire detection and alarm equipment. Design and construction for fire safety is also controlled by local codes, the local fire marshal, and the building's insurance provider.

Codes often relax prescriptions when an active fire suppression system is designed into the building. They may allow size limits to be exceeded in a sprinklered building, or when a building is divided by firewalls into separate areas, each of which does not exceed the size limit. Detailed computer analysis of fire spread and occupant evaluation for a given design

TABLE 18.2 SELECTED NFPA CODES AND STANDARDS

Title	Description
NFPA 101® Life Safety Code®	Minimum requirements for means of egress, fire alarm systems, fire and smoke detection equipment.
NFPA 70: National Electrical Code (NEC) Article 760 Fire Alarm Systems	Fire alarm system wiring including fire detection and alarm notification, sprinklers, safety functions, damper control, fan shutdown. Includes elevator capture and intercom.
NFPA 72®: National Fire Alarm and Signaling Code	Regulations for protective signaling systems and their components.
NFPA 80: Standard for Fire Doors and Other Opening Protectives	Regulates installation and maintenance of assemblies and devices used to protect openings in walls, floors, and ceilings against the spread of fire and smoke.
NFPA 99: Health Care Facilities Code	Fire and life safety issues in healthcare facilities.
NFPA 220: Standard on Types of Building Construction	Defines construction assemblies based on combustibility and fire resistance rating of their structural elements.

may allow greater distances to exits, larger open floor areas, and alternative construction methods. This performance-based analysis approach to design requires close cooperation between building designers and fire code enforcement officials.

The design professional in charge of a project is ultimately responsible for making sure that the design meets all applicable codes. Interior designers often check building codes for fire safety requirements. Whether specifying a fabric for a commercial project or checking a floor plan for the number, size, and location of exits, interior designers rely on applicable code requirements. The interior designer must be familiar with the codes for each project's location, and must make sure that the design complies. It is important to verify the most recent versions of applicable codes. Failure to do so can result in costly mistakes, delays in construction, and a very unhappy client.

Construction Types

To be classified a certain construction type, a building must meet the minimum standards for every structural element in that type. The construction type can come into play when a designer changes existing interior structural elements or adds new ones. If changes are not consistent with the existing building materials, they could reduce the entire building classification. This would reduce building safety and could affect building insurance and liability.

Interiors projects that are affected by construction types and building sizes include ones that relocate walls or add a stairway or other structural work. The primary structural element that an interior designer may be involved with is an interior wall, which could be either load-bearing or nonload-bearing. The designer's work may also affect firewalls and party walls, smoke barriers, and shaft enclosures. Interior designers sometimes also work with columns, floor and ceiling assemblies, and roof/ceiling assemblies.

Each construction type assigns structural elements a minimum fire protection rating. This is the number of hours the structural element must be able to resist fire without being affected by flame, heat, or hot gases. It is essentially a fire endurance rating. Where construction is required to be fire-resistant, it must typically have 1-hour fire-resistant construction throughout. It is essential to verify fire-resistance requirements with current applicable codes.

Occupancy Hazard Classifications

Building codes classify various occupancies according to fire hazard. (See Table 18.3) These classifications are used to determine the design of sprinkler systems. Generally, sprinkler systems are required for Factory, Hazardous, and Storage occupancies, or where large groups of people are present, as in Assembly, Institutional and large Mercantile and Residential occupancies. The requirements are based on the number of occupants, the mobility of the occupants, and the types of hazards present. Verify occupancy classifications with authorities with jurisdiction.

Many codes now require sprinklers in residential occupancies. Residential sprinklers that are tested and listed for protection of dwelling units are designed for fast response. They are sensitive to both smoldering and rapidly developing fires. They are designed to have one or two heads open quickly to prevent smoke and toxic gases from filling small rooms. Residential sprinklers are designed to deliver water high enough on walls and ceilings to prevent the fire from getting above the sprinklers. By cooling the ceiling, they reduce the number of sprinklers that open, limiting water damage.

Residential systems are designed for buildings that do not ordinarily have sufficient water supply capacity for standard sprinkler systems. Most codes that require residential sprinkler systems exempt bathrooms up to 55 square feet (5.1 meters^2); closets with a smallest dimension up to 3 feet (0.9 m); open

TABLE 18.3 OCCUPANCY HAZARD CLASSIFICATIONS

Classification	Contents	Conditions Similar To:	Maximum Sprinkler Spacing
Light Hazard	Low quantity, combustibility, and rate of heat release. Relatively easy fire protection.	Residential, churches, auditoriums, hospitals, museums, offices, restaurant seating areas, educational, institutional	15 feet (4.6 m). Protection area per sprinkler head maximum 200 square feet (18 m²). Sprinklers do not need to be staggered.
Ordinary Hazard	Moderate to high quantity and rate of heat release, relatively low to high combustibility. Materials may cause rapidly developing fires.	Bakeries, laundries, dry cleaners, manufacturing facilities, large library stack areas, mercantile, post offices, restaurant service areas, stages	15 feet (4.6 m). Protection area per sprinkler head maximum 130 square feet (12 m²) for noncombustible ceiling and 120 square feet (11 m²) for combustible ceiling. Sprinklers must be staggered if over 12 feet (3.7 m) between heads.
Extra Hazard	Very high quantity and combustibility, rapid fire; severe hazard. Fire development with high heat release rates may occur.	Aircraft hangars, plywood and particle board manufacturing, textile processing, upholstering with plastic foams, paint shops	12 feet (3.7 m). Protection area per sprinkler head maximum 90 square feet (8.4 m²) for noncombustible ceiling, 80 square feet (7.4 m²) for combustible ceiling. Sprinklers staggered if over 8 feet (2.4 m) between heads.

porches, garages, and carports; uninhabitable attics and crawl spaces not used for storage; and entrance foyers that are not the sole means of egress.

MEANS OF EGRESS

A **means of egress** is a continuous and unobstructed path of travel from any point in the building to its exit or a public way. The means of egress is also the path of travel an occupant uses to obtain a safe area of refuge within the building. A means of egress must provide safe and adequate access from any point in a building to protected exits leading to a place of refuge. There are three components to an egress system: exit access, exit, and exit discharge. (See Figure 18.5)

The threat of panic behavior in a fire drives many code requirements, although panic behavior is actually rare. In a fire, building occupants typically make decisions in a predictable manner. First, they detect the smell of smoke; the sounds of breaking glass, sirens, or alarm bells; or more rarely, the sight of flames. Next, the occupants define how severe they believe the fire to be. How other people act is influential, and can even result in a refusal to evacuate in the early stages of a fire. This is followed by coping behavior, the decision whether to flee or fight the fire. Clear exit pathways and access to firefighting equipment are critical to the decision. The NFPA *Fire Protection Handbook* discusses human behavior in fires.

Typical building code requirements indicate the number of exits that must be provided for fire egress, the width of doors, passageways, and fire stairs, and the acceptable length of travel. Specific requirements vary with building type and occupancy.

Building Types

Means of egress requirements differ with various building types. Low-rise buildings are typically easier to evacuate than high-rise buildings. As we have observed, death and injury by fire occurs more often in residential buildings.

LOW-RISE AND HIGH-RISE BUILDINGS

In low-rise buildings, it is a reasonable goal to evacuate all occupants between the detection of the fire and the arrival of the firefighters. (See Table 18.4) This is not the case with high-rise buildings.

Firefighting equipment is ordinarily limited to seven floors, or around 90 feet (27 m). Fire codes classify buildings with an occupiable floor more than 75 feet (23 m) above the lowest fire department access as high-rise, with special design considerations that apply.

Frequently, only two exit stairways are provided in a high-rise building. A 15-story building housing 60 persons per

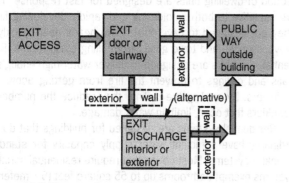

Figure 18.5 Means of egress components

TABLE 18.4 EVACUATION CODE PROVISIONS FOR LOW-RISE BUILDINGS

Egress Component	Provisions
Exit access	Clearly defined pathways to exits. Keep relatively clear of smoke. Minimum clear wheelchair access 32" (813 mm).
Vertical egress exits	Smokeproof towers, exterior and interior stairs and ramps, some escalators.
Horizontal egress exits	Doors leading directly to outside, 2-hour fire-rated enclosed hallways, moving walks. Special horizontal exits have internal firewalls with two fire doors swinging in opposite directions.
Exit discharge	Area to outside exits leading to a public way.

floor per stair can be evacuated in around 9 minutes. A 50-story building with 240 persons per floor per with the same stair size takes at least 2 hours, 11 minutes, to evacuate. When doors are held open, smoke enters the stairwell. Consequently, people are increasingly advised, but may refuse, to evacuate.

RESIDENTIAL EGRESS

A single means of egress is generally permitted when there is a maximum of 10 occupants and where the length of travel within a dwelling unit does not exceed 75 feet (23 m); this applies even if an individual residence is multistory. The exit must be directly to the outside or, if within a multifamily structure, to a floor that has the prescribed number of exits required by the floor area and/or occupancy load exiting on that floor (typically at least two). Exits must be located according to appropriate remoteness and travel distance requirements.

The 2015 IRC requires that the means of egress provide a continuous and unobstructed path of vertical and horizontal egress travel from all portions of the dwelling to the required egress door without requiring travel through a garage. At least one side-hinged egress door with a clear opening width of not less than 32" (813 mm) is required to be provided for each dwelling unit. The clear height of this required egress door opening shall be at least 78" (1981 mm). Egress doors are required to be readily operable from inside the dwelling without the use of a key or special knowledge or effort.

The 2015 IRC requires that basements, habitable attics, and every sleeping room have at least one operable emergency escape and rescue opening that opens directly into a public way, or to a yard or court that opens to a public way. Where basements contain one or more sleeping rooms, such an opening is required in each sleeping room. The 2015 IRC details requirements for emergency egress through an opening below grade, where either a window well or bulkhead must be provided.

Always verify means of egress requirements with current editions of the codes recognized by the authorities having jurisdiction for a specific project.

Means of Egress Components

As indicated earlier, any means of egress has three components: the exit access, the exit, and the exit discharge. A means of egress is comprised of both vertical and horizontal passageways including doorways, corridors, stairs, ramps, enclosures, and intervening rooms. The international construction codes and *Life Safety Code* set most egress requirements. The Americans with Disability Act (ADA) should also be reviewed for related requirements.

All areas or spaces within a building should have at least one exit door or exitway that cannot be locked against egress. At least two of the lockable exits in all rooms, spaces, and areas that can be occupied by more than 50 people must not be lockable against egress.

To calculate the minimum egress widths per floor, calculate the net or gross floor area, as specified in the applicable code table. Then divide the calculated floor area by the occupant load to get the number of occupants for whom exits must be provided for that floor.

The next step is to calculate the exit capacity based on its clear width. Although requirements vary, generally the required egress width per person is 0.3" (7.6 mm) for stairs and 0.2" (5.1 mm) for other elements. A design is usually based on a 22" (559 mm) wide unit of exit. For an additional fraction under one-half, add 12" (305 mm), or a full unit for over one-half.

Doors along an escape route in most cases must open in the direction of travel from indoors to outdoors; this may not be required for occupancies of fewer than 50 people. A door swinging outward can interfere with an exit passageway and is best recessed, but most codes allow a door to project into a corridor as long as does not reduce the width of the passageway by more than half.

EXIT ACCESS

The **exit access** is that portion of a means of egress that leads to an exit. An exit access leads from the room or space to the exit, and can include doors, stairs, ramps, corridors, aisles, and intervening rooms. (See Figure 18.6) The exit access does not necessarily require a fire rating nor need to be fully enclosed. When a fire rating is required, it is usually one hour.

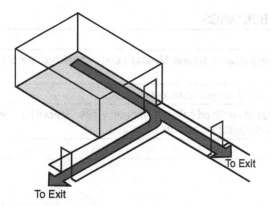

To Exit

To Exit

Figure 18.6 Exit access

Source: Redrawn from Francis D.K. Ching and Corky Binggeli, *Interior Design Illustrated* (3rd ed.), Wiley 2012, page 245

Doors in an exit access are regulated as to type, size, and swing direction, depending upon where they are located. Doors along a corridor must be fire-rated. A minimum height of 80"(2032 mm) is typically required. Large spaces with high occupancies may require two or more exit access doorways.

An **exit access corridor** is any corridor leading to an exit in a building. The typical corridor may require a 1-hour rating. The width and maximum length of travel of the corridor are limited by codes and by accessibility requirements. Exit access corridors must be enclosed by walls of fire-resistive construction in order to serve as required exits.

An **aisle** is a pathway created by furniture, equipment, merchandise, or other obstructions with a maximum wall height of 69" (1753 mm). If the furniture or equipment is any higher, it is considered a corridor. The pathways between movable panel systems in offices are considered aisles, as are the paths between tables and chairs in restaurants and display racks in stores. Rules for aisles that are part of an exit access are similar to those for corridors.

The 2015 International Building Code (IBC) requires an **aisle accessway** that provides a path to an adjacent aisle or aisle accessway to be 30" (762 mm) in Groups B and M where not required to be accessible. Display areas for merchandise (merchandise pads) must have a 30" (762 mm) minimum aisle accessway on at least one side. Check the Americans with Disabilities Act (ADA) for specific dimension requirements for accessible spaces.

According to the 2015 IBC, the required clear width of aisle access ways adjacent to seating at a table or counter is measured to a line 19" (483 mm) away from and parallel to the edge of the table or counter. Where tables or counters are served by fixed seats, the aisle or aisle accessway width is measured to walls, edges of seating, and tread edges. (See Figure 18.7) Other requirements also apply.

Assembly occupancies sometimes have fixed seating for large numbers of people. Codes restrict aisle widths depending upon the size of the occupancy, the number of seats served by each aisle, and whether the aisle is a ramp or a stair. The minimum distances between seats and where the aisles terminate are also regulated.

An exit access should be as direct as possible. In some projects, the access path may need to pass through an adjoining room or space before reaching a corridor or exit. This may be allowed if the path is a direct, unobstructed, and obvious means of travel toward the exit.

Codes may allow smaller rooms to empty through larger spaces to access a corridor. Reception areas, lobbies, and foyers are allowed as long as they meet code requirements. Kitchens, storerooms, restrooms, closets, bedrooms, and other spaces subject to locking are generally not allowed as part of an exit access except in a dwelling unit or where there is a limited number of occupants. Rooms with a higher susceptibility to fire are also restricted.

No single floor should have steps or stairs within a means of egress. Ramps can be used wherever there is a change of elevation and access is required for people in wheelchairs. Ramps have width and clearance requirements similar to those of corridors. Ramps require landings at certain intervals and with certain dimensions depending on the length of the ramp and the number of changes in direction. Landings are also required at the top and bottom of the ramp. The ramp and landings require specific edge details and a nonslip surface. Handrails are typically required when ramps exceed a certain angle or rise, and guards may also be required.

For more information on ramps, see Chapter 5, "Floor/Ceiling Assemblies, Walls, and Stairs."

EXITS

The **exit** is the portion of a means of egress that is separated from all other spaces of a building. The exit leads from the exit access to the exit discharge, and must provide an enclosed, protected means of evacuation for occupants of the building in the event of fire. There are specific requirements for the quantity, location, and size of exits, along with other code and accessibility requirements similar to those for the exit access. The exit must be fully enclosed and fire-rated with minimal penetrations. Exits are typically required to have a 2-hour fire rating, as compared to the exit access and exit discharge that usually require one hour.

Vertical exits include smokeproof towers, exterior and interior stairs and ramps, escalators meeting specific requirements, and in some cases, elevators. Horizontal exits comprise doors leading directly to the outside, 2-hour fire-rated enclosed hallways, and moving walks. Special horizontal exits consist of internal firewalls penetrated by two fire doors, with one door swinging in each direction.

The exit must either open into another exit, an exit discharge, or directly onto a public way. A fire exit is generally a 2-hour-rated fire stair, a fire-protected horizontal passage

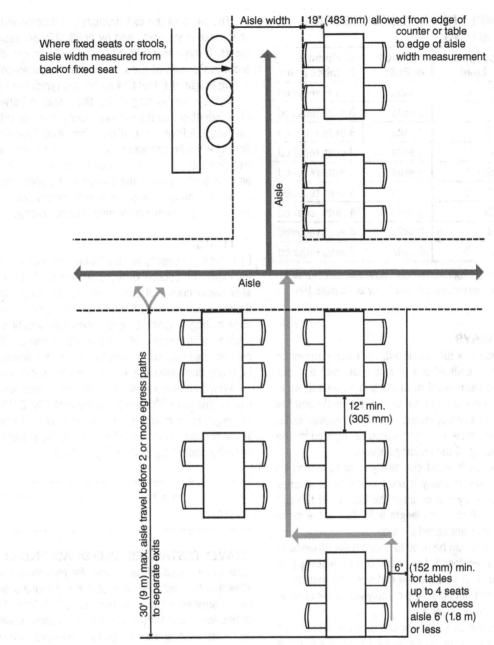

Aisle width

19" (483 mm) allowed from edge of counter or table to edge of aisle width measurement

Where fixed seats or stools, aisle width measured from backof fixed seat

Aisle

Aisle

30' (9 m) max. aisle travel before 2 or more egress paths to separate exits

12" min. (305 mm)

6" (152 mm) min. for tables up to 4 seats where access aisle 6' (1.8 m) or less

Figure 18.7 Aisle access

Source: Redrawn from Francis D.K. Ching and Steven R. Winkel, *Building Codes Illustrated* (2nd ed.), Wiley 2007, page 163

across an acceptable fire barrier, or a door opening directly to the outside from a ground floor room or corridor. In some cases, it may be an exit passageway with walls of fire-rated construction.

One exit is required for a room with fewer than 50 occupants. Two remotely located doors are required per room with 50 or more occupants. The **one-half diagonal rule** usually applies. It states that, for individual rooms, floors, and overall buildings, two exits must be separated by distance at least one-half the dimension of a diagonal line drawn across the area under consideration.

Codes usually require that each door be sized to handle at least half of the occupants, even if there are more than two doors. This typically also applies to exit doors and fire stairs. Some exceptions may apply for places of assembly.

The maximum distance permitted from the door of any room to the farthest protected exit is usually 150 to 200 feet (46 to 61 meters). The floor with the largest number of occupants requires the highest number of exits. Four exits are required by code for any occupant load over 1000. This load carries down to all the floors below the fourth, which everyone must pass to leave the building. (See Table 18.5)

TABLE 18.5 SAMPLE NUMBER OF EXITS FOR MULTISTORY BUILDING

Floor Number	Occupancy Load	Number of Exits	Minimum Number Exits
Floor 8	450	2 exits	2 exits required
Floor 7	825	3 exits	3 exits required
Floor 6	495	2 exits	3 exits required
Floor 5	800	3 exits	3 exits required
Floor 4	1020	4 exits	4 exits required
Floor 3	982	3 exits	4 exits required
Floor 2	905	3 exits	4 exits required
Floor 1	400	2 exits	4 exits required
Basement	51	2 exits	2 exits required

Source: Data from Sharon Koomen Harmon and Katherine E. Kennon, *The Codes Guidebook for Interiors* (5th ed.), Wiley 2011, page 166

EXIT PASSAGEWAYS

An **exit passageway** is a fully enclosed, fire-rated corridor or hallway connecting a required exit or exit court with a public way. It provides the same level of protection as an exit stair. The exit passageway consists of the surrounding walls and the doors leading into it, has no openings other than required exits, and is enclosed by fire-resistive construction as required for the walls, floors, and ceiling of the building served.

An exit passageway is most commonly used to extend an exit. If an enclosed exit stairway is not located at an exterior wall, the exit passageway can connect the bottom of the exit stair to an exterior exit door. Its length is limited to the maximum dead-end corridor permitted by code.

An exit passageway can be used to shorten the distance to an exit by adding an enclosed, fire-rated corridor leading to a door at an exit stair. This gives a new end point for measuring the distance to the exit, and can help comply with code requirements for travel distances to an exit.

An exit access such as a corridor can also exit into an exit passageway. This may occur on the ground floor of a building when secondary exits are required, as in malls and office buildings with central building cores. The exit passageway is created at the building perimeter between two tenants, so that an exterior door can be reached off the common corridor.

EXIT DISCHARGES

All exits must discharge to a safe place of refuge outside the building, such as an exit court or public way at ground level. An **exit discharge** may be a courtyard, patio, or exterior vestibule connecting the exterior exit door to the public way. A **public way** is a street, alley, or similar parcel of land open to the sky and permanently available for free passage and use of the general public. Small alleys or sidewalks less than 10 feet (3 meters) wide are not considered public ways, and become exterior exit discharges connecting the exterior exit door to a larger alley, sidewalk, or street.

The width of the exit discharge is determined by the width of the exit it supports and by applicable accessibility requirements. If more than one exit opens into an exit discharge, the width is the sum of the various exits' requirements. A minimal ceiling height of 8 feet (2.4 meters) is typically required.

In the main building lobby, the distance between the door of an exit stair and the exterior door is considered to be an exit discharge. A foyer or vestibule (a small enclosure on the ground floor of a building between the end of a corridor and the exterior exit door) can be part of an exit discharge. If the size of such an enclosure is kept to the minimum, the codes may not require a high fire rating. A larger enclosure may be considered an exit passageway, which may require a higher rating.

EXIT SIGNS

Exit signs are usually required wherever two or more exits are mandated by code. Exit signs are located at the doors of all stair enclosures, exit passageways, and horizontal exits on all floors. An exit sign is placed at an exterior exit door and at any door exiting a space or area when the direction of egress is unclear, with directional signs at other places. Some smaller occupancies may not require exit signs. It is advisable to clearly label any door, passage, or stair that is "Not an Exit."

Within an exit access, the maximum distance from an exit sign is limited to 100 feet (30.5 meters). The 2015 IBC requires exit signs to be mounted flush to the door or wall near the floor in guest rooms in Group R-1 occupancies, in addition to ceiling- or wall-mounted signs. (See Figure 18.8)

For more information on exit signs, see Chapter 17, "Lighting Systems."

TRAVEL DISTANCES AND DEAD-END CORRIDORS

Most tables used to determine the maximum travel distance allowable to reach the nearest fire exit make a distinction between sprinklered and unsprinklered buildings. However, this distinction is currently under review as some code officials believe that existing allowances for sprinklers to have raised the maximum distance above reasonably safe levels. Check current applicable codes.

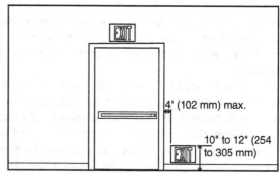

Figure 18.8 Floor-level exit sign location

A dead-end corridor is typically a branch off a main corridor that links at least two fire stairs. This arrangement provides only one means of egress until it meets the main corridor. The 2015 IBC requires that where more than one exit or exit access doorway is required, the exit access shall be arranged such that there are not dead ends in corridors more than 20 feet (6 meters) in length. Length limits for other conditions and other requirements also apply.

EXIT STAIRS

An **exit stair** is the most common type of exit and comprises a protected enclosure. (See Figure 18.9) The exit stair includes the stair enclosure, any doors opening into or exiting out of the stairway enclosure, and the stairs and landings inside the enclosure. The enclosure of an exit stair must be constructed of rated assemblies.

An exit stair may lead to an exit passageway, an exit court, or public way. The doors of an exit stair must swing in the direction of the exit discharge. Exit stairs must allow firefighters to move up while occupants move down. They are sized to meet requirements that allow people in the stairwell to continue down without interference from doors. Unenclosed access stairs do not count as required exit stairs.

Enclosed stairs may fill with smoke. Two separate exit stairs within the same shaft (called scissor stairs) are not recommended as their entrances are too close together; this arrangement could create confusion, and a fire could block both entrances. The safest fire stairways are in **smokeproof towers**, with stairs having direct access to outdoor air at each floor.

Code requirements for the number and size of fire stairs consider both economic and safety issues. People use stairs more where there are more and larger stairways. Making them more visible with code-approved fire-resistant glazing increases their appeal.

Most of the stairs that we deal with in buildings are exit stairs. Another category covered by codes is the **exit access stair**, which is not as common as a typical exit stair. Exit access stairs are usually used within a space when one tenant occupies more than one floor of a building or where there is a mezzanine. Exit access stairs usually do not need a fire-rated enclosure unless they connect more than two floors.

For more information on exit stairs and ramps, see Chapter 5, "Floor/Ceiling Assemblies, Walls, and Stairs."

OCCUPANT LOADS

Building codes use the **occupant load** to establish the required number and width of exits for a building. The occupant load determines the maximum number of people allowed in a specific occupancy at any one time.

Note that the occupant load is not the same as an occupancy classification. An occupancy classification indicates the use of the space, rather than the number of people occupying it.

The code assigns a predetermined amount of space or square feet that is required per occupant within specific occupancies and building uses. This amount of space is called the **occupant load factor**.

The occupant load of a space is the total number of persons that may occupy a building or portion of a building at any one time. It is determined by dividing the floor area assigned to a particular use by the square feet (in the United States) per occupant permitted in that use.

The occupant load is calculated using the area inside the exterior walls. Some load factors are calculated in gross square feet, including interior wall thicknesses and all miscellaneous spaces in the building as a whole. Others are based on net square feet of actual occupied space, not including corridors, restrooms, utility closets, and other unoccupied areas. Sometimes fixed items that take up space, such as interior walls, columns, built-in counters, and shelving are deducted from the number of square feet used in the calculations. Once again, check the specific current code for the jurisdiction of your project.

AREAS OF REFUGE

Areas of refuge (refuge areas) are provided in high-rise buildings and for wheelchair users in multistory buildings. In large

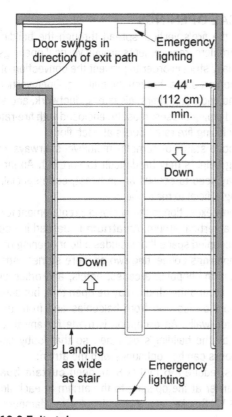

Figure 18.9 Exit stair

(within figure)
Door swings in direction of exit path
Emergency lighting
44" (112 cm) min.
Down
Down
Landing as wide as stair
Emergency lighting

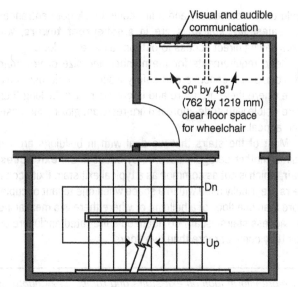

Figure 18.10 Area of refuge near stairway

Source: Redrawn from Sharon Koomen Harmon and Katherine E. Kennon, *The Codes Guidebook for Interiors* (5th ed.), Wiley 2011, page 159

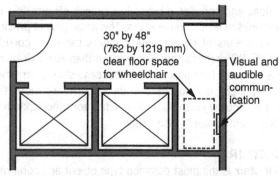

Figure 18.11 Area of refuge near elevator lobby

Source: Redrawn from Sharon Koomen Harmon and Katherine E. Kennon, *The Codes Guidebook for Interiors* (5th ed.), Wiley 2011, page 159

buildings, not everyone may be able to evacuate in time, and refuge areas provide a place to wait that is protected from smoke. Ideally, refuge areas should remain free of smoke, gases, heat, and fire throughout the fire and until rescue. The Americans with Disabilities Act (ADA) sets minimum requirements for accessible spaces in refuge areas.

Areas of refuge are commonly located adjacent to a protected stairway, and are protected from smoke. (See Figure 18.10) They are provided with communications devices to summon firefighters to rescue the people in the refuge.

When a stairway is used as a refuge area, it is designed to hold all the building occupants, allowing 3 square feet (0.28 meters^2) per person. Horizontal exits can be areas of refuge, as can smoke-protected vestibules or enlarged landings adjacent to exit stairways.

In an exit stairwell, the landings at the doors entering the stairs can be enlarged so that one or more wheelchairs can wait for assistance without blocking the means of egress. An alternative arrangement provides an alcove for wheelchairs in a portion of an exit access corridor located immediately adjacent to the exit enclosure. Another possibility is to provide space immediately adjacent to an exit enclosure by creating an enclosed exit discharge such as a vestibule or foyer. (See Figure 18.11)

ELEVATORS AND ESCALATORS

Elevators are not normally used as part of a means of egress because of their inherent unreliability in a fire. When a fire alarm is set off, all elevators are immediately recalled to the ground floor. Firemen with special keys can manually operate the elevators so that firefighters with respirators can take fire equipment quickly to the floor below the fire.

Currently, code and fire safety officials are reconsidering the use of elevators for occupant evacuation in very tall buildings. They are already used in some Asian buildings to aid evacuation, and are being investigated for use in the United States. The use of elevators by occupants during emergency situations has been employed in the Stratosphere Tower in Las Vegas, Nevada.

For more information on elevators and fire safety, see Chapter 19, "Conveyance Systems."

VERTICAL OPENINGS

Because the fire's vertical spread through the building is the biggest problem, requirements around vertical openings tend to be especially strict in order to prevent the convection of fire and combustion products through the building. Open vertical shafts of any kind, including stairs, elevators, ductwork, and electrical wiring and piping chases, must be enclosed with fire-rated walls with self-closing fire-rated doors at each floor.

The code status of complex multilevel stairways with walls of varying heights may be difficult to sort out. An interior designer may need to consult an architect, code specialist, and/or building official to make the call.

The only exception to the enclosure requirement for vertical shafts is a vertical atrium. An **atrium** is defined in codes as a roofed, occupied space that includes a floor opening or a series of floor openings connecting two or more stories. Atriums are often found in shopping arcades, hotels, and office buildings. Balconies around the atrium may be open to it, but surrounding rooms must be isolated from balconies and from the atrium by fire-rated walls. An exception is made for any three floors selected by the building's designer, so that lobby spaces on several floors can be continuous with the atrium.

Codes require a 6 foot (1.8 m) deep **curtain board** as a smoke barrier at the opening to the atrium at each floor. (See Figure 18.12) Smoke detectors and motorized dampers in ducts are required.

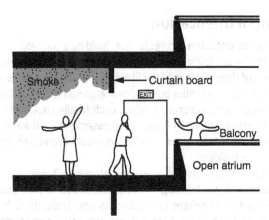

Figure 18.12 Curtain board

Source: Redrawn from Walter T. Grondzik and Alison G Kwok, *Mechanical and Electrical Equipment for Buildings* (12th ed.), Wiley 2015, page 1154

A building with an atrium must have sprinklers throughout, with sprinklers located 6 feet (1.8 m) on center at the lobby level, the atrium floor level, and where other floors open to an atrium, to create a water curtain. The frames for glazing at the perimeter of the atrium should be designed for thermal expansion so that they will not cause their glass to break when they get hot. The atrium must be provided with fans with dampers that open and turn on automatically in case of fire to bring fresh air into the space at ground level and exhaust smoke at the ceiling level.

HORIZONTAL EXITS

Horizontal exits are passages through a wall constructed as required for an occupancy separation, protected by an automatic-closing fire door, and leading to an area of refuge in the same building or on approximately the same level in an adjacent building. (See Figure 18.13) Occupants escape from a fire on one side by moving horizontally through self-closing fire doors to the other side to the safety of an area of refuge.

Doors in a horizontal exit must be fire-rated and swing in the direction of travel to the exit. If the horizontal exit has an area of refuge on either side, it must have two doors together, swinging in opposite directions, so that occupants can push through the doorway in either direction.

When you pass through the door of a horizontal exit, the whole space beyond is considered an area of refuge where you can either wait for assistance or use another exit to leave the building. Horizontal exits are used to provide refuge for large building populations in healthcare, detention, and educational buildings.

Horizontal exits may reduce the number of other types of exits required, but codes place limits on the total number of horizontal exits a building can have. Horizontal exits are only allowed when two or more exits are required.

PROTECTING THE BUILDING

The structure of a building is protected to prevent collapse of the building within the time the fire runs its course or to delay the collapse of low buildings until all occupants have escaped and firefighters have had a reasonable chance to save the building. The building may survive to be salvaged rather than being demolished after the fire. Protecting the structure protects the occupants, firefighters, and neighboring buildings. Tall buildings present a significant danger if all or part of the building falls.

The most important elements of the structure to be protected are the columns. Next in importance are the girders and the beams, and lastly, the floor slabs.

Most large buildings are constructed of either reinforced concrete or protected steel. Steel does not burn but loses much of its structural strength in a fire and will sag or collapse at the sustained temperatures frequently reached by ordinary building fires. Concrete is more resistant to fire than steel, but its cement binder can disintegrate, potentially resulting in serious structural damage if the fire lasts long enough.

Steel beams and columns are encased in concrete, lath, and plaster, or surrounded with multiple layers of gypsum wallboard (drywall) for protection. (See Figure 18.14) They are also sometimes sprayed with lightweight mineral insulation in cementitious

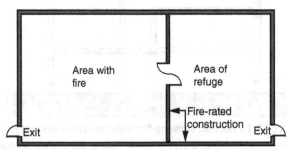

Figure 18.13 Horizontal exit

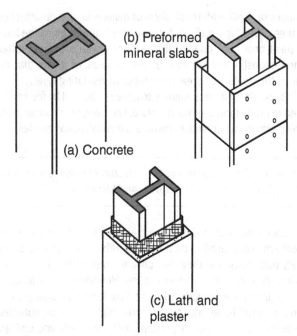

Figure 18.14 Fire-resistant steel column assemblies

Figure 18.15 Column with spray-on fire protection
Source: Courtesy of Herb Fremin

binders or have preformed slabs of mineral insulation attached to them. (See Figure 18.15) **Intumescent coatings**, in the form of paint or a thick coating that is put on with a trowel, soften when exposed to heat and release bubbles of a gas that expands the coating to create a protective insulating layer.

Brick, tile, and mineral fibers that are unaffected by fire can be used to protect the building structure. However, their mortar joints may disintegrate, potentially causing the construction can fail.

For more information on building structural systems, see Chapter 4, "Building Forms, Structures, and Elements."

Low industrial and commercial buildings of unprotected steel are considered to be noncombustible, but there is an unlikely possibility that they may collapse rapidly in a hot fire before occupants have time to escape. Buildings constructed of heavy timber are considered to be slow burning buildings, and are permitted to be one to two stories higher than unprotected steel buildings. Plaster or gypsum wallboard walls and ceilings offer one-half hour of protection for smaller wooden buildings.

Compartmentation

Compartmentation protects the building's occupants and property by confining the fire, heat, smoke, and toxic gases to the area of their origin until the fire is extinguished or burns itself out completely. An entire building or a large space can be divided into two or more separate spaces, each totally enclosed within a fire barrier envelope of floor/ceiling assemblies and walls. This prevents the spread of fire, smoke, and heat beyond a restricted area of the building.

Compartmentation is required between different types of functions within a building. Compartmentation is also used to provide areas of refuge for occupants and firefighters. In row houses, walls separate dwellings into separate compartments.

Compartmentation is used to isolate spaces where fires often originate or where they are especially unacceptable. Code required fire-resistance levels for fire-rated construction range from 30 minutes to 4 hours. Construction assemblies are tested for structural adequacy, integrity, and insulation.

Openings in firewalls must be protected by fire-rated doors and fire dampers in forced-air systems. Fire-resistant barriers to limit vertical fire spread are required by codes. Roll-down shutters and accordion doors can also be used to compartmentalize a building.

FIRE BARRIERS

Fire barriers are fire-rated structural elements. They include wall, ceiling, or floor systems that prevent the spread of flame and heat through the use of fire-rated structural materials with fire-resistant (FR) ratings. Fire barriers can be divided into three types. **Firewalls** have the highest fire ratings and are usually part of the building shell. **Fire separation walls** are used to create fire-rated compartments within a building. The fire ratings of the third category, **floor/ceiling assemblies**, depend on the walls they surround.

Firewalls, also called party walls, are often used to subdivide a building into two separate types of construction. Firewalls combine with fire-resistant floors to contain fires both horizontally and vertically. (See Figure 18.16) They are also used to separate one occupancy from another in a mixed-use building.

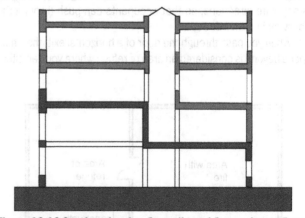

Figure 18.16 Section showing firewalls and fire-resistant floors
Source: Redrawn from Francis D.K. Ching, *Building Construction Illustrated* (5th ed.), Wiley 2014, page 2.07

Firewalls provide continuous protection from the foundation of the building to the roof and to each exterior wall. They are built so that if one side of the wall falls, the other side would remain standing. Firewalls typically have 3- to 5-hour ratings.

Building codes limit the number of penetrations in a fire-rated wall. Interior designers generally are not involved in designing firewalls, but our work may involve penetrations. All openings in a firewall are limited to a certain percent of the wall's length. Openings may require protection by self-closing firewalls, fire-rated window assemblies, or fire and smoke dampers. Openings for penetrations must be carefully sealed to keep the compartment airtight and prevent fire spread to adjacent spaces. Any access panel doors to utilities must maintain fire barrier ratings.

Fire separation walls, which include tenant separation walls (demising walls), corridor walls, vertical shafts, and room separations, are more likely than firewalls to be added or changed during an interiors project. Tenant separation walls create fire rated compartments within a building that separate two tenants or dwelling units. They typically require 1-hour ratings, depending on the occupancy and whether sprinklers are used.

Corridor walls have ratings of from one to two hours depending upon how corridors are used, the occupancy, and whether sprinklers are used. Corridor walls that are used as exits usually have a 2-hour fire rating, and corridors used as exit accesses generally require 1-hour ratings. Typically, codes require that corridor walls be continuous from floor slab to floor slab and penetrate suspended ceilings. Some corridor walls may also act as demising walls, and then stricter requirements apply.

The walls that create vertical shaft enclosures for stairwells, elevators, and dumbwaiters are usually continuous from the bottom of the building to the underside of the roof deck. Stairs used as part of an exit have requirements for fire ratings, can only have limited penetrations, and may require that the enclosure be smokeproof. Stair enclosures are required to have a 1-hour fire rating for up to three stories, and 2-hour ratings for four or more stories. Where stairs connect only two floors within a single occupancy, the space may be considered to be an atrium, and the enclosure restrictions may be less restrictive.

Most rooms within a space do not require fire rated walls. Where the contents of the room may be hazardous, codes may specify that they be separated from the rest of the building by a fire rated wall.

Fire rated floors and ceilings are rated as either floor/ceiling or roof/ceiling assemblies. The assembly consists of everything from the bottom of the ceiling material to the top of the floor or roof above. This includes all the ducts, piping, and wires between the finished ceiling and the finished floor above it. The required ceiling rating is determined by the ratings of the surrounding walls.

CONCEALED SPACES

Fire can spread quickly in concealed spaces over suspended ceilings, behind walls, within pipe chases, in attics, and under raised floors. Noncombustible materials should be specified wherever possible in these spaces. Automatic fire detection and suppression systems and oxygen deprivation systems can be used in concealed spaces. Compartmentalization with fire-stops or firewalls can break up continuous concealed spaces. Automatic fire detection and suppression equipment, including oxygen deprivation approaches, can be used in unoccupied concealed spaces.

Construction Assemblies and Elements

ASTM E-119—*Standard Test Methods for Fire Tests of Building Construction and Materials* establishes 1-hour, 2-hour, 3-hour, and 4-hour ratings for construction assemblies. Assemblies that are tested according to this standard include permanent partitions, shaft enclosures for stairways and elevators, floor/ceiling constructions, doors, and glass openings. Doors and other opening assemblies also receive 20, 30, and 45-minute ratings.

Any opening that pierces the entire thickness of a construction assembly is referred to as a through-penetration. Codes require that penetrations in fire-rated assemblies be protected with fire assemblies in the form of fire doors, fire windows, firestops, and fire dampers. These opening protectives or through-penetration protection systems must have fire protection ratings. The combined width of all openings must not exceed 25 percent of the length of the wall. Usually, no opening greater than 120 sq. ft. (11 square meters) is permitted. Any assembly that passes the required tests must have a permanent label attached to it to prove it is fire-rated.

FIRE DOORS

Fire doors are actually entire door assemblies. The typical fire door includes the door itself, the frame, the hardware, and the doorway (wall opening). Fire door assemblies are required to protect the openings in fire-rated walls. The whole assembly is tested and rated as one unit.

Exit access corridor enclosure walls require a 1-hour construction assembly rating, and doors in those walls require at least a 20-minute rating. For a firewall with a required 4-hour assembly rating, a door with a 3-hour rating is required.

Many types of doors are regulated as fire doors by building codes, including swinging and vertical sliding doors, along with accordion folding, roll-down, and bi-parting doors among others.

Fire doors are typically flush, either solid-core wood or metal, with mineral composition cores. A few panel doors may meet fire door requirements, and some fire doors may have applied finishes to improve their appearance. Frames are wood, hollow metal, steel, or aluminum. The doorframe and hardware must have a fire rating similar to that of the door itself. The maximum fire door size is 4 by 10 feet (1.2 by 3 m).

Hardware for fire doors includes hinges, latches and locksets, and pulls and closers. Hinges, latchsets, and closing devices are the most stringently regulated as they must hold the door closed securely during a fire, and must withstand the pressure and heat the fire generates. The door must be self-latching and equipped with a closer. (See Figure 18.17)

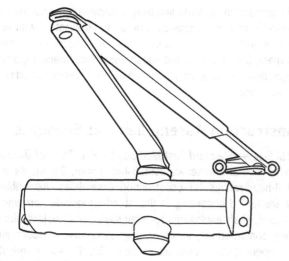

Figure 18.17 Door closer

Door closers are hydraulic or pneumatic devices that automatically close doors quickly but quietly. Building codes require the use of self-latching, self-closing doors with UL-rated hardware to protect openings in firewalls and occupancy separations. For doors that are normally kept in an open position, an automatic closing device uses a fusible link that is triggered by heat or activated by the smoke detector to close the door. Other doors with lower ratings may require self-closing devices to close the door after each use.

Fire-rated exit doors also require a specific type of latch. The most common is called **fire exit hardware**, and consists of a door-latching assembly that disengages when pressure is applied on a horizontal bar spanning the interior of an emergency exit door at waist height. (See Figure 18.18)

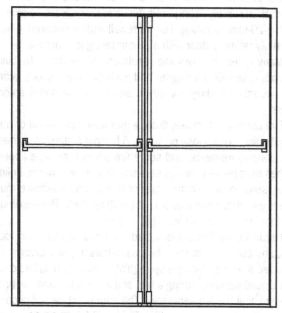

Figure 18.18 Fire exit hardware

Fire exit hardware is tested and rated. The term "panic hardware" is often used, but technically panic hardware is not tested and should not be used on fire-rated doors.

Fire exit hardware is typically required in Assembly and Educational occupancies, and is often used on other exit doors. The codes also regulate the width, direction of swing, and location of required exit doors, according to the use and occupancy of the building. The 2010 ADA Standards for Accessible Design requires that the force necessary to push open or pull open a door be no greater than five pounds.

Fire exit hardware may need to be locked for security reasons. Anyone attempting to use the door will set off an alarm—not a problem during a fire, but very conspicuous and embarrassing otherwise.

For more information on fire doors and windows, see Chapter 6, "Windows and Doors."

WINDOWS

The types of windows that are covered by building code fire regulations include casement, double hung, hinged, pivot, and tilting windows. Stationary windows, sidelights, transom lights, view panels, and borrowed lights are also included. Glass block walls are also covered by fire regulations.

Building codes regulate the clear opening of any operable window that serves as an emergency exit for a residential sleeping space. Typically, the minimum area permitted is 5.7 square feet (0.53 m²), with a minimum clear width of 20" (508 mm) and clear height of 24" (610 mm). The sill must be no more than 44" (1118 mm) above the floor. Codes may restrict the location of glazing.

Windows with fire ratings typically consist of a frame, wired glass, and hardware. They are used for openings in corridors, room partitions, and smoke barriers. Window ratings are similar to those of doors, with hour classifications usually not greater than one hour.

Wired glass has traditionally been used as fire-rated glazing. (See Figure 18.19) It consists of a wire mesh embedded in the middle of a glass sheet. The wire distributes heat and increases the strength of the glass. Wired glass has relatively low impact resistance. Codes are increasingly eliminating the use of wired glass in hazardous locations where it is susceptible to impact and breakage, such as doors, sidelites, and windows near the floor, in all types of buildings.

An alternative to traditional wired glass is ceramic glass with up to 3-hour ratings in doors and up to 90 minutes in other locations. Another fire-rated option is transparent wall panels with ratings of up to two hours.

A final category of fire-rated glazing is specially tempered glass. However, these products only carry ratings of 20 or 30 minutes, and cannot withstand the thermal shock of water thrown from sprinklers or a fire hose. Such products are sometimes used in 20-minute rated doors.

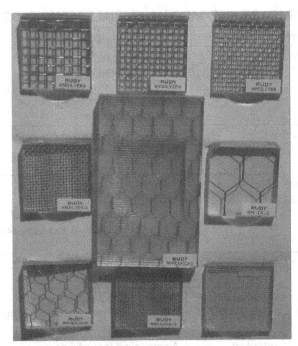

Figure 18.19 Wired glass

Figure 18.20 Glass block

Glass block typically has a 45-minute rating, with newer types available with 60- to 90-minute ratings. (See Figure 18.20) Codes limit the number of square feet of glass block permitted as an interior wall. They also limit glass block used as a view panel in a rated wall, where it is required that it be installed in steel channels.

Safety glazing is required for a glazed panel where there is a walking surface within 36" (916 mm) horizontally inside or outside of the plane of the glazing. (See Figure 18.21) Safety

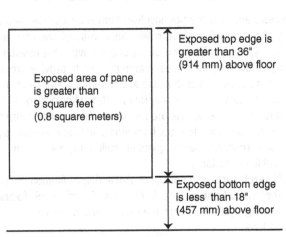

Figure 18.21 Safety glazing limits

glazing is not required when there is a protective bar at least 1½" (38 mm) wide is installed on the accessible side of the glazing 34" to 38" (864 to 965 mm) above the floor. Other requirements also apply to panes beyond defined limits.

FIRE DAMPERS AND DRAFT STOPS

Fire dampers are used in HVAC ductwork to automatically interrupt the flow of air through the duct system during an emergency. The fire damper restricts the passage of smoke, fire and heat. A fire damper includes a fusible link on either side of the assembly the duct penetrates. This link melts during a fire, causing the fire damper to close and seal the duct. Fire dampers must be installed whenever a duct passes through a wall, ceiling, or floor that is part of a fire-rated assembly.

Draft stops are required in combustible construction to close off large concealed spaces. They are not required to be noncombustible themselves. Draft stops are placed between the ceiling and the floor above, in attic spaces, and in other concealed spaces to create separate spaces and prevent the movement of air.

FIRESTOPS

Firestops are required at through penetrations in fire and smoke barriers. Firestops may also be required in concealed spaces between walls and in connections between horizontal and vertical planes. Firestops restrict the passage of smoke, heat, and flames in concealed spaces. They seal and protect openings for plumbing pipes, electrical conduit and wire, HVAC ducts, cables, and so forth passing through walls, floors and ceilings.

The most common way to create a firestop system is to fill the open space with a fire-rated material and finish it with a sealant. Factory-built firestop devices are typically installed as part of a penetration through a wall or ceiling/floor assembly.

MATERIALS AND FIRE PROTECTION

Fire-resistant construction involves both reducing the flammability of a material and controlling the spread of fire. Fire-rated materials, assemblies, and construction have a fire-resistance rating required by their uses. Materials used to provide fire protection must be noncombustible and able to withstand very high temperatures without disintegrating. They should also be low conductors of heat to insulate the protected materials from the heat generated by a fire. Such materials include concrete, gypsum or vermiculite plaster, gypsum wallboard, and a variety of mineral fiber products.

Control starts with careful design and specification of building materials, finishes, and furnishings. As this book focuses on building systems rather than materials, we cover only the basics here.

The process of identifying the appropriate finish classification begins with identifying the occupancy classification of the building or space, and whether it is a new or existing building. Building code finish chapters include finish tables specifying required finishes for different means of egress components and types of buildings.

Some occupancies that have special requirements include healthcare, detention, or correctional institutions; hotels or dormitories; and apartment buildings. Buildings with unusual structures may also have special requirements. Stricter finish requirements usually apply where occupants are immobile or have security measures imposed on them restricting their freedom of movement, or in overnight accommodations. Sprinklers throughout a building may change the finish class ratings.

Fire-safety-related terms used for building materials include noncombustible, combustible, flammable, inflammable, fire-rated, flame-or fire-resistant, and flame retardant. (See Table 18.6)

Smoke causes around 80 percent of all fire casualties, and often originates from burning or smoldering plastic materials.

Codes and Standards

Organizations publish lists of tested assemblies for walls and partitions, floor/ceiling systems, and roof/ceiling systems. They set standards for protection of beams, girders and trusses, columns, and window and door assemblies.

Interior wall finishes that are subject to code provisions include most of the surfaces applied over fixed or moveable walls, partitions, and columns. Interior finishes for ceilings are covered in codes. Coverings applied over finished or unfinished floors, stairs, and ramps are also included. Standards and testing requirements for interior finishes and furnishings change often. Codes set minimum requirements, and it may be prudent to be more stringent.

TABLE 18.6 FIRE-SAFETY MATERIALS TERMINOLOGY

Term	Description
Noncombustible	Materials will not ignite and burn when subjected to fire. Includes steel, iron, concrete, and masonry. Actual performance depends on how they are used.
Combustible	Materials will ignite and continue to burn when a flame source is removed.
Flammable and inflammable	Both terms mean the same thing: tending to ignite easily and burn rapidly. Both are the equivalent of highly combustible, a less confusing term.
Fire-rated	A product that has been tested to obtain an hourly fire rating.
Flame-resistant or fire-resistant	Building components or systems with specified fire resistance ratings based on fire resistance tests.
Flame retardant	A compound that inhibits, suppresses, or delays the production of flames to prevent the spread of fire.
Fireproof or flameproof	Nothing is actually fireproof or flameproof; all construction materials, components and systems have limits where they will be irreparably damaged by fire.

Built-in cabinetry and seating with continuous expanses of plastic laminates and wood veneers are considered interior finishes by many jurisdictions. High-back upholstered restaurant booths may be restricted in certain jurisdictions. In general, only relatively small amounts of foam plastics and cellular materials can be used as wall or ceiling finishes.

The *Life Safety Code* has a table for finish materials. UL certifies materials, systems, and assemblies used for fire resistance and separation of adjacent spaces to safeguard against the spread of fire and smoke. The Gypsum Association's *Fire Resistance Design Manual (GA-600-12)* depicts over 600 systems that may be used for fire-rated walls and partitions, floor/ceiling systems, roof/ceiling systems, and to protect columns, beams, and girders. Be aware that test results and ratings of materials and assemblies can become invalid if the products are not used and maintained properly.

It is the interior designer's responsibility to check requirements and to select furnishings with knowledge of codes and standards. Codes affecting furnishings cover exposed finishes found in furniture and window treatments, such as fabrics, wood veneers, and laminates. Also included are nonexposed finishes like the foam in upholstered seating and the linings in draperies. Furniture includes whole pieces of furniture and upholstered

TABLE 18.7 FINISH AND FURNITURE TESTS

Test	Description
Mattress Test	Pass/fail test used to determine heat release, smoke density, generation of toxic gases, and weight loss when a mattress is exposed to a flame
Methenamine Pill Test	Pass/fail flammability test required for all carpets and certain rugs manufactured for sale in the United States
Radiant Panel Test	Measures tendency to spread a fire, and minimum energy required to sustain a flame for carpet, resilient and hardwood flooring, wall base
Room Corner Test	For napped, tufted, or looped textiles used as coverings on walls and ceilings
Smolder Resistance Test	Tests how new upholstered furniture smolders before either flaming or extinguishing (Cigarette Ignition Test)
Steiner Tunnel Test	Tests flame spread and smoke developed for interior finishes applied to walls, ceilings
Toxicity Test	Measures the amount of toxicity a material emits when it is burned (LC50 or Pitts Test)
Upholstered Seating Test	Pass/fail flame-resistance test for entire piece of furniture
Vertical Flame Test	Pass/fail test for vertical treatments (window treatments, large wall hangings, and decorative plastic films)

seating as well as panel systems. Considerations when specifying materials include ease of ignition; rates of flame spread, heat release, and smoke release; and toxicity of combustion products.

All exits and paths of travel to and from exits must be clear of furnishings, decorations, or other objects. No drapes or mirrors are allowed to obscure exit doors, and mirrors are prohibited next to exit doors. Attention must not be drawn away from exit signs.

Wood columns, heavy timber beams, and girders are typically allowed to remain exposed, because they are spaced relatively far apart and do not provide a continuous surface for flame spread.

Intumescent materials that expand rapidly when touched by fire create air pockets to insulate the surface from the fire, or swell material to block openings through which fire and smoke could travel. Intumescent paints, caulks, and putties are available, as are ¼" (6 mm) thick sheets with a variety of facing materials.

Finish Classes and Test Ratings

Tests for finishes and furniture look at the potential of the material to contribute to the overall fire and smoke growth and spread. (See Table 18.7) Small-scale tests are performed on a small sample of the finish or furnishing. Larger scale tests use a larger sample or full assembly including the finish, substrate, adhesive, fasteners, and another parts; they may include an entire room or a whole piece of furniture.

According to the 2015 IRC, wall and ceiling finishes for a residence shall have a flame spread index of not greater than 200. Exceptions to this include trim, door and window frames, or materials less than ⅟₂₈" (0.91 mm) thick cemented to the surface that meet specified conditions. Wall and ceiling finishes shall have a smoke-developed index of not greater than 450. There are also limitations on the use of foam plastics.

Chapter 7 of the 2015 IRC deals with the design and construction of interior and exterior wall coverings for buildings. It sets requirements for interior coverings or wall finishes including gypsum plaster, cement plaster, gypsum board and panel products, ceramic tile, and other finishes.

The International Code Council (ICC) recognizes three classes of interior finish ratings. (See Table 18.8) Class A is the strictest.

NONTESTED FINISHES AND RETARDANTS

Sometimes finishes geared to residential use or from smaller manufacturers with specialty items have not been tested. The

TABLE 18.8 INTERIOR FINISH RATINGS

Rating	Description
Class A	Includes any material classed at a flame-spread rating of less than 25 with a smoke developed rating below 450.
Class B	Includes materials with flame-spread ratings between 25 and 75, and smoke test ratings below 450.
Class C	Includes flame spreads from 76 to 200, and limits smoke ratings to below 450.

interior designer may then need to have the finish tested or make sure it is properly treated to meet code requirements. Depending on the situation, testing companies can be very costly to use, as they may have to simulate actual installations.

Flame-retardant chemicals inhibit or resist the spread of fire. Their widespread use is undergoing scrutiny, as they appear to be less effective as fire-safety elements than previously believed, and more dangerous to the environment and human health. **Polybrominated diphenyl ethers (PBDE)** are used as flame retardants in building materials, furnishings, polyurethane foams, and textiles. PBDE and some other flame retardants have been associated with fertility problems in humans. They are banned by the European Union and in some US states.

FIREFIGHTING

Building design and building systems are critical to the safety and effectiveness of firefighters. Smoke management is a major part of this effort.

Smoke Management

Smoke kills more people in building fires than heat or structural collapse. Even if a person is not killed by smoke, smoke inhalation can result in memory loss and lingering physical effects. Fires in modern buildings usually last less than 30 minutes, but smoke problems can remain present for hours. Smoke inhalation can lead to unclear thinking, and hot smoke washing over people can result in panic as logic vanishes and fear overwhelms them.

Smoke control is required by most building codes. The goals of **smoke management** are to reduce deaths and property damage and to provide for continuity of building operations with minimum smoke interference. A barrier's effectiveness in limiting smoke movement depends on how smoke is able to leak through it and the pressure differences on each side of it. Pressure depends on the complex interactions of the fire, stack effect, wind, building geometry, and the HVAC system.

Traditional passive methods of modifying smoke movement are used to protect building occupants and firefighters and reduce property damage. They include compartmentation using fire barriers, smoke vents, and smoke shafts. The major flow paths are open or closed doors and windows. Smoke control doors are normally held open magnetically, and released to close when the fire alarm system is activated. Smoke also moves due to airflow through cracks in partitions, floors, exterior walls, or roofs. Leakage occurs where pipes penetrate walls and floors, through cracks where walls meet floors, and around doors.

The effectiveness of smoke vents and smoke shafts depends on their proximity to the fire, the buoyancy of the smoke, and the presence of other driving forces. The HVAC system can be shut down in a fire to limit its contribution to smoke spread. Current practice uses fans to control the movement of smoke.

CONFINEMENT

Smoke should be confined to the area of the fire and excluded from refuges. Firewalls and smoke barriers confine smoke. A large open space above dividing walls can hold a great deal of smoke while the building occupants evacuate.

As introduced earlier, curtain boards are partial-depth smoke partitions suspended from the ceiling to trap hot air and smoke. They help to set off fire detection and suppression systems more quickly. Curtain boards lose their effectiveness quickly when the smoke layer becomes too thick to contain, or as air pressure forces smoke below the boards.

Water curtains can inhibit the flow of smoke but only delay its spread and do not eliminate it. They allow time to evacuate occupants and help firefighters control the fire.

A smokeproof enclosure is made by enclosing an exit stairway by walls of fire-resistive construction, accessible by a vestibule or by an open exterior balcony. The smokeproof enclosure must be ventilated by natural or mechanical means to limit penetration of smoke and heat. Stairs in smokeproof towers have direct access to outdoor air and to firefighting equipment at each floor, and are therefore the safest stairs. Building codes usually require one or more of the exit stairways of a high-rise building to be protected by a smokeproof enclosure.

Wall assemblies that are continuous from outside wall to outside wall and from floor slab to floor slab are good smoke barriers. In tall buildings, vertical shafts for stairs, elevators, and waste and linen chutes can be designed with ventilation or pressurization systems to be smokeproof.

SMOKE CONTROL SYSTEMS

Smoke control systems are required in buildings of over six floors. The exit access corridors must be continually pressurized. All doorways and openings in stairways, corridors, and exit passageways in fire-resistant construction must be protected with doors, fire shutters, or dampers with equivalent fire ratings. Stairway doors opened during evacuation and other doors are sometimes accidentally left open or propped open throughout fires; pressurization keeps smoke out.

Diluting smoke with outdoor air early in a fire may help people evacuate a burning building. Dilution is not enough to control smoke alone, especially when toxic fumes are present. Smoke dilution is usually combined with confinement and an early detection and suppression system.

Special exhaust systems that function only in fires are becoming more common. They use a combination of air velocity and air pressure to control smoke movement. Smoke exhaust systems work well in large-volume atriums, removing smoke at the ceiling and supplying fresh air below. Smoke exhaust systems help keep toxic gases out of refuge areas, and help reduce concentrations of dangerous gases. They also aid in removing smoke after the fire is extinguished.

Automatic ventilation hatches vent heat and smoke without fans. They are suitable for smaller and one-story buildings. Heat and smoke trigger the controls and the hatches open

Figure 18.22 Automatic smoke ventilation hatch

individually. Ventilation hatches improve conditions near the fire for firefighters, and help firefighters on the roof locate the fire inside the building. (See Figure 18.22)

The coordination of the HVAC, fire detection and suppression, and smoke exhaust systems is essential to smoke management. The fire detection and suppression system must activate the smoke exhaust fans and override conventional HVAC controls.

FIRE DETECTION

Fire protection systems are primarily designed to protect life, and secondarily to prevent property loss. They must be tailored to the needs of the specific facility. They are designed as part of the overall fire protection system of a building. There are three basic parts to a fire protection system.

- Signal initiation: manual or automatic equipment
- Signal processing control equipment
- Alarm indication: audible plus visual

Fire protection systems are designed to detect the existence of a fire as early as possible, to sound an alarm, and to extinguish the fire or at least contain it and its effects until firefighters can bring it under control. Sprinkler systems are designed to simultaneously start to put out the fire and send out an alarm when water flows through a sprinkler head.

Fire alarms are often connected to private regional supervisory offices that call the municipal fire department. All public buildings and some other buildings are required to have fire detection and alarm systems with an indicator of the location of the fire.

The **alarm-initiating device** is a signal source that senses fire or smoke. Occupants can initiate an alarm by using a manual pull station. Automatic alarm equipment can be a fire detector, smoke detector, or water flow switch that works whether the building is occupied or unoccupied. Automatic equipment can initiate an audible and/or visible alarm locally, remotely, or both, and can also actuate an automatic fire suppression system. Many state and local building codes require automatic fire and smoke detectors.

Building fire alarm control panels for commercial and light industrial applications are complete systems that can detect a fire, sound an alarm, and activate extinguishing functions. (See Table 18.9) They can be programmed to close fire doors, close

TABLE 18.9 COMMERCIAL FIRE ALARM SYSTEMS

Alarm System Type	Description
Multiple-dwelling	Audible/visual alarms wake all sleepers, may include living room. Smoke detection in corridors, service, utility, storage rooms. Hardwired with standby power. Alarm light over door of each apartment, plus emergency voice/alarm system in high-rises.
Protected premises	Sounds alarm only in protected premises, not in outside central location.
Auxiliary	Local system with direct connection to municipal fire alarm box. Used for public buildings such as schools, government offices, museums.
Remote station protective signaling	Similar to auxiliary system, with line to manned 24-hour service, then phoned to fire department. Private buildings unoccupied for extended periods.
Proprietary	Large multibuilding facilities such as universities, manufacturing facilities. Facility's personnel monitor center on site.
Central station	Similar to proprietary, but owned and operated by service company.

fire dampers in air conditioning and heating ducts, provide auxiliary power for exhaust fans to remove smoke, shut down building fan system, and turn off other machinery.

Emergency power is required for a fire alarm system. Most detection and alarm systems convert power from 24V DC batteries to AC.

A fire progresses through four stages: incipient, smoldering, flame, and heat. Different types of fire and smoke detectors are designed to indicate problems at each of these stages. (See Table 18.10) Incipient stage detectors are the most sensitive— too sensitive for many interior spaces where they produce false alarms. Most fires are detected at the smoldering stage.

Types of detectors include smoke, flame, and heat detectors, among others. (See Figures 18.23, 18.24, and 18.25)

Wilson cloud chamber type detectors are sensitive to microscopic particles in the early stages of a fire but insensitive to dust. They continuously sample air in the protected space and give few false alarms. Wilson cloud chamber detectors require piping and are expensive in small installations. These detectors are used in high-value installations like museums, data processing spaces, libraries, clean rooms, and facility control rooms.

Problem areas for smoke detectors include kitchens, laundries, boiler rooms, shower rooms, and other spaces with high humidity and steam as well as repair shops and laboratories where open flames are used, and garages and engine test facilities where exhaust gases affect sensors. Smoking rooms and

TABLE 18.10 FIRE AND SMOKE DETECTORS

Stage	Description	Detectors
Incipient	Ignition temperature reached and invisible particles and combustion gases released but hardly noticeable	Ionization detectors, gas sensing detectors
Smoldering	Soot plus gases visible mostly as smoke	Smoke detectors, photoelectric detectors, air sampling systems
Flame	Fire emits visible light and infrared (IR) radiation plus smoke and gases	Flame detectors for fires with little smoke; ultraviolet (UV) and IR radiation detectors, plus combined UV/IR
Heat	Very hot fire expands air, spreads fire vertically by stack effect and in all directions by radiation	Spot or linear heat detectors for rapidly spreading fires with little smoke; thermister used to sense heat

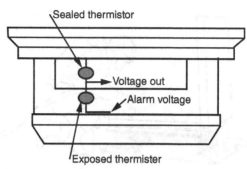

Figure 18.25 Heat detector

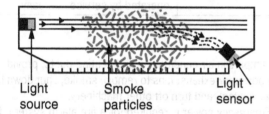

Figure 18.23 Smoke detector

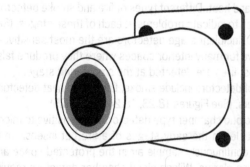

Figure 18.24 Flame detector

areas near designated smoking areas can be a problem, as can areas with heavy accumulations of dust and dirt. High volumes of air movement near loading docks, exit doors, and discharging ducts and registers are also problems. Avoid locations in kitchens where normal cooking processes will activate the alarm.

Residential Detectors

A basic residential smoke detection system places a listed smoke detector outside and adjacent to each sleeping area, in each sleeping room, at the head of every stair, and at least one on every level including the basement. Combined smoke and heat detectors are recommended in the boiler room, kitchen, garage, and attic. An alarm in any detector should set off an alarm in all audible and visible units. The system should have an annunciated central control unit, backup power, and wiring on supervised circuit with a fault trouble alarm.

Smoke detector units are usually placed 6" to 12" (152 to 305 mm) from the ceiling when mounted on a wall. If the alarm is too close to the intersection of the wall and ceiling or too near a doorway, air currents may carry smoke and heat past the unit.

Smoke detectors are subject to false alarms from moisture and particles in the air. In general, the greater sensitivity of the detector, the more false alarms. Choosing the appropriate type of alarm and avoiding placement where conditions cause problems both limit false alarms. If a smoke alarm must be located in a poor location, more than one type of detector should be used, with provisions for extra maintenance and verification of alarms.

Most jurisdictions require installation with hard wiring for smoke detectors in residential occupancies and in hotel or motel units. Interconnected detectors tied into the building electrical system and with a battery backup are required in many new homes and homes with additions or alterations. Other homes are required to have at least battery-operated units. Residences are usually required to have smoke detectors outside each sleeping area and on all habitable floors. Townhouses may have even stricter requirements.

MULTIPLE-DWELLING DETECTORS

Smoke detectors should be located in the corridors of multiple dwelling buildings, and in service spaces, and utility and storage rooms. Battery powered detectors are not permitted in multiple dwellings. All fire alarm circuits should have standby power.

FIRE ALARM SYSTEMS

The goals of a fire alarm system are first to protect life and secondly to prevent property loss. Systems are tailored to specific building types and uses. A fire alarm system includes equipment for signaling that there is a problem, for processing the signal, and for alerting people as to the situation. A fire alarm system can initiate fan controls, smoke venting, smoke door closers, rolling shutters, and elevator controls as part of an overall fire protection plan.

The architect or designer of the fire alarm system must ascertain which current regulations have jurisdiction before designing the system. The codes generally specify where manual or automatic fire signaling systems or fire alarm systems are required. The codes specify required systems and provide testing data. An electrical engineer will be involved in the design of an extensive fire alarm system.

Alarms are initiated by detectors or by a manual pull station that may include a handset for two-way communication. Depending on the building type, an alarm system can alert occupants with bells, horns, or sirens; the ADA includes requirements for visible alarms. Systems may alert a central building station control panel to the location of a fire and alert local fire and police departments as well.

Sometimes a person first notices the fire, and gives the alarm by using a pull station or telephone. **Manual pull stations** are available in various forms. (See Figure 18.26) The most common has a glass rod or window that must be broken to move a handle and activate the switch. An alternative design has a handle behind a cover that must be opened, or direct access to a handle restrained by spring. Manual fire alarm

initiation stations must be placed in the normal path of egress to be used by a person exiting the building. Manual stations must be well marked and easily found.

Interior designers should never specify painting over smoke detectors or other fire safety equipment, as this may hamper their effectiveness by keeping fusible links from melting.

Handicapped-accessible types of pull boxes are available that are pushed rather than pulled and that take minimal effort to operate. Both regular pull boxes and accessible boxes must be red.

Some systems automatically close fire doors and shutters, as well as fire dampers in ducts. They may provide auxiliary power to operate fire safety systems such as exit signs, egress lighting, and smoke exhaust fans. Controls can turn off specified machinery including air handlers to prevent the spread of smoke by the HVAC system. Systems typically also return elevators to the ground floor and keep them there.

Residential Alarm Systems

Residential occupancies include single-family residences, multi-family apartment buildings, townhouses, and condominiums. Their alarm systems should provide sufficient time for evacuation of residents, and for initiation of appropriate countermeasures in larger buildings. System components include alarm-initiating devices, wiring and control panel, and audible alarm devices.

MULTIPLE-DWELLING ALARM SYSTEMS

Multiple-dwelling alarm systems are used in apartment houses, dormitories, hotels, motels, and boarding houses. (See Table 18.11) Alarm systems are designed to provide early warning and orderly egress at times when the building occupants may be asleep, potentially including in living rooms. Audible and visual alarms are positioned so that all sleeping persons, including those with sight or hearing impairments, will be wakened. There should be an alarm light over the door of each apartment or suite to indicate the alarm location, especially if the central panel only shows a zone location. In high-rise residential buildings, an emergency voice alarm communication system should be provided.

All alarms must be identifiable by addressing or annunciation that indicates the location of the alarm. **Annunciator panels** that have a map and lights can be located at a system control panel in the building management office or at the lobby desk of a hotel or dormitory to help firefighters.

In apartments, false alarms are common from kitchen smoke and excessive dust. Some apartment building alarm systems give only a local alarm for evacuation of the apartment. A separate central heat detector system sounds a remote alarm. This reduces the number of false alarms but increases the risk of a fire growing before activation of the fire suppression system or before firefighting crews are dispatched.

Figure 18.26 Fire alarm manual pull station

TABLE 18.11 BASIC MULTI-UNIT RESIDENTIAL ALARM SYSTEMS

Equipment	Description
Listed smoke detectors	Outside and adjacent to each sleeping room, at head of every stair, at least one on every level including basement. Combined smoke and heat detectors in boiler room, kitchen, garage, and attic.
Alarms	An alarm in any detector produces alarm in all audible and visual units.
Annunciated control panel	Panel shows device location, shuts off oil and gas lines, and attic fan to help prevent spread of smoke.
Backup power	Supervised storage battery with trickle charger.
Wiring	On supervised circuits that sound distinctive trouble alarm if fault occurs.

The *NFPA 101—Life Safety Code* and *NFPA 72: National Fire Alarm and Signaling Code* contain detailed requirements for residential fire alarm systems.

Commercial and Institutional Systems

Commercial and institutional building alarm systems vary greatly. Where permitted, **presignaling** that alerts only key personnel may be advantageous for buildings where an evacuation alarm would not be readily tolerated. In schools (especially primary grades), rapid, orderly evacuation is the primary requirement. Public buildings should have an auxiliary connection to the fire department.

High-rise office buildings are required to be equipped with emergency voice/alarm communication systems because of their configuration and the fact that it is not practical to evacuate all occupants at the same time.

Alarm System Operation

An alarm signal can tie into an annunciation panel placed in normal path of egress from building. An annunciation control panel has red indicator light for each detection zone. The panel may be located at a fire-department command station directly accessible from the street. It must be well marked and easily found, and should not be camouflaged or hidden.

When the alarm initiating device is directly connected to indicating devices in a remote fire station or police headquarters, the situation usually requires additional alarm indicators on site to signal building occupants to evacuate. Alarms are generally located by location coding. (See Table 18.12)

The type and placement of audible and visual alarm devices in a public building must meet NFPA 72 and ADA requirements. Codes specify minimum levels and locations of audible signals

TABLE 18.12 ALARM SYSTEM CODING

System Type	Description
Noncoded systems	May be zoned and with annunciation. Continuously ringing bells, horns, and lights.
Master-coded systems	Generate four rounds of code sounded and flashed on all building alarm devices when any signal-initiating device operates.
Zone-coded systems	Identifiable by alarmed zone. Zone lights must go to panel or annunciator, or coding sounds on all gongs in building.
Dual-coded systems (combination coded and noncoded)	Sends identifying coded alarm to maintenance office, with separate continuous ring evacuation alarm throughout building. Requires continuously staffed office.
Selective-coded systems	Fully coded system with all manual devices individually coded. All automatic devices trip code transmitters at the panel. Usually large systems with sprinkler transmitters and smoke detectors as subsystems.
Presignaling systems	Alert only key personnel at their work locations. Personnel investigate and turn in alarm if necessary. Used only in buildings where evacuation difficult and staff sufficient to investigate cause of alarm.

for sleeping areas and mechanical equipment rooms. Visible signal requirements are covered in NFPA 72 and the ADA as well as other codes. Combination audiovisual fire alarm devices are available.

The 2010 ADA Standards require accessible warning systems to be both audible and visual, including in existing facilities where the system is upgraded, replaced, or newly installed. The ADA sets requirements for the type and specific locations. Where required, alarms must be provided in each restroom, hallway, and lobby, and in other common use areas such as meeting rooms, break rooms, examination rooms, and classrooms. In occupancies with multiple sleeping units, a percentage of the units must be equipped with a visible alarm as well as an audible alarm. (See Figure 18.27)

High-rise office buildings may require emergency **voice alarm communication systems**. Voice fire alarms allow specific instructions to be issued to occupants of each part of building regarding safe areas of refuge and rescue efforts in progress. This is especially important in the upper floors of high-rise buildings, and also benefits visitors in hotels and convention centers who tend to ignore or misunderstand bells or horns without specific verbal instructions.

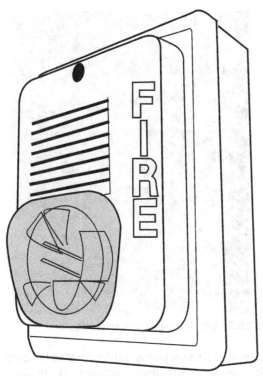

Figure 18.27 Audible and visual fire alarm

FIRE SUPPRESSION

Automatic sprinkler systems extinguish incipient fires before they have a chance to get out of control. Sprinkler heads are so efficient that one to two heads can usually put out a fire.

As indicated earlier, building codes commonly allow sprinklered buildings to have greater distances between exits, eliminating one or more stairways in a large building. By allowing larger floor areas between fire separations, some fire-resistant walls and doors may be eliminated. Buildings may be allowed to have greater overall areas and heights. Some structural elements may need less fire protection, and the building may be able to contain greater amounts of combustible building materials.

Water cools, smothers, emulsifies, and dilutes the fire, but it also damages building contents, and can conduct electricity when used as a stream. Water will not put out burning oil; the flammable oils will float and burn on the surface. When water hits a hot fire, the steam can harm firefighters. Despite these disadvantages, water remains one of the main ways to suppress a fire.

There are other methods used to put out building fires. Carbon dioxide, other gases, foaming agents, and dry chemicals extinguish flames by chemically inhibiting flame propagation, suffocating flames by excluding oxygen, interrupting the chemical action of oxygen uniting with fuel, or sealing and cooling the combustion center.

Due to complex spacing and sizing of supply pipes, most sprinkler systems are designed by professional contractors or engineers working for sprinkler manufacturers. (See

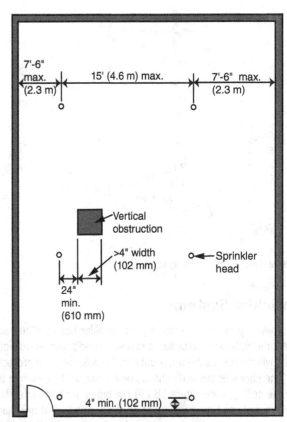

Figure 18.28 Sprinkler clearances

Figure 18.28) *NFPA 13—Standard for the Installation of Sprinkler Systems* lists detailed requirements for sprinkler systems.

The interior designer should work closely with the sprinkler system designer to verify sprinkler head locations and provide adequate clearance at each sprinkler. Typically, a minimum of 18″ (457 mm) must be left open below the sprinkler head deflector. The interior designer should be especially observant of this requirement where wall cabinets or shelving are used, as in storage rooms, kitchens, and libraries.

The distribution system for fire protection must be immediately identifiable to firefighters. It is rarely treated as a visually integrated design element for this reason.

As we previously indicated, in high-rise buildings and buildings with large areas there are places that cannot be reached by firefighters' ladders and hoses. While most fire deaths occur in smaller, often residential buildings, larger commercial, industrial, and institutional buildings create a potential for many deaths and injuries from a single fire. High-rise buildings require an inordinate length of time to evacuate. Stack effects can be created in high-rise buildings over 75 feet (23 meters) tall. Such buildings must have their own firefighting system, which is usually an automatic sprinkler system.

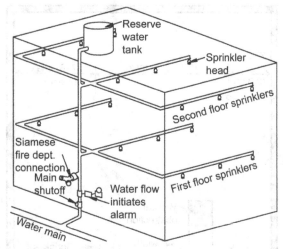

Figure 18.29 Sprinkler system

Sprinkler Systems

Automatic sprinkler systems rely on sprinkler heads. When activated, the water stream strikes a sprinkler head's shaped deflector plate, which spreads the water out over the area the head protects.

The sprinkler system consists of a network of pipes in or below the ceiling. (See Figure 18.29) The pipes are connected to a water supply and have valves or sprinkler heads that are made to open automatically at a certain temperature. Each sprinkler head is controlled by a plug or link of fusible metal that melts at a temperature of around 150°F (66° C).

A sprinkler system may require large supply pipes and valves, fire pumps, and access for monitoring and maintenance. (See Figure 18.30) They are generally considered unsightly, and must be accommodated in the architectural design.

Figure 18.30 Sprinkler supply pipes

Figure 18.31 Interior siamese connection

A **siamese connection** is usually installed for larger buildings for use by fire department pumper trucks to pump water from a hydrant to the sprinkler system. (See Figure 18.31) It provides two or more connections through which firefighters can pump water to a standpipe or sprinkler system.

SPRINKLER HEADS

The most common type of sprinkler head keeps the water in the system by a plug or cap held tightly against the orifice (opening) by levers or other restraining devices. The restraining device is typically a glass bulb containing colored liquid and an air bubble. The heat of a fire causes the liquid to expand until it absorbs the air bubble. Continued expansion ruptures the glass bulb, releasing water through the orifice in steady stream. Pendant, upright, and sidewall heads are often used. (See Figures 18.32 and 18.33) There are a variety of other types of sprinkler heads as well. (See Table 18.13)

Sprinkler heads are available with various finishes. It may be possible to paint the coverplate on a concealed sprinkler or the escutcheons on recessed and flush types, but you should check the applicable code and be sure not to seal the cover.

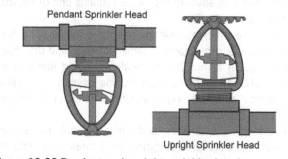

Pendant Sprinkler Head

Upright Sprinkler Head

Figure 18.32 Pendant and upright sprinkler heads

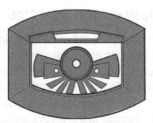

Figure 18.33 Sidewall sprinkler head

RESIDENTIAL SPRINKLERS

The 2015 IRC covers requirements for dwelling unit fire sprinkler systems. Some codes now require sprinklers in all residential occupancies. Most codes make an exception for bathrooms up to 55 square feet (5.1 m²), closets with their least dimension not greater than 3 feet (0.9 m), and open porches, garages, and carports. Uninhabitable attics and crawl spaces not used for storage and entrance foyers that are not the sole means of egress are other common exceptions.

A **residential sprinkler** is a fast response device, listed for protection of dwelling units. It is sensitive to both smoldering and rapidly developing fires. A residential sprinkler is designed to open quickly to fight a fire with only one or two heads operating as residences have a smaller water supply and toxic gases and smoke quickly fill small spaces.

Residential sprinklers have special water distribution pattern. They are designed to put water on walls high enough to prevent a fire from getting above the sprinkler openings. This strategy also helps cool gases at the ceiling, so fewer sprinklers open, reducing water damage.

SPRINKLER SYSTEM PIPING

Water sprinkler system types include wet pipe, dry pipe, pre-action, and deluge systems. (See Table 18.14)

TABLE 18.13 SPRINKLER HEAD TYPES

Type of Head	Description
Upright	Installed with orifice facing upward and deflector on top of exposed piping.
Pendant	Hangs down from pipe with orifice facing downward and deflector below.
	Flush heads have only heat-detecting element below ceiling.
	Concealed heads are entirely above ceiling. Cover plate falls away in fire.
Sidewall	Throw spray of water to cover entire small room with one sprinkler head. Typically one per hotel or apartment room adjacent to one wall.
Quick-response (fast response)	Required in light hazard occupancies (office buildings, hotels, motels). More sensitive to heat, opening sooner than ordinary sprinkler heads, so fewer heads needed. May open from high heat not related to fire.
Early suppression fast-response (ESFR)	For special fire hazards encountered in high-piled storage. Sprinkler's higher pressure and flow penetrate the fire's base faster.
Extra large orifice	Deliver large quantities of water where water pressures are relatively low.
Multilevel	Used where other sprinklers are at higher plane within same space.
Extended coverage sprinklers	Allowed for unobstructed construction with flat, smooth ceilings with limited slope and luminaires, and grills flush or recessed, other uses.

TABLE 18.14 WATER SPRINKLER SYSTEM TYPES

Type	Description
Wet pipe	Most common, contains water, connected to water supply under pressure at all times, fast acting, most reliable type.
Dry pipe	Pipes filled with air under pressure. When sprinkler head opened, water fills piping and flows from open sprinkler. Slower operation than wet pipe. Normally installed upright and only where freezing could be a problem.
Pre-action	Eliminates water damage from accidental discharge of automatic sprinklers. Deluge valve holds water back until opened by fire-detection system. Water not discharged until individual sprinkler opens from heat from fire or manually. Somewhat unreliable.
Deluge	Delivers most water in least time. Water admitted to sprinklers or spray nozzles open at all times, supplied through valve opened by automatic detection system. Extra-hazard occupancies where flammable liquids are stored, flash fire risk.

A space with sprinklers should have adequate water drainage during and after the fire. Floor drains safely carry water away from the building. Salvage covers can protect sensitive objects and direct water toward drainage points. A readily accessible outside valve that controls all the normal sources of supply to the system can cut off water promptly when it is no longer needed.

STANDPIPES AND HOSES

Standpipe and hose systems help fight fires within buildings. They are classified by their intended use. System components may include standard hose racks or hose racks and fire extinguishers in cabinets.

Standpipes are water pipes that extend vertically through the building to supply fire hoses at every floor. Today, standpipes are often located within a fire stair and without a hose. They are not intended for use by amateurs, which can result in delay calling firefighters. Locating fire hose cabinets on stairwell landings allows their use both upstairs and down, but shortens the length of the total run beyond the stairwell. (See Figure 18.34)

Standpipes and hoses are supplied either by a separate water reserve, upfeed pumping, or fire department connections. Wet standpipes contain water under pressure and are fitted with fire hoses for emergency use by building occupants. Dry standpipes do not contain water, but are used by the fire department to connect fire hoses to a fire hydrant or pumper truck.

Figure 18.34 Standpipe access panel

Other Fire Suppression Systems

A variety of other methods are used when water damage threatens the structure or contents of a building almost as much as a fire would. These include intumescent materials, mist systems, foams, carbon dioxide (CO_2), dry chemicals, and clean agent gases. (See Table 18.15) The production of Halon has been phased out due to its damage to the earth's ozone layer.

Portable Fire Extinguishers

A portable fire extinguisher can save lives and property by putting out a small fire at an early stage or containing it until the fire department arrives. However, portable extinguishers have limitations.

Portable fire extinguishers are movable and do not require access to plumbing lines. They are rated for the class of fire they are designed to fight. How many are required and where

TABLE 18.15 OTHER FIRE SUPPRESSION SYSTEMS

System Type	Description
Mist systems	Fast alarm initiation, rapid response. Smaller volumes of water, less damage. Mist easily moves around obstructions.
Carbon dioxide	Displaces oxygen, can suffocate people; used for sealed-off closed spaces without people or animals. Automatic gas system most common. No cleanup required. Purging gas to atmosphere adds to global warming.
High-expansion foam systems	Used in confined areas. Foam generated by blowing air through screen sprayed with detergent to completely cover area. Little water, but soapy film or residue requires cleanup. Used for large spaces with flammable liquids. Obscures vision but does not suffocate people.
Low-expansion foams	Introduced into water in sprinkler system. Smothers fire with minimum water use. When foam exhausted, system can become deluge sprinkler.
Dry chemicals	Commercial kitchens where flash fire in cooking appliance can ignite grease in plenum or duct. Dry chemical with sodium bicarbonate base can extinguish fire in seconds from nozzles over cooking areas, in ducts.
Clean agent gases	Replacements for Halon include a potassium-based aerosol and FM-200® (heptafluoropropane). FM-200® is greenhouse gas, but does not cause ozone depletion, with shorter lifetime, leaves no residue.

they must be located depend on the hazard classification of the occupancy. They must be located in conspicuous places along ordinary paths of egress.

Fire extinguishers may be surface-mounted or recessed within the wall using a special cabinet with a vision panel. The extinguisher must be visible at all times, be tested regularly, and have an approved label. This presents a challenge to the interior designer, since fire extinguishers and related equipment are bright red and in highly visible locations. Showing this equipment on interior elevations helps designers and their clients become aware of the final appearance of the room.

The extinguisher must have adequate force to fully extinguish the fire. The typical residential fire extinguisher is not designed to fight large or spreading fires, and may run out in eight or fewer seconds. Extinguishers must be located where they are both quick and safe to reach in case of fire. The person who handles the fire extinguisher must have the strength and knowledge to use it properly and without hesitation.

Building codes specify which occupancies and types of building uses require fire extinguishers. Most occupancies require extinguishers, and some specific areas within buildings have special requirements. Commercial kitchens and smaller kitchens and break rooms in commercial spaces require extinguishers. NFPA 10— *Standard for Portable Fire Extinguishers* establishes requirements.

Fire extinguishers are classified by types represented by letters. (See Table 18.16) They also have force ratings indicated by numbers. The higher the rating number, the more extinguishing agent the unit contains, and therefore, the larger the fire it should be able to put out. A higher force number also means a heavier extinguisher.

A fire extinguisher should be located in or near a residential kitchen, but not next to the cooking surface or oven, where it may not be accessible in the event of a fire. It should be placed close to an exit, within the universal 15" to 48" (381 to 1219 mm) reach range.

Interior designers need to be familiar with the codes and related ADA requirements for portable fire extinguishers. If a fire suppression system is to be used by the building's occupants, it must be mounted at accessible heights and located within accessible reaches from a front or side wheelchair approach. Fire suppression equipment may not protrude more than 4" (102 mm) into the path of travel. This requirement may eliminate bracket-mounted fire extinguishers and surface mounted fire protection cabinets in some areas.

In this chapter, we have explored how to prevent injury to people and property from building fires. In Chapter 19, we look at elevator, escalator, and materials handling equipment that conveys people and materials within the building.

TABLE 18.16 PORTABLE FIRE EXTINGUISHERS

Class	Contents	Use	Effect
Class 1A to 40A	Water, aqueous film-forming foam (AFFF), film-forming fluoroprotein foam (FFFP), multipurpose dry chemical	Ordinary combustibles such as wood, cloth, paper, rubber, many plastics	Water: Heat-absorbing, cooling; dry chemicals: coating, interruption of combustion chain reaction
Class 5B to 40B	CO_2, dry chemicals, AFFF, FFFP	Flammable or combustible liquids, flammable gases, greases, similar materials	Exclude oxygen, inhibit release of combustible vapors, or interrupt combustion chain reaction
Class C	CO_2 or dry chemicals	Fires in live electrical equipment	Do not conduct electricity
Class A:B:C	Dry chemicals, primarily ammonium phosphate.	Multipurpose	Ammonium phosphate leaves hard reside if not thoroughly cleaned up immediately
Class D	Dry powders: copper or graphite compound or sodium chloride	Combustible metals or metal alloys	Designed and labeled for a specific metal
Class K	Potassium-acetate based low PH agent mist	Cooking fires involving vegetable oils and animal oils and fats	Helps prevent fires from grease splashes while cooking appliance cools

Conveyance Systems

19

Elevators travel vertically to carry passengers, equipment, and freight from one level of a building to another. Escalators move large numbers of people efficiently and comfortably among a limited number of floors.

> Of the many decisions that must be made by the designer of a multistory building, probably none is more important than the selection of the vertical transportation equipment—that is, the passenger, service, and freight elevators and the escalators. Not only do these items represent a major building expense. . . but the quality of elevator service is also an important factor in a tenant's choice of space in competing buildings. (Walter T. Grondzik and Alison G. Kwok, *Mechanical and Electrical Equipment for Buildings* (12th ed.), Wiley 2015, page 1445)

INTRODUCTION

Conveyance systems include both horizontal (moving walks, horizontal conveyors) and vertical transportation systems (elevators, escalators, and dumbwaiters). Vertical transportation is a determining factor in a building's shape, core layout, and lobby design. Vertical transportation comprises 10 to 15 percent of the construction budget for tall buildings, plus operating costs.

History

The elevator has a longer history than we might assume. Vertical shafts in the Coliseum contained lifts operated by ropes and pulleys to transport scenery, animals, and sometimes, gladiators from underground to the arena. During Europe's Middle

Ages, lifting devices were also used, although generally avoided for conveying people due to frequent failures. (See Figure 19.1)

In the eighteenth century, a few elevators were installed in palaces, including a passenger elevator for Louis XV of France in 1743. A safer screw-powered elevator was built by Ivan Kulibin in the czar's Winter Palace in St. Petersburg, Russia, in 1793.

Figure 19.1 Medieval elevator design
Source: Redrawn from design by Konrad Kyeser, 1405, public domain (Wikipedia)

Historically, skyscrapers and elevators evolved together. The first electric elevator was built by Werner von Siemens in Germany in 1880. In 1852, Elisha Otis introduced the safety elevated with a governor device that locks the elevator to its guides in the event that it descends at an excessive speed.

Codes and Standards

Building codes heavily regulate elevator design, installation, and signals. (See Table 19.1) These codes affect the interior designer's choices for elevator cabs and lobbies.

The American Society of Mechanical Engineers (ASME) Standard A17.1—*Safety Code for Elevators and Escalators* sets strict installation requirements for vertical transportation equipment. Some states and cities have their own stricter codes. Other localities and states may have additional requirements.

Most codes require emergency power in specific building types to operate at least one elevator at a time, plus power (usually a generator and batteries) for lights and communication.

ELEVATORS

Any multistory building needs ways to get people and objects from one floor to another. Stairs are the most basic means of vertical transportation, of course, and are included even in very

TABLE 19.1 ELEVATOR AND ESCALATOR CODES AND STANDARDS

Title	Description
ASME Standard A17.1—*Safety Code for Elevators and Escalators*	American Society of Mechanical Engineers (ASME) Installation requirements, including Limited-Use/Limited Application (LU/LA) elevators
ASME A17.3-2011—*Safety Code for Existing Elevators and Escalators*	Requirements for electric and hydraulic elevators and escalators
ASME A17.4-1999—*Guide for Emergency Personnel*	Safety of elevators, escalators and related conveyances.
NFPA 101 *Life Safety Code*	Fire safety requirements
NFPA 70 (NEC)	Electrical requirements
2010 ADA Standards for Accessible Design	Signage, car controls, doors, cars, LU/LA elevators
ANSI A117.1—*Accessible and Useable Buildings and Facilities*	Accommodations for persons with disabilities
Building Transportation Standards and Guidelines	National Elevator Industry, Inc. (NEII) online reference

tall buildings as secure exits in the event of fire. However, nobody wants to walk up 20 flights of stairs or carry furniture and supplies up them, which is where elevators and escalators come in. There are four basic types, including passenger, service, freight, and residential elevators.

For information on stairs, see Chapter 5, "Floor/Ceiling Assemblies, Walls, and Stairs."

Elevator Design

Architects work with engineers and elevator consultants or manufacturers to design the complexities of an elevator installation. Elevator design involves decisions about the number of elevators, speed, and capacity of the system, which are determined by the number of people served and the building height.

Although interior designers are not usually responsible for deciding how many elevators will be in the building or where they will be located, these decisions affect space planning, as elevators take up a great deal of space at critical locations and are focal points for circulation paths.

Interior designers are often involved in selecting the finishes for elevator cabs and lobbies, and for the appearance of buttons and indicators in the cab and at each floor landing. Because people congregating at elevator lobbies are often forced to stand around waiting for an elevator, the design of these areas can have a great impact on the comfort of building occupants and visitors, and on the impression they have of the building and the businesses within it. This is especially important for people who have to use the elevators every day, when unpleasant, unsafe, or uncomfortable surroundings become a dreaded part of the daily routine. The design of elevators and their lobbies also has implications for security, fire safety, and maintenance of these areas.

The ground floor elevator lobby is also called the lower terminal, and is usually located close to the main entrance, with a building directory, elevator indicators, and possibly a control desk nearby. Lobbies are designed to be large enough for the peak load of passengers, with 5 square feet (0.5 m²) of floor space allowed per person waiting for one or more elevators. The same allowance should be made for hallways approaching the lobby.

General rules for designing an elevator system allow one elevator for every 250 to 300 people. The size of the elevator car and the frequency of trips determine the car's capacity. This is independent of the number of cars in the elevator bank. According to actual counts in many existing installations during peak periods, cars are not usually loaded to maximum capacity but are typically only eighty percent full.

Manufacturers and elevator consultants supply standard layouts for elevators, including dimensions, weights, and structural loads. The average trip time is determined by the time spent waiting in the lobby plus the time it takes to travel to a median floor stop. For a commercial elevator, a trip of less than one minute is highly desirable, with 75 seconds considered acceptable. A trip time of 90 seconds becomes annoying, and anything over 120 seconds exceeds the limits of toleration. For residential elevators, users often spend a minute or more of the trip time just waiting for the elevator.

The manufacturer sets structural requirements. Structural columns support the elevator from the foundation all way up to the penthouse, with the main beams supporting the penthouse floor.

Originally DC electricity was used for elevators to allow fine speed control, but now AC motors that can be finely controlled are used. This results in energy savings and smaller space requirements.

A power outage sets the car brake immediately and the car remains stationary and does not descend to the nearest landing. Hydraulic cars can be lowered by operation of manual valve. Small traction cars can be cranked to a landing by hand, but large cars are fixed in position.

ELEVATOR ACCESSIBILITY

At a minimum, elevators are required to meet 2010 ADA Standards for Accessible Design, with additional accommodations to meet a specific building intent or local codes. The ADA mandates self-leveling car features and sets standards for car controls and illumination. The ADA also addresses elevator door operation and signals. Visible and audible signals are required at each hoistway entrance to indicate which car is answering a call and the car's direction of travel.

Elevator cars are required to provide a clear floor space 36" (915 mm) by 48" (1220 mm) minimum. ADA minimum dimensional requirements vary with the door location. (See Table 19.2 and Figures 19.2 and 19.3)

Accessible elevator cars with doors opening to one side must have a minimum width of 68" (1727 mm). Cars with center opening doors must be a minimum of 80" (2032 mm) wide. The minimum clear depth is 51" (1295 mm).

Accessibility requirements for elevators are complex and detailed. Refer to the current applicable ADA standards to verify requirements for a specific project.

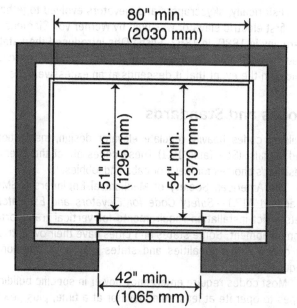

Figure 19.2 Accessible elevator cab dimensions, centered door

Source: Redrawn from 2010 ADA Standards for Accessible Design, Figure 407.4.1

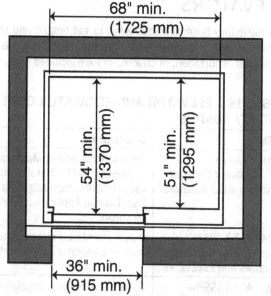

Figure 19.3 Accessible elevator cab dimensions, side (off-centered) door

Source: Redrawn from 2010 ADA Standards for Accessible Design, Figure 407.4.1

TABLE 19.2 2010 ADA MINIMUM ELEVATOR CAR DIMENSIONS

Door Location	Door Clear Width (mm)	Inside Car Side to Side (mm)	Inside Car, Back Wall to Front Return (mm)	Inside Car Back Wall to Inside Face of Door (mm)
Centered	42" (1065)	80" (2030)	51" (1295)	54" (1370)
Side (off-centered)	36" (915)	68" (1725)	51" (1295)	54" (1370)
Any	36" (915)	54" (1370)	80" (2030)	80" (2030)
Any	36" (915)	60" (1525)	60" (1525)	60" (1525)

Source: Redrawn from 2010 ADA Standards for Accessible Design, Table 407.4.1 Elevator Car Dimensions

PARTS OF AN ELEVATOR

The main parts of an elevator include the car, cables, elevator machine, control equipment, counterweights, hoistway, guide rails, penthouse, and pit. (See Figure 19.4 and Table 19.3) An elevator is essentially a cage of fire-resistant material supported on a structural frame, with lifting cables attached to its top. The car is guided in its vertical travel within the shaft by guide shoes on its side members.

An elevator is provided with safety doors, operating-control equipment, floor-level indicators, illumination, emergency exits, and ventilation. It is designed for long life, quiet operation, and low maintenance.

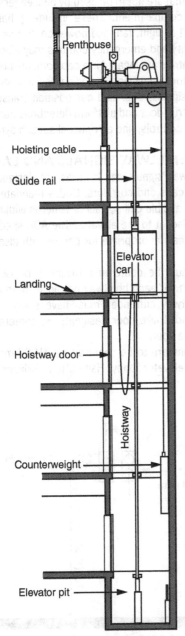

Figure 19.4 Electric elevator section

Source: Redrawn from Francis D.K. Ching and Corky Binggeli, *Interior Design Illustrated* (3rd ed.), Wiley 2012, page 212

TABLE 19.3 ELEVATOR PARTS

Part	Description
Car (cab)	Carries freight or passengers up and down in elevator shaft
Hoistway (shaft)	Vertical space for travel of one or more elevators
Guide rails	Vertical steel tracks on side walls of shaft that control travel of car
Cables	Connected to the top beam of the elevator; lift cab in the shaft
Counterweights	Rectangular cast iron blocks mounted in a steel frame to counterbalance the elevator cab
Elevator machine	Driving motor on heavy structural frame turns sheave to lift and lower car, along with other equipment.
Penthouse	Houses elevator machine on roof.
Control equipment	Drive (motion) control: velocity, acceleration, position determination, leveling of car
	Operating control: car door operation and functioning of car signals, including floor call buttons and indicating device
	Supervisory control: group operation of multiple-car installations; indicating and control devices: car and hallway buttons, lanterns, etc.
Elevator pit	Extends from the level of the lowest landing to the floor of the shaft

Elevators can be noisy. Noise-sensitive areas such as sleeping rooms should be located away from elevator shafts and machine rooms. Solid-state equipment eliminates the clatter and whirring sound of older machine rooms.

ELEVATOR DOORS

Both car and hoistway doors should be designed to coordinate with the overall architecture of the building. Doors may be single, double, or four-panel, opening in the center or to one side. A center-opening door is the fastest. Car and hoistway doors for a given elevator are the same size.

Code requirements indicate that the clear opening for elevator doors should be at least 42" (1067 mm), with 48" (1219 mm) preferred. Smaller doors are appropriate only in residential or small, light traffic commercial buildings. For elevators with small cars and a short rise, a swing-type manual corridor door may be permitted. Larger cars need power operated sliding doors.

With a door only 36" (914 mm) wide, two people are not able to pass each other at the same time. This delays loading until unloading of passengers is complete, affecting the speed and quality of service.

Doors must have delayed door-closing capacity with detection beams that reopen a door without contact when they sense a passenger. Delayed door closings increase travel time, so in buildings with traffic peaks, one or more elevators may be designated for use by people with disabilities during busy periods.

Doors can be equipped with an electronic sensing device that detects passengers in wide area on the landing in front of the car, rather than only in the door's path. The device often has an audible signal, and the door stays open a predetermined length of time. This arrangement is especially useful where passengers cannot approach the entrance or enter the car quickly, as when they are towing baggage or holding children, using wheelchairs, or moving bulky objects.

ELEVATOR CABS

The elevator cab interior is a virtually inescapable and intimate place. It is important for an elevator cab in a commercial or institutional building to create a positive impression. Interiors must deal with physical abuse, gravitational stress from rapid acceleration and deceleration, and shifting and vibration through constant movement. In addition, people in elevators are sometimes uneasy about traveling in a confined space and in close contact with strangers.

Pre-engineered, pre-manufactured elevator systems are completely engineered systems with known performance and cost. They offer rapid delivery and lower cost, with minimal architectural and owner supervision required.

The interior designer is likely to be involved in the décor of elevator cabs and the styling of hallway and cab signals. The normal elevator specification describes the intended operation of the equipment, and includes an amount to cover the basic finishes of the cabs. The type and function of signal equipment specified, along with finishes and styling, are options that the architect and interior designer specify.

Standard original equipment manufacturer choices may not be very compelling. However, a custom designed elevator interior can be costly, time consuming, and subject to cancellation.

Elevator cab interiors may be finished in wood paneling, plastic laminate, stainless steel, and other materials. Floors are typically tile, wood, or carpet. The choice of material depends on the architectural style of the building, the budget available, and the practicality of the material for the elevator's intended use. One set of protective wall mats is usually provided for each bank of elevators, especially if there is no separate service car.

Ceiling coves, ceiling fixtures, or completely illuminated luminous ceilings provide lighting for the cab. Lighting fixtures may be standard or special designs. The goal in lighting the cab should be to provide pleasant, even illumination from sources that are resistant to vandalism and abuse.

Cab finishes should be appropriate to use by people with disabilities. Many people with vision problems are able to see with sufficient, nonglaring lighting. Sturdy handrails and nonslip finishes help people who have mobility problems. Well-designed signals and call buttons avoid confusion for everyone, including people with perceptual problems.

CAB OPERATING PANELS AND SIGNALS

Within the elevator cab, signals indicating the travel direction and present car location are either part of the cab panel or separate fixtures. A voice synthesizer may announce the floor, direction of travel, and safety or emergency messages inside the car. Voice synthesizers are very helpful for people with vision problems.

The car's operating panel must have full-access buttons for call registry, door opening, alarm, emergency stop, and firefighters' control. An intercom connected to the building control office provides added security. Sometimes a door-closing button is provided if hand operation is anticipated.

Controls that are not to be used by passengers are grouped in a locked compartment. These include a hand operation switch as well as light, fan, and power control switches. Other special security and emergency controls may also be included. Still other controls are located in a cab compartment accessible only to elevator technicians, including devices controlling door motion, car signals, door and car position transducers, load-weighing control, door and platform detection beam equipment, visual display controls, and an optional speech synthesizer.

CAB AND HALLWAY SIGNALS AND LANTERNS

Cab and hallway signals and lanterns are designed to fit with the décor of cabs and corridors. Codes mandate the location of visible and audible call signals or lanterns within sight of the floor area adjacent to the elevator. The ADA specifies requirements for signals appropriate for people with disabilities. (See Figure 19.5)

Signals must be centered a minimum of 72" (1829 mm) above the floor at each hoistway entrance. Both jambs of the elevator hoistway entrances must have signage with raised characters and Braille floor designations, centered 60" (1524 mm) above the floor.

Call buttons are to be centered 42" (1067 mm) above the floor in each elevator lobby. Hall buttons indicate the desired

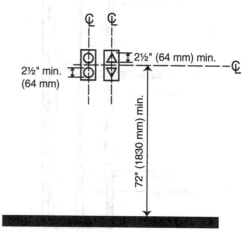

Figure 19.5 ADA 2010 elevator hall signals

Source: Redrawn from 2010 ADA Standards for Accessible Design, Figure 407.2.2.2

direction of travel, and confirm visually that the call has been placed. A hall lantern at each car entrance must visually indicate the direction of travel of the arriving elevator, and preferably, its designation.

An audible signal of the car's imminent arrival encourages people to move to the arriving car and speeds up service. Hall stations can be equipped with special switches for fire, priority, and limited access service as required.

Elevator Machines

The two most common types of elevators are **electric (traction) elevators** and **hydraulic elevators**. (See Table 19.4)

Passenger Elevators

Spatial requirements and traffic patterns are determining factors in interior space planning for passenger elevators. Performance goals include minimizing waiting time for a car at any floor level, comfortable acceleration, rapid transportation, and smooth, rapid braking. The elevator should provide accurate automatic leveling at landings, rapid loading and unloading at all stops, and quick, quiet door operation. All mechanical equipment should operate smoothly, quietly, and safely under all conditions of loading. Emergency and security equipment should be reliable.

There should be good floor status and travel indication both in cars and at landings, with easily operated car and landing call buttons or other devices. Comfortable lighting and a generally pleasant car atmosphere are also important. Elevator shaftways and lobbies should be integrated into the building layout and design. Cars and shaftway doors should be treated in a manner consistent with the building's architectural design.

TABLE 19.4 ELEVATOR MACHINE TYPES

Type	Description
Geared traction	Uses electric motor with gear box to adjust rotation speed of sheave. Medium-rise buildings.
Gearless traction	Traction sheave rotated directly by electric motor, faster than geared. High-rise buildings. Provides a very smooth, high-speed ride.
Machine-room-less (MRL)	Newer technology. Smaller electric motor fits inside elevator shaft; no machine room. More efficient motors use least energy of any type.
Hydraulic	Supported by hydraulic mechanism (plunger) extending down into deep well. Plunger attached to bottom of car raises and lowers car.

FIRE SAFETY CONSIDERATIONS

A firefighter's return emergency service in the elevator is required by American National Standards Institute (ANSI) and additional local fire codes. Emergency personnel should have a means of two-way communication with cars and the control center.

Some building codes require elevator shafts to have smoke vents at the top, allowing the hoistway to become a smoke evacuation shaft in an emergency. If there is a fire on a lower floor, the shaft fills with smoke, which helps clear smoke from the area of the fire. However, this prevents firefighters and other people from using the elevator. Codes have required that in the event of a fire, all elevator cars close their doors and return nonstop to the lobby or another designated floor, where they park with their doors open. They can then only be operated in manual mode with a firefighter's key in the car panel.

For more information on elevator fire safety see Chapter 18, "Fire Safety Design."

RESIDENTIAL AND LU/LA ELEVATORS

Small private-residence elevators can double as wheelchair lifts. Because they require overhead equipment space, standard traction elevators are uncommon in private residences, and hydraulic elevators must have a plunger bore hole below. Instead, residential elevators often rely on winding-drum units, roped hydraulics, or worm and screw units. Building codes consider residential elevators a separate class if they have a maximum size of 18 square feet (1.7 m²), a load of 1400 pounds (635 kg), a rise of 25 feet (7.6 m), and a speed of 30 fpm (0.15m/s).

Residential elevators are available with laminate or wood car interiors. The doors, which can be designed to look like residential wood doors, have concealed safety locks. Car sizes vary to allow for more headroom or more platform area as needed. (See Figure 19.6)

Residential elevators are marketed for people with mobility problems. However, the small cab size may not accommodate a wheelchair or another person providing assistance. Cabs can have a single opening or two openings opposite each other or at right angles. Separate machine space is required.

Since the enactment of the ADA, there has been a need for another type of vertical transportation, the **limited use/limited application (LU/LA) elevator**. A LU/LA elevator is defined as a power passenger elevator where the use and application is limited by size, capacity, speed and rise, intended primarily to provide vertical transportation for people with physical disabilities. LU/LA elevators provide high-quality access for people in wheelchairs and their companions.

LU/LA elevators were created to fill the void between the commercial elevator and the vertical platform or wheelchair lift. Typical applications include schools, libraries, small businesses, churches, and multifamily housing.

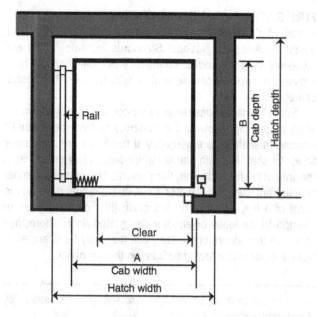

Cab size	A (width)	B (depth)
3 x 3	3' (914 mm)	3' (914 mm)
3 x 4	3' (914 mm)	4' (1219 mm)
Special	12 sq. ft. (1 sq. m) max. platform area	

Figure 19.6 Residential elevator
Source: Redrawn from Walter T. Grondzik and Alison G. Kwok, *Mechanical and Electrical Equipment for Buildings* (12th ed.), Wiley 2015, page 1526

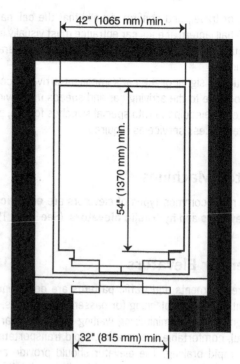

Figure 19.7 LU/LA elevator (new construction)
Source: Redrawn from 2010 ADA Standards for Accessible Design, Figure 408.4.1

LU/LA elevator cars are required by the 2010 ADA Standards to have a minimum clear width of 42" (1065 mm) and a clear depth of 54" (1370). (See Figure 19.7) Exceptions are provided for cars with a clear depth of 51" (1295 mm) and 51" (1295 mm) depth with a clear door opening of 36" (915 mm). An exception is also made for existing cars with 36" (915 mm) clear width and 54" (1370 mm) clear depth and a minimum net clear platform area of 15 square feet (1.4 m²).

WHEELCHAIR LIFTS

Inclined wheelchair platform lifts and chair lifts are available in a variety of designs. (See Figures 19.8 and 19.9) They are covered by the elevator code and must be installed in accordance with code requirements, including safety elements and controls.

Chair lifts can be used on most stairs, even ones with turns. Wheelchair lifts are available with up to a 14-foot (4.3-meter) rise. An inclined chair lift uses the same space as the stairs, and is much less expensive than an elevator.

Vertical platform lifts are safe, economical, and space conserving ways to overcome architectural barriers up to twelve feet high. They are manufactured with a stationary enclosure, including gates and doors, as needed for each application.

Figure 19.8 Inclined platform lift on stair

A vertical platform lift takes up a significant amount of space, unless it can be tucked in next to a stair landing. (See Figure 19.10)

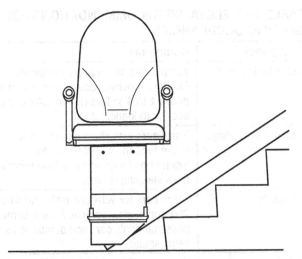

Figure 19.9 Chair lift

Source: Redrawn from Walter T. Grondzik and Alison G. Kwok, *Mechanical and Electrical Equipment for Buildings* (12th ed.), Wiley 2015, page 1527

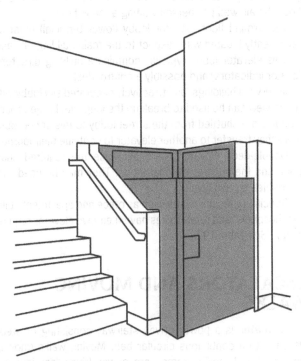

Figure 19.10 Vertical platform lift

Freight Elevators

Design factors for **freight elevators** include the amount of weight that must be transported per hour, the size of each load, the method of loading, and the distance of travel. The type of load, type of doors, and speed and capacity of cars is also considered.

For low rises below 60 feet (18 meters), hydraulic elevators provide accurately controlled, smooth operation, and accurate automatic leveling. Cabs are made of heavy-gauge steel with a multiple-layer wood floor designed for hard service. Ceiling lighting fixtures must have guards against breakage. Freight elevator gates slide up vertically and are at least 6 feet (1.8 meters) high. Hoistway doors lift vertically or are center opening, and are manually or power operated.

Hydraulic and mechanical vertical lifts are available for warehouse and industrial use. They are open frameworks custom-fit to the application, ranging from simple two-level applications to sophisticated multilevel, multidirectional systems.

Service Cars and Special Elevators

In office buildings, one **service car** is typically provided for every 10 passenger cars. Service cars can serve as passenger cars at peak times. A service car has a door 48" to 54" (1219 to 1372 mm) wide for furniture, and should have access to a truck door or freight entry, plus to the lobby.

Because hospital elevators must accommodate gurneys, wheelchairs, beds, linen carts, and laundry trucks, the cabs are much deeper than normal. A hospital elevator can hold more than 20 people, and service is slow.

OBSERVATION CARS

An **observation car** comprises a glass-enclosed car attached to a traction lifting mechanism behind the car. (See Figure 19.11) The back is treated as a screen to hide the equipment. Observation cars can also be designed with hydraulic lift mechanisms and cantilevered cars.

Figure 19.11 Glass-enclosed elevator

INCLINED AND RACK AND PINION ELEVATORS

Inclined elevators are cars that ride up a diagonal path on inclined rails, pulled by a traction cable. The St. Louis Gateway Arch has a 10-passenger inclined elevator on each side.

Rack and pinion elevators ride up or down a rack. A rotating cogged-wheel pinion is attached to a vertical rack, moving the attached car up and down the rack. They are simple and safe for an unlimited rise with low maintenance and operating costs, and use little space. A rack and pinion system was used for a 210-foot (64-m) rise in the 1986 renovation of the Statue of Liberty in New York to evacuate heart attack victims. Rack and pinion elevators are used indoors and outdoors in industrial environments for vertical transport of passengers and materials.

Elevator Security

If someone is being attacked in an elevator, the attacker can render the enclosed space of the elevator cab inaccessible by pressing the emergency stop button. The attacker can then restart the elevator and escape at any floor. To attempt to counter this danger, alarm buttons are provided that alert building occupants and any available security personnel.

Elevators must be equipped with communications equipment by code. A two-way communication system with hands-free operation is best for security in the car. Security improves with a closed circuit television monitor with a wide-angle camera in each car, with continuous monitoring at a building security desk.

Sometimes it is necessary to restrict access to or from a given floor or elevator car. Pushbutton combination locks and coded cards may work, but do not keep an unauthorized person from following the user inside the elevator. The best systems combine automatic monitoring and access devices with continual supervision by persons who know the appropriate action to take in an emergency.

Elevator Systems

Single-zone elevator systems in which all cars serve all floors are generally used for buildings under 15 stories. Multizone systems split into two or more zones are used for buildings over 20 stories. Buildings with between 16 to 19 stories can use either type of system.

Large elevator systems use very sophisticated controls. The controls for small systems may be much simpler. Solid-state systems are universal now on new elevators.

Elevator Lobbies

The elevator lobby on each floor is a focal point from which corridors radiate for access to all rooms, stairways, service rooms, and other spaces. Elevator lobbies must be located above one another.

The elevator lobby is usually the first place people see on each floor. The lobby serves as a waiting area, and needs to be kept clear of other circulation. Lobbies must be large enough

TABLE 19.5 ELEVATOR RECOMMENDATIONS FOR SPECIFIC OCCUPANCIES

Occupancy	Comments
Office buildings	Where used as service elevator for furniture moving, recommend oversized doors 4 to 4-½ feet (1.2 to 1.4 m) and access to loading dock.
Apartment buildings	Small cars with short rise may have swing-type manual corridor door. Isolate hoistways and machine rooms from sleeping rooms.
Hospitals	Large cars for vehicular traffic hold over 20 people, slow service. Adding some passenger-only cars and dumbwaters helps speed.
Retail stores	One or two elevators for staff, people with disabilities.

to allow the peak number of passengers to wait comfortably. Approximately 4 to 5 square feet (0.4 to 0.5 m²) of floor area should be allowed per person waiting at peak time.

The ground floor elevator lobby (lower terminal) must be conveniently located with respect to the main building entrances. This elevator lobby typically contains a building directory, elevator indicators, and possibly a control desk.

In very tall buildings, an attractively decorated sky lobby with a good view can be used to break up the long trip. Large groups of people are shuttled from the street lobby to this upper lobby where they transfer to another elevator to continue their journey.

Double-deck elevators cars decrease the required shaft space and the number of local stops. They may be used with sky lobbies on two levels.

Specific types of spaces such as office and apartment buildings, hospitals, and retail stores have their own elevator requirements. (See Table 19.5)

ESCALATORS AND MOVING WALKS

An **escalator** is a power driven stairway consisting of steps attached to a continuous circular belt. Moving walks (moving sidewalks) and moving ramps are power-driven, continuously moving surfaces, similar to a conveyor belt, used for carrying pedestrians horizontally or along low inclines.

Escalators

Escalators (also called electric stairways or moving stairs) move more people faster than elevators. They move large numbers of people efficiently and comfortably through up to six floors, although they are most efficient for connecting two to three floors, with elevators preferred for over three floors. Their decorative design allows users to observe panoramic views. In addition to the standard straight escalator, special escalator designs include curved escalators.

You cannot be trapped on an escalator in a power failure, and escalators do not require emergency power, because you can simply walk up or down the stationary escalator as though it were a stairway.

Escalators require space for floor openings and for circulation around the escalator. Because escalators move at a constant speed, there is practically no waiting period, but there should be adequate queuing space at each loading and discharge point.

Escalators may not be used as required fire exits. They cannot carry wheelchairs or freight, so the building still needs at least one elevator.

The earliest working moving stair was designed in 1896 by Jesse Reno and installed as a novelty ride at Coney Island in Brooklyn, New York. The first commercial escalator, designed by Charles Seeberger and Elisha Graves Otis, won the first prize at the 1900 *Exposition Universelle* in France.

ESCALATOR COMPONENTS

An escalator is structurally supported on a truss. (See Table 19.6) The balustrade beside the steps has a moving handrail. (See Figure 19.12)

Escalators are preferred over elevators by storeowners, as the customers see merchandise while changing levels. They are located on the main line of traffic so users can see them readily and identify the escalator's destination. Customers should be able to move toward the escalator easily and comfortably. When laying out a retail space, avoid blocking the line of sight to the escalator with large displays.

TABLE 19.6 ESCALATOR COMPONENTS

Component	Description
Truss	Welded steel frame supports escalator on both ends, and in middle if rise is over 18' (5.5 m). Provides space for mechanical equipment.
Tracks	Steel angles attached to truss guide step rollers and control step motion.
Drive system	Sprocket assemblies, chains, and machine work like bicycle chain drive.
Emergency stop button	Located at both ends of escalator. Stops the drive machine and applies brake.
Elongated newels	ADA requires elongated newels with minimum of two horizontal treads before landing plate to allow people to adjust before leaving escalator.
Handrails	Synchronized with tread motion for passenger stability and support.
Balustrade assembly	Designed for maximum passenger safety stepping on and off. Side panels of the escalator made of fiberglass, wood, or plastic.

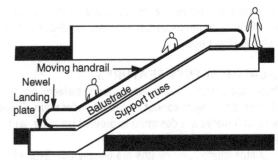

Figure 19.12 Parts of an escalator

In the United States, all escalators rise at an angle of 30 degrees from the horizontal. Standard escalators rise 10 to 25 feet (3 to 7.6 m). There must be 7 feet (2.1 m) clearance overhead.

ESCALATOR DESIGN

A tread 40" (1016 mm) wide could theoretically accommodate two people. (See Table 19.7) In reality, for psychological and physical reasons, one person per tread in an alternating diagonal pattern is the most common use. One half of the treads on an escalator with 24" (610 mm) treads are typically unused.

ASME A17.1 currently defines the width of an escalator as the width of stair tread (in inches). Previously the width was the distance between balustrades; that measure now called "size."

An escalator needs adequate queuing space at each loading and discharge point. Backups are dangerous when people are constantly exiting the escalator, especially in theaters and stadiums during peak traffic flows. Backups can be avoided by providing well-marked escalators with enough capacity. The landing space in front of escalator should be a minimum of 6 to 8 feet (1.8 to 2.4 m) deep, and more for higher speed escalators.

The length of an escalator base on a 30-degree angle incline is equal to 1.782 times the floor-to-floor height. Added to this is the length of the horizontal landing platforms at the top and bottom.

Collecting space at intermediate landings relieves pressure. Physical divisions at intermediate landing turnaround points guide riders away from the discharge points. A setback for the next escalator eases a 180-degree turn, so that people do not have to bunch up where they step onto the next escalator.

TABLE 19.7 STANDARD ESCALATOR WIDTHS AND SIZES

Tread Width	Tread Size	Number of Persons
24" (610 mm)	32" (813 mm)	One adult and one child (1¼ persons)
32" (813 mm)	40" (1016 mm)	2 adults per step
40" (1016 mm)	48" (1219 mm)	2 adults per step

Escalators should exit to an open area with no turns or change of direction necessary. If turns are needed, large clear signs should direct users.

ESCALATOR SAFETY FEATURES

Escalators are extremely safe, with smooth surfaces and handrails designed so fingers cannot be caught under them. Top and bottom comb plates are designed to prevent jamming, and are trip-proof.

Escalator injuries and deaths are usually the result of tripping or falling rather than equipment malfunction. Items (including body parts) can be caught in toothed escalator treads and/or between treads and the escalator sides. Safety mechanisms sometimes do not stop movement as intended.

ESCALATOR FIRE PROTECTION

Escalator mechanical equipment (trusses, return treads, motors) must be enclosed in fire-rated construction. The escalator opening between floors must be designed to prevent a fire from spreading.

There are four methods of fire protection for escalators. (See Table 19.8) The code requires one or more when an escalator pierces more than two floors to prevent a fire from spreading through the escalator opening.

Escalators are typically not allowed as a means of egress. Some exceptions are made for escalators in existing buildings if they are fully enclosed within fire-rated walls and doors. Codes may also require specific sprinkler system configurations.

TABLE 19.8 ESCALATOR FIRE PROTECTION

Method	Description
Rolling fire shutter	Activated by temperature and smoke detectors. Entirely closes off wellway at a given level, prevents drafts and fire spread. Uncommon in the United States, more common in the United Kingdom and Europe.
Smoke guard	Fireproof baffles surround wellway and extend down about 20" (508 mm) from ceiling to deflect smoke and flame. Automatic water curtain from sprinkler heads on ceiling then isolates the escalator.
Spray-nozzle curtain	Similar to a smoke guard. Closely spaced, high-velocity water nozzles create compact water curtain, prevent smoke and flames from rising. All nozzles open simultaneously.
Sprinkler-vent	Fresh air intake on roof. Blower drives air down through wellway, and roof exhaust fan creates strong draft upward through exhaust duct, drawing air from just under ceilings. Includes spray nozzles.

ESCALATOR LIGHTING

Escalators must have adequate lighting for safety, especially at the landings. Supplement general lighting on the ceiling above the escalator, especially on the comb plates. Balustrade lighting is also possible. In addition, as a featured part of the décor, lighting should enhance the visual focus on the escalator.

ARRANGEMENT OF ESCALATORS

The crisscross arrangement is the most common arrangement for escalators. It locates the entrances and exits to their upper and lower ends at opposite ends of the escalator. Crisscross escalators take up the minimum amount of floor space and have the lowest structural requirements.

Separation of escalators creates a longer walk for the customer to view merchandise. Separation allows easier mixing of entering riders with continuing riders. About 10 feet (3 m) is the maximum feasible separation.

Separated crisscross arrangements consist of only an up or a down escalator in one location. They allow space at the end of the run for merchandise. If the distance to the next run is too long or if the next escalator is not in sight, users become annoyed. This is made worse when the floor space for the trip between escalators is inadequate, resulting in crowding, pushing, and delays.

Walk-around crisscross escalators reverse the direction of second-level stairs, forcing the passenger to walk around the entire length of the stair to continue up. This arrangement requires extra floor space around escalators. It makes store displays highly visible, but is potentially annoying.

Spiral crisscross escalators provide an uninterrupted trip with an up spiral and a down spiral. (See Figure 19.13) Stairs nest into each other to economize space. Spiral crisscross escalators can be used for up to 5 floors without annoying the rider.

Parallel escalators face in the same direction. They are less efficient, more expensive, and use more floor space than crisscross arrangements, but have an impressive appearance. Parallel escalators can use spiral arrangements or stacked parallel

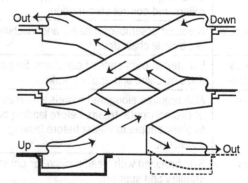

Figure 19.13 Spiral crisscross escalators
Source: Redrawn from Walter T. Grondzik and Alison G. Kwok, *Mechanical and Electrical Equipment for Buildings* (12th ed.), Wiley 2015, page 1537

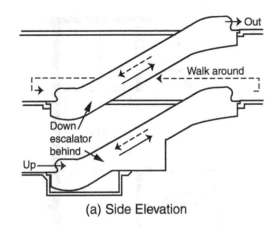

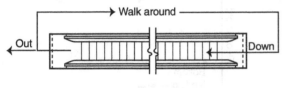

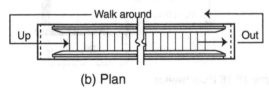

(b) Plan

Figure 19.14 Parallel stacked escalators

Source: Redrawn from Walter T. Grondzik and Alison G. Kwok, *Mechanical and Electrical Equipment for Buildings* (12th ed.), Wiley 2015, page 1539

arrangements. (See Figure 19.14) Banks of three or four parallel units are often used in transportation terminals, with all but one in a bank traveling in the same direction; their direction can be reversed to accommodate the heaviest traffic direction.

Moving Walks and Ramps

Moving walks and ramps are very similar in construction and operation, but differently applied. Their components are like those of an escalator, but with a flattened pallet in place of steps. Moving walks and ramps are limited to about 1000 feet (305 meters) in length. The depth of the supporting truss is typically 3'-6" (1 m), which is likely to impinge on the ceiling of the floor below.

The dimensions and speed of moving walks and ramps are not as standardized as escalators, and are usually designed for the specific application. A variety of widths are made, with 55" (1397 mm) wide units in transportation terminals allowing walkers to pass other passengers.

MOVING WALKS

Moving walks (also called moving sidewalks or autowalks) are high-capacity continuous people movers designed to reduce

congestion and force movement along a designated path. They eliminate or accelerate the need to walk long distances in a building. A moving walk consists of a power driven continuously moving surface similar to a conveyor belt.

Moving walks can be used to transport large, bulky objects easily. Their most common use is in air transport terminals and other transportation facilities. (See Figure 19.15)

Moving walks are also used to move people past display window or other points where congestion or stopping is undesirable. In addition, they are useful to people with mobility problems.

Although a moving walk was first introduced at the 1893 World's Columbian Exposition in Chicago, the first commercial installation in the United States did not occur until the 1950's, when the Speedwalk was built by the Goodyear Company for Hudson & Manhattan's Eire Station in Jersey City, New Jersey.

Moving walks move large numbers of people horizontally for distances of at least 100 feet (30.5 m). Distances over 300 feet (91 m) use multiple units end to end with on/off space between. They are available in widths from 2 to 9 feet (0.6 to 2.8 m), and are commonly 32", 40", or 56" (813, 1016, or 1422 mm) wide.

A moving walk must not incline more than five degrees from the horizontal. The maximum speed for horizontal moving walkways is 2 miles per hour (mph) (3.2 kmh).

MOVING RAMPS

A **moving ramp** is a moving walkway that inclines between five and fifteen degrees. Moving ramps offer a way for wheeled vehicles and large, heavy packages to move vertically and

Figure 19.15 Moving walk

horizontally through a building. They are also an option for people who would have trouble using an escalator. Multilevel stores use moving ramps to transport shopping carts to rooftop parking lots. Transportation terminals use them to carry luggage carts that cannot easily negotiate escalators.

MATERIALS HANDLING

Until the late 1970s, materials were transported within commercial and institutional buildings primarily by hand, with some mechanical assistance. Office messengers carried mail. Hospitals used dumbwaiters, service elevators, conveyors, or chutes. Large stores used pneumatic tubes to carry money. Today, these tasks can be done automatically and usually much more rapidly. The initial cost of automatic systems is high, but the reduction in labor and increased speed result in a short payback period and a rise in efficiency.

Materials handling systems include elevator and conveyor types and pneumatic systems, among others. (See Table 19.9)

Dumbwaiters

Manual load-unload dumbwaiters are used in department stores to transport merchandise from stock areas to selling and pickup centers. Hospitals use dumbwaiters to transport food, drugs, and linens. Restaurants with more than one floor carry food from the kitchen and return soiled dishes in dumbwaiters.

Dumbwaiter designs include traction or drum styles. (See Figure 19.16) The car is often divided by shelves. Dumbwaiters can be designed to load at floor, counter, or other specified heights. Maximum platform area for a dumbwaiter is 9 square feet (0.8 m²), and maximum height is 4 feet (1.2 m). Many dumbwaiters are limited to 50 feet (15.2 m) in height and 300 to 500 pounds (136 to 227 kg) per load.

EJECTION LIFTS

Ejection lifts are automated dumbwaiters that move relatively large items in carts or baskets vertically and rapidly. Institu-

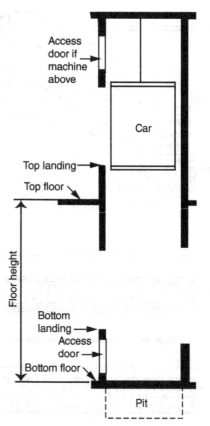

Figure 19.16 Dumbwaiter
Source: Redrawn from Walter T. Grondzik and Alison G. Kwok, *Mechanical and Electrical Equipment for Buildings* (12th ed.), Wiley 2015, page 1530

tional and other facilities use them for rapid vertical movement of relatively large items. They can deliver food carts, linens, dishes, and bulk-liquid containers, for example. Each load is carried in a cart or basket, which is manually or automatically loaded. At delivery, the item must be picked up and horizontally transferred to its final destination.

Conveyors

Industrial facilities and commercial buildings like mail-order houses use horizontal conveyors. They are relatively low cost and can carry large quantities of merchandise. However, they demand an inflexible right-of-way, are noisy, and may be dangerous if misused. Horizontal conveyors are used in airports for check-in baggage, in cafeterias for soiled dishes, and in post offices and mail-order houses for parcels.

A selective vertical conveyor picks up and delivers tote boxes (also called trays) that are carried along a continuous chain. The operator puts the item in a tote box, addresses the box, and places it at a pickup point. The next empty carriage on the chain picks up the box and delivers it.

TABLE 19.9 MATERIALS HANDLING SYSTEMS

System Type	Description
Elevator-type systems	Vertical-lift cars including dumbwaiters and ejection lifts
Conveyor-type systems	Horizontal, vertical, or inclined
Pneumatic systems	Sophisticated pneumatic tube systems and pneumatic trash and linen systems
Other systems	Automated messenger carts, automatic track-type container delivery systems, others

Pneumatic Systems

Pneumatic systems were invented in the nineteenth century and continue to be used for the physical transfer of items, although they have been replaced by digital media for data transmission. They are reliable, rapid, and efficient. Pneumatic tube systems consist of single or multiple loops of tubes 2½" to 6" (64 to 152 mm) in diameter, and are also available in special shapes. In the past pneumatic tube systems relied on large, noisy compressors. Newer computer-controlled systems are relatively quiet.

Pneumatic trash and linen systems provide for rapid movement of bagged or packaged trash and linen from numerous outlying stations to a central collection point. Health codes require separate systems for trash and linen. Linen systems are commonly used in hospitals. Trash systems are found in many types of buildings, often along with trash compactors.

A pneumatic trash or linen system consists of large pipes 16", 18", or 20" (406, 457, or 508 mm) in diameter. They carry one load at a time at 20 to 30 feet (6 to 9 meters) per second. These pneumatic systems do rely on large and very noisy compressors that require acoustical isolation. However, they perform their specific task efficiently and cheaply.

Automated Container Delivery Systems

In an automated container delivery system, containers are locked onto a motorized carriage, which is in turn locked onto the track system. Containers come in a variety of sizes. Automated container delivery systems are easy to retrofit into a building, due to their small size and flexible track layout, but tend to be expensive.

In another variation, passive guidance tapes installed below the carpet invisibly route robotic battery-powered vehicles that can connect to elevators for vertical transport. They are used for pickup and delivery of parts in industrial settings, for food and supplies distribution in hospitals, and for mail and document pickup and delivery in offices.

With this chapter's look at conveyance systems, we have nearly finished our survey of building systems. In Chapter 20, we look at how technology is changing communications, security, and control equipment in today's buildings.

20

Communications, Security, and Control Equipment

Buildings are being designed and operated with computer software that integrates building systems, combining communications, security, and control systems. Today's residential client may have a home alarm system, fire monitoring, weather monitoring, and other security systems, as well as wireless and mobile control of appliances, lighting, and other building functions. The emerging field of interior technology design can aid in creating a home lighting and media audiovisual experience controlled with a remote or touch screen.

INTRODUCTION

No area of equipment design and application in buildings has seen such rapid and sustained changes as that of signal equipment. Signal equipment encompasses all communication and control equipment, the function of which is to assist in ensuring proper building operation. Included are: surveillance equipment such as that for fire and access control; audio and visual communication equipment such as telephone, intercom, and television (both public and closed-circuit); and timing devices such as clock and program equipment and all types of time-based controls. (Walter T. Grondzik, Alison G. Kwok, *Mechanical and Electrical Equipment for Buildings* (12th ed.), Wiley 2015, page 1409)

In this chapter, we look at the design of communications, security, and control equipment for residential and commercial

buildings. We do not pretend to know what the future will bring, so our focus is on basic design principles.

Signal Systems

Signal systems send and receive electronically coded information. They include all communication and control equipment, security, music and sound, intercom, clock and program, paging, and building automation systems. (See Table 20.1.) These functions were previously separate, but are now frequently combined and multipurpose. Signal systems are designed by the electrical consultant, or by special fire protection, audiovisual, or acoustical consultants.

For fire detection and alarm systems, see Chapter 18, "Fire Safety Design." For HVAC controls, see also Chapter 14, "Heating and Cooling."

Public buildings such as hotels, motels, hospitals, schools, and museums typically have special service requirements beyond those of office buildings with normal data processing and telephone services, and may require public address, piped music, and **closed-circuit television (CCTV)**.

All signal systems consist of a source, a means of conveying the signal, and indicating equipment at the destination. (See Table 20.2) Symbols for signal devices appear on drawings. (See Figure 20.1)

TABLE 20.1 SIGNAL SYSTEM TYPES

Type	Description
Surveillance equipment	Security, fire, and access control
Audiovisual communication equipment	Telephone
	Broadcast TV with very high frequency (VHF) and ultra high frequency (UHF) reception
	Closed-circuit TV for security or educational purposes
	Paging and sound systems with AM/FM tuners, intercom
Timing devices	Clock and program equipment, time-based controls; incorporated into building mechanical control systems
	Master clock system with interconnected clocks and bells
HVAC controls	From simple thermostats to computerized energy management systems

TABLE 20.2 BASIC SIGNAL SYSTEM COMPONENTS

Component	Description
Source	Sensor to pick up signal, plus CCTV camera, telephone or intercom device to process and transmit it. TV or radio signal antenna.
Means of conveying signal	Usually low-voltage wiring or radio airwaves. TV antenna cables and close-circuit connections must be shielded, generally not grouped with phone lines due to possible signal interference.
Indicating equipment	Audible, visual, printed hard copy. Signal indicators include loudspakers, computer monitors, bells, horns, sirens, and flashing lights.

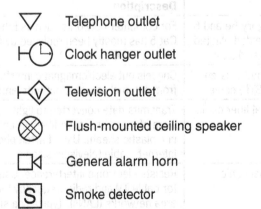

- ▽ Telephone outlet
- Clock hanger outlet
- Television outlet
- Flush-mounted ceiling speaker
- General alarm horn
- ⓢ Smoke detector

Figure 20.1 Signal device symbols

Source: Redrawn from Vaughn Bradshaw, *The Building Environment* (3rd ed.), Wiley 2006, page 336

COMMUNICATION SYSTEMS

Communications wiring supports computers, telephones, videos, televisions, audio microphones and speakers, and signal devices such as clocks. Separate wiring circuits are required for the sound and signal equipment of telephone, cable, intercom, and security or fire alarm systems.

The building's electrical system has close relationships with its communications systems. The installation of communications wiring also affects walls, partitions, and finishes.

Communication and building control wiring uses low-voltage wires to carry information within buildings. Other information is carried on powerline carrier (PLC) systems, fiber optics, or wireless systems. Power line carrier (PLC) signals are high frequency signals impressed on electric power wiring.

See Chapter 16, "Electrical Distribution," for more information on power line carrier (PLC) signals.

Residential Communications

Residences today are intimately linked with the rest of the world through mobile and wireless devices. Sophisticated automated residential systems control security, fire alarms, time functions, thermostats, blinds, lighting, and door locks among other items. These systems use dedicated wiring, control bus, or PLC signals, which all show a trend toward consolidated control, with a single control panel serving multiple residential systems.

Residential telephone service normally follows the overhead or underground route of the electric service, but with a separate service entrance. Telephone wiring installed after a residence is constructed requires surface-mounted cable that is unsightly and often objectionable. Telephone service can be prewired by running cables within the wall framing into empty device boxes for later connection. Wireless telephone service is decreasing the need for wired telephone and data lines.

Wireless communications allow occupants to work anywhere in the home. A home planning center with a cell phone charging station, often in the kitchen, can act as the communication center for household members.

The kitchen or another space may function as a home office, with a computer and Internet connection, laptop docking station, electronic charging station, and other equipment such as a printer and shredder. Plan for cord management with cable channels or raceways if necessary. Internet wiring also provides support for Internet-ready appliances.

Refrigerators today connect wirelessly to the Internet to help track expiration dates or order food online through a smartphone. A screen embedded into a digital-wall backsplash allows the user to cook with TV chefs or check a security camera feed. Software allows smart devices to talk to one another, and voice and gesture-enabled controls are on the horizon.

Provide good ventilation for computer components; they should not be enclosed in cabinets. Computers may need one or more dedicated circuits with surge protection appropriate to the equipment.

Televisions in the kitchen and elsewhere may link to a home cable or satellite system. Connections may need to be provided for a DVD player plus speakers and a sound system; wireless systems are increasingly available. Equipment can be located elsewhere with ceiling, recessed, or wall-mounted speakers.

The home may have internal communication networks intended for small children and older adults needing assistance. These should be accessible to all family members and located where all users can easily reach controls. Help systems that use telephone connections for emergency calls can be lifesavers.

A basic residential intercom system has one or more masters and several remote stations, and can allow the front door to be answered from various points in the home. Voice communication and CCTV may be added for identification. Leaving the intercom system in the open position allows remote monitoring. Low voltage multiconductor intercom cable is generally run concealed in walls, attics, and basements. Other systems eliminate separate wiring and make remote stations portable by imposing voice signals on the house power wiring, with plug-in connections to a power outlet.

Office Building Communications

Office building communications systems frequently combine four functions into a single network. These include intra-office voice communication (intercom), interoffice and intra-office data communication using telephone and communication cabling, outside communication via telephone company or data lines, and a paging function. Office building communications systems are often purchased or leased from a private company. The same instruments and switching equipment can be used for both intercom and outside connection.

Office building communications planning may require setting aside large amounts of space in critical locations for the service entrance room (equipment room), vertically stacked riser spaces (shafts) and riser closets, and satellite closets where required. Horizontal distribution between closets and devices may use conduit, boxes and cabinets, underfloor raceways, and over-the-ceiling systems. Here again, wireless communications promise savings in space and equipment.

It is important for interior designers to consider the location of data and phone receptacles when laying out desks. Receptacles should be located on walls where desks will be placed, rather than in locations that would require running extension cords.

School Communication Systems

An integrated sound-paging-radio system designed for school use provides a means of distributing signals from recordings, broadcasts, or live sound to selected areas. The simple system can provide a CD player and single microphone with a single channel to all the speakers in the school. More complex systems can distribute multiple input signals to different areas of the school. A small system can be installed in a compact desktop console. A large system often requires a separate console, frequently built as a desk with adequate space for the equipment and the person operating it. An alcove of 30 to 50 square feet (28 to 4.7 m²) may be reserved for it and a library of recordings.

Electronic teaching equipment for schools is evolving so fast that it is not feasible to predict future needs accurately. Both passive and interactive modes are used.

Passive-mode equipment encompasses all recorded material in any format that is available to the student via a form of information retrieval. Passive-mode usage includes printed, audio, and video material in conventional and electronic library forms.

Interactive mode involves a student using a computer for individual study at his/her own pace. Building designers must accommodate rapid developments are occurring in this field at all educational levels with provisions for electrical power, cable raceways, lighting, and HVAC.

Data and Communications Wiring

Interior designers are often responsible for showing the location of electrical power, data, and telephone connections, both in homes and in offices. The design goal is flexibility and ease of access. Plugging in equipment should not require the user to crawl around on the floor. Rewiring for new technology should not involve ripping up walls and floors. The special design needs for data and communications may make it worthwhile to hire a design consultant familiar with the latest technology.

Many small low-voltage wires may be twisted in pairs or coaxial form. (See Table 20.3 and Figures 20.2 and 20.3) Some are shielded to prevent signal interference between circuits.

TABLE 20.3 COMMUNICATIONS CABLES

Type	Description
Category 5e and 6 unshielded twisted pair (UTP) cable	For computer networks such as Ethernet. Cat 5 has mostly been replaced by Cat 5e and Cat 6 cable.
Shielded twisted pair (STP) cable	Cancels out electromagnetic interference from external sources; outdoor landlines.
Optical fiber cable	Transmits data converted to light pulses. Core of glass or plastic filament encased in a plastic sheath. Used for telephone, Internet, cable television signals.
Coaxial cable	Resists electronic interference. Used for cable TV and radio frequency local area networks (LANs). Typically a single wire conductor with a layer of insulation, outer shielding conductor, and second layer of insulation.

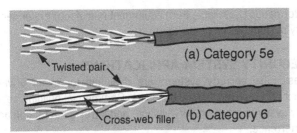

Figure 20.2 Category 5e and 6 unshielded twisted pair (UTP) cable

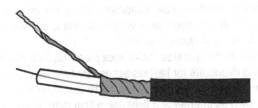

Figure 20.3 Coaxial cable

Figure 20.4 Fiber optic cable

Figure 20.5 Wiring on ceiling

Flat cables are used for signal and communication wiring and both electrical and fiber-optic cables and accessories for data transmission. Fiber optic cables are used instead of copper cabling in installations with very heavy data transmission loads, video systems, or high-security, low-noise, and broadbandwidth requirements. (See Figure 20.4)

As long as we continue to use wires to carry information, wires will be part of our interior environments. A great deal of the wiring currently in existing buildings may be inactive, left over from a previous tenant or use. (See Figure 20.5) As wireless technology expands our options, more and more wiring will become obsolete.

Access flooring raised only a few inches can house wiring that remains easily accessible and almost invisible. (See Figure 20.6) Height changes may need to be ramped or otherwise made accessible.

See Chapter 16, "Electrical Distribution," for more information on access flooring and wiring.

Premise Wiring

The system of raceways, boxes, and outlets dedicated to all types of communications systems (with the general exception of audio signals) is known as **premise wiring**. The term often does not include the wiring itself. Premise wiring raceways are often surface mounted to allow frequent access and to accommodate

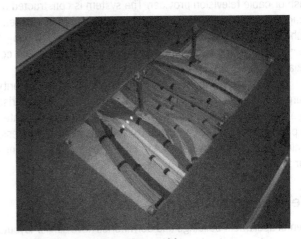

Figure 20.6 Access floor data cables

preterminated data cables that can be difficult to pull through recessed raceways. (See Figure 20.7) The larger premise wiring raceways are easier and less expensive to install than other surface-mounted units. Premise wiring devices and raceways are also referred to as **wire management**.

Television

Cable television systems may receive their signals from an outdoor antenna or satellite dish, a cable company, or a closed-circuit system. If several outlets are required, a 120V outlet is supplied

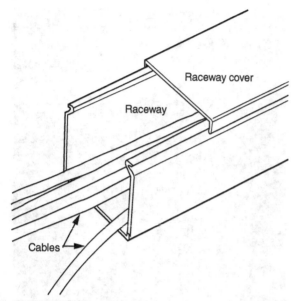

Figure 20.7 Signal cabling raceway

to serve an amplifier. Coaxial cables in a nonmetallic conductor raceway transmit the amplified signal to the various outlets.

In today's multiple-dwelling residences, each room has one or more cable/Internet "jack" outlet for signals from a satellite dish or cable television provider. The system is constructed as empty conduits connecting to cable-pulling points in cabinets; subcontractors later install the wiring. In residential low-budget construction, only floor and wall sleeves are supplied and co-axial cables are run exposed.

Closed-circuit television (CCTV) cameras used for security purposes are very common in banks, retail stores, high-rise apartment buildings, and industrial complexes. They are located in parking garages, elevators, and all possible means of access, including doors and windows and exterior ventilation openings for fans and ducts.

Telecommunications

Rental apartment buildings and dormitories have similar layouts on each floor, and cable can be run in risers through sleeves in vertically aligned closets within apartments. Individual rooms can be prewired without conduit or with only a few short sleeves. Telephone closets require adequate lighting and electrical service for maintenance.

Large installations require a service connection, terminal enclosures, riser spaces, and other equipment similar to electrical systems. Large systems are usually designed, furnished, and installed by a telecommunications company.

Security Systems

Building security systems should be part of the initial stages of building design. Today, security hardware often requires wiring.

Information is available from equipment manufacturers or security system design consultants.

SECURITY SYSTEM APPLICATIONS

Residential security systems are designed for either private residences or multifamily buildings. Commercial building security systems focus on surveillance, and intrusion detection and deterrence.

RESIDENTIAL SECURITY SYSTEMS

Residential alarm systems can be as simple as wireless, battery-operated do-it-yourself equipment. More complex options include additional sirens, temperature sensors, and water sensors installed just above floor level.

Smartphone-operated locks today can replace traditional locks and deadbolts for keyless access that allows residents to monitor and customize who gains entry and keep track of activities such as Internet and light use within their house. Smartphone controlled security alarms can cause lights to change color or flash to alert the homeowner of an intruder. Products also allow homeowners to video-chat with someone ringing the doorbell, even when they are not themselves home.

Residences have most commonly used magnetic door and window switches as well as passive infrared (PIR) and/or motion detectors. Providing a switch at the end of a long cord allows a resident to set of the alarm manually. Intrusion alarm systems can be monitored by a security company's continuously supervised central station for direct response or notification of police.

Residential solutions can combine networking, computing, automation, and entertainment integrated into a central control platform. Lighting control, climate control, whole-house music and video, and energy monitoring are also available.

In apartment buildings, the security and doorbell functions are often combined. A two-way intercom in the building entrance allows apartment tenants to screen callers before opening the lobby entrance door.

MULTIPLE-DWELLING SECURITY SYSTEMS

Apartment buildings frequently combine security and doorbell functions. Using a two-way intercom between the building entrance and each apartment, occupants can screen callers and push their release button to open the lobby entrance door. A closed-circuit television is often added to the system.

Emergency call buttons within apartments are often added to apartment buildings for intruder alarms. In luxury apartment buildings, apartment doors can be monitored from a central security desk.

Housing designed for older people often includes provisions to unlock an apartment door to allow helpers summoned by lights and alarms. Emergency call systems for elderly, handicapped, or other residents alert people outside a closed apartment to an emergency inside due to illness or distress. Many construction and housing codes describe required equipment, including a call initiation button in each bedroom and bathroom that is monitored 24 hours per day and will register an audible alarm and visible annunciated signal. Additional signals in each floor's corridor and at each apartment alert immediate neighbors to a distress call.

Magnetic cards and electronic combination locks are especially useful for residential facilities catering to transients as they allow easy code changes.

HOTEL AND MOTEL SECURITY SYSTEMS

Security systems for hotels and motels include room access security and equipment security. Most modern hotels use electronic room locks with opening device codes changed with each guest. These locks may have coding changed from a central console, comprise magnetically or electronically coded cards, or use a programmable electronic lock and coded key device.

Equipment security for guest rooms and meeting rooms is typically designed by a specialty theft control consultant. A system applied in hotels, schools, office buildings, and industrial facilities senses the disconnection of equipment from its power connection and transmits an alarm to an annunciator at a control location.

SCHOOL SECURITY SYSTEMS

Security equipment for schools includes door and window sensors that can both trip local alarm devices and notify police headquarters. A perimeter alarm detection system helps prevent after-hours entry. Exterior lighting can be activated to deter vandals. An exit-control alarm is used for doors locked on the outside that must be operable from inside in an emergency.

School clock and program systems are today combined into a single system that also provides timing for all programmable switches and controllers. A clock and program device controls clock signals, audible devices, and other optional devices. A tone produced on a classroom speaker is often preferred to bells, gongs, buzzers, or horns.

OFFICE BUILDING SECURITY SYSTEMS

Office building security systems normally use some type of manual watchman's tour system with key and clock stations so that surveillance of unoccupied areas is conducted on a regular basis. A simple manual system consists of small cabinets containing a key that are placed at intervals around the building. The watchman uses the key to operate a special wall-mounted or portable clock that records the time he checked each specific location. A computerized version of this system automatically records data and provides a printout. Systems are also available that permit constant supervision from a central location.

There is a current trend toward integration of security systems with communications, fire safety, and emergency management systems. Some CCTV, fire, mass notification systems, and burglar alarms systems have been integrated for access control.

Building security affects the interior layout. It may be necessary to reach a compromise between providing as many fire exits as possible versus as few as required for security purposes.

Exit doors may not be locked against the direction of exit travel, but may be locked to entry. Enclosed exits are especially likely to create security problems as they provide easy escape routes for thieves, offer a way to travel to prohibited floors or avoid being questioned by reception desk, and can be inconspicuously used to gain unauthorized entry.

Security equipment electronically extends the surveillance abilities of a limited security force to remove or reduce the incidence of theft, assault, or vandalism. Security personnel are still needed to monitor security devices and apprehend criminals. Detection devices can automatically alert the police, or private security guards can both alert authorities and apprehend criminals.

Security Equipment

Security equipment for buildings begins with intrusion detection. Public emergency reporting systems (PERS) provide access to aid people in public spaces.

INTRUSION DETECTION

Intrusion detectors at doors and windows and motion detectors within the building trigger an alarm if someone enters an unauthorized area. (See Figure 20.8 and Table 20.4) Intrusion detection begins when a sensor detects the problem. Detecting devices may be CCTV, motion detectors, intrusion detectors, or smoke/fire detectors, as separate systems or as parts of an integrated security system. The signal is then processed and appropriate measures are taken. These may include sounding alarms, turning on lights, and/or sending signals to central proprietary or private surveillance services or police.

EMERGENCY COMMUNICATIONS SYSTEMS

Public emergency reporting systems (PERS) may be located in key egress and public gathering areas. Fire, police, or medical emergencies can be reported to qualified operators within facility. Pulling fire handle sends an audio alarm and visible location indication to the building control room. Lifting the handset from its cradle sounds a visual and audio alarm in the central control room in direct contact with the PERS station. The medical handle summons medical/paramedical service automatically.

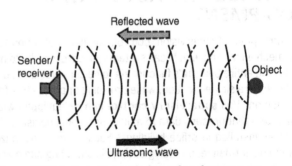

Figure 20.8 Active ultrasonic intrusion detector

TABLE 20.4 INTRUSION DETECTORS

Type	Description
Simple normally closed (NC) contact sensors	Transmit alarm signal. Used in closed, supervised circuits. Include magnetic contacts and spring-loaded plunger contacts for doors and windows, window foil, and pressure/tension devices.
Mechanical motion detectors	Spring-mounted contact suspended inside second contact surface. Motion on surface on which device is placed produces alarm. Very sensitive, most units have sensitivity adjustment.
Photoelectric devices	When beam is received, contact in receiver is closed. Beam interruption causes contact to open, producing alarm. Laser or infrared (IR) beams. Laser beam signal can make perimeter fence.
Passive infrared (PIR) presence detector	Lens or mirror concentrates IR radiation on sensor. Rapid change in IR reading indicates object entering or leaving space, triggers alarm. Also used as occupancy sensors to turn lights off.
Motion detectors	Detect changes in frequency of signal and initiates alarm if moving body reflects signal. Ultrasonic and microwave units.
Acoustic detectors	Alarm when increase in noise level or frequencies (breaking glass, forced entry). Can also be occupancy sensors for lighting.

Firefighter communication systems use phones that plug into special jacks in stairwells, elevator lobbies, and other key locations for communication among firefighters and with the command station. In large high-rise buildings, central control centers often control the HVAC and electrical services as well as monitor fire alarms from a single point.

CONTROL AND AUTOMATION EQUIPMENT

A central point of supervision, control, and data collection for the mechanical and electrical systems of an office building allows the entire building's functioning to be surveyed and controlled from a single location, while providing opportunities for automation. This **supervisory control center** equipped with computers that process data to make operational decisions is routinely installed in office buildings. Such systems optimize system performance, providing savings in operating and maintenance costs.

Automation

In general, supervisory control center systems are referred to as **building automation systems (BAS)**. Spaces that house BAS centers require good lighting and ventilation as well as extensive raceway space, but take up little area.

A remote control system, as opposed to an automated system, employs a technique by which an action can be performed manually at the device being controlled from a remote location by some intermediate means, such as low-voltage wiring or a wireless signal.

An **automated** system uses an automatic signal from a timing device or a programmable device such as a microprocessor or computer. An automated system can simply use an automated signal to control a single action. Alternatively, it may be very complex yet control only a single function, such as an automated lighting system with sensors activating or overriding scene presets that activate dimmers and switches. Such a system is referred to as a **stand-alone (automated) system**. Interconnecting and supervising several stand-alone systems produces an integrated control system which, when applied to building systems, is called a building automation system (BAS).

Standalone control and automation systems serve residences, multidwelling buildings, hotels and motels, and schools. They can also control lighting systems along with window shades.

Building automation is today economically feasible and makes detailed multipoint monitoring and control in real time practical. This promotes efficiency, environmental benefits, and cost savings. It is becoming increasingly cost effective to retrofit building automation systems to existing buildings.

Control wiring supports fire alarms, security, lighting, and HVAC control systems. Powerline carrier (PLC) systems are used mainly for building control functions. They use line voltage cables to carry information signals as well as power.

Intelligent Buildings

According to the Intelligent Buildings Institute 1987 definition, an **intelligent building** is "a building which provides a productive and cost-effective environment through optimization of its four basic elements—structure, systems, services, and management—and the interrelationships between them.... Optimal building intelligence is the matching of solutions to occupant needs." (quoted in Walter T. Grondzik and Alison G. Kwok, *Mechanical and Electrical Equipment for Buildings* (12th ed.), Wiley 2015, page 1439)

INTELLIGENT RESIDENCES

Intelligent residences, also called smart or automated homes, vary in the amount of automation with which they are equipped. The simplest may have a low-voltage control system

TABLE 20.5 HOME AUTOMATION SYSTEM COMPONENTS

Device	Description
Sensors	Measure or detect temperature, humidity, daylight, motion, etc.
Controllers	Computer or dedicated home automation controls
Actuators	Motorized valves, light switches, motors, etc.
Buses	For wired or wireless communication
Interfaces	For human-machine and/or machine-to-machine interaction running on smartphone or tablet

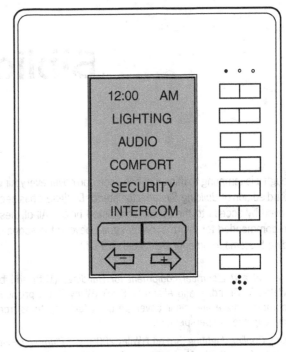

Figure 20.9 Residential control system keypad

with time-based programming using a relatively simple micro-processor. Others use a touch-screen computer for control. Devices may communicate over dedicated wiring, over a wired network, or wirelessly. Today, the user's telephone can control any portion of the system directly.

Residential systems should be laid out early in the design process, with consideration of specific ways to monitor and reduce energy use. (See Table 20.5) Including an electronic systems professional in the design process is a good idea.

Private residential automated systems handle security, fire alarm, time functions, lighting, and other equipment. They use dedicated wiring, control bus, or PLC signals. A single control panel (annunciator) can serve multiple residential systems.

Building Controls

Control systems can include occupancy sensors set to respond to the occupants' habits and to sunrise and sunset times. Light level sensors respond to daylighting. Wireless battery-powered automatic shades aid both daylighting and security. Pressure sensors on stairs can control lighting.

Controls no longer have to be installed in unattractive wall boxes. Thermostats and lighting controls are available with appealing designs, or can be customized with screen backgrounds or cover plates that coordinate with the interior design.

Residential HVAC system controls are zoned for the areas used most. (See Figure 20.9) They can be equipped with "away" and vacation buttons, and respond to remote commands to turn up the heat when the occupant is on their way home. Thermostats are now available that can learn users' preferences and set temperatures to reduce heating and cooling bills. Wall clutter is reduced when the thermostat is replaced by a dime-sized disk that can be painted over.

LIGHTING CONTROL SYSTEMS

Computerized lighting control establishes lighting levels and settings, reacts to daylight and occupancy sensors, and provides window shade control. Conference room control of lights and shades can be done through the speaker's lectern.

ENERGY MANAGEMENT CONTROL SYSTEMS

Energy management control systems (EMCSs) turn off equipment that is not in use, optimize HVAC operation, and cycle electrical loads to limit overall demand. They can include a security monitoring system. An EMCS requires lighting, ventilation, and space for raceways.

This brings the final chapter of *Building Systems for Interior Designers, Third Edition*, to a close. Still to come are the Bibliography and Index, which are presented to assist you, the reader, in finding the information you need.

Bibliography

Although it is tempting to think that one book can cover everything, the third edition of *Building Systems for Interior Designers* has been improved by access to the publications listed below. All of these are recommended for information that goes beyond the scope of *Building Systems for Interior Designers* (3rd ed.).

Mechanical and Electrical Equipment for Buildings (12th ed.) by Walter T. Grondzik and Alison G. Kwok (Wiley 2015) provides comprehensive and clear coverage of building systems from the engineering perspective.

Heating, Cooling, Lighting (3rd ed.) (Wiley 2009) and *Plumbing, Electricity, Acoustics* by Norbert M. Lechner (Wiley 2012) combines good historical information with clear explanations.

The Building Environment: Active and Passive Control Systems (3rd ed.) by Vaughn Bradshaw (Wiley 2006) addresses sustainable design in the context of architects and builders.

Building Construction Illustrated (5th ed.) by Francis D.K. Ching (Wiley 2014) presents basic concepts with exceptionally fine illustrations.

Environmental Issues for Architecture by David Lee Smith (Wiley 2011) explains basic principles in a way that facilitates communication among design professionals, and has good historical information.

Kitchen Planning (2nd ed.) and *Bath Planning* (2nd ed.), both by Julia Beamish, Kathleen Parrott, JoAnn Emmel, and Mary Jo Peterson (Wiley and National Kitchen & Bath Association [NKBA] 2013), supply guidelines, codes, and standards for residential design.

Other books written by Corky Binggeli for interior designers that are available from Wiley include:

Interior Graphic Standards (2nd ed.) (Editor-in-Chief)

Interior Graphic Standards (2nd student ed.) (Editor-in-Chief)

Materials for Interior Environments (2nd ed.)

Interior Design Illustrated (2nd ed.), with Francis D.K. Ching

Interior Graphic Standards Field Guide to Commercial Interiors

Index

Page references in boldface indicate illustrations.